An Integrated Approach to Biotechnology

An Integrated Approach to Biotechnology

Edited by Wade Walsh

SYRAWOOD
PUBLISHING HOUSE
New York

Published by Syrawood Publishing House,
750 Third Avenue, 9th Floor,
New York, NY 10017, USA
www.syrawoodpublishinghouse.com

An Integrated Approach to Biotechnology
Edited by Wade Walsh

International Standard Book Number: 978-1-64740-084-2 (Hardback)

Cataloging-in-Publication Data

An integrated approach to biotechnology / edited by Wade Walsh.
    p. cm.
Includes bibliographical references and index.
ISBN 978-1-64740-084-2
1. Biotechnology. 2. Genetic engineering. 3. Chemical engineering. I. Walsh, Wade.
TP248.2 .I58 2022
660.6--dc23

TABLE OF CONTENTS

PREFACE

The purpose of the book is to provide a glimpse into the dynamics and to present opinions and studies of some of the scientists engaged in the development of new ideas in the field from very different standpoints. This book will prove useful to students and researchers owing to its high content quality.

Biotechnology is a broad area of biology that involves the use of organisms and living systems to make or develop products for a specific use. Biotechnology branches out into green biotechnology, blue biotechnology, bioinformatics, red biotechnology, white biotechnology, industrial biotechnology, etc. The modern-day biotechnology includes new and diverse sciences such as applied immunology, genomics, development of pharmaceutical therapies and recombinant genes. Its applications are prominent in four major industries-crop production and agriculture, non-food uses of crops and other products, health care, and environmental uses. It strives to provide a fair idea about this discipline and to help develop a better understanding of the latest advances within this field. As this discipline is emerging at a rapid pace, the contents of it will help the readers understand the modern concepts and applications of the subject. This book elucidates the concepts and innovative models around prospective developments with respect to biotechnology.

At the end, I would like to appreciate all the efforts made by the authors in completing their chapters professionally. I express my deepest gratitude to all of them for contributing to this book by sharing their valuable works. A special thanks to my family and friends for their constant support in this journey.

Editor

Ginseng Berry, a Promising Anti-Aging Strategy: Recent Opinions on the Biological Effects of a Traditional Korean Ingredient

Juewon Kim[1,2#], Si Young Cho[1#], Su Hwan Kim[1], Sunmi Kim[1], Chan-Woong Park[1],
Hyun Woo Park[1], Dae Bang Seo[1*], Song Seok Shin[1*]

[1]R&D Unit, Amore Pacific Corporation, Yongin-si, Gyeonggi-do 446-729, Republic of Korea, Japan
[2]Department of Integrated Biosciences, University of Tokyo, Chiba 277-8562, Japan

*Corresponding authors: Dae Bang Seo, Beauty Food Research Institute, R&D Unit, AmorePacific Corporation, Yongin-si, Gyeonggi-do 446-729, Republic of Korea; E-mail: sdbang@amorepacific.com

Song Seok Shin, Beauty Food Research Institute, R&D Center, AmorePacific Corporation, Yongin-si, Gyeonggi-do 446-729, Republic of Korea; E-mail: ssshin@amorepacific.com

Abstract

A recent effort in the development of new medications and immune modulatory agents is to search for candidates among natural products because they have relatively low toxicities in clinical applications. Ginseng root has been used as a traditional medicine in Korea, Japan and China and has demonstrated efficacy against various human diseases, such as cancer, viral infectious diseases, diabetes, and atherosclerosis. Recent observations and clinical studies have elevated the interest in the potential health effects of the ginseng berry, an association that appears to be due to the phytochemical content of this fruit. The ginseng berry has various bioactivities, such as anti-diabetic, anti-cancer, anti-inflammation, anti-oxidation, anti-neuro degeneration, and enhancement of sexual function bioactivities. Moreover, the effective anti-aging component of the ginseng berry, syringaresinol, has the ability to stimulate longevity via gene activation. Further molecular and clinical studies are necessary to elucidate the numerous bioactive substances in the ginseng berry that contribute to public health.

Keywords: Ginseng Berry; Ginsenoside Re; Syringaresinol; Panax Ginseng; Bioactive; Phyto Chemical

Introduction: Ginseng and the Ginseng berry

Ginseng (*Panax ginseng*) is a popular herbal medicine that has been used in Asia for 5,000 years [1]. Ginseng is classified as fresh, white, or red ginseng, depending on the processing method. As a traditional herb, red ginseng is known as an adaptogen that restores and improves normal well-being. The use of this herbal plant has been widespread throughout the world because of its therapeutic effects. The well-known biochemical and pharmacological effects include anti-cancer [2], anti-fatigue [3], and anti-diabetic effects [4], along with promoting the synthesis of DNA, RNA and proteins [5]. The herb is used as a tea, an extract, or raw directly from powdered root [6]. The representative bioactive compounds are widely considered to begin senosides,

which are ginseng-specific saponins [7]. Currently, more than 100 naturally occurring saponins of various types and products of enzymatic conversion have been isolated from the roots, stems, leaves, flowers, berries, and seeds of ginseng. The different parts of ginseng contain distinct ginsenoside profiles, and thus, different parts probably possess different pharmacological effects [8].

Recently, many health reports have recommended an increase in fruit intake as part of a healthy dietary pattern [9,10]. These reports allow for various forms of fruit, including fresh, frozen, and dried, as well as juices, and recommend fruits such as oranges, apples, bananas, grapes, raisins, and berries. Whereas berries are known as a good source of potassium or fiber, recent studies suggest that berry fruits are a rich source of many phytochemicals that have a broad spectrum of bioactivity and a positive impact on general health. Several berry fruits, including blackberries, blueberries, cranberries, raspberries, and strawberries, have recently received attention as a result of their effects *in vitro* and their associations with lowered risks for some chronic diseases, which were found in recent observational research [11,12]. As a perennial herb, ginseng develops flowers and fruits that bloom in its third and fourth year. Unlike the widely used ginseng root, the ginseng berry is preserved for planting and has not been used by general populations. A recent study reported that the ginseng leaf and berry have higher levels of ginsenosides than ginseng root, and their pharmacological activities have also been reported [13]. In this review, we will summarize the research on the role of dietary ginseng berries in delaying aging, as well as evidence suggesting positive biological effects to prevent age-related diseases.

Ginseng Berry Bioactive Content and Composition

The bioactive components of ginseng are triterpene glycosides or saponins, which are commonly regarded as ginsenosides.

It has been reported that ginsenosides are the most effective agents in ginseng in the treatment and prevention of cancer and the regulation of blood glucose and blood pressure [14]. Ginsenosides are divided into three major groups based on the triterpene aglycones panaxadiol, panaxytriol, and olenolic acid derivatives [15]. Other chemical compounds from *Panax ginseng* include alkenes, alkynes, sterols, fatty acids, mono-triterpenes, phenyl propanoids, kairomones, carbohydrates (sugars and polysaccharides), amines, flavonoids, organic acids and vitamins. In addition to amino acids, nucleic acids, various enzymes and inorganic compounds are obtained from ginseng [16]. More than 60 different types of ginsenoside have been identified that are contained in the plant roots, leaves and fruits [17,18]. Because different parts of the plant contain distinct ginsenoside content, the pharmacological activity of the various parts of the plant may be different. Recent studies have demonstrated that the ginseng berry has a different ginsenoside profile and higher ginsenoside content than the root [19]. Interestingly, among the ginsenosides, ginseng berry extract contains high levels of ginsenoside Re, amounting to almost more than 30-40 times that of ginseng root, indicating that the ginseng berry may be a superior form to ginseng root extract for ingesting a large amount of ginsenoside Re [20]. In addition, ginseng berry extract contains larger amounts of vitamin E, vitamin K, folic acid, and potassium than the raw materials (i.e., skin, flesh, juice) of ginseng. Currently, ginseng berry extract is being evaluated in clinical and preclinical trials because its components are more efficacious as compared to ginseng root extract.

Chemistry and Pharmacological Effects of Ginsenoside Re

Ginsenosides are glycosides that contain an aglycone with a dammarane (except Ro). They are divided into two groups based on the type of aglycone: the proto panax adiol ginsenoside group and the proto panaxa triolginsenoside group. Ginsenosides possess different chemical structures due to variations in the type of sugar moiety and the number and site of attachment. Ginsenoside Re belongs to the proto panaxatriol group and is a major component in ginseng leaf and berry, occurring in much higher quantities than in root [13]. Previous studies have shown that ginsenoside Re exhibits multiple pharmacological activities via different mechanisms both *in vivo* and *in vitro*. First, ginsenoside Re has anti-inflammatory effects, and it ameliorates inflammation by inhibiting macrophage activation [21,22] and regulating auto phagy [23]. Ginsenoside also has anti-diabetic activities. Ginsenoside Re reverses insulin resistance in the muscles of high-fat-diet-fed rats [24], and this effect is most likely due to the inhibition of NFκB [25]. Moreover, ginsenoside Re lowers blood glucose and lipid levels [26,27], as well as exhibits an anti-diabetic effect in ob/ob mice [28]. It also reduces the oxidative stress level in pancreatic beta-cells and diabetic rats [29,30] and, interestingly, attenuates diabetes-associated cognitive deficits in rats [31]. Several studies have suggested that ginsenoside Re has protective effects and beneficial functions on the cardiovascular system, such as contractive and electromechanical alternans [32-35], anti-arrhythmic effects [33,36] nit-ischemic activity [37-

39], angio genic regeneration [40,41], and electrophysiological activities of cardiac cells [42-45]. Ginsenoside Re also exhibits neuro protective effects, and the beneficial effects of ginsenoside Re on Alzheimer's disease [46-48], Parkinson's disease [49], and depression [50] have been reported. The neuro protection of ginsenoside Re is mediated by an anti oxidative effect [38,51], the regulation of inflammatory mediators [52], and nitric oxide signaling [53]. Some reports have proposed that ginsenoside Re can promote sperm capacitation [54] and motility [55] and also has an estrogenic effect [56]. Ginsenoside Re has demonstrated angio genic effects in *in vitro* [41,57] and *in vivo* [40,41] models. The multitude of pharmacological activities of ginsenoside Re can be obtained by dietary ingestion of the ginseng berry. Oral ingestion of ginseng berry extract results in significantly higher absorption (0.33-0.75%) compared to the low oral bioavailability of ginsenoside Re from ginseng root (0.19-0.28%) [20]. In herbal or alternative medicine, a whole herbal extract might be advantageous compared to isolated natural ingredients.

Ginseng Berry: Preventive and Therapeutic Roles

Because the ginseng berry has more abundant ginsenoside content than the root parts [58] (Table 1), the ginseng berry not only exhibits ginseng root-like effects but also has many other specific biological activities. Moreover, in addition to ginsenoside Re, the ginseng berry contains other bioactive components that can be efficiently absorbed from dietary ginseng berry extract [20] .Here, we review the pharmacological activities of a whole extract of the ginseng berry as well as evidence suggesting the potential of a novel anti-aging compound.

Anti-Diabetic Activity

Ginseng has received increasing attention as a complementary and alternative medicine for the treatment of diabetes. Ginseng extract treatment has been reported to have hypoglycemic effects in animal models of type 1 and 2 diabetes [4, 59]. A previous study reported that ginseng berry extract exhibited greater hypoglycemic activity as compared to the same dosage of a root extract [60]. And the consumption of ginseng berry extract increased insulin secretion and ameliorated hyperglycemia in diabetic mice [61,62]. The anti-diabetic effects of the ginseng berry that have been discussed focus on effective components [63], reduced blood glucose levels [64-67] and administration [68,69]. A recent study revealed that ginseng berry extract improved insulin sensitivity in aged mice by increasing protein levels of tyrosine phosphorylated insulin receptor substrate-1 and insulin resistance-related protein AKT [70]. According to these results, the ginseng berry may ameliorate age-related metabolic disorders, such as diabetes.

Anti-Cancer Activity

The reports from early studies demonstrated that ginseng has strong immune-stimulatory properties, such as macrophage and dendritic cell activation, proliferation, and viability of spleen cells [71,72]. Recent studies have reported that the ginseng berry also exhibits anti-cancer activity in *in vitro* [2, 73-77] and *in vivo* [2, 78,79], as well as the ability to attenuate chemotherapy-induced side effects [80,81]; these effects result from the promotion of dendrite cell maturation. Interestingly, the ginseng berry induced a higher degree of co-stimulatory molecule up regulation than the root extract at the same concentrations [82]. These studies indicate that the ginseng berry is an intense tumor therapeutic vaccine adjuvant that can be used in investigations and clinical research.

Anti-Inflammation and Anti-Oxidative Activity

Recent studies have reported that anti-inflammatory compounds prevent the progression of atherosclerosis without altering the blood lipid profiles in hyper lipidemic mice [83,84], indicating that anti-inflammatory compounds may be used as therapeutic agents for the treatment of inflammatory diseases. The ginseng berry has been shown to suppress reactive oxygen species production [85,86], NF-κB activation [19] and inflammatory gene expression [87] in vitro and in vivo. The ginseng berry suppressed atherosclerotic lesion development by inhibiting NF-κB-mediated atherogenic inflammatory gene expression through the induction of antioxidant enzymes without lowering serum lipid levels in a hyper lipidemic mouse model [19]. Moreover, chronic pretreatment with ginseng berry attenuated oxidative stress in cardio myocytes [88] and up regulated human umbilical vein endothelial cell proliferation and migration [89]. These studies provide insight into the therapeutic potential of the ginseng berry for the treatment of oxidative stress and inflammation-related diseases.

Anti-Sexual Dysfunction

Sexual dysfunction has a severe impact on the quality of life of affected individuals. Previous studies reported that more than half of the male population has some degree of erectile dysfunction [90] and one-third of the global male population, across all ages, has some degree of premature ejaculation [91]. To treat these symptoms, PDE5 inhibitors and selective serotonin reuptake inhibitors are used; however, these drugs can produce negative side effects, including headache, gastrointestinal disorder, muscle pain and blurred vision and may have dangerous interactions with other medications [92,93]. To avoid the risks of side effects, people often turn to dietary ingredients, such as ginseng. Ginsenosides have been shown to enhance nitric oxide production by inducing nitric oxide synthase activity [94,95]. Recently, ginsenoside Rg1, which is abundantly present in the ginseng berry, was also shown to improve male copulation behavior via the nitric oxide/cGMP pathway [96]. Clinical observation of patients after 8 week oral treatment indicated that ginseng berry improved all domains of sexual function including erectile dysfunction and premature ejaculation [97]. The ginseng berry had a greater relaxation effect on rabbit corpus cavernosum smooth muscle than did ginseng root extract and increased intra cavernosal pressure in a rat model in both a dose- and duration-dependent manner. This relaxing effect might be mediated by nitric oxide production [98]. According to these studies, the ginseng berry can be used as an alternative medicine for men with sexual dysfunction.

Anti-neuro Degeneration Activity

Continued research indicates the occurrence of neuronal and behavioral deficits during aging, even in the absence of neurodegenerative diseases such as Alzheimer's diseases and Parkinson's diseases. There has been a growing interest in a number of pharmacological approaches to help slow the rate of both cognitive and functional declines associated with aging. Recently, several dietary supplements with either straw berry or blue berry extracts have been reported to reduce some neurological deficits in animal models of aging [99,100]. Fruits are beneficial in both forestalling and reversing the deleterious effects of aging on neuronal communication and behavior [101]. Some of the actions reported to be elicited by ginseng include an ability to induce effects within the central nervous system that control functions related to stamina, fatigue, physical stress, and the functions of memory, learning, and behavior [102]. A number of different cognitive tests have indicated that ginseng exerts potential positive effects on memory and learning performance in a variety of animal species [103-105]. In addition, herbal mixtures that contain ginseng have been proven to improve cognitive performance [106]. The neuro active effects of free amino acids in ginseng seed and berries have also been proposed [107]. It has been suggested that one potential mechanism by which the ginseng berry improves various neurological functions is via an interaction with the cholinergic and serotoninergic neurotransmitter systems. The suggestion of this pathway is supported by reports that have shown that selective damage to serotonergic neurons affects certain aspects of memory functions, specifically, spatial working memory [108,109]. Moreover, one of the behavioral paradigms found to be improved by ginseng and ginseng berry supplementation was that of electroconvulsive shock, which is known to modulate the cholinergic neurotransmitter system, especially within brain areas, such as the hippocampus [110-112]. It has also been proposed that ginseng and the ginseng berry enhance the components of cholinergic systems, such as choline acetyl transferase, which is also thought to be important in the formation of memory [113-115]. Although there have been a number of studies emphasizing the potential helpful effects of ginseng on cognitive performance in animal models, few epidemiological reports have been performed. In fact, a comprehensive investigation of the literature found few studies exploring the effects of ginseng on human cognitive performance, in which significant improvement in mental arithmetic and abstraction tests were reported [116-118].

Exploring other Bioactive Constituents in the Ginseng Berry and Syringaresinol

Because of its numerous potent biological activities, there have been many efforts to discover other useful components in the ginseng berry besides ginsenoside. The ginseng berry has many unique bioactive constituents compared to the generally used ginseng root. Through various assays [13,119-122], a new triterpenoid saponin, isoginsenoside-Rh3 [123], alkaloid ginsenine [124] and a dammarane-type triterpene ketone, panaxadione

[125,126], were isolated. Interestingly, ginsenoside-free molecules promote ethanol metabolism [127]. Recently, a lignin compound, syringaresinol(4,4'-(1S,3aR,4S,6aR)-tetrahydro-1H,3H-furo[3,4-c]furan-a,4-diylbis(2,6-dimethoxyphenol)), was isolated from panax ginseng pulp and found to activate SIRT1 gene expression, leading to delayed cellular senescence and improved endothelial cell function in endothelial cells [128]. Syringaresinol treatment induced the binding of FOXO3 to the SIRT1 promoter in a sequence-specific manner, leading to the induction of SIRT1 expression. Syringaresinol exists either exclusively as one enantiomer or as enantiomeric mixtures in plant foods. Recent studies revealed that (+)-syringaresinol, but not (-)-syringaresinol, up regulates SIRT1 gene expression; thus, the ginseng berry, with a predominantly high content of (+)-syringaresinol, exhibits higher activity in inducing SIRT1 gene expression. Syringaresinol has enantio selective effects upon biological activity [129]. Syringaresinol also has protective effects against hypoxia/re oxygenation-induced injury. Syringaresinol caused the destabilization of hypoxia-inducible factor 1 following hypoxia/re oxygenation and then protected cellular damage and death in a FOXO3-dependent mechanism [130]. These findings strongly suggest that the ginseng berry has potential as an effective anti-aging reagent.

Concluding Remarks

The ginseng berry is a rich source of dietary bioactivities and has various biological activities in addition to that of the ginseng root. It possesses higher ginsenoside content than its root, which has been traditionally used in herbal medicine for many human diseases and age-related attenuates. In this study, we reviewed the biological and pharmacological activities of the ginseng berry, including anti-diabetic, anti-cancer, anti-inflammation, anti-neuro degeneration, and also sexual function effects. In addition, an anti-aging component of the ginseng berry, syringaresinol, has the potential for activating the longevity genes sirt1 and foxo. According to numerous reports, the ginseng berry has the potential to be widely used as an anti-aging reagent for many age-related human diseases and to increase vitality. Basic research has suggested a number of potential mechanisms of action for ginseng berry bioactive substances, although further molecular research is necessary. Furthermore, the optimal dose of ginseng berry bioactive substances has not been determined for urinary tractor cardiovascular health. Another major deficiency concerning the evaluation of existing clinical reports is the lack of quantification of ginseng berry bioactive substances or assessment of their concentration in blood or urine. There is potent experimental evidence that ginseng berry bioactive substances have favorable effects on blood glucose metabolism, blood pressure, oxidative stress, inflammation, cancer, and endothelial function. As noted, the average daily fruit consumption is substantially less than what is recommended. In part, encouraging consumption of a greater proportion of plant foods, including fruit, to achieve a healthy dietary pattern will help to attain the recommended dietary intake of micronutrients. Although reference intake values have yet to be developed for phytochemicals, there is a growing consensus that their bioactivities importantly contribute

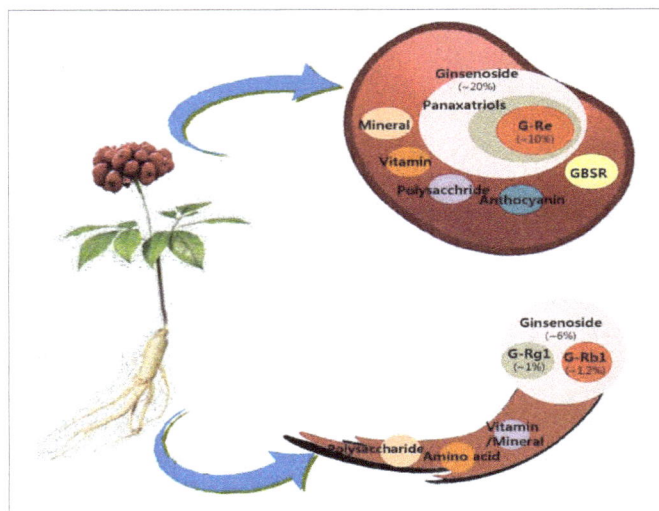

Figure 1: The different constitutions and contents of ginseng berry- and root. Ginseng berry comprises approximately 20% ginsenoside compared to 6% ginsenoside of root part. Especially, ginseng berry has more than 30 to 40 times amount of ginsenoside Re than root (0.1%) whereas root contains 1% of ginsenoside Rg1 and Rb1. Moreover ginseng berry contains 2-3 times content of crude saponin and 20-30 times of ginsenoside than root part. In addition, ginseng berry contains many of vitamins, minerals, and polyphenols as well.

Table 1: Amount and contents of ginsenoside in ginseng berry, leaf and root part.

| Part | Ginsenoside (mg/g) | | | | | | | |
| | PPD | | | | PPT | | | PPT/PPD |
	Rb1	Rb2	Rc	Rd	Re	Rg1	Rg2	
Berry	8	20	21	18	100	19	9	1.9
Leaf	5	3	2	4	42	8	4	3.8
Root	9	3	10	3	4	5	1	0.4

Ginsenoside contents and amount of ginseng berry, leaf and root [58]. PPD: protopanaxadiol; PPT: protopanaxatriol.

to promoting public health and reducing the risk of chronic diseases. Berry fruit, including the ginseng berry, represents an especially rich source of many phenolic acids and flavonoids that have been associated with these benefits. Additional research that clarifies specific dietary guidance with regard to the type of berry should help elevate our intake of these bioactive moieties.

Acknowledgements

All authors read and approved the final manuscript. I thank all lab members for the discussions. In addition to their contribution to the writing, D.B.S, S.S.S, and S.Y.C. outlined and co-edited this article. S.H.K., S.K., C.P., and H.W.P depicted the Figure data of the ginseng berry and Table 1.

References

1. Kennedy DO, Scholey AB. Ginseng: potential for the enhancement of cognitive performance and mood. Pharmacol Biochem Behav. 2003; 75(3): 687-700.

2. Xie JT, Wang CZ, Zhang B, Mehendale SR, Li XL, Sun S, et al. In vitro and in vivo anticancer effects of American ginseng berry: exploring representative compounds. Biol Pharm Bull. 2009; 32(9): 1552-8.

3. Christensen LP. Ginsenosides: chemistry, biosynthesis, analysis, and potential health effects. Adv Food Nutr Res. 2009; 55: 1-99. doi: 10.1016/S1043-4526(08)00401-4.

4. Xie J, Mehendale S, Yuan C. Ginseng and diabetes. Am J Chin Me. 2005; 33: 397-404.

5. King ML, Murphy LL. Role of cyclin inhibitor protein p21 in the inhibition of HCT116 human colon cancer cell proliferation by American ginseng and its constituents. Phytomedicine 2010; 17(3-4): 261-8. doi: 10.1016/j.phymed.2009.06.008.

6. Bucci LR. Selected herbals and human exercise performance. Am J Clin Nutr. 2000; 72(2 Suppl): 624S-36S.

7. Choi J, Kim TH, Choi TY, Lee MS. Ginseng for health care: a systematic review of randomized controlled trials in Korean literature. PLoS One. 2013; 8(4): e59978. doi: 10.1371/journal.pone.0059978.

8. Attele AS, Wu JA, Yuan CS. Multiple pharmacological effects of ginseng. Biochem Pharmacol. Biochem Pharmacol. 1999; 58(11): 1685-93.

9. USDA; U.S. Department of Health and Human Services. Dietary guidelines for Americans, 2010. 7th ed. Washington: U.S. Government Printing Office. 2010.

10. Kiefte-de Jong JC, Mathers JC, Franco OH. Nutrition and healthy ageing: the key ingredients. Proc Nutr Soc. 2014; 73(2): 249-59. Doi: 10.1017/S0029665113003881.

11. Paredes-López O, Cervantes-Ceja ML, Vigna-Pérez M, Hernández-Pérez T: Berries: improving human health and healthy aging, and promoting quality life-a review. Plant Foods Hum Nutr. 2010; 65(3): 299-308. doi: 10.1007/s11130-010-0177-1.

12. Basu A, Lyons TJ. Strawberries, blueberries, and cranberries in the metabolic syndrome: clinical perspectives. J Agric J Agric Food Chem. 2012; 60(23): 5687-92. doi: 10.1021/jf203488k.

13. Wang CZ, Zhang B, Song WX, Wang A, Ni M, Luo X et al. Steamed American ginseng berry: ginsenoside analysis and anticancer activities. J Agric Food Chem. 2006; 54(26): 9936-42.

14. Yin J, Zhang H, Ye J. Traditional Chinese medicine in treatment of metabolic syndrome. Endocr Metab Immune Disord Drug Targets. 2008; 8(2): 99-111.

15. Kim TH, Lee SM. The effects of ginseng total saponin, panaxadiol and panaxatriol on ischemia/reperfusion injury in isolated rat heart. Food Chem Toxicol. 2010; 48(6): 1516-20. doi: 10.1016/j.fct.2010.03.018.

16. Chang YS, Seo EK, Gyllenhaal C, Block KI.C. Panax ginseng: a role in cancer therapy? Integr Cancer Ther. 2003; 2(1): 13-33.

17. Nakamura S, Sugimoto S, Matsu. Chem Pharm da H, Yoshikawa M. Medicinal flowers from flower buds of American ginsen. Panaxquinquefolium L Chem Pharm Bull (Tokyo). 2007; 55(9): 1342-8.

18. Chenling Q, Yuping Bai, Xiangqun Jin, Yutang Wang, Kun Zhang, Jingyan You, et al. Study on ginsenosides in different parts and ages of Panaxquinquefolius. 2009; 115(1): 340–6.

19. Kim CK, Cho DH, Lee KS, Lee DK, Park CW, Kim WG, et al. Ginseng berry extract prevents atherogenesis via anti-inflammatory action by upregulation phase II gene expression. Evid Based Complement Alternat Med. 2012; 2012: 490301. doi: 10.1155/2012/490301.

20. Joo KM, Lee JH, Jeon HY, Park CW, Hong DK, Jeong HJ, et al.

Pharmacokinetic study of ginsenoside Re with pure ginsenoside Re and ginseng berry extracts in mouse using ultra performance liquid chromatography/mass spectrometric method. J Pharm Biomed Anal. 2010; 51(1): 278-83. doi: 10.1016/j.jpba.2009.08.013.

21. Lee IA, Hyam SR, Jang SE, Han MJ, Kim DH. Ginsenoside Re ameliorates inflammation by inhibiting the binding of lipopolysaccharide to TLR4 on macrophages. J Agric Food Chem. 2012; 60(38): 9595-602.

22. Paul S, Shin HS, Kang SC. Inhibition of inflammations and macrophage activation by ginsenoside-Re isolated from Korean ginseng (Panax ginseng C.A.Meyer). Food Chem Toxicol. 2012; 50(5): 1354-61. doi: 10.1016/j.fct.2012.02.035.

23. Son YM, Kwak CW, Lee YJ, Yang DC, Park BC, Lee WK, et al. Ginsenoside Re enhances survival of human CD4+ T cells through regulation of autophagy. Int Immunopharmacol. 2010; 10(5): 626-31. doi: 10.1016/j.intimp.2010.03.002.

24. Han DH, Kim SH, Higashida K, Jung SR, Polonsky KS, Klein S, et al. Ginsenoside Re rapidly reverses insulin resistance in muscles of high-fat diet fed rats. Metabolism. 2012; 61(11): 1615-21. doi: 10.1016/j.metabol.2012.04.008.

25. Zhang Z, Li X, Lv W, Yang Y, Gao H, Yang J, et al. Ginsenoside Re reduces insulin resistance through inhibition of c-Jun NH2-terminal kinase and nuclear factor-kappaB. Mol Endocrinol. 2008; 22(1): 186-95.

26. Quan HY, Yuan HD, Jung MS, Ko SK, Park YG, Chung SH. Ginsenoside Re lowers blood glucose and lipid levels via activation of AMP-activated protein kinase in HepG2 cells and high-fat diet fed mice Int J Mol Med. 2012; 29(1): 73-80. doi: 10.3892/ijmm.2011.805.

27. Lee OH, Lee HH, Kim JH, Lee BY. Effect of ginsenoside Rg3 and Re on glucose transport in mature 3T3-L1 adipocytes. Phytother Res. 2011; 25(5): 768-73. doi: 10.1002/ptr.3322.

28. Xie JT, Mehendale SR, Li X, Quigg R, Wang X, Wang CZ, et al. Anti-diabetic effects of ginsenoside Re in ob/ob mice. Biochim Biophys Acta. 2005; 1740(3): 319-25.

29. Lin E, Wang Y, Mehendale S, Sun S, Wang CZ, Xie JT, et al. In pancreatic beta-cells. Am J. Antioxidant protection by American ginseng. Am J Chin Med. 2008; 36(5): 981-8.

30. Cho WC, Chung WS, Lee SK, Leung AW, Cheng CH, Yue KK. Ginsenoside Re of Panax ginseng possesses significant antioxidant and antihyperlipidemic efficacies in streptozotocin-induced diabetic rats. Eur J Pharmacol. 2006; 550(1-3): 173-9.

31. Liu YW, Zhu X, Li W, Lu Q, Wang JY, Wei YQ, et al. Ginsenoside Re attenuates diabetes-associated cognitive deficits in rats. Pharmacol Biochem Behave. 2010; 101(1): 93-8. doi: 10.1016/j.pbb.2011.12.003.

32. Jin ZQ. The action of ginsenoside Re on inotropy and chronotropy of isolated atria prepared from guinea pigs. Planta Med. 1996; 62(4): 314-6.

33. Wang Y, Yuan CS, Lipsius S. Ginsenoside Re experts anti-arrhythmic effects in cat ventricle myocytes. Exp Biol Abstr. A277.

34. Kang SY, Schini-Kerth VB, Kim ND. Ginsenosides of the protopanaxatriol group cause endothelium-dependent relaxation in the rat aorta. Life Sci. 1995; 56(19): 1577-86.

35. Wang YG, Zima AV, Ji X, Pabbidi R, Blatter LA, Lipsius SL. Ginsenoside Re suppresses electromechanical alternans in cat and human cardiomyocytes Am J Physiol Heart Circ Physiol. 2008; 295(2): H851-9. doi: 10.1152/ajpheart.01242.2007.

36. Chen CX, Zhang HY. Protective effect of ginsenoside Re on isoproterenol-induced triggered ventricular arrhythmia in rabbits.

Zhongguo Dang Dai Er Ke Za Zhi. 2009; 11(5): 384-8.

37. Liu Z, Li Z, Liu X. Effect of ginsenoside Re on cardiomyocyte apoptosis and expression of Bcl-2/Bax gene after ischemia and reperfusion in rats. J HuazhongUnivSciTechnolog J Huazhong Univ Sci Technolog Med Sci. 2002; 22(4): 305-9.

38. Chen LM, Zhou XM, Cao YL, Hu WX. Neuroprotection of ginsenoside Re in cerebral ischemia-reperfusion injury in rats. J Asian Nat Prod Res. 2008; 10(5-6): 439-45 doi: 10.1080/10286020801892292.

39. Scott GI, Colligan PB, Ren BH, Ren J. Ginsenosides Rb1 and Re decrease cardiac contraction in adult rat ventricular myocytes: role of nitric oxide. Br J Pharmacol. 2001; 134(6): 1159-65.

40. Huang YC, Chen CT, Chen SC, Lai PH, Liang HC, Chang Yal. A natural compound isolated from Panax ginseng as a novel angiogenic agent for tissue regeneration. Pharm Res. 2005; 22(4): 636-46.

41. Yu LC, Chen SC, Chang WC, Huang YC, Lin KM, Lai PH, et al. Stability of angiogenic agents, ginsenoside Rg1 and Re, isolated from Panax ginseng: in vitro and in vivo studies. Int J Pharm. 2007; 328(2): 168-76.

42. Bai CX, Sunami A, Namiki T, Sawanobori T, Furukawa T. Electrophysiological effects of ginseng and ginsenoside Re in guinea pig ventricular myocytes. Eur J Pharmacol. 2003; 476(1-2): 35-44.

43. Bai CX, Takahashi K, Masumiya H, Sawanobori T, Furukawa T. Nitric oxide-dependent modulation of the delayed rectifier K+ current and the L-type Ca2+ current by ginsenoside Re, an ingredient of Panax ginseng, in guinea-pig cardiomycytes. Br J Pharmacol. 2004; 142(3): 567-75.

44. Kim HS, Lee JH, Goo YS, Nah SY. Effects of ginsenosides on Ca2+ channels and membrane capacitance in rat adrenal chromaffin cells. Brain Res Bull. 1998; 46: 245-251.

45. Jin ZQ, Liu CM. Effect of ginsenoside Re on the electrophysiological activity of the heart. Planta Med. 1994; 60(2): 192-3.

46. Shi J, Xue W, Zhao WJ, Li KX. Pharmacokinetics and dopamine/acetylcholine releasing effects of ginsenoside Re in hippocampus and mPFC of freely moving rats. Acta Pharmacol Sin. 2013; 34(2): 214-20. doi: 10.1038/aps.2012.147.

47. Ji ZN, Dong TT, Ye WC, Choi RC, Lo CK, Tsim KW. Ginsenoside beta-amyloid Re attenuate and serum-free induced neurotoxicity in PC12 cells. J Ethnopharmacol. 2006; 107(1): 48-52.

48. Chen F, Eckman EA, Eckman CB. Reductions in levels of the Alzheimer's amyloid beta peptide after oral administration of ginsenosides. FASEB J. 2006; 20: 1269-1271.

49. Xu BB, Liu CQ, Gao X, Zhang WQ, Wang SW, Cao YL. Possible mechanisms of the protection of ginsenoside Re against MPTP-induced apoptosis in substantia nigra neurons of Parkinson's disease mouse model. J Asian Nat Prod Res. 2005; 7(3): 215-24.

50. Lee B, Shim I, Lee H, Hahm DH. Effects of ginsenoside Re on depression- and anxiety-like behaviors and cognition memory deficit induced by repeated immobilization in rats. J MicrobiolBiotechnol.2012; 22: 708-720.

51. López MV, Cuadrado MP, Ruiz-Poveda OM, Del Fresno AM, Accame ME. Neuroprotection effect of individual ginsenoside on astrocytes primary culture. Biochim Biophys Acta. 2007; 1770(9): 1308-16.

52. Lee KW, Jung SY, Choi SM, Yang EJ. Effects of ginsenoside Re on LPS-induced inflammatory mediators in BV2 microglial cells. BMC Complement Altern Med. 2012; 12: 196. doi: 10.1186/1472-6882-12-196.

53. Kim KH, Song K, Yoon SH, Shehzad O, Kim YS, Son JH. Rescue of PINK1 protein null-specific mitochondrial complex IV deficits by ginsenoside Re activation of nitric oxide signaling. J Biol Chem. 2012; 287(53): 44109-20. doi: 10.1074/jbc.M112.408146.

54. Zhang H, Zhou Q, Li X, Zhao W, Wang Y, Liu H, Li N, et al. Ginsenoside Re promotes human sperm capacitation through nitric oxide-dependent pathway. Mol Reprod Dev. 2007; 74: 497-501.

55. Zhang H, Zhou QM, Li XD, Xie Y, Duan X, Min FL, et al. Ginsenoside Re increases fertile and asthenozoospermic infertile human sperm motility by induction of nitric oxide synthase. Arch Pharm Res. 2006; 29(2): 145-51.

56. Bae EA, Shin JE, Kim DH. Metabolism of ginsenoside Re by human intestinal microflora and its estrogenic effect. Biol Pharm Bull. 2005; 28(10): 1903-8.

57. Leung KW, Leung FP, Huang Y, Mak NK, Wong RN. Non-genomic effects of ginsenoside Re in endothelial cells via glucocorticoid receptor. FEB S Lett. 2009; 581(13): 2423-8.

58. Zhao YQ, Yuan CL. Chemical constituents of the fruit of Panax ginseng C. A. Meyer. Zhongguo Zhong Yao Za Zhi. 1993; 18(5): 296-7, 319.

59. Kimura M, Waki, Chujo T, Kikuchi T, Hiyama C, Yamazaki K, et al. Effects of hypoglycemic components in ginseng radix on blood insulin level in alloxan diabetic mice and on insulin release from perfused rat pancreas. J Pharmacobiodyn. 1981; 4(6): 410-7.

60. Dey L, Xie JT, Wang A, Wu J, Maleckar SA, Yuan CS. Anti-hyperglycemic effects of ginseng: Comparison between root and berry. Phytomedicine. 2003; 10(6-7): 600-5.

61. Park EY, Kim HJ, Kim YK, Park SU, Choi JE, Cha JY, et al. Increase in insulin secretion induced by panax ginseng berry extracts contributes to the amelioration of hyperglycemia in streptocin induced diabetic mice. J Ginseng Res. 2012; 36(2): 153-60. doi: 10.5142/jgr.2012.36.2.153.

62. Xie JT, Wu JA, Mehendale S, Aung HH, Yuan CS. Anti-hyperglycemic effect on the polysaccharides fraction from American ginseng berry extract in ob/ob mice. Phytomedicine. 2004; 11(2-3): 182-7.

63. Attele AS, Zhou YP, Xie JT, Wu JA, Zhang L, Dey L, et al. Antidiabetic effects of Panax ginseng berry extract and the identification of an effective component. Diabetes. 2002; 51(6): 1851-8.

64. Xie JT, Zhou YP, Dey L, Attele AS, Wu JA, Gu M,et al. Ginseng berry reduces blood glucose and body weight in db/db mice. Phytomedicine. 2002; 9(3): 254-8.

65. Xie JT, Aung HH, Wu JA, Attel AS, Yuan CS. Effects of American ginseng berry extract on blood glucose levels in ob/ob mice. Am J Chin Med. 2002; 30(2-3): 187-94.

66. Xie JT, Wang CZ, Ni M, Wu JA, Mehendale SR, Aung HH, et al. American ginseng berry juice intake reduces blood glucose and body weight in ob/ob mice. J Food Sci. 2007; 72(8): S590-4.

67. Kim ST, Kim HB, Lee KH, Choi YR, Kim HJ, Shin IS, et al. Steam-dried ginseng berry fermented with lactobacillus plantarum controls the increase of blood glucose and body weight in type 2 obese diabetic db/db mice. J Agric Food Chem. 2012; 60(21): 5438-45. doi: 10.1021/jf300460g .

68. Dey L, Zhang L, Yuan CS. Dey L, Zhang L, Yuan CS. Anti-diabetic and anti-obese effects of ginseng berry extract: comparison between intraperitoneal and oral administrations. Am J Chin Med. 2002; 30(4): 645-7.

69. Yuan CS, Tanaka H. Bioactivity of American ginseng by knockout extract preparing using monoclonal antibody. Curr Drug Discov

Technol. 2011; 8(1): 32-41.

70. Seo E, Kim S, Lee SJ, Oh BC, Jun HS. Ginseng berry extract supplementation improves age-related decline of insulin signaling in mice. Nutrients. 2015; 7(4): 3038-53. doi: 10.3390/nu7043038.

71. Zhang G, Huihua G, Yi L. Stability of halophilic proteins: from dipeptide attributes to discrimination classifier. Int J Biol Macromol. 2013; 53: 1-6. doi: 10.1016/j.ijbiomac.2012.10.031.

72. Byeon SE, Lee J, Kim JH, Yang WS, Kwak YS, Kim SY, et al. Molecular mechanism of macrophage activation by red ginseng acidic polysaccharide from Korean red ginseng. Mediators Inflamm. 2012; 2012: 732860. doi: 10.1155/2012/732860.

73. Wang W, Zhao Y, Rayburn ER, Hill DL, Wang H, Zhang R. In vitro anti-cancer activity and structure-activity relationships of natural products isolated from fruits of Panax ginseng. Cancer Chemother Pharmacol. 2007; 59(5): 589-601.

74. Wang CZ, Xie JT, Fishbein A, Aung HH, He H, Mehendale SR, et al. Antiproliferative effects of different plant parts of Panaxnotoginseng on SW480 human colorectal cancer cells. Phytother Res. 2009; 23(1): 6-13. doi: 10.1002/ptr.2383.

75. Li XL, Wang CZ, Sun S, Mehendale SR, Du W, He TC, et al. American ginseng berry enhances chemopreventive effect of 5-FU on human colorectal cancer cells. Oncol Rep. 2009; 22(4): 943-52.

76. Xie JT, Du GJ, McEntee E, Aung HH, He H, Mehendale SR, et al. Effects of triterpenoid glycosides from fresh ginseng berry on SW480 human colorectal cancer cell line. Cancer Res Treat. 2011; 43(1): 49-55. doi: 10.4143/crt.2011.43.1.49.

77. Jang HJ, Han IH, Kim YJ, Yamabe N, Lee D, Hwang GS, et al. Anticarcinogenic effects of products of heat-processed ginsenoside Re, a major constituent of ginseng berry, on human gastric cancer cells. J Agric Food Chem. 2014; 62(13): 2830-6. doi: 10.1021/jf5000776.

78. Hao M, Wang W, Zhao Y, Zhang R, Wang H. Pharmacokinetics and tissue distribution of 25-hydroxyprotopanaxadiol, an anti-cancer compound isolated from Panax ginseng, in athymic mice bearing xenografts of human pancreatic tumors. Eur J Drug Metab Pharmacokinet. 2011; 35(3-4): 109-13. doi: 10.1007/s13318-010-0022-9.

79. Lee S, Kim MG, Ko SK, Kim HK, Leem KH, Kim YJ. Protective effect of ginsenoside Re on acute gastric mucosal lesion induced by compound 48/80. J Ginseng Res. 2014; 38(2): 89-96. doi: 10.1016/j.jgr.2013.10.001.

80. Mehendale SR, Aung HH, Yin JJ, Lin E, Fishbein A, Wang CZ et al. Effects of antioxidant herbs on chemotherapy-induced nausea and vomiting in a rat-pica model. Am J Chin Med. 2004; 32(6): 897-905.

81. Mehendale S, Aung H, Wang A, Yin JJ, Wang CZ, Xie JT, et al. American ginseng berry extract and ginsenoside Re attenuate cisplatin-induced kaolin intake in rats. Cancer Chemother Pharmacol. 2005; 56(1): 63-9.

82. Zhang W, Cho SY, Xiang G, Min KJ, Yu Q, Jin JO.Ginseng berry extract promotes maturation of mouse dendritic cells. PLoS One. 2015; 10(6): e0130926. doi: 10.1371/journal.pone.0130926.

83. Nam KW, Kim J, Hong JJ, Choi JH, Mar W, Cho MH, et al. Inhibition of cytokine-induced IkB kinase activation as a mechanism contributing to the anti-atherogenic activity of tilianin in hyperlipidemic mice. Atherosclerosis. 2005; 180(1): 27-35.

84. Brand K, Page S, Rogler G, Bartsch A, Brandl R, Knuechel R, et al. Activated transcription factor nuclear factor-kappa B is present in the atherosclerotic lesion. J Clin Invest. 1996; 97(7): 1715-22.

85. Shao ZH, Xie JT, Vanden Hoek TL, Mehendale S, Aung H, Li CQ,

et al. Antioxidant effects of American ginseng berry extract in cardiomyocytes exposed to acute oxidant stress. Biochim Biophys Acta. 2004; 1670(3): 165-71.

86. Xie JT, Shao ZH, Vanden Hoek TL, Chang WT, Li J, Mehendale S, et al. Antioxidant effects of ginsenoside Re in cardiomyocytes. Eur J Pharmacol. 2006; 532(3): 201-7.

87. Bae HM, Cho OS, Kim SJ, Im BO, Cho SH, Lee S, et al. Inhibitory effects of ginsenoside Re isolated from ginseng berry on histamine and cytokine release in human mast cells and human alveolar epithelial cells. J Ginseng Res. 2012; 36(4): 369-74. doi: 10.5142/jgr.2012.36.4.369.

88. Mehendale SR, Wang CZ, Shao ZH, Li CQ, Xie JT, Aung HH, et al. Chronic pretreatment with American ginseng berry and its polyphenolic constituents attenuate oxidant stress in cardiomyocytes. Eur J Pharmacol. 2006; 553(1-3): 209-14.

89. Lei Y, Gao Q, Chen KJ. Effects of extracts from Panaxnotoginseng and Panax ginseng fruit on vascular endothelial cell proliferation and migration in vitro. Chin J Integr Med. 2008; 14(1): 37-41. doi: 10.1007/s11655-008-0037-0.

90. Feldman HA, Goldstein I, Hatzichristou DG, Krane RJ, McKinlay JB. Impotence and its medical and psychosocial correlates: results of the Massachusetts male aging study. J Urol. 1994; 151(1): 54-61.

91. Laumann EO, Nicolosi A, Glasser DB, Paik A, Gingell C, Moreira E, et al. Sexual problems among women and men aged 40-80y: prevalence and correlates identified in the global study of sexual attitudes and behaviors. Int J Impot Res. 2005; 17(1): 39-57.

92. Moreira SG Jr, Brannigan RE, Spitz A, Orejuela FJ, Lipshultz LI, Kim ED. Side-effect profile of sildenafil citrate (Viagra) in clinical practice. Urology. 2000; 56 (3): 474-6.

93. Ferguson JM. SSRI antidepressant medications: adverse effects and tolerability. Prim Care Companion J Clin Psychiatry. 2001; 3(1): 22-27.

94. Li Z, Niwa Y, Sakamoto S, Shono M, Chen X, Nakaya Y. Induction of inducible nitric oxide synthase by ginsenosides in cultured porcine endothelial cells. Life Sci. 2000; 67(24):2983-9.

95. Kim ND, Kim EM, Kang KW, Cho MK, Choi SY, Kim SG. Ginsenoside R3 inhibits phenylephrine-induced vascular contraction through induction of nitric oxide synthase. Br J Pharmacol. 2003; 140(4): 661-70.

96. Wang X, Chu S, Qian T, Chen J, Zhang J. Ginsenoside Rg1 improves male copulatory behavior via nitric oxide/cyclic guanosine moniphosphate pathway. J Sex Med. 2010; 7(2 Pt 1): 743-50. doi: 10.1111/j.1743-6109.2009.01482.x

97. Choi YD, Park CW, Jang J, Kim SH, Jeon HY, Kim WG, et al. Effects of Korean ginseng berry extract on sexual function in men with erectile dysfunction: a multicenter, placebo-controlled, double-blind clinical study. Int J Impot Res. 2013; 25(2): 45-50. doi: 10.1038/ijir.2012.45

98. Cho KS, Park CW, Kim CK, Jeon HY, Kim WG, Lee SJet al. Effects of Korean ginseng berry extract on penile erection: evidence from in vitro and in vivo studies. Asian J Androl. 2013; 15(4): 503-7. doi: 10.1038/aja.2013.49.

99. Gemma C, Mesches MH, Sepesi B, Choo K, Holmes DB, Bickford PC, et al. Diets enriched in foods with high antioxidant activity reverse age-induced decreases in cerebellar beta-adrenergic function and increases in proinflammatory cytokines. J Neurosci. 2002; 22(14): 6114-20.

100. Wang Y, Chang CF, Chou J, Chen HL, Deng X, Harvey BK, et al. Dietary supplementation with blueberries, spinach, or spirulina reduces

ischemic brain damage. Exp Neurol. 2005; 193(1): 75-84.

101. Joseph JA, Shukitt-Hale B, Casadesus G. Reversing the deleterious effects of aging on neuronal communication and behavior: beneficial properties of fruit polyphenolic compounds. Am J Clin Nutr. 2005; 81(1 Suppl): 313S-316S.

102. Petkov V. Effects of ginseng on the brain biogenic monoamines and 3', 5'– AMP system. Experiments on rats. Arzneimittelforschung. 1978. 28(3): 388-93.

103. Jaenicke B, Kim EJ, Ahn JW, Lee HS. Effects of Panax ginseng extract on passive avoidance retention in old rats. Arch Pharm Res. 1991; 14(1): 25-9.

104. Petkov VD, Mosharrof AH. Effects of standardized ginseng extract on learning, memory and physical capabilities. Am J Chin Med. 1987; 15(1-2): 19-29.

105. Petkov VD, Kehayov R, Belcheva S, Konstantinova E, Petkov VV, Getova D, et al. Memory effects of standardized extracts of Panax ginseng, Ginkgo biloba and their combination Gincosan. Planta Med. 1993; 59(2): 106-14.

106. Nishiyama N, Chu PJ, Saito H. An herbal prescription, S-113m, consisting of biota, ginseng and schizandra, improves learning performance in senescence accelerated mouse. Biol Pharm Bull. 1996; 19(3): 388-93.

107. Kuo Y, Ikegami F, Lambein F. Neuroactive and other free amino acids in seed and young plants of Panax ginseng. Phytochemistry. 2003; 62(7): 1087-91.

108. Lehmann O, Jeltsch H, Lehnardt O, Pain L, Lazarus C, Cassel JC. Combined lesions of cholinergic and serotonergic neurons in the rat brain using 192 IgG-saponin and 5, 7-dihydroxytryptamine: neurochemical and behavioural characterization. Eur J Neurosci. 2000; 12(1): 67-79.

109. Balse E, Lazarus C, Kelche C, Jeltsch H, Jackisch R, Cassel JC. Intrahippocampal grafts containing cholinergic and serotonergic fetal neurons ameliorate spatial reference but not working memory in rats with fimbria-fornix/cingular bundle lesions. Brain Res Bull. 1999; 49(4): 263-72.

110. Mingo NS, Cottrell GA, Mendonca A, Gombos Z, Eubanks JH, Burham WM. Amygdala-kindled and electroconvulsive seizures alter hippocampal expression of the m1 and m3 muscarinic cholinergic receptor genes. Brain Res. 1998; 810(1-2): 9-15.

111. Vann SD, Brawn MW, Erichsen JT, Aggleton JP. Fos imaging reveals differential patterns of hippocampal and para-hippocampal subfield activation in rats in response to different spatial memory tests. J Neurosci. 2000; 20(7): 2711-2718.

112. Zhao W, Chen H, Xu H, Moore E, Meiri N, Quon MJ, et al. Brain insulin receptors and spatial memory. Correlated changes in gene expression, tyrosine phosphorylation, and signaling molecules in the hippocampus of water maze trained rats. J Biol Chem. 1999; 274(49): 34893-902.

113. Stancampiano R, Cocco S, Cugusi C, Sarais L, Fadda F. Serotonin and acetylcholine release response in the rat hippocampus during a spatial memory task. Neuroscience. 1999; 89(4): 1135-43.

114. Darnaudery M, Koehl M, Piazza PV, Le Moal M, Mayo W. Pregnenolone sulfate increases hippocampal acetylcholine release and spatial recognition. Brain Res. 2000; 852(1): 173-9.

115. Meck WH, Williams CL. Choline supplementation during prenatal development reduces proactive interference in spatial memory. Brain Res Dev Brain Res. 1999; 118(1-2): 51-9.

116. D'Angelo L, Grimaldi R, Caravaggi M, Marcoli M, Perucca E, Lecchini S, et al. A double-blind placebo-controlled clinical study on the effect of a standardized ginseng extract on psychomotor performance in healthy volunteers. J Ethnopharmacol.1986; 16(1): 15-22.

117. Sorensen H, Sonne J. A double-masked study of the effect of ginseng on cognitive functions. CurrTher Res. 1996; 57: 959-968.

118. Winther K, Ranlov C, Rein E, Mehlsen J. Russian root improves cognitive functions in middle-aged people whereas Ginkgo biloba seems effective only in the elderly. J Neurol Sci. 1997; 150: S90.

119. Wang CZ, Wu JA, McEntee E, Yuan CS. Saponinscompositon in American ginseng leaf and berry assayed by high-performance liquid chromatography. J Agric Food Chem. 2006; 54(6): 2261-6.

120. Sritularak B, Morinaga O, Yuan CS, Shoyama Y, Tanaka H. Quantitive analysis of ginsenosides Rb1, Rg1, and Re in American ginseng berry and flower samples by ELISA using monoclonal antibodies. J Nat Med. 2009; 63(3): 360-3. doi: 10.1007/s11418-009-0332-x.

121. Kim YK, Yoo DS, Xu H, Park NI, Kim HH, Choi JE, et al. Ginsenoside content of berries and roots of three typical Korean ginseng cultivars. Nat Prod Commun. 2009; 4: 903-6.

122. Morinaga O, Uto T, Yuan CS, Tanaka H, Shoyama Y. Evaluation of a new eastern blotting technique for the analysis of ginsenoside Re in American ginseng berry pulp extracts. Fitoterapia. 2010; 81(4): 284-8. doi: 10.1016/j.fitote.2009.10.005.

123. Wang JY, Li XG, Zheng YN, Yang XW. Isoginsenoside-Rh3, a new triterpenoid saponin from the fruits of Panax ginseng C. A. Mey. J Asian Nat Prod Res. 2004; 6(4):2 89-93.

124. Wang JY, Li XG, Yang XW. Ginsenine, a new alkaloid from the berry of Panax ginseng C. A. Meyer. J Asian Nat Prod Res. 2006; 8(7): 605-8.

125. Sugimoto S, Nakamura S, Matsuda H, Kitagawa N, Yoshikawa M. Chemical constituent from seeds of Panax ginseng: structure of new dammarane-type triterpene ketone, panaxadione, and hplc comparisons of seeds and flesh. Chem Pharm Bull (Tokyo). 2009; 57(3): 283-7.

126. Zhao JM, Li N, Zhang H, Wu CF, Piao HR, Zhao YQ. Novel dammarane-type sapogenins from Panax ginseng berry and their biological activities. Bioorg Med Chem Lett. 2011; 21(3): 1027-31. doi: 10.1016/j.bmcl.2010.12.035.

127. Lee do I, Kim ST, Lee DH, Yu JM, Jang SK, Joo SS.. Ginsenoside-free molecules from steam-dried ginseng berry promote ethanol metabolism: an alternative choice for an alcohol hangover. J Food Sci. 2014; 79(7): C1323-30. doi: 10.1111/1750-3841.12527.

128. Cho SY, Cho M, SEO DB, Lee SJ, Suh Y. Identification of a small molecule activator of SIRT1 gene expression. Aging (Albany NY). 2013; 5(3): 174-82.

129. Park HW, Cho SY, Kim HH, Yun BS, Kim JU, Lee SJ, et al. Enantioselective induction of SIRT1 gene by syringaresinol from Panax ginseng berry and Acanthopanaxsenticosus harms stem. Bioorg Med Chem Lett. 2015; 25(2): 307-9. doi: 10.1016/j.bmcl.2014.11.045.

130. Cho SY, Cho M, Kim J, Kaeberlein M, Lee SJ, Suh Y. Syringaresinol protects against hypoxia/reoxygenation-induced cardiomyocytes injury and death by destabilization of HIF-1a in a FOXO3-dependent mechanism. Oncotarget. 2014; 16(1): 43-55.

Biochemical profiling of antifungal activity of betel leaf (*Piper betle L.*) extract and its significance in traditional medicine

Sarika Pawar[1#], Vidya Kalyankar[2], Bela Dhamangaonkar[1], Sharada Dagade[3],
Shobha Waghmode[2]* and Abhishek Cukkemane[1]

[1]*Bijasu Agri Research Laboratory LLP, Sr. No. 37, Kondhwa Industrial Estate, Khadi Machine Chowk, Kondhwa, Pune-411048, India*
[2]*Department of Chemistry, M. E. S Abasaheb Garware College, Pune - 411 007, Maharashtra, India*
[3]*Department of Chemistry, Y. M. College, Bharathi Vidyapeeth, Pune-411038, Maharashtra, India*
[#]*Present Address: Microbial Diversity Research Centre, Dr. D. Y. Patil Biotechnology and Bioinformatics Institute, Dr. D. Y. Patil Vidyapeeth, Tathawade, Pune-411033, India*

Corresponding author: *ShobhaWaghmode, Department of Chemistry, M. E. S Abasahaeb garware College, Pune - 411 007, Maharashtra, India;*
; E-mail: shobhawaghmode@yahoo.co.in

Abstract

Piper betle (Linn) commonly called as betel leaf is a widely cultivated plant in the Indian subcontinent. The traditional Indian ayurvedic document describes several of its medicinal properties including as an effective antifungal agent. The present study was conducted to evaluate the secondary metabolite that contributes to its antifungal activity. *Invitro* studies were performed on molds and yeasts on antifungal using fractions obtained from ethyl acetate, hexane and ethanol-methanol extracts by well-diffusion technique. Ethyl acetate extracts showed highest anti-fungal activity. Using biophysical techniques such as Nuclear Magnetic Resonance and Fourier transform infrared spectroscopy techniques; we identified the molecule as derivative of the phenyl propanoid family akin eugenol. The molecule can be readily purified using a 2 step solvent extraction procedure along with silica column chromatography. These findings reveal the antifungal and possible commercial potential of the plant extract and its potential in agriculture against pest management and food spoilage.

Aim: To determine the antifungal activity of *Piper betle* L. extracts.

Keywords: Antifungal; Ayurveda; Fourier transform infra-red spectroscopy; Nuclear magnetic resonance spectroscopy; Food spoilage; Agriculture.

Introduction

In recent years, the increase in resistance in known fungal pathogens to the available antifungal drugs has raised enormous challenges to public health issues [1-3]. In addition, conventional antifungal drugs have undesirable side effects and are very toxic such as chlorhexidine, imidazole and amphotericin B [4]. One important demand of critical significance in this context is to search for novel antifungal agents that would be less toxic and more effective. Interestingly; several medicinal plants have been extensively investigated in order to find novel bioactive compounds [5]. Moreover, several studies have suggested that a number of plant species possesses promising antimicrobial compounds [6-9]. *Piper betle* Linn. (*Piperaceae*), a slender creeping plant, is widely distributed in India, Sri Lanka, Thailand and other tropical countries. This plant has deep green heart shaped, smooth, shinning and long stalked leaves, with pointed apex. Betel leaf possess strong aromatic flavor and have been long in use for the preparation of traditional Indian ayurvedic herbal remedies. It has been reported for the treatment of various diseases such as conjunctivitis, boils and abscesses, cuts and injuries etc. [5,10]. In addition, it also acts as a breath freshener, a digestive and pancreatic lipase stimulant and a pain killer in joint pain [11,13]. Even though all these positive effects of betel leaf are known, the biochemicals of these favorable effects remain obscure. Moreover, betel leaf extract has previously been shown to have strong antimicrobial effects in review [14-19]. A couple of research articles [20,21] have demonstrated the potential of the leaf extract on dermatophytes. Here in, we show that the leaf extract possess antifungal activity against various plant pathogens. The present study was sought to investigate the effects of ethyl acetate, hexane and ethanol-methanol extracts of this plant leave on fungal pathogens.

Here in, we employed biophysical techniques to identify the active metabolite that provide the betel leaf extract with its anti-fungal and other pharmacological properties that have been valued in traditional Indian ayurveda.

Materials and methods

Betel leaf extract

The crude extract and isolation was done using Silica Gel (100-200 mesh) for column chromatography and HPLC grade solvents. Betel leaf was collected from Western Maharashtra farm (Indapur tehsil), India. It was cleaned first with distilled water and dried in shadow for a week. The dried leaves were powdered and 100 g was used for extraction. The powder was transferred into 1 L conical flask with 500 mL ethanol to completely soak the powder, which was incubated at room temperature for 24 hours.

The sample was filtered using ordinary filter paper directly into the clean round bottom flask and set for distillation. We obtained up to 30 mL of distillate from a single run under controlled heating at 55 ° C under reduced pressure. We repeated the procedure 3 times and pulled the collected distillate containing crude extract. The crude extract was concentrated on a water bath at 55 ° C under reduced pressure.

During the extraction optimization step, we added 100 mL ethyl acetate/ethanol-methanol/ hexane to the concentrated 100 mL crude extract and mixed well. The mixture was separate during separating funnel and the ethyl acetate/n-hexane fraction was collected in another conical flask. All 3 extracts were concentrated in procedure mentioned above and was further purified using column chromatography using the respective solvent .The eluent was pooled and analyzed by thin-layered chromatography (TLC), Silica gel 60 F254, preloaded Silica gel on Alumina sheets with ethyl acetate as the solvent and observed under UV chamber and Iodine vapors.

Fungal strains and growth conditions

We tested the effect of betel leaf extract on opportunistic fungal pathogen *Aspergillus niger,* which is the causative agent of black mold in several fruits and vegetables [22]. We also isolated saprophytic and opportunistic pathogenic fungi *Rhizopus sp.* [23] from the leaves of *Murraya koenigii.* (curry leaves). Lastly, we isolated a mold from the leaf of figs infested with rust, which was not obligate parasite *Cerotelium (Physopella) fici* [24] but we have tentatively identified it as wild *Aspergillus sp.* These cultures were cultured and maintained on Potato Dextrose Agar (PDA- HIMEDIA) at 4 ° C. A stock inoculums spore suspension of each fungal culture was prepared from fresh, mature (3-daysold) cultures grown on potato dextrose agar plates at 28 ° C.

In-vitro antifungal assay

Antifungal activity of betel leaf extract was tested against *A.niger,* wild *Aspergillus sp.* and *Rhizopus sp.* of the ethyl acetate, hexane and ethanol-methanol extracts of betel leaf sample was tested by well diffusion method. In brief, 500 µL of fungal spore suspension was added to 20 mL PDA medium and poured in petri dish. After solidification, wells of 5mm in diameter were made on this plate. Each well was filled with 50 µL of ethyl acetate, hexane and ethanol-methanol herbal extract. Potassium tellurite was used as positive control and ethyl acetate, hexane and ethanol-methanol solvent used as negative control. The antifungal assay plates were incubated at 30 ° C for 36h. The antifungal activities of the extracts were determined by measuring the diameter of the inhibition zone in millimetres (mm).

NMR and FTIR

The IR spectra of neat sample were recorded on Nicolette iD5, Thermo scientific at room temperature. Standard[1] H NMR spectra was recorded on Jeol 200MHz using CDCl$_3$solvent and TMS (Euriso-top) as a reference at room temperature.

Results

Isolation of active ingredient from *Piper betle* extract

The beetle leaf extract was purified to homogeneity. During the solvent extraction steps, we observed three spots on the TLC plates. After solvent extraction the crude extract was concentrated and run on a column containing 100-200 mesh Silica Gel and Sodium Sulfate for further purification. In all three solvent types that we tested we observed a single spot from there fraction.

Antifungal susceptibility assay

In the next step, we studied the effect of the organic extract against a range of bacterial and fungal cultures. In all tested cases, we observed antimicrobial effect of the extract against fungal cultures only. This is in stark contrast to preciously reported data where few groups have observed anti-bacterial effect of the *Piper betle* extract [14,17,25,26]. Results obtained in the present study revealed that the ethyl acetate extract possess effective antifungal activity against all the tested fungal cultures (Table-1, Figure-1).The highest antifungal activity of ethyl acetate extract was observed for *A.niger* and Black rust followed by *Rhizopus sp.* Antifungal activity of hexane extract was also significant against *A.niger* (5 mm) and *Aspergillus* sp. (8 mm), while ethanol-methanol extract was ineffective against any of the tested fungal cultures. In addition to the molds, we also tested the effects of wild yeasts that were isolated from leaves. We again noticed a strong anti-fungal activity against them. From all the preparations, ethyl acetate fraction provided the best results. This could be because the extraction process is more efficient and perhaps also stable over other solvent.

Biophysical characterization of the extract

We performed NMR and FTIR spectroscopy in order to understand the functional groups associated with the isolated active compound. In the ^{1}H-NMR spectrum, we observed three

Table 1

Fungal culture	Inhibition zone (mm)			
	Ethyl acetate	Hexane	Ethanol-Methanol	K-tellurite (+ control)
Aspergillus niger	28	5	-	6
Aspergillus sp.	28	8	-	7
Rhizopus sp.	23	-	-	-

Antifungal activity of ethyl acetate, hexane and ethanol-methanol herbal extract against indicated fungi.

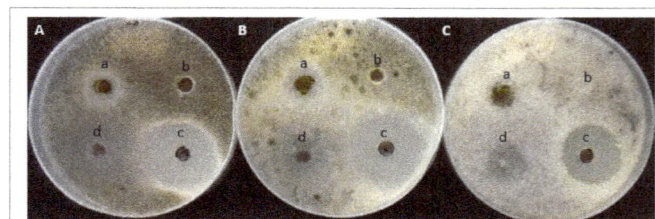

Figure 1: Antifungal activity of ethyl acetate, hexane and ethanol-methanol extracts of herbal sample against A) *A.niger* B) *Aspergillus* sp. and C) *Rhizopus.* Extract preparation in a) Hexane, b) Ethanol-Methanol, c) Ethyl acetate and positive control d) Potassium tellurite

types downfield peaks at 6.7, 7.2 and 7.25 ppm for aromatic ring. Additionally plenty of peaks were observed for protons associated with polar and aliphatic groups in the 3.25-5.5 and 1-2.25 ppm, respectively. From the characteristics of the spectra and the results published by other groups, the extracted compound is a derivative of the phenyl propanoid family to which antimicrobials eugenol-chavicol belong [27]. The FTIR spectra well corroborated with the NMR spectrum. We observed a broad peak in the range of 3700 to 3000, which corresponds to absorption caused by N-H, C-H and O-H single bonds. We did not observe any peaks characteristic of triple bonds in the range of 2,500 to 2,000. A wide range of double bond specific groups in C=O and C=C was observed in the region from 2,000 to 1,500. Lastly, the region from 1700-600 show finger print that is reminiscent of phenyl propanoid eugenol [27,28].

Discussion

The antifungal activity of the various extract highlights that ethyl acetate based extraction process provided the best for antifungal property. Surprisingly, unlike other results where *Piperbetle* extract that showed both anti- bacterial and fungal activities, we did not observe any anti-bacterial effect against gram positive *Staphylococcus aureus* and *Bacillus* sp.; few gram negative bacteria such as *Escherichia coli, Pseudomonas aeruginosa* and *Xanthomonas campestris*. This may be due to the fact that our purification method is different to the ones reported before (Table 2). In all previous studies crude extracts have been used hence many groups have reported anti-bacterial, anti-fungal and in some cases anti-cancer effects as well. In our methodology, we have enriched the phenyl propoanoid derivative in the extract by first ethanolic extract and then treating with other organic solvents like ethyl acetate/ hexane/

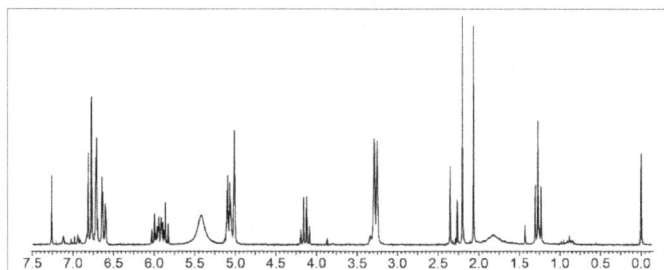

Figure 2: 1D ^{1}H-NMR spectrum of the active ingredient of *Piper betle* extract is suggestive of molecule from the phenyl propanoid- eugenol family

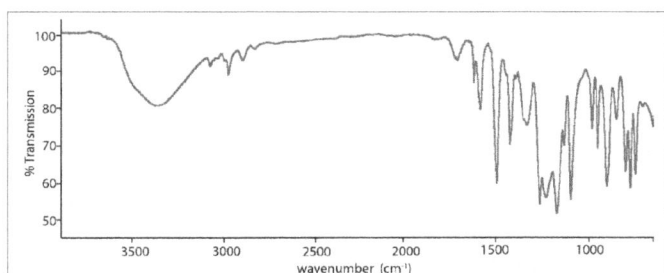

Figure 3: FTIR spectrum of the active ingredient of *Piper betle* extract.

Table 2

Active compound	Extraction method	Result
Hydroxychavicol	choloroform extract	anti-fungal (Ali et al., 2010); antibacterial (Sharma et al 2009)
	hydrodistillation	anti-bacterial (Basak and Guha, 2015)
	Concoction- boiling extraction	anti-cancer (Gundala et al., 2014)
Eugenol	ethanol or water-soxhlet extraction	*Streptococcus mutans* (Deshpande and Kadam, 2013)
	solvent extraction: methanol, ethyl acetate and petroleum ether	*Streptococcus mutans* (Deshpande and Kadam, 2015)
	hydro-distillation	anti-bacterial (Sugumaran et al., 2011)
Hydroxychavicol and Eugenol	methanolic extract	anti-cancer (Paranjpe et al., 2013)
	liquid-liquid and supercritical fluid extraction	Comparison of extraction methods(Singtongratana et al., 2013)

Summary of bio-active extraction procedures from *Piper betle*

ethanol-methanol. Furthermore, the large zone of clearance that we observe highlights the better extraction and stability of the phenyl propanoid in the ethyl acetate fraction.

Conclusion

Our finding sheds light on one of the important biochemical metabolite that contributes to the significance of betel leaf in traditional Indian ayurvedic medicine. The metabolite can be readily extracted in a two-step process using organic solvents and silica column chromatography. The active ingredient belongs to the phenyl propanoid family belonging to eugenol-chavicol group have been shown to possess strong antimicrobial properties [15,26,29,30]. More importantly, our results highlight the potential of using betel leaf extract as a potent anti-fungal agent for farming and perhaps also food storage against different types of molds. Food spoilage is a major agricultural problem accounting for heavy losses; therefore it is necessary to make the process commercially viable and efficacious against various pathogens and food spoilage organisms [31,32].

Acknowledgements

We thank the following institutes like. M. E. S Abasaheb Garware college, Vijay Chemicals and Neeti Developer for support; and Dr. Nivedita Cukkemane for critically reading the manuscript.

References

1. Alexander BD, Perfect JR, Antifungal resistance trends towards the year 2000. Implications for therapy and new approaches. Drugs. 1997;54(5):657–678.

2. Ghannoum MA, Rice LB, Antifungal agents: mode of action, mechanisms of resistance, and correlation of these mechanisms with bacterial resistance. Clin Microbiol Rev. 1999;12(4):501–517.

3. Neely MN, Ghannoum MA. The exciting future of antifungal therapy.

Eur J Clin Microbiol Infect Dis. 2000; 19(12):897–914.

4. White TC, Marr KA, Bowden RA, 1998. Clinical, cellular and molecular factors that contribute to antifungal drug resistance. Clin Microbiol. 1998;11(2):382–402.

5. Peter KV. Handbook of herbs and spices. 1st Edition Vol 1 and Vol 2. Sawston, Cambridge: Woodhead Publishing Limited;2004.

6. Nakamura CV, Ishida K, Faccin LC, Filho BPD, Cortez DAG, Rozental S, et al. In vitro activity of essential oil from Ocimum gratissimum L. against four Candida species. Res Microbiol. 2004;155(7):579–586. doi:10.1016/j.resmic.2004.04.004

7. Wannissorn B, Jarikasem S, Siriwangchai T, Thubthimthed S. Antibacterial properties of essential oils from Thai medicinal plants. Fitoterapia. 2005;76(2):233–236.

8. Khan R, Islam B, Akram M, Shakil S, Ahmad A, Ali SM, et al. Antimicrobial Activity of Five Herbal Extracts Against Multi Drug Resistant (MDR) Strains of Bacteria and Fungus of Clinical Origin. Molecules. 2009;14(2):586–597. doi: 10.3390/molecules14020586

9. Jagtap SD, Deokule SS, Pawar PK, Harsulkar AM, Kuvalekar A A. Antimicrobial activity of some crude herbal drugs used for skin diseases by Pawra tribes of Nandurbar district. J Nat Prod. 2010;1(2):216–220.

10. Guha P. Betel leaf: The Neglected Green Gold of India. J Hum Ecol. 2006;19(2):87–93.

11. Norton SA. Betle: consumption and consequences. J Am Acad Dermatol. 1998;38(1):81–88.

12. Prabhu MS, Patel K, SaraawathiG, Srinivasan K, Effect of orally administered betel leaf (Piperbetel leaf Linn.) on digestive enzymes of pancreas and intestinal mucosa and on bile production in rats. Indian J Exp Biol. 1995;33(10):752–756.

13. Pradhan D, Suri KA, Pradhan DK, Biswasroy P. Golden Heart of the Nature: Piper betle L. Journal of Pharmacognosy and Phytochemistry. 2013;1(6).

14. Ramji N, Ramji N, Iyer R, Chandrasekaran S, 2002. Phenolic antibacterial from Piper betle in the prevention of halitosis. J Ethnopharmacol. 2002;83(1-2):149–152.

15. Sugumaran M, Suresh Gandhi M, Sankarnarayanan K, Yokesh M, Poornima M, Sree Rama rajasekhar, Chemical composition and antimicrobial activity of vellaikodivariety of Piper betle Linn Leaf oil against dental pathogens. Int J PharmTech Res. 2011;3(4):2135–2139.

16. Datta A, Ghoshdastidar S, Singh M. Antimicrobial Property of Piper betel leaf against Clinical Isolates of Bacteria. IJPSR. 2011;2(3):104–109.

17. Kaveti B, Tan L, Antibacterial Activity of Piper betle Leaves. International Journal of Pharmacy Teaching and Practices. 2011;2(3):129–132.

18. Chakraborty D, Shah B. Antimicrobial, anti-oxidative and anti-hemolytic activity of Piper betel leaf extracts. IJPPS. 2011;3(3):192–199.

19. Das S, Parida R, Sriram Sandeep I, Nayak S, Mohanty S. Biotechnological intervention in betelvine (Piper betle L.): A review on recent advances and future prospects. Asian Pac J Trop Med. 2016;9(10):938–946. doi:10.1016/j.apjtm.2016.07.029

20. Trakranrungsie N, Chatchawanchonteera A, Khunkitti W. Ethnoveterinary study for anti-dermatophytic activity of Piper betle, Alpinia galangal and Allium ascalonicum extracts in vitro. Res Vet Sci. 2008;84(1):80–84.

21. Sharma KK, Saikia R, Kotoky J, Kalita JC, Das J. Evaluation of Anti-dermatophytic activity of Piper betle, Allamanda cathertica and their combination: An in vitro and in vivo study. Int J PharmTech Res. 2011;3(2):644–651.

22. Sharma R. Pathogenecity of Aspergillus niger in plants. Cibtech Journal of Microbiology. 2012;1(1):47–51.

23. Wilson CL, Wisniewski ME. Biological control of postharvest diseases of fruits and vegetables: an emergency technology. Annual Review of Phytopathology. 1989;27:425–441. doi: 10.1146/annurev. py.27.090189.002233

24. Laundon GF, Rainbow AF. Cerotelium fici. CMI Descriptions of Pathogenic Fungi and Bacteria. 1971;281:1–2.

25. Deshpande SN, Kadam DG. GCMS analysis and antimicrobial activity of Piper betle (Linn) leaves against Streptococcus mutans. Asian J Pharm Clin Res. 2013;6(5):99–101.

26. Basak S, Guha P. Modelling the effect of essential oil of betel leaf (Piper betle L.) on germination, growth, and apparent lag time of Penicillium expansum on semi-synthetic media. Int J Food Microbiol. 2015;215:171–178. doi:10.1016/j.ijfoodmicro.2015.09.019

27. Carrasco AH, Espinoza CL, Cardile V, Gallardo C, Cardona W, Lombardo L, Catalán MK, Cuellar FM, Russo A. Eugenol and its synthetic analogues inhibit cell growth of human cancer cells (Part I.). Journal of the Brazilian Chemical Society. 2008;19:543–548.

28. Paranjpe R, Gundala SR, Lakshminarayana N, Sagwal A, Asif G, Pandey A, et al. Piper betel leaf extract: anticancer benefits and bio-guided fractionation to identify active principles for prostate cancer management. Carcinogenesis. 2013;34(7):1558–1566. doi: 10.1093/carcin/bgt066

29. Prakash B, Shukla R, Singh P, Kumar A, Mishra PK, Dubey NK. Efficacy of chemically characterized Piper betle L. essential oil against fungal and aflatoxin contamination of some edible commodities and its antioxidant activity. Int J Food Microbiol. 2010;142(1-2):114–119.

30. Saxena M, Khare NK, Saxena P, Syamsundar KV, Srivastava SK. Antimicrobial activity and chemical composition of leaf oil in two varieties of Piper betle from northern plains of India. JSIR. 2014;73(2):95–99.

31. Rawat S. Food Spoilage: Microorganisms and their prevention. Asian J. Plant Sci. Res. 2015;5(4):47–56.

32. Adeyeye SAO. Fungal mycotoxins in foods: A review. Cogent Food & Agriculture. 2016;2(1). doi: 10.1080/23311932.2016.1213127

33. Ali I, Khan FG, Suri KA, Gupta BD, Satti NK, Dutt P, et al. In vitro antifungal activity of hydroxychavicol isolated from Piper betle L. Annals of Clinical Microbiology and Antimicrobials. 2010;9:1–7. doi: 10.1186/1476-0711-9-7

34. Sharma S, Khan IA, Ali I, Ali F, Kumar M, Kumar A. et al. Evaluation of the Antimicrobial, Antioxidant, and Anti-Inflammatory Activities of Hydroxychavicol for Its Potential Use as an Oral Care Agent. Antimicrobial Agents and Chemotherapy. 2009;53(1):216–222. doi:10.1128/AAC.00045-08

35. Gundala SR, Yang C, Mukkavilli R, Paranjpe R, Brahmbhatt M, Pannu V. et al. Hydroxychavicol, a betel leaf component, inhibits prostate cancer through ROS-driven DNA damage and apoptosis. Toxicol Appl Pharmacol. 2014;280(1):86–96. doi: 10.1016/j.taap.2014.07.012

36. Deshpande SN, Kadam DG. Evaluation of antimicrobial activity of extracts of Piper betle (Linn) leaves against Streptococcus mutans. World Journal of Pharmacy and Pharmaceutical Sciences. 2015;4(11):1040–1050.

37. Singtongratana N, Vadhanasin S, Singkhonrat J. Hydroxychavicol and Eugenol Profiling of Betel Leaves from Piper betle L. Obtained by Liquid-Liquid Extraction and Supercritical Fluid Extraction. Kasetsart J (Nat Sci). 2013;47:614–623.

38. Naik PM, Al-Khayri JM. Impact of Abiotic Elicitors on In vitro Production of Plant Secondary Metabolites: A Review. J Adv Res Biotech. 2016;1(2):7.

3

Effective Microbial Consortium of Bacteria Isolated from Hydrocarbon Polluted Soils of Gujarat, India

Mandalaywala HP, Ratna Trivedi*

Department of Environmental Sciences, Shree Ramkrishna Institute of Computer Education & Applied Sciences
M.T.B. college campus, Athwalines, Surat-395001, Gujarat, India

***Corresponding author:** Ratna Trivedi, Department of Environmental Sciences, Shree Ramkrishna Institute of Computer Education & Applied Sciences, M.T.B. College campus, Athwalines, Surat-395001, Gujarat, India. E-mail: mandalaywalahetal@gmail.com; drratnatrivedi@gmail.com*

Abstract

Hydrocarbons are widely distributed in environment owing to its extensive use as pesticides, petroleum products or other organic compounds. Hydrocarbons being mutagenic and carcinogenic in nature, it has led to various serious threats to living form. Pollution can ensue due to accidental spillage or leakages while handling and transportation of such compounds. Thus it becomes necessary to ensure safe removal and disposal of pollutants, to avoid further dispersal in different environmental layers. Bioremediation is one such effective treatment methods which render the pollutants harmless. Thus studying hydrocarbon utilizing bacteria becomes an essential step to formulate an effective bioremediation process. Hydrocarbon utilizing bacteria were isolated from a hydrocarbon based polluted site and scrutinized by series of tests. Among the isolates, five *Pseudomonas species* have been found to be effective in hydrocarbon utilizing capability. The isolates were identified using 16 sRNA sequencing procedure.

Keywords: Hydrocarbons; Hydrocarbon utilizing bacteria; Effective microbial consortium; 16s rRNA sequence

Introduction

Microorganisms play a vital role in maintaining a healthy ecological balance. Their ability to transform and degrading many types of pollutants have been widely recognized [1]. Several microorganisms may be involved in the reactions of biogeochemical cycles and in some cases they are the only living forms that are capable of degrading the complex elements of nature and regenerating a form that can be consumed by other organisms [2]. Hydrocarbons are organic compounds that lack functional groups and thus make it a polar. They are chemically less reactive at room temperature [3]; and insoluble in water. In this study petroleum derivatives are taken into consideration as hydrocarbon source. Various community play vital role in degradation of hydrocarbon. Petroleum is a heterogeneous mixture of hydrocarbons including aliphatic, alicyclic and aromatic in varying concentration depending upon the origin and nature. Henceforth, hydrocarbons term is used in general for the ease of discussion here. Hydrocarbons differ in their susceptibility to the microbial degradation, and are in general ranked as follows:

Linear alkanes > branched alkanes > small aromatics > cyclic alkanes i.e., the simpler the structure, more the susceptibility to microbial attack [4]. Hydrocarbons are hazardous to plants and animals, are also known to be carcinogenic, mutagenic and potent immuno-toxicants posing a serious threat to human and animal health [4]. Thus incorporation of hydrocarbons in natural environment is not desirable, makes its treatment and proper disposal inevitable. Various important processes influence the fate of hydrocarbons in nature, like sorption, volatilization, a biotic transformation (chemical or photochemical), and biotransformation. Since microorganisms play an important role in biogeochemical cycles, the biotransformation is of major concern. The other method mentioned either just fixes or transports the contaminants or is not very effective in nature [2]. Biotransformation is a natural process, which can be used to solve environmental issues, the process then known as Bioremediation.

Bioremediation and organisms degrading hydrocarbons

Bioremediation can be defined as the use of living organisms to detoxify or remove pollutants owing to their diverse metabolic capabilities. The process is considered as non-invasive and cost effective which makes it favorable to practice. [4]. Bioremediation is carried out by living forms. Among others, microorganisms are also involved in biologically degrading complex matter into simpler fragments, which renders it harmless and also makes it available to other organisms for their consumption. Bioremediation is achieved by existing metabolic potential of indigenous microorganisms or by introducing microorganisms through selection on basis of their catabolic activities or even by introduction of genetically enhanced microorganisms which are incorporated with such functions [1]. Its effectiveness depends on the extent to which the microbial population or consortium can be enriched and maintained in environment, when few or no indigenous degradative microorganisms exists in the contaminated area. Hydrocarbons are insoluble in water, thus in order to consume hydrocarbon derivatives, microorganisms need to emulsify it first in the solution or medium. For this, they are known to produce surface active agents i.e., biosurfactants, thus rendering the hydrocarbons susceptible to biodegradation [5].

Microbial community

Various microorganisms have hydrocarbon degradative potential. They consume hydrocarbons either individually or in consortia. Many aerobic bacterial genus viz., *Pseudomonas*, *Mycobacterium*, *Rodococcus*, *Arthobacter*, *Acinetobacter*, *Nocardia* and *Bacillus* are known for their degradative properties. Some fungi also have the ability to degrade hydrocarbon based pollutants like *Planerochaete chrysosporium* (White rot fungus) is an example of ligninolytic fungi capable of degrading polyaromatic hydrocarbons and other harmful environmental pollutants. *Cunninghamella echinulata* and Mycorrhizal fungi have also been used for the remediation of PHC-polluted soil [6].

Effective microbial consortium

The advantages of employing mixed cultures as opposed to pure cultures in bioremediation have been well documented. The degradative capacity of any microbial consortium is not necessarily the result of merely adding together of the capacities of the individual strains forming the association. It could be attributed to the effects of synergistic interactions among members of the association [7].

Mutagenesis to enhance degradation

Newly arisen mutations can have very different impacts on the fitness of the organism, ranging from deleterious through neutral to beneficial. However, they appear at very different rates [8]. There are various ways to induce mutation. One such method is use of UV radiation. UV mutagenesis has also come up as a tool,

to enhance degradation by indigenous bacteria [9]. Indigenous microorganisms play a major role in bioremediation, as they are adapted to the extremities of the polluted site. They are able to consume the pollutant as substrate and convert it to simpler forms which are safe to handle by the environment. Studying the characteristics of the polluted site and physiology of the indigenous microbes can thus help to develop a relation between the two; thus can give better understanding and insight on the bioremediation processes. The phylogenetic study helps one to infer the interrelationship between organisms. Thus phylogenetic study of such indigenous microbes can help us understand the relation between them better. And this can also help us to infer the synergistic relationship that may prevail between them to carry out the degradation process. The purpose of the study is to collect the sample from hydrocarbon contaminated site, to identify microbial communities and prepare microbial indices and to determine relationship between physiological study in community development.

Materials and Methods

Sample collection

Hydrocarbon based polluted samples were collected from industrial area located near south Gujarat (Figure 1).

Physico-chemical & organic analysis of the sample

The collected samples were scrutinized for their physic-chemical & organic properties according to the standard prescribed methods [10].

Enrichment: The samples were then enriched by suspending in Mineral Salt Medium (MSM).

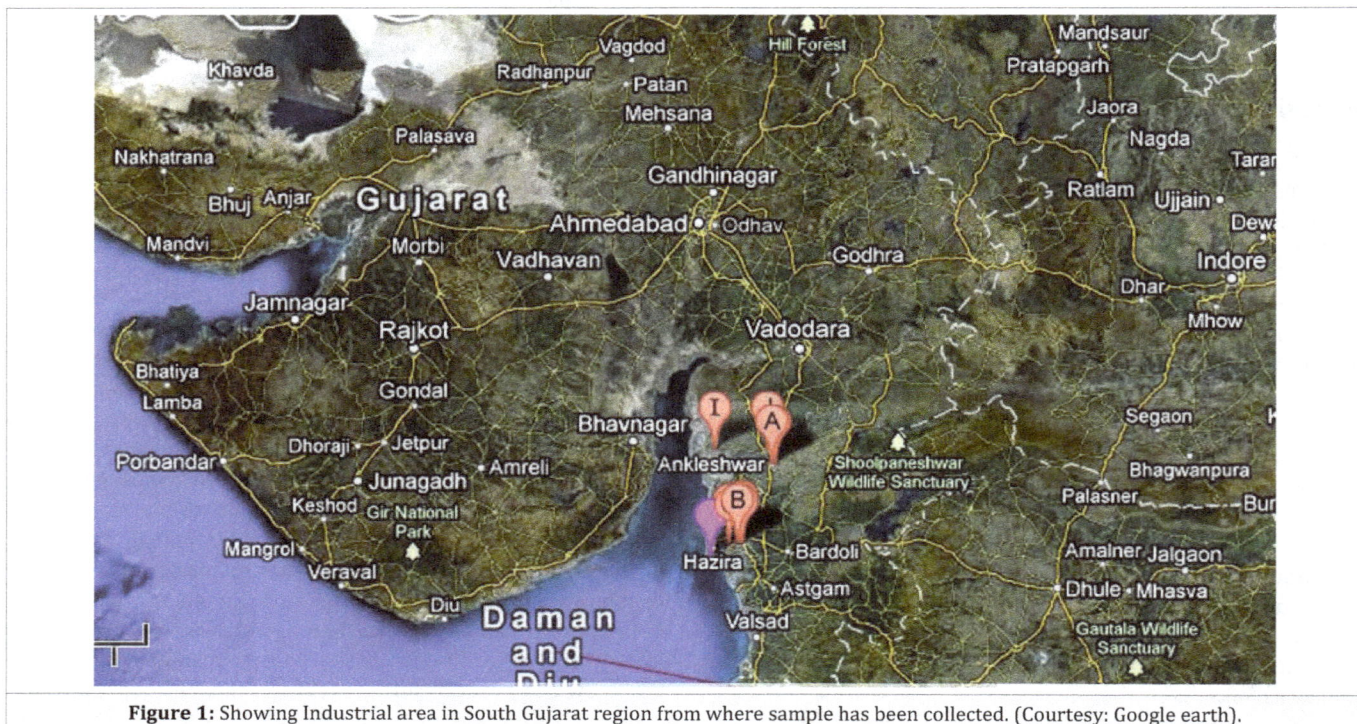

Figure 1: Showing Industrial area in South Gujarat region from where sample has been collected. (Courtesy: Google earth).

Microbial analysis: The samples after enrichment was scrutinised further by inoculating in different mediums: Mineral Salt Medium (MSM) and Tributyrin Agar medium (TBA); in addition to petroleum derivatives so as to culture only hydrocarbon degrading microbes [11]. Thus a synthetic medium was formulated using hydrocarbon based substrate i.e., petroleum derivatives along with the hydrocarbon extracted from the collected sample. Tributyrin Agar showed colonies development within 48 hours as compared to more than 72 hours in the former, thus TBA was used further in the study.

Tests for hydrocarbon degrading activity: Series of preliminary tests were carried out to check the hydrocarbon degrading capacity of the isolates, viz., drop collapse test, oil spread method, Blue plate agar, Haemolytic activity was checked using blood agar and then confirmatory test-Phenol sulphuric acid test [12].

Effects of various factors on degradation: Various factors like temperature, pH, nitrogen sources and substrate sources and concentration were tested for their effect on degradation rate.

Effective microbial consortium: Microbial consortium was formulated from the isolates, to evaluate the degradation properties of individual isolates as compared to different consortium so formulated.

Mutagenesis: Induced mutation was carried out by exposing 24 hours old culture to UV radiation for 30 mins. The degradation carried out by mutated strain was compared to that of carried out by wild type strain [9].

Result and Discussion

Analysis of samples

Sample collected from polluted sites were characterized for their physical, chemical and organic properties. The results obtained are as mentioned in the tables below (Table 1; Figure 2(a,b,c,d,e)).

These samples were suspended in MSM media, as mentioned above for enrichment of indigenous microorganisms. Figure 3 shows the enrichment flasks.

Then the medium was streaked on MSM and TBA, with petrol spread on top so as to grow only hydrocarbon consuming colonies. In table 2 & 3, the results of microbial community count and calculation of microbial diversity Indices is shown.

Only bacterial community was observed in the sample, no fungal or actinomycete growth was observed throughout the study. Bacterial colonies so formed were checked for varies characteristics, viz., Gram's reaction, motility, Capsule formation

and so on. 11 isolates were obtained and were studied further. The results are mentioned in the table 4.

Morphological characterization of isolates

Out of 11 above mentioned isolates, 8 isolates viz., A,B,C,D,F,G,H & K were found to be Gram Negative. The remaining three isolates; E, I& J were found to be gram variable. Their microscopic study showed that all of the isolates were short rods. The isolates were checked for capsule formation by carrying out capsule staining method: Maneval's method of capsule staining. It was found that isolates were non- capsulated. Motility was checked by standard method, which indicated that all isolates were motile.

Oil spreading assay

This method was developed by Morikawa. It is based on the oil displacement activity [12]. All the isolates were assessed by this method. But none of the isolates gave positive result.

Drop collapse method

This assay relies on the destabilization of liquids by secondary metabolites produced by hydrocarbon consuming microorganisms, i.e., Biosurfactants [12]. None of the isolates gave positive result in this method.

Blue agar plate

This method was developed by [13]. The microbes of interest are inoculated on a light blue mineral salt agar plate containing cationic surfactant Cetyl-Trimethyl-Ammonium Bromide (CTAB) and the basic dye methylene blue. If microbes are able to degrade hydrocarbons, they secret anionic surfactants which form a dark blue, insoluble ion pair with cetyl-trimethyl-ammonium bromide and methylene blue. The positive isolates will thus form dark blue coloured colonies [12]. Isolates A, C, D, F, G & H gave positive results in this test. Other colonies showed negative results (Figure 4).

Blood agar haemolysis

This method was developed by [14]; and is based on the fact that hydrocarbon degrading microbes are able to haemolyse the red blood; observe for zone of hemolysis [12]. Isolates A, C, D, F & G showed zone of hemolysis (Figure 5; Table 5).

Confirmatory test

Phenol- sulphuric acid test: The positive isolates were further scrutinized by confirmatory test [12] (Figure 6). Only positive results are shown here. Isolates A, C, D, F& G showed positive results in this test. Isolate H gave positive result only with blue agar plate; still it was checked further for the hydrocarbon degrading property.

Table 1: Result of physical analysis of sample.				
Parameters	Sample-A	Sample-B	Sample-C	Sample-D
Physical state	Viscous Liquid	Viscous Liquid	Viscous Liquid	Solid
Appearance	Dark Brown	Dark Brown	Dark Brown	Black
Flammability	Flammable	Flammable	Slight Flammable	Flammable

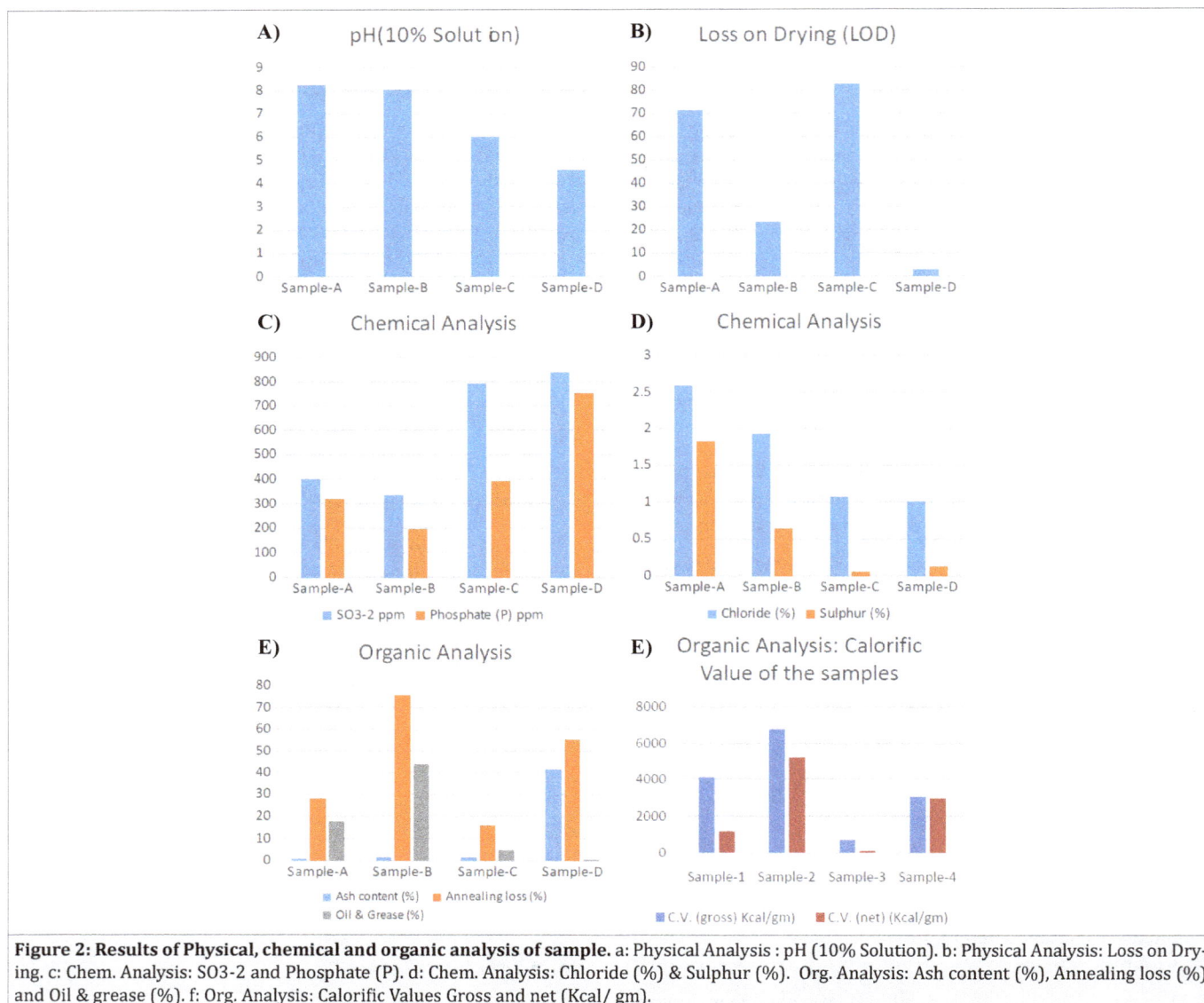

Figure 2: Results of Physical, chemical and organic analysis of sample. a: Physical Analysis : pH (10% Solution). b: Physical Analysis: Loss on Drying. c: Chem. Analysis: SO3-2 and Phosphate (P). d: Chem. Analysis: Chloride (%) & Sulphur (%). Org. Analysis: Ash content (%), Annealing loss (%) and Oil & grease (%). f: Org. Analysis: Calorific Values Gross and net (Kcal/ gm).

Table 2: Microbial Community.

Sample	Bacteria (CFU/ml)	Colonies
A	1.2×10^5	3
B	1.1×10^5	2
C	1.1×10^5	2
D	1.3×10^5	4

Table 3: Microbial diversity Indices (Using *Past-3 Software*).

Index	Bacterial diversity
Taxa_S	4
Individuals	4
Dominance_D	0.2512
Simpson_1-D	0.7488
Shannon_H	1.384
Evenness_e^H/S	0.9975
Brillouin	0.6034
Menhinick	1.845
Margalef	2.164
Equitability_J	0.9982
Fisher_alpha	12.73
Berger-Parker	0.2128
Chao-1	4

Results of effects of various factors on degradation

Nitrogen Sources

The effect of different nitrogen sources on the degradation capacity was also studied. The nitrogen sources taken into consideration were peptone, Sodium Nitrate (NaNO₃), potassium

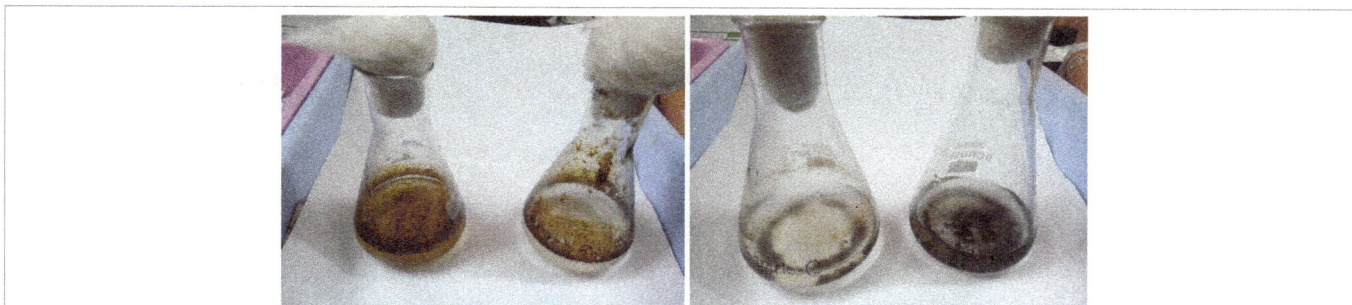

Figure 3: Samples A, B, C & D suspended in MSM media.

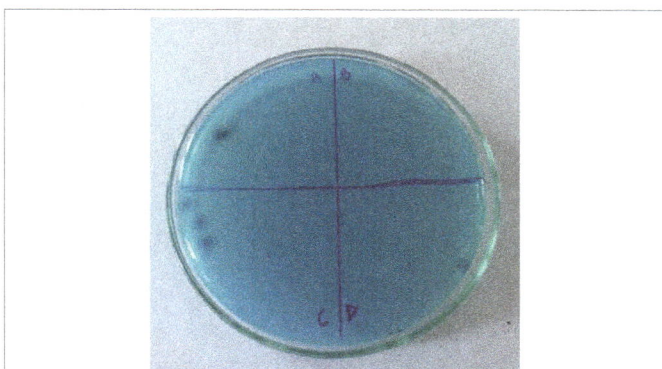

Figure 4: Blue Agar Plate.

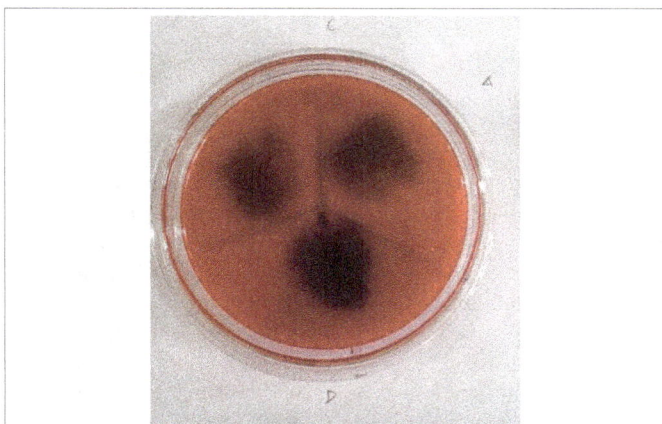

Figure 5: Blood Agar Plate.

Nitrate (KNO_3), Calcium Nitrate Tetra Hydrate ($Ca(NO_3)_2 . 4H_2O$), Ammonium Sulphate (NH_4SO_4), Ammonium Chloride (NH_4Cl) and NH_4NO_3). Among these, peptone gave better results (Figure 7a). Here a too exceptionally high result with isolate G was observed with Ammonium Sulphate was due to pigment formation.

Different substrate concentration

A synthetic medium was prepared by using petroleum derivatives. The concentration and type of petroleum derivative is varied to study its effect on the growth and degradative properties with varying substrates Figure 7b to 7g. shows results of different concentration of petrol, kerosene, diesel and combinations of these petroleum derivatives at different concentration.

Degradation by formulated consortium

Formulated consortium was studied for their degradative properties, and the result of the same are as shown in figure 7h & figure 7i. In Figure 7h, OD of consortium formulated by combination of two isolates: Consortium I and by combination of three isolates: Consortium II. The relative population size of the isolates for these consortiums were kept equal i.e., in 1:1 ratio. The results of consortium I as been shown in Figure 7h; combination of CF and CG gave highest OD among others. And in figure 7i, OD of combination of Consortium II has been shown. Among which ACF showed maximum OD i.e., 0.4, which is higher than CF & CG as well which is nearly 0.25 (Figure 8).

Enhancing degradative property by mutation

The hydrocarbon degrading property of isolates was seen to be improved when isolates were mutated by UV radiation. The figure below shows that mutation enhanced the degradative properties of the isolates as compared to that of wild type strains (Figure 9).

Identification of microorganisms

Isolated microorganisms has been identified by 16s RNA gene sequencing (Saffron's Gene Laboratory) and further submitted to NCBI. The submitted sequences got accession numbers as below (Table 6):

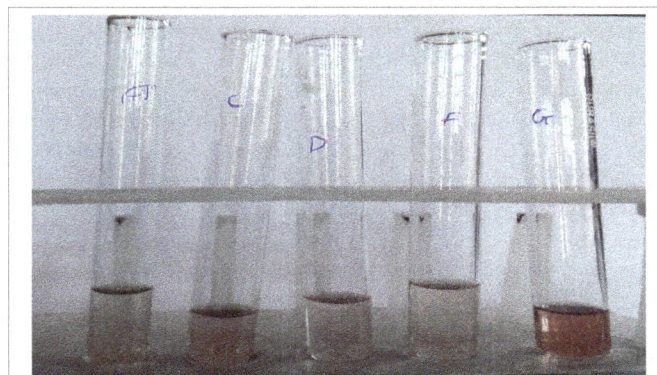

Figure 6: Phenol – Sulphuric Acid test (Confirmatory test).

Table 4: Characterization of isolates.

Isolate	Gram's Reaction	Capsule formation	Motility	Oil Spreading Assay	Drop Collapse	Blue Agar Plate	Haemolytic reaction	Confirmatory test (Phenol- Sulphuric Acid test)
A	Negative	No	Yes	No	No	Yes	Yes: ß	Yes
B	Negative	No	Yes	No	No	No	No	No
C	Negative	No	Yes	No	No	Yes	Yes: ß	Yes
D	Negative	No	Yes	No	No	Yes	Yes: ß	Yes
E	Variable	No	Yes	No	No	No	No	No
F	Negative	No	Yes	No	No	Yes	Yes : ß	Yes
G	Negative	No	Yes	No	No	Yes	Yes: ß	Yes
H	Negative	No	Yes	No	No	Yes	No	No
I	Variable	No	No	No	No	No	No	No
J	Variable	No	No	No	No	No	No	No
K	Negative	No	No	No	No	No	No	No

Table 5: Diameter of zone of haemolysis.

Colony	Isolate A	Isolate C	Isolate D	Isolate F	Isolate G
Zone diameter	1 mm	2 mm	2 mm	1 mm	1 mm

Table 6:

Sample	Isolate	Strain	Accession number
A	Isolate A	*Pseudomonas aeruginosa* strain RRLP1	KU314415
A	Isolate G	*Pseudomonas aeruginosa* RRLP1	KU314419
B	Isolate C	*Pseudomonas aeruginosa* Strain AS-1	KU314416
B	Isolate H	*Pseudomonas aeruginosa* Strain JQ-41	KU314420
C	Isolate D	*Pseudomonas aeruginosa* Strain SI5(1)3	KU314417
D	Isolate F	*Pseudomonas spp.* Strain 14-1	KU314418

Phylogenetic study

Phylogram was prepared in MEGA 7 software Figure 10.

Conclusion

The samples extracted from the hydrocarbon based waste matter showed majorly Gram negative to gram variable microorganisms. Out of eleven, six *Pseudomonas* species were isolated with Hydrocarbon degrading properties. These dominated *Pseudomonas* species were identified by 16s RNA sequence analysis. The Pseudomonas species identified are Isolate A: *Pseudomonas aeruginosa* strain RRLP1, Isolate C: *Pseudomonas aeruginosa* Strain AS-1, Isolate D: *Pseudomonas aeruginosa* Strain SI5 (1)3, Isolate F: *Pseudomonas spp.* Strain 14-1, Isolate G: *Pseudomonas aeruginosa* RRLP1, Isolate H: *Pseudomonas aeruginosa* Strain JQ-41. On assessing effects of various parameters on their hydrocarbon degradation properties, it was found that at the isolates were able to grow under various physical conditions since their origin physical condition was also extreme. It was observed that, at temperature 30°C, consistent growth was observed as compared to 37°C, which showed varied growth response. When isolates were exposed to varied pH values, they gave growth in pH range of 4.5 to pH 9.0. However, at pH 4 no growth was observed with any isolate and at pH 9, comparatively less growth was observed considering lesser OD values. Consistent growth was seen in pH range of 5.5 to 6.5. It was also observed that isolates were able to grow in different N sources but among those, peptone gave consistent results. And in carbon concentration, petrol 0.005% gave consistently good growth than Kerosene and diesel at different concentrations. High growth among the isolates was observed by isolate D: *Pseudomonas aeruginosa* Strain SI5(1)3, at petrol concentration of 0.005% (v/v), with peptone as Nitrogen source, with pH of 6.0, when incubated at temperature 30°C for 72 hours. And isolate F: *Pseudomonas spp.* Strain 14-1 was able to show pigmentation in different physical conditions which was not observed in any other isolate. The study represented that *Pseudomonas* species isolated from hydrocarbon contaminated waste matter were able to degrade various petroleum derivatives at different concentration and also under the influence of various physical conditions. Moreover different strains of *Pseudomonas* should be further studied for their hydrocarbon degradative properties at large scale. This work can be helpful to design consortia to biodegrade hydrocarbon pollution in natural environment and to have more insight in this area of research which needs more attention considering the prevailing pollution rate.

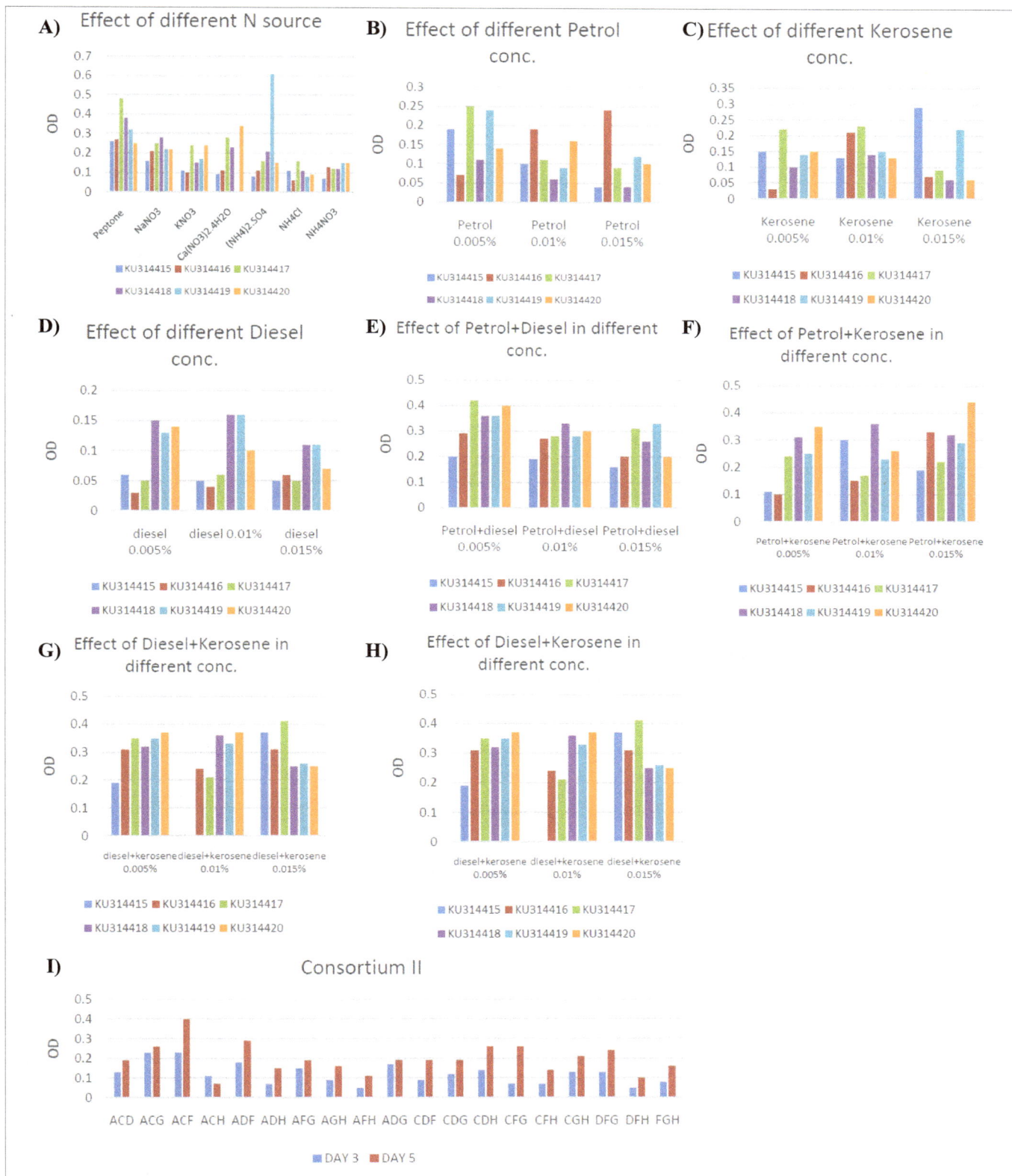

Figure 7: Results of effects of different parameters (a, c, d, f, g, h and i are the isolates as mentioned below). a: Effect of different Nitrogen (N) source. b: Effect of different Petrol concentration. c: Effect of Different Kerosene Concentration. d: Effect of Different Diesel Concentration. e: Effect of Petrol+Diesel in different conc. f: Effect of Petrol+Kerosene in different conc. g: Effect of Diesel+kerosene different conc. h: Degradation by consortium prepared by combining two isolates. i: Degradation by consortium prepared by combining three isolates.

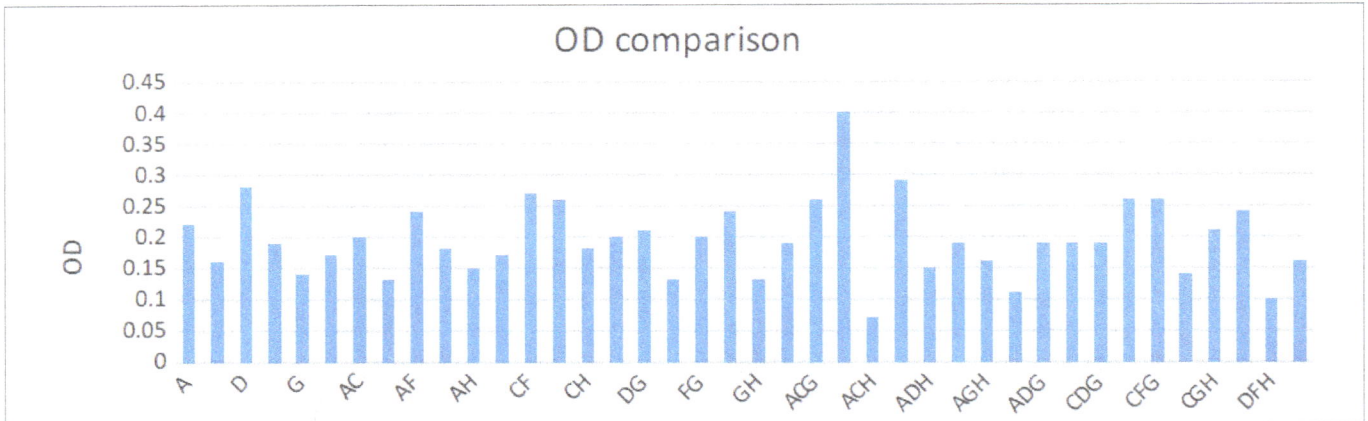

Figure 8: Comparison between individual isolate and consortium.

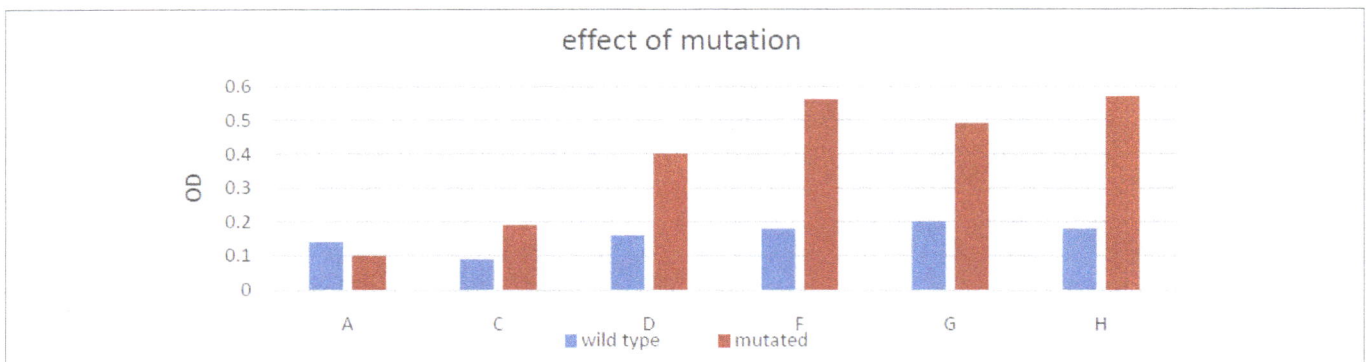

Figure 9: Effect of Mutation on cell growth.

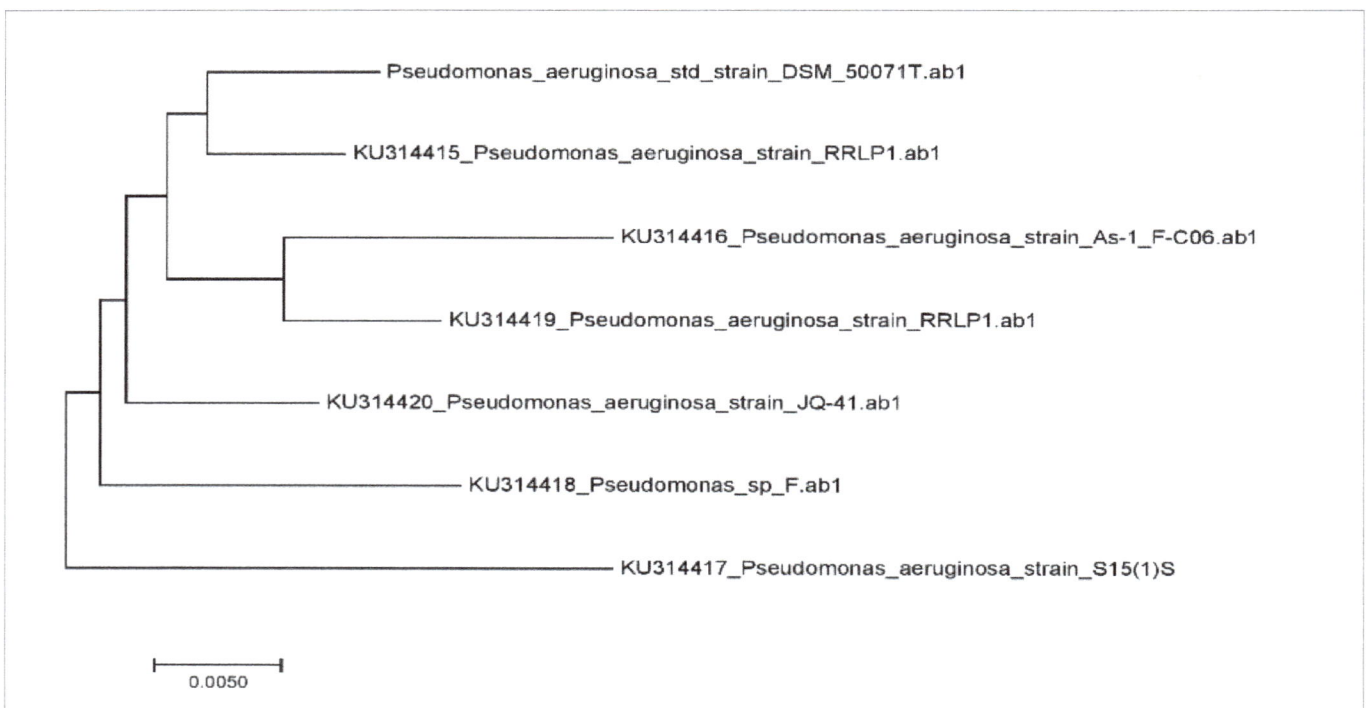

Figure 10: Phylogram of the isolates.

References

1. Ilyina A, Castillo S, Villarreal S, Ramirez E, et al. Isolation of soil bacteria for bioremediation of hydrocarbon contamination. вестн. Моск. Ун-та. 2003:44(1):88-91.

2. Peixoto RS, Vermelho AB, Rosado AS. Petroleum-Degrading Enzymes: Bioremediation and New Prospects. Enzyme Res. 2011;2011:475193. Doi: 10.4061/2011/475193.

3. Oliveira I. et al. Microbial Hydrocarbon Degradation: Efforts to Understand Biodegradation in Petroleum Reservoirs. DOI: 10.5772/55920.

4. Das N, Chandran P. Microbial Degradation of Petroleum Hydrocarbon Contaminants: An Overview. Biotechnol Res Int. 2011;2011:941810. Doi: 10.4061/2011/941810.

5. Panda SKKRN, Panda CR. Isolation and identification of petroleum hydrocarbon degrading microorganisms from oil contaminated environment. Int J Environ Sci (IJEST). 2013;3(5):1314-1324. doi:10.6088/ijes.2013030500001.

6. Chibuike, GU, Obiora, SC. Bioremediation of hydrocarbon-polluted soils for improved crop performance. Int J Environ Sci (IJEST).2014;4(5):840-858. Doi: 10.6088/ijes.2014040404524.

7. Farinazleen MG, Raja Noor ZAR, Abu Bakar S, Mahiran B. Biodegradation of hydrocarbons in soil by microbial consortium. Int Biodeterior Biodegradation.2004;54(1):61-67. Doi:10.1016/j.ibiod.2004.02.002.

8. Denamur E, Matic I. Evolution of mutation rates in bacteria. MolMicrobiol.2006;60(4):820-7. Doi:10.1111/j.

9. Naveenkumar SMN, Ganesan S, Manivannan SP, Velsamy G. Isolation, Screening and In Vitro Mutational Assessment of Indigenous Soil Bacteria for Enhanced Capability in Petroleum Degradation. Int J Environ Sci (IJEST) 2010;1(4):493-518.

10. APHA, AAWI. Standard Methods for the Examination of Water and Wastewater. In: Eugene W. Rice, Laura Bridgewater. 22nd Edition. 2012.

11. Anupam Mittal, Padma Singh. Isolation of hydrocarbon degrading bacteria from soils contaminated with crude oil spills. Indian J Exp Biol. 2009;47(9):760-5.

12. Rehman Naziya NMA, SMB, Dixit PP and Deshmukh AM. Screening Of Biosurfactant Producing Microorganisms from Oil Contaminated Soils of Osmanabad Region, Maharashtra, India. International Science Journal. 2014;1(1):35-39.

13. Inka Siegmund and Fritz Wagner. New method for detecting rhamnolipids excreted by Pseudomonas species during growth on mineral agar. Biotechnology Techniques. 1991;5(4):265-268. Doi: 10.1007/BF02438660.

14. Catherine NM, Terry YK. Chow, Bernard F. Gibbs. Enhanced biosurfactant production by a mutant Bacillus subtilis strain Applied Microbiology and Biotechnology. 1989;31(5):486-489. Doi: 10.1007/BF00270781.

4

Microbial Community Structure of Activated Sludge as Investigated with DGGE

Shah MP*

Division of Applied & Environmental Microbiology, Enviro Technology Limited, Industrial Waste Water Research Laboratory, India

*****Corresponding author:** *Shah MP, Division of Applied & Environmental Microbiology, Enviro Technology Limited, Industrial Waste Water Research Laboratory, India, E-mail: shahmp@beil.co.in*

Abstract

Microbial activity and structure of the bacterial community of activated sludge reactors, which treated industrial wastewater, were studied. Microbial communities, including ammonia oxidizing bacteria, Eubacterium and actinomycete communities were studied in two different systems with the polymerase chain electrophoresis gradient denaturing gel reaction using amplified gene fragments, 16S rRNA of bacteria. Both systems, which used an anoxic-aerobic process and anaerobic-anoxic-aerobic process, respectively, received the same industrial wastewater, operating under the same conditions and showed similar processing performance. Oxidizing bacterial communities of ammonia from two systems showed almost identical structures corresponding to ammonia removal, while the actinomycete bacterial community showed obvious differences. FISH results showed that the ammonia-oxidizing bacterial cells in the anaerobic-anoxic-aerobic system increased by 3.8 ± 0.2% of the total bacterial population, while those in the anoxic-aerobic system represented 1.7 ± 0.2%. Thus, the existence of an anaerobic-anoxic environment in the anaerobic-aerobic system resulted in a marked increase in biodiversity.

Keywords: Industrial waste water; 16S rRNA; Anoxic; Anaerobic; Aerobic; FISH

Introduction

The growth of the world population, the development of various industries, and the use of fertilizers and pesticides in modern agriculture has overloaded not only the water resources but also the atmosphere and the soil with pollutants [1]. The degradation of the environment due to the discharge of polluting wastewater from industrial sources is a real problem in a number of countries. This situation is even not as good as in developing countries like India where little or no treatment is carried out before the discharge [2]. Moter and Gobel developed the first activated sludge system for purification of wastewater in Manchester [3]. However, the role of microbial consortia in this process is still not completely understood. Culture-based techniques were found to be too selective to give a comprehensive and authentic picture of the entire microbial community as it has been estimated that the majority (over 99%) of bacteria in nature cannot be cultivated by using traditional techniques [4]. Activated sludge is a very thorny system, comprised of a variety of populations including heterotrophic and autotrophic bacteria, fungi and protozoa [5]. The relationship between the microbial composition and the treatment performance of activated sludge processes has long attracted the attention of microbial ecologists and environmental engineers, as this information might be useful for the proper design and operation of biological wastewater treatment systems. Protozoa have been studied and utilized as an important indicator for judging process performance and effluent quality since the 1970s, because these large sized microorganisms can be directly observed and identified under a microscope [6]. The development of DNA-based techniques has revolutionized the ability to characterize and identify the diversity and taxonomy of environmental organisms in a wide variety of niches [7,8], such as food [9], soil [10], water [11] and the human body [12]. A major advantage of this approach is that it allows monitoring, exposure and investigation of the genetic targets of interest directly from environmental samples, lacking the additional steps of cultivation and recovery [13,14], which are known to be inefficient in recuperating symbiotic, facultative, stationary, slow growing, pH sensitive and various other fastidious microorganisms [15,16]. In spite of its attractiveness, many molecular studies applied to soil and water have indicated that the choice of processing method and the design of extraction protocols may affect the degree of lysis of the microorganisms present in the sample (and hence the recovery of their template DNA) [17], the integrity and size of DNA obtained [18] and the extent of co-extraction of both organic and inorganic impurities which may interfere with PCR amplification [19]. These factors may also affect the usefulness and applicability of the DNA for further molecular analysis [20] and drastically affect the recovery of molecular diversity, leading to mistakes in the interpretation of the true diversity and taxa present [21,22]. Despite the knowledge that microbial communities evolving in wastewater treatment plants contribute on handling processes, there are only a few reports concerning the study of bacterial communities [23]. Until now, studies were carried out using mainly traditional microbiological schemes. The emergence of molecular techniques allowed the conquering of the problems associated with culture-dependent methods that lead to an underestimation of the true diversity. Molecular methods such as 16S rDNA clone libraries [24], ribosomal intergenic spacer analyses [25], 16S-restriction fragment length polymorphism [26], repetitive extragenic palindrome polymerase chain reaction [27] and Fluorescent In Situ Hybridization [FISH] [27] have already been applied to the study of wastewater-associated microbial communities. The combination of PCR amplification of 16S rRNA genes with Denaturing Gradient Gel Electrophoresis [DGGE] analysis has also provided a useful means to directly characterize bacterial populations within many samples. Polymerase chain reaction –denaturing gradient gel electrophoresis has been successfully used in many fields of microbial ecology to assess the diversity of microbial communities and to determine the community dynamics in response to environmental variations. Studies concerning bacterial diversity in waste waters using a DGGE-based approach have been performed for reactor systems [28] and activated sludge [28], revealing the presence of highly complex bacterial communities. However, petty work has been done in order to apply this methodology to assess the bacterial diversity in industrial wastewater where the organic matter degradation takes place. In the present study, microbial communities of two different systems, an anoxic-anaerobic-aerobic process and an anoxic-aerobic process, respectively, receiving identical sewage and having similar treatment performance, were determined using group- specific PCR-DGGE and subsequent sequence analysis of rRNA genes. The community diversity of eubacteria and ammonia oxidizing bacteria, were investigated in the two systems to evaluate the effects of different designs on microbial populations. FISH was used for the determination of the ratio of ammonia oxidizing bacteria in each system with a probe.

Materials & Methods

DNA Isolation from samples of activated sludge

Activated sludge samples were collected from the biological system of the industrial treatment plant, pelleted by centrifugation (5000 x g, 10 min, 4°C) and stored at -45°C until isolation of DNA. Total genomic DNA was extracted from 0.3 g of activated sludge samples according to the mechanical method [6]. The samples were washed three times with 1 x PBS buffer and disrupted with bead beating in lysis buffer [Tris-HCl 100 mM, 100 mM EDTA, 1.5 M NaCl; pH = 8.0). The samples were incubated 20 minutes at 1400 revolutions per minute and 200 ul of 10% SDS was added. After 30 minutes incubation at 65°C the samples were centrifuged twice at 13,000 rpm and placed on spin filters. DNA fixed on the filter was washed twice with a solution A1. The amount of DNA was measured spectrophotometrically, using qubit and stored at -20°C until PCR amplification.

Chemical analysis

Water temperature and dissolved oxygen were determined in situ with a WTW model 330i/ SET and a model WTW OXI 96, respectively. Influent characteristics, namely the biochemical oxygen demand, chemical oxygen demand, suspended solids and pH were determined by standard methods [29] and portrayed in Table 1.

Analysis of microbial community diversity by means of 16S rRNA using PCR-DGGE

The DNA of the bulk community was extracted from 1.0 mL of sludge using a Fast DNA Spin Kit for Soil. The extracted DNA was then subjected to PCR touchdown, using primers 341F and 534R [30]. The primer 341F contained a 44 bp GC clamp. Amplification was performed in a thermal cycler. PCR products were separated using a code-D and 1 mm thick polyacrylamide gel system containing 8% (w/ v) acrylamide-bisacrylamide (37.5:1), TAE IX buffer, and a denaturing gradient of 30% to 70% (v/ v). Electrophoresis was performed in TAE buffer at 60°C and constant voltage for 14 hours. Gels were stained with 1:10 000 (v/ v) SYBR Green I and photographed using Gel Doc 2000 equipped with MULTIANALYST software. The central parts of DGGE bands were excised with a razor blade and soaked overnight in purified water. A portion of this (10 ul) was then removed and reamplified as described above. The re-amplified DNA fragments from the DGGE bands were sequenced directly or cloned into the pGEM-T Easy vector system prior to sequencing. The sequences were checked for possible chimeras using CHIMERA_CHECK program on the website of ribosomal database project. To determine the phylogenetic position of microorganisms detected in DGGE, the sequences of 16S rRNA genes analyzed were compared with databases of sequences available via BLAST search. The band patterns and intensities of the scanner gels were analyzed using Gel Compar software. After applying subtraction drive working capital, an analysis of each channel, acquiring densitometric curves was carried out by the software. A DNA band was identified if the tape represented more than 1.0% of the total intensity channel. A matrix was then constructed using this information, and has been used to calculate a set of digital values to describe the diversity of bacterial communities. As a parameter to the structural diversity of the microbial community, the Shannon index [31] of overall diversity, H, was calculated with the following formula: $H = -\Sigma Pi \bullet \ln (Pi)$ wherein Pi is the probability that major bands in a track: H was calculated on the basis of the bands on the gel lanes, using the intensities of the bands after the peak heights in the densitometric curve. The probability of material, Pi, Pi was calculated as $= ni/ N$, where ni is the height of the peak i, and N is the sum of all peak heights in the densitometric curve. For the analysis of ammonia-oxidizing bacteria and activated sludge actinomycete populations, a nested PCR technique was used to increase the sensitivity [37]. In the first round, 1 μL of the extracted DNA was added to Mastermix PCR and different primers were used, each with their own corresponding PCR protocol. During the second round of PCR, 1 μL amplified product in the first round was added to 49 ul of PCR mixture and then reamplified using their own protocol and corresponding PCR primers shown in Table 2.

Table 1: Influent and effluent characteristics

Type	CODCr (mg/ l)	BOD (mg/ l)	NH4+-N (mg/ l)	TN (mg/ l)	TP (mg/ l)	SS (mg/ l)
Raw Sewage	450 - 800	250 - 400	35.6 – 47.5	52 - 68	49.0 – 62.0	5.2 – 6.6
A₂O effluent	38.6 - 48.8	12.4 - 18.6	11.0 – 25.0	38 - 44	36.0 – 41.0	0.6 – 3.8
AO effluent	32.4 - 40.2	10.12 - 14.84	8.6 – 17.8	9.4 – 18.2	33.6 – 38.5	5.3 – 5.9

Sequence Analysis of denaturing gradient gel electrophoresis Profiles

The denaturing electrophoresis gel gradient profiles were analyzed with the fingerprint software database TM diversity. On the basis of the presence [1] or absence (0) of individual bands in each lane, a binary matrix was constructed. Binary data representing the banding patterns were used to generate a Dice pair wise distance matrix. A dendrogram was obtained by unweighted pair group analysis of the mean of the cluster means. The distance matrix was also used for the construction of a multi-scale diagram scaling, on a two dimensional plane with artificial x- and y-axis where each denaturing gradient electrophoresis gel fingerprint is placed at a certain time, so that similar samples are represented together. Clustering analysis and MDS were performed using the software Primer 5 [32]. Denaturing gradient electrophoresis gel patterns were also examined using two indices to field many aspects of microbial diversity. The Shannon-Weaver index of diversity, H [33] and the index of equitability, E [34] were calculated for each sample as follows:

$$H = -\sum (n_i / N) \log((n_i / N)),$$

$$E = H / \log S,$$

Where, ni is the intensity of the relative area of each band electrophoresis denaturing gradient gel, S is the number of bands in denaturing gradient electrophoresis gel and N is the sum of all surfaces of all bands in a given sample [35]. The statistical significance of the variance in the index was assessed by a two-way analysis of variance. A canonical correspondence analysis was used to determine the multiple relationships between each variant denaturing gradient gel electrophoresis banding patterns and environmental parameters. The analysis was performed using CANOCO for Windows Version 4.5 [36] and its significance was evaluated by the Monte Carlo test with 1000 permutations.

GC-MS analysis

The Gas Chromatograph - Mass Spectrometry (GC-MS) analysis was performed by a MP5890GC/ MS. Chromatography was conditioned as follows: SE- 54 capillary column (25 m × 0.32 mm); the column temperature was maintained at 40°C for 2 minutes, then heated to 250°C, with an increment of 3 ~ 5 ° C/ min and held at 250°C for 30 min. Mass conditions were as follow: temperature of the MS ion source was 250°C; the voltage multiplier is 2400 V; the electron energy is 70 eV.

Fluorescence in situ hybridization analysis

Prior to hybridization, the samples were dispersed into single cells by sonication, and then treated immediately and fixed in 4% paraformaldehyde for 3 hours at 4°C. After that, the biomass was washed with phosphate buffered saline (pH 7.4) and stored at a 1:1 ratio of phosphate buffered saline and 100% ethanol at 0° - 20°C. All Hybridizations were performed at 46°C for 120 min as described by Manz, et al. (1992). The oligonucleotide probes of the 16S rRNA target - and the stringency used in this study are listed in Table 3.

After hybridization, the slides were mounted with Citifluor

prevent money laundering and examined with an epi-fluorescence microscope Axio skop 2. All processing and image analysis were performed with the standard software provided by Zeiss. Three probes are listed in Table 3.

Results

Physico-chemical characterization

Sampling in the industrial wastewater treatment plant was done from two biological systems of aeration tanks. Influent physical and chemical characteristics during the sampling period are shown in Figure 1. Parameters such as biochemical oxygen demand, chemical oxygen demand and total suspended solids showed a high variation with pronounced peaks. The pH ranged from 6.7 to 7.6. Water temperature and DO concentration values registered in both systems were also rather unstable, with temperatures ranging from 11.3 to 29.7 ˚C and DO varying between 0.10 and 3.14 mg/ l (Figure 1).

Bacterial community analysis using 16S rRNA gene PCR-DGGE

To follow the evolution of the microbial community during operation of the reactor, 16S rRNA PCR amplified gene fragments were analyzed using DGGE (Figure 2). DGGE profiles show changes in the microbial population due to progressive reductions in HRT. The average band number per lane in each reactor used for diversity analysis was 19.7 (from 17 to 22) Al, 18.8 (from 15 to 22) in the A- II and 19.7 (from 17-22) in AI + II. The number and thickness of the bands observed in the DGGE profiles provide an estimate of species richness. The Shannon diversity index, H, from the DGGE band pattern of each sample was calculated to

Table 2: PCR Primers used in this study

Primers	Number of cycles	PCR conditions						References
		Denaturation		Annealing		Elongation		
		˚C	min	˚C	min	˚C	min	
F243, R1378r	35	95	1	63	1	72	2	[38]
CTO189fAB, CTO189Fc, CTO654r	35	95	1	57	1	72	2	[38]
GC-P338f, P518r	30	95	1	55	1	72	2	[38]

Table 3: Analysis of wastewater components

Sample	Hydrocarbons	Alcoholic aldehydes and ketone acids	Aromatic hydrocarbons	Phenols	Halogenated hydrocarbons	Others
Influent	40	25	16	1	4	25
A-I	58	28	16	4	18	58
A-II	48	18	22	2	12	10
A-I+II	52	26	20	3	15	14

determine the diversity of the microbial community. Figure 3 shows changes in the Shannon diversity index H, occurring at the same time than changes in HRT. After HRT reduction of 48 to 24 h, the values of H in A-I and A-II increased slightly (2.80 and 2.63, respectively). At HRT 12 h, microbial diversity recovered slightly in A-I. The final reduction of HRT decreased to 6 h H for the three reactors. 16S rRNA gene sequences of 28 major groups (9 to A-I bands, 10 bands to A-II and A-9 bands I + II). Most of the sequences were found to be clustered in the Proteobacteria (17 bands) and *Bacteroidetes* (6 bands). The other band sequences were found to be clustered in the *Actinobacteria* (2 strips), phylum TM7 (2 bands) and *Acidobacteria* (1 band). In the Proteobacteria, most sequences were combined in the β-Proteobacteria, in particular in the control *Burkholderiales* (10 bands). In A-I, bands F4, F5 and F7 were present in all periods of operation and have become widespread as HRT has been reduced. F1, F2, F3 made minor bands, but was observed in almost all periods. F6 became widespread on days 40 and 46 (6 pm HRT).

In A-II, the profile does not change significantly during days 15-30 (HRT of 24 h and 12 h) or days (40-50 HRT of 6 hours). Major groups, including Z3, Z4 and Z8, who were present during long periods (HRT of 24 h and 48 h) persisted, but became minor after 30 days (HRT 12 h). Z6, Z7 and Z9 have become large groups after 40 days (HRT) 6 h of operation. In A-I + II, Group N4, which is a minor component of the community during the first period, gradually became dominant after the operation of the

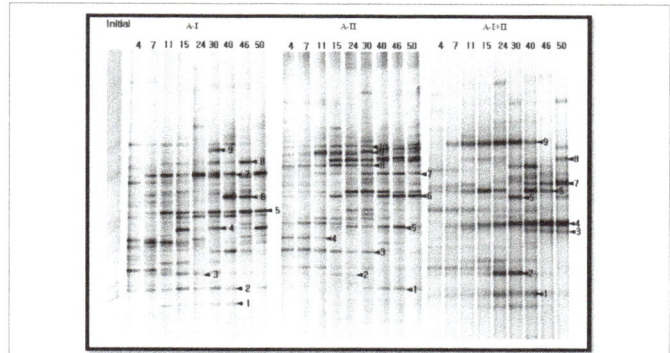

Figure 2: Denaturing gradient gel electrophoresis profiles of 16S rRNA gene of bacterial communities.

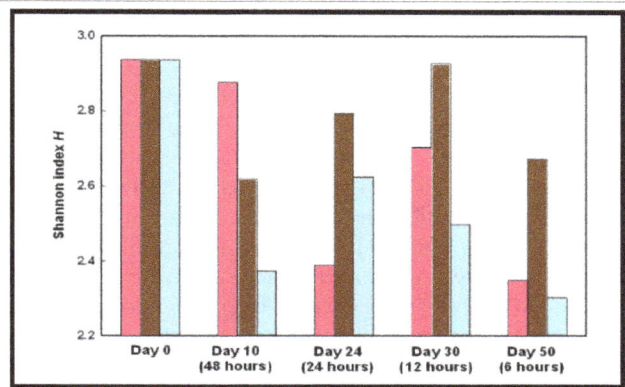

Figure 3: Change of Shannon index values throughout reactor operation (■: A-I ; □ A II; □ :A-I-II) Shannon index values (H) were calculated on the basis of the number and intensity of bands on the gel tracks. HRT is indicated in parenthesis.

reactor supported (with decreasing HRT). N1, N2, N5, N6 and N9 became dominant between days 15 and 40. Finally, days 40-50, new groups such as N7 and N8 became dominant. Only a limited number of bands with greater than 98% similarity with each other were recovered from sludge from all three aeration tanks. One contained F3, Z2, and N2, which are associated with *Zoogloea*, and the other contained F6, Z6, and N5, which are associated with *Acidovorax*. Sludge A-I and A-II had two sequences in common: F2 and Z1, which were associated with *Microbacterium*, and the other contained F9 and Z10, which were associated with the TM7 branch. These results demonstrate that the bacterial community is significantly different between the reactors, depending on the type of cyanide used.

GC-MS analysis

Analysis of samples in different places of the aeration tank was performed. Total ion chromatograms are shown in Figure 4. It showed that the type and amount of organic matter in wastewater have a decreasing trend in the flow direction. A-I + II, the areas of ICT crest of the wave were almost the same. This demonstrates that the disposal capacities in the last three organic compartments were extremely limited and most organics were removed in A-I and A-II. Further analysis of the types and relative

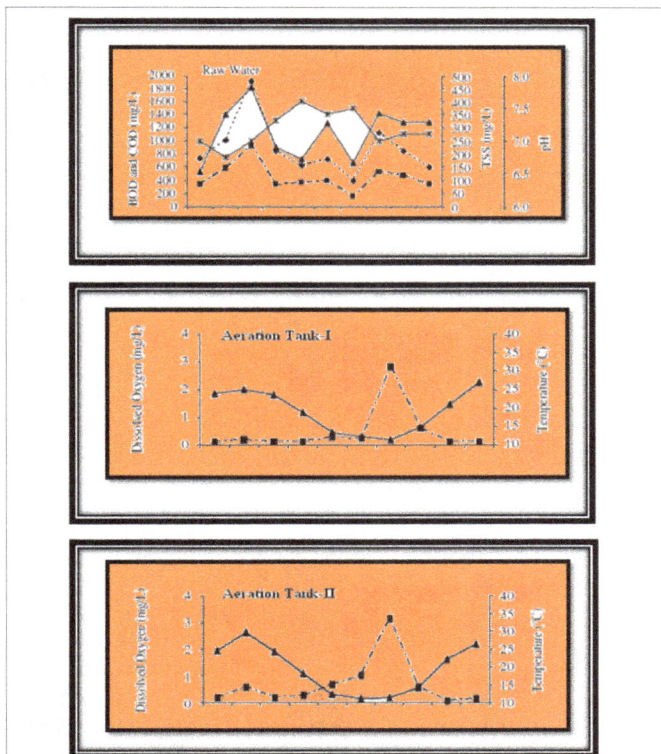

Figure 1: Waste water treatment systems operations conditions & performance (a) BOD (■), COD (♦), TSS (▲), pH (✳) in raw water ; (b) temperature (▲), Dissolved oxygen (■)

amounts of the organic phases are illustrated in Table 3. It was shown that there were seven major types of organic matter in the influent. The number of organic matter in the influent was 112 and increased to 185 after the influent mixed with the return sludge in A-I. Then the number fell by 113 and 48, but rose again to 116 in the secondary settling tank. The number of organic matter in the secondary settling tank was close to that of the tributary. Thus, along the direction of flow of water, types of organic materials initially increased and then gradually decreased, and finally increased sharply. The sudden increase in substances was mainly those refractory organic as alcoholic aldehyde, ketone acid, hydrocarbons and halogenated hydrocarbons. It was deduced that these refractory organics were initially adsorbed by the activated sludge in an aeration condition. And in anoxic conditions of the secondary settling tank, the activity of aerobic bacteria and *Zoogloea* significantly reduced.

FISH analysis

In this investigation, several probes were used for the analysis of ammonia oxidizing bacteria in both systems. Probes β-AO233, Nsp436 and Nmo254 were respectively used for the detection of bacteria belonging to *Nitrosospira* cluster, and the cluster *Nitrosomonas* halophilic and halo tolerant, and total ammonia oxidizing bacteria (Table 4). It was found that in both systems, the genus *Nitrosomonas* shows about 1.8 % (the system A2O) and 1.6% (AO system) respectively. The *Nitrosospira* proportions in the two systems were 2.2% (the system A2O) and 1.2% (AO system), respectively. Hybridizations with Nmo254 probe showed that the total number of ammonia oxidizing bacteria in the system represented A2O average 3.6 ± 0.2% of the total bacterial population while the AO system was only 1.9 ± 0.2%.

Discussion

This study was undertaken to promote our knowledge of

Table 4: Oligonucleotide Probes used in this study

Probe	Sequence (59 - 39)	Target group
b-AO233	AGCTAATCAGRCATCGG	All b-subgroup ammonia oxidizers
Nsp436	TTTCGTTCCGGCTGAAAG	All *Nitrosospira* spp.
Nmo254	GTAGGCCSTTACCCYACC	All *Nitrosomonas* spp.

how microbial communities in wastewater are important in governing the settlement patterns of the bacterial community. An excellent strategy to assess these patterns is ensured by bacterial inoculation experiments. Microbial communities, obtained by activated sleds, are added to the pre-sterilized effluent selected and the implementation of certain community structures is compared. This study provides data that support an analysis of the foundations of the creation of the bacterial community. The study of the composition, structure and dynamics of microbial communities in aerated lagoons is essential to understand and ensure the proper functioning of the treatment system, a valuable tool for improving the design of aerated ponds. Since the ecological function of microorganisms depended on its community structure, operational performance and degrading treatment system efficiency could be reflected by changes in the microbial community structure. Each agency has its inherent niche and optimal substrates, and the microbial community would adjust its structure in response to the changing environment. Whereas little information is available on microbial communities that inhabit these ecosystems, the approach based on PCR-DGGE, which was applied here, has shown to be effective in obtaining new data on the structure and dynamics of these communities. In addition, the constant changes in DO concentration, the dramatic reduction of biodegradability and recycling of sludge also contributed to the constant changes in environmental conditions, which led to repetitive structural changes in bacterial community. As shown in Table 1, both systems show nearly identical performance in the removal of chemical oxygen demand, biochemical oxygen demand and suspended solids, while the anoxic-aerobic system had a slightly higher removal of ammonia and the anaerobic-anoxic-aerobic system had a higher phosphate removal.

Although wastewater systems received identical, despite fairly similar constitutions, operating conditions, and treatment performance, both systems have shown quite different structures of microbial communities, except for ammonia oxidizing bacteria. It is interesting to note that the anoxic-aerobic-anaerobic system had far richer compositions of bacterial populations, actinomycetes and yeast. The richest community structures of these populations have been clearly linked to the creation of the anaerobic compartment. Most species that increased in the anoxic-anaerobic-aerobic system are perhaps those that tend to thrive in anaerobic environments. Temperature, DO and pH were the parameters that have shown to exert more influence on DGGE profiles. Yuz, et al. [23] previously reported temperature and DO as decisive parameters that affect community structure. Gilbride and colleagues [28] also found a significant correlation between the temperature and the structure of the bacterial communities of the activated sludge, as well as other influential

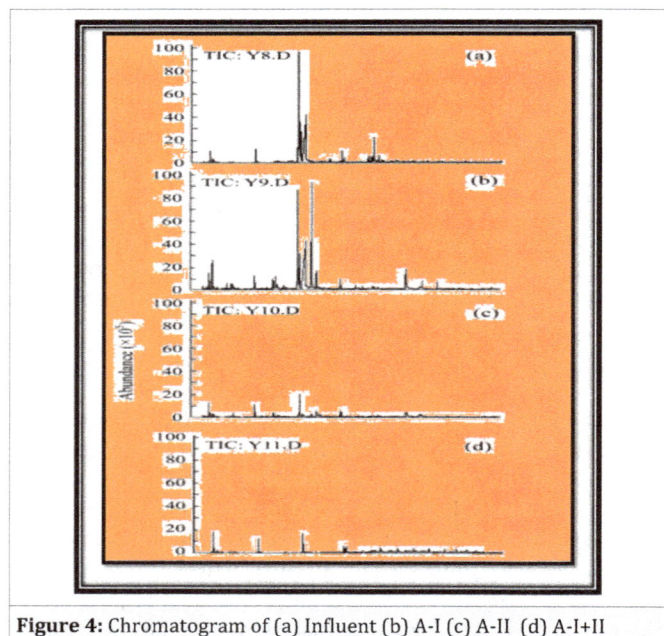

Figure 4: Chromatogram of (a) Influent (b) A-I (c) A-II (d) A-I+II

parameters, such as COD and BOD, which showed no significant correlation. Table 1 indicates that the AO system has almost no phosphate removal capability. The A2O system, on the other hand, had shrinkage in TP of about 50%, which can be attributed to the possible accumulation of ODP. The results of bacterial DGGE bands sequencing indicated that the dominant population that appear only in the A2O system belongs to the gamma Proteobacteria. Gamma proteobacteria are related to the elimination of phosphates [41]. To stabilize the structure of the bacterial community and the purification efficiency for industrial wastewater process A/ O, the following strategies can be considered: (a) to strengthen the pre-treatment units to reduce the fluctuation of the waste water and avoiding shock loads to the activated sludge system; (b) transforming the original O3 compartment in the anoxic tank and packed compartment A/ S with the immobilized carriers in different places. Thus, after most of the organic materials responsible for the aerobic bacteria are depleted, anaerobic bacteria and facultative bacteria could become dominant for the degradation of pollutants by the alteration of the oxygen concentration. The relatively low removal of TP in the A2O system, on the other hand, might be related to the relatively high residual nitrate from the anoxic compartment. By this biodegradation step by step, the diversity of the microbial population and the stability of the community structure would be improved. The nitrate concentration of the mixed liquor in the anoxic compartment was as high as 12.4 mg / L, which could be used as sufficient electron acceptors for the denitrifying bacteria that use organic substrates. Activated sludge actinomycetes have recently become the research focus because they are believed to play an important role in sludge bulking and foaming in activated sludge plants [43]. Competition between PAO and denitrifying bacteria for organic compounds could be the main reason for the weak removal of TP [42]. Microorganisms in different biological compartments would view its unique functions and the removal efficiency of pollutants will be improved. The average of the two IVR systems was 150 l / g (A2O) and 100 l/ g (AO) respectively. Aside from these negative roles, actinomycetes are active in the decomposition of organic matter. The diversity of actinomycetes in both systems suggests that some actinomycetes could also play an important role in the elimination of the organic substances. Further investigation, however, is needed to draw a final conclusion. High temperatures and low DO levels with the presence of microorganisms previously associated with anaerobic ecosystems found in this study, may explain the reduced efficacy of treatment. We hypothesize that the increase of the temperature and the depletion of DO levels create anoxic micro niches, promoting the growth of anaerobic bacteria, such as sulfate-reducing bacteria [44]. Among the four populations analyzed, only ammonia oxidizing bacterial communities have demonstrated a clear similarity (77.5%) between the two processes, suggesting that the introduction of anaerobic compartment has not changed the wealth of AOB populations significantly. However, when analyzing the DGGE data from complex environmental samples through several related to DNA extraction and purification methods [45], the relative efficiency of gene amplification [46] or the PCR inhibition due to the presence of humic acids and heavy metals [47], or the impact of

artifact bands due to excessive cycles of amplification should be taken into account. Nevertheless, analysis of bands DGGE profiles focuses on the numerical analysis of ability to be applied to the results obtained by molecular techniques. However, as shown by the results of FISH analysis, the proportion of the total AOB bacterial numbers in the two systems was very different [3.6 ± 0.2% for the system A2O and 1.9 ± 0.2% for the AO system). Obviously, the A2O system had many more cells than AOB in the AO system. A previous study [48] demonstrated that sludge retention time mainly influenced the total AOBs in activated sludge systems. But in this study, the two systems had similar SRT and MLSS. Further studies are necessary in order to explain this phenomenon completely.

References

1. Maulin P Shah, Patel KA, Nair SS, Darji AM, Shaktisinh Maharaul. Optimization of Environmental Parameters on Decolorization of Remazol Black B Using Mixed Culture. American Journal of Microbiological Research. 2013;1(3):53-56.

2. Maulin P Shah. Microbial Degradation of Azo Dye by *Pseudomonas* spp. MPS-2 by an Application of Sequential Microaerophilic and Aerobic Process. American Journal of Microbiological Research. 2013;43:105-112.

3. Moter Annette, Gobel Ulf B. Fluorescence in situ hybridization (FISH) for direct visualization of microorganisms. Journal of Microbiological Methods. J Microbiol Methods. 2000;41(2):85-112.

4. Rondon Michele R, August Paul R, Bettermann Alan D, Brady Sean F, Grossmann Trudy H, Liles Mark R et al. Cloning the soil metagenome: a strategy for accessing the genetic and functional diversity of uncultured microorganisms. Applied and Environmental Microbiology. 2000;[6]66:2541-2547.

5. Madoni P, Davol D, Gibin G. Survey of filamentous microorganisms from bulking and foaming activated-sludge plants in Italy. Water Res. 2000;34(6):1767–1772.

6. Chen S, Xu M, Cao H. the activated-sludge fauna and performance of five sewage treatment plants in Beijing China. Eur J Protistol(2004);40(2) 30:147–152.

7. Nocker A, Burr M, Camper AK. Genotypic microbial community profiling: a critical technical review. Microb Ecol. 2007;54(2):276-89.

8. Holben, W.E ,Harris. D. DNA-based monitoring of total bacterial community structure in environmental samples Mol Ecol. 1995;4(5):627-31.

9. Giraffa G, Neviani E. DNA-based, culture-independent strategies for evaluating microbial communities in food-associated ecosystems. Int J Food Microbiol. 2001;67(1-2):19-34.

10. Ranjard L, Poly F, Nazaret S. Monitoring complex bacterial communities using culture-independent molecular techniques: application to soil environment. Res. Microbiol. 2000;151(3):167-177.

11. Kawai M, Matsutera E, Kanda H, Yamaguchi N, Tani K, Nasu M. 16S ribosomal DNA-based analysis of bacterial diversity in purified water used in pharmaceutical manufacturing processes by PCR and denaturing gradient gel electrophoresis. Appl Environ Microbiol. 2002;68(2):699-704.

12. Zhou X, Bent SJ, Schneider MG, Davis CC, Islam MR, Forney LJ. Characterization of vaginal microbial communities in adult healthy women using cultivation-independent methods. Microbiology. 2004;150(Pt 8):2565-73. doi: 10.1099/mic.0.26905-0.

13. Amann RI, Ludwig W, Schleifer KH. Phylogenetic identification and in situ detection of individual microbial cells without cultivation. Microbiol Rev. 1995;59(1):143-69.

14. Dahllöf I. Molecular community analysis of microbial diversity. Curr Opin Biotechnol. 2002;13(3):213-7.

15. Colwell R, Grimes J. Semantics and strategies. In Non-culturable Microorganisms in the Environment; Colwell R, Grimes J. Eds. ASM: Washington. DC, USA, 2000;1-5.

16. Kazuhiro K. Nonculturable bacterial populations that control environmental processes. Biosci. Ind. 1999;57:731-736.

17. Frostegard A, Courtois S, Ramisse V, Clerc S, Bernillon D, Le Gall F, et al. Quantification of bias related to the extraction of DNA directly from soils. Appl Environ Microbiol. 1999;65(12):5409–5420.

18. Wu L, Li F, Deng C, Xu D, Jiang S, Xiong Y. A method for obtaining DNA from compost. Appl Microbiol Biotechnol. 2009;84(2):389-95. doi: 10.1007/s00253-009-2103-8.

19. Abu Al-Soud W, Râdström P. Capacity of nine thermostable DNA polymerases to mediate DNA amplification in the presence of PCR-inhibiting samples. Appl Environ Microbiol. 1998;64(10):3748-53.

20. Weiss A, Jérôme V, Freitag R. Comparison of strategies for the isolation of PCR-compatible, genomic DNA from a municipal biogas plants. J Chromatogr B Analyt Technol Biomed Life Sci. 2007;853(1-2):190-7.

21. Stach JE, Bathe S, Clapp JP, Burns RG. PCR-SSCP comparison of 16S rDNA sequence diversity in soil DNA obtained using different isolation and purification methods. FEMS Microbiol Ecol. 2001;36(2-3):139-151.

22. McIlroy SJ, Porter K, Seviour RJ, Tillett D. Extracting nucleic acids from activated sludge which reflect community population diversity. Antonie Van Leeuwenhoek. 2009;96(4):593-605. doi: 10.1007/s10482-009-9374-z.

23. YuZ, Mohn W. Bacterial diversity and community structure in an aerated lagoon revealed by ribosomal intergenic spacer analysis and 16S rDNA. Appl Environ Microbiol. 2001;67(4):1565–1574. doi: 10.1128/AEM.67.4.1565-1574.2001

24. Otawa K, Asano R, Oba Y, Sasaki T, Kawamura E, Koyama F, et al. Molecular analysis of ammonia-oxidizing bacteria community in intermittent aeration sequencing batch reactors used for animal waste water treatment. Environ Microbiol 2006;8(11):1985–96. DOI: 10.1111/j.1462-2920.2006.01078.x.

25. Baker CJ, Fulthorpe RR, Gilbride KA. An assessment of variability of pulp mill waste water treatment system bacterial communities using molecular methods. Water Qual Res J Can 2003;38:227–42.

26. Gilbride KA, Fulthorpe RR. A survey of the composition and diversity of bacterial populations in bleached kraft pulp-mill waste water secondary treatment systems. Can J Microbiol. 2004;50(8):633-44. DOI: 10.1139/w04-031.

27. Baker CJ, Fulthorpe RR, Kimberley, A. Gilbride. An assessment of variability of pulp mill waste water treatment system bacterial communities using molecular methods. Water Qual Res J Can 2003;38:227–42.

28. Gilbride KA, Frigon D, Cesnik A, Gawat J, Fulthorpe RR. Effect of chemical and physical parameters on a pulp mill biotreatment bacterial community. Water Res. 2006;40(4):775-87. DOI: 10.1016/j.watres.2005.12.007.

29. Casserly C, Erijman L. Molecular monitoring of microbial diversity in an UASB reactor. Int Biodeterior Biodegrad 2003;52(1):7–12.

30. Ingvorsen, K., B. Hjer-Pedersen, and S. E. Godtfredsen. Novel cyanide-hydrolyzing enzyme from Alcaligenes xylosoxidans subsp. denitrificans. Appl Environ Microbiol. 1991;57(6):1783-9.

31. Kim, B. S., H. M. Oh, H. J. Kang, S. S. Park, and J. S. Chun. Remarkable bacterial diversity in the tidal flat sediment as revealed by 16S rDNA analysis. J. Microbiol. Biotechnol. 2004;14:205-211.

32. Clarke K, Gorley R. PRIMER v5: user manual/tutorial. Plymouth, UK: PRIMER-E. 2001.

33. Shannon CE, Weaver W. The mathematical theory of communication. Urbana, IL: University of Illinois Press. 1963.

34. Pielou EC. Ecological diversity. New York: Wiley. Polz MF, Cavanaugh CM. Bias in template-to-product. 1975.

35. Fromin N, Hamelin J, Tarnawski S, Roesti D, Jourdain Miserez K, Forestier N. Statistical analysis of denaturing gel electrophoresis finger printing patterns. Environ Microbiol. 2002;4(11):634-43.

36. ter Braak CJF, Verdonschot PFM. Canonical correspondence analysis and related multi variate methods in aquatic ecology. Aquat Sci 1995; 57(3):255–89.

37. Boon N, Windt WD, Verstraete W, Top EM. Evaluation of nested PCR-DGGE with group-specific 16S rRNA primers for the analysis of bacterial communities from different wastewater treatment plants. FEMS Microbiol Ecol. 2002;39(2):101-12. doi: 10.1111/j.1574-6941.2002.tb00911.x.

38. Cocolin L, Bisson LF, Mills DA. Direct profiling of the yeast dynamics in wine fermentations. FEMS Microbiol Lett. 2000;189(1):81-7.

39. YuZ, Mohn W. Bacterial diversity and community structure in an aerated lagoon revealed by ribosomal intergenic spacer analysis and 16S rDNA. Appl Environ Microbial. 2001;67:1565–74.

40. Gilbride KA, Frigon D, Cesnik A, Gawat J, Fulthorpe RR. Effect of chemical and physical parameters on a pulp mill biotreatment bacterial community. Water Res. 2006;40(4):775-87.

41. Kavanaugh RG, Randall CW. Bacterial populations in a biological nutrient removal plant. Water Sci Technol 1994;29(7):25-34.

42. Kuba T, Wachtmeister A, van Loosdrecht MCM, Heijnen JJ. Effect of nitrate on phosphorus release in biological phosphorus removal systems. Water Sci Technol 1994;30(6):263-269.

43. Davenport RJ, Curtis TP, Goodfellow M, Stainsby FM, Bingley M. Quantitative use of fluorescent in situ hybridization to examine relationships between mycolic acid-containing actinomycetes and foaming in activated sludge plants. Appl Environ Microbiol. 2000;66(3):1158-66.

44. Schramm A, Santegoeds C, Nielsen H, Ploug H, Wagner M, Pribyl M. On the occurrence of anoxic microniches, denitrification and sulfate reduction in aerated activated sludge. Appl Environ Microbiol 1999;65(9):4189–96.

45. Lemarchand K, Berthiaume F, Maynard C, Harel J, Payment P, Bayardelle P. Optimization of microbial DNA extraction and purification from raw wastewater samples for downstream pathogen detection bymicroarrays . J Microbiol Methods. 2005;63(2):115-26. DOI: 10.1016/j.mimet.2005.02.021.

46. Suzuki MT, Giovannoni SJ. Bias caused by template annealing in the amplification of mixtures of 16Sr RNA genes by PCR. Appl Environ Microbiol 1996; 62(2):625–30.

47. Tebbe CC, Vahjen W. Interference of humic acids and DNA extracted directly from soil in detection and transformation of recombinant DNA from bacteria and yeast. Appl Environ Microbiol. 1993; 59(8): 2657–2665.

48. Limpiyakorn T, Shinohara Y, Kurisu F, Yagi O (2005) Communities of ammonia-oxidizing bacteria in activated sludge of various sewage treatment plants in Tokyo. FEMS Microbiol Ecol. 2005;54(2):205-17.

5

Development and Evaluation of a Multiplex PCR for Simultaneous Detection of Five Foodborne Pathogens

Thuy Trang Nguyen[1,3#], Vo Van Giau[2,3#*], Tuong Kha Vo[4*]

[1]Department of Pharmacy, Ho Chi Minh City University of Technology (HUTECH), Ho Chi Minh City, Vietnam
[2]Deparment of Faculty of Food Technology, Ho Chi Minh City University of Food Industry (HUFI), 140 Le Trong Tan, Tan Phu district, Ho Chi Minh City, Vietnam
[3]Department of Bionano Technology, Gachon Medical Research Institute,Gachon University, Seongnam, South Korea
[4]Vietnam Sports Hospital, Ministry of Culture, Sports and Tourism, Do Xuan Hop Road, My Dinh I Ward, Nam Tu Liem District, Hanoi City, Vietnam
#Authors contributed equally to this article

*Corresponding author: Vo Van Giau, Deparment of Faculty of Food Technology, Ho Chi Minh City University of Food Industry, 140 Le Trong Tan, Tan Phu district, Ho Chi Minh City, Vietnam, E-mail: vovangiau911@gmail.com

Abstract

Foodborne pathogens present serious concerns to human health and can even lead to fatalities. The gold standard for pathogen identification – bacterial culture – is costly and time consuming. A cheaper and quicker alternative will benefit in controlling food safety. In this study, we developed a multiplex-PCR protocol for simultaneous detection of five Foodborne pathogens including *Escherichia coli* O157:H7, *Staphylococcus aureus*, *Salmonella* spp., *Listeria monocytogenes*, and *Vibrio cholerae*, based on five genes stx1, femA, invA, iap, và ctxA, respectively. Specific primers for multiplex PCR amplification of the stx (Shiga-like toxin), nuc (thermo nuclease), inv A (invasion protein A), iap (invasive associative protein), and ctx A (cholera toxin A) genes that were established to amplify simultaneous detection of the target pathogens. The assay was also validated for its specificity, sensitivity, and applied to test some spiked food samples. The results showed the products expected multiplex PCR fragments of approximately 112, 244, 301, 453, 518 and 720bp for *S. aureus*, *Salmonella* spp. *V. cholera*, *L. monocytogenes*, *E. coli* O157:H7 and 16S rRNA, respectively. The assay was specific to the targeted pathogens and was sufficiently sensitive and robust to effectively analyze market samples. The whole process took less than 24 h to complete indicating that the assay is suitable for reliable and rapid identification of these five foodborne pathogens, which could be suitable in microbial epidemiology investigation.

Keywords: Foodborne pathogens; Multiplex-PCR; Five genes; Detection; Simultaneous

Introduction

The incidence of foodborne diseases has increased over the years and is a serious health hazard in both developing and developed countries. *Escherichia coli* O157:H7 (*E. coli* O157:H7), *Salmonella* spp., and *Vibrio cholera* (*V. cholera)* are likely the most common cause of foodborne disease [1, 2]. The well-known *E. coli* bacteria that produce Shiga toxin (STEC) is *E. coli* O157:H7 strains are foodborne infectious agents that cause a number of life-threatening diseases, including hemorrhagic colitis (HC) and hemolytic uremic syndrome (HUS) [3]. According to recent reports by the Center for Disease Control and Prevention from 2011-2014, there were 11 multistate outbreaks of STEC in the United States of America with six of them attributed to *E. coli* O157:H7. With low infectious dose, an inoculation of fewer than 10–100 CFU of *E. coli* O157:H7, is sufficient to cause infection [4]. Shiga toxin (Stx) is one of the major virulence factors involved in *E. coli* O157:H7 pathogenesis [5]. The stx gene is well associated with a prophage and including a variety of subtypes shiga toxin that are described as stx1, stx1c, stxfc, stx2, stx2e, stx2d and stx2g [6]. With capable of producing enterotoxins and coagulase enzyme, *Staphylococcus aureus* (*S. aureus*) has already been involved in a number of food-poisoning outbreaks. Producing of an extracellular thermostable nuclease and coagulase with the same frequencies that were important phenotypic identifying markers of *S. aureus* [7-9]. The nuc gene encodes for the production of a thermostable endonuclease enzyme, which has been used for the correct identification of *S. aureus* in previous studies [10,11]. Approximately 20,000 hospitalizations and 378 deaths per year in the United States, *Salmonella* spp. are the leading bacterial cause of acute gastrointestinal illness. There are some serotypes as *S. enteritidis*, *S. typhi*, *S. paratyphi* A, *S. paratyphi*B, *S. paratyphi*C, and *S. choleraesuis* that can cause foodborne illnesses [12,13]. The invA gene is a good candidate gene to invade mammalian cells and subsequently cause disease [14,15], and it presents in all pathogenic serovars as a maker has been the most frequently used for *Salmonella* spp. Detection [16-18]. *Listeria monocytogenes* (*L. monocytogenes*) has been found from dairy, frozen aquatic and meat products [19,20] and is associated with listeriosisis, a servere disease with morbidity and of high mortality of 20–30% [21-23]. One of the main virulence genes, the iap gene encodes p60 protein, is associated with the presence of the invasion-associated as the mechanism of pathogenicity in *L. monocytogenes*. Similarity, cholera has been one of the most feared diseases for human. The vast majority of strains associated with epidemic cholera are attributed to toxigenic *Vibrio cholera* (*V. cholerae*) with the O1 serotype [24]. The ctxAB operon that encodes cholera toxin resides in filamentous bacteriophage CTXΦ genome confer the ctx operon to *V. cholerae* strains as a prophage that carries the *ctxA* and *ctxB* genes.

Illnesses resulting from the consumption of foods contaminated with pathogens and/ or their toxins have a wide range of economic and public health impact worldwide [25]. The current gold-standard method for detecting foodborne pathogen in food encompasses enrichment with subsequent plating on selective media, biochemical reactions, and serological tests, which are time-consuming and labor-intensive [26]. Currently,, the official procedure for detection of pathogenic bacteria also used to a cultural method, and this procedure could take from 3 to 5 days for confirmation, which is a disadvantage when the results are needed promptly [27,28]. Hence, faster technologies have been applied to develop rapid and enhance sensitive analytical protocol for the foodborne pathogens. The polymerase chain reaction (PCR) is still the most commonly used for detection of the targets bacterial, which based on the identification of the target gene of specific bacteria present after the exponential application with the high sensitivity and specificity. It has become an important tool for detecting and identifying pathogenic organisms in various foods [29-33]. Consequently, since multiplex PCR assay has been able to simultaneously amplify multiple gene targets by using several sets of target specific or degenerated primers in a single tube [34], it has greatly improved the sensitivity, specificity, and speed of detecting pathogenic organisms [35]. Furthermore, multiplex PCR assay, in comparison with uniplex PCR assays, could save considerable time and workload, and improve efficiency [26,30,31-36].

In this study, we developed a multiplex PCR assay for the rapid and simultaneous detection of five epidemic foodborne pathogens, namely *Escherichia coli O157:H7*, *Staphylococcus aureus*, *Salmonella* spp, *Listeria monocytogenes*, and *Vibrio cholera*. The performance of the multiplex assay, including its sensitivity, specificity, and precision in quantitative analyses, was comprehensively evaluated in comparison with the traditional methods. The capacity of the proposed assay to detect multiple target pathogens simultaneously was also tested, and the effect of non-target interference on the assay performance was evaluated. The results obtained with artificially contaminated food samples and real samples demonstrate that the multiplex PCR assay can simultaneously detect these five target foodborne pathogens in foods with high sensitivity and reliability.

Materials and Methods

Bacterial strains and their cultivation

The strains used for specificity testing are listed in Table 1. For the identification of five food-borne pathogens and the sensitivity of the multiplex PCR assay experiments, the following strains were used: three strains *E. coli* O157:H7, one produces both Stx1 and Stx2 (NLU), and one produces Stx1 only (NIHE), the last one has Stx2 only (HCMUS), all were obtained from previously worked [37,38], *S. aureus* ATCC6538, *S. enterica* ATCC 14028,

L. monocytogenes A TCC15313, and *V. cholera* ATCC 17802. All strains were grown in tryptic soy broth (TSB) or brain heart infusion (Merck, Germany) at37°C for 24 h. The simultaneous enrichment broth (SEB), which used for simultaneous enrichment of five pathogenic bacteria in this study, was described previously by Kobayashi et al. [39].Then, the culture broth was used for DNA extraction and subjected to the multiplex PCR assay.

Pathogen detection by the conventional culture method

For detection of *E. coli* O157:H7, each 25 g of each sample was diluted in 225 mL of Modified Tryptone Soya broth (mTSB-Oxoid, UK) added with Novobiocin, homogenized for 2 min at 260 rpm using a Stomacher (Model 400 circulator, Seward, Norfolk, England) and incubated for 18–24 h at 41.5°C according to ISO 16654 (2001) method, as well as the remaining steps. After enrichment and immune magnetic concentration steps, the selective and differential isolation of enterohemorrhagic *E. coli* O157:H7 was carried out on MacConkey Agar with Sorbitol, Cefixime, and Tellurite (CT-SMAC—Oxoid, UK) and incubated overnight at 42°C. From each sample one well isolated suspected colony was transferred to tryptone soy agar (Oxoid) and incubated for 24 h at 37°C. Subsequently, one isolate from the subculture was further tested for agglutination with an *E. coli*

Table1. Bacterial strains and their sources were employed in this study

No	Bacteria	Serovar	Source
	The target strains		
1	*E. coli*	O157:H7	NIHE
2	*E. coli*	O157:H7	HCMUS
3	*E. coli*	O157:H7	NLU
4	*S. aureus*		ATCC6538
5	*Salmonella enterica*		ATCC14028
6	*L. monocytogenes*		ATCC15313
7	*V. cholerae*		ATCC17802
	The non-target strains		
1	*E. coli*		ATCC 11775
2	*E. coli*		ATCC 25922
3	*E. coli*(I1)		Clinical isolate
4	*E. coli*(I2)		Clinical isolate
5	*E. coli*(I3)		Chicken isolate
6	*E. coli*(I4)		Beef isolate
7	*E. coli*(I5)		Salad isolate
8	*Clostridium perfringens*		ATCC13124
9	*Bacillus cereus*		ATCC11778
10	*Shigella sonnei*		ATCC 9290

ATCC: American Type Culture Collection;
NIHE: National Institute Of Hygiene And Epidemiology;
HCMUS: HCM University of Science; NLU: Nong Lam University

O157:H7 latex test kit (Becton–Dickinson, USA) for sero group O157:H7 confirmation. For detection of *S. aureus*, each 10 g food sample was diluted with 90 mL of sterile Saline Petone Water (Merck, Germany) and pummeled in a Stomacher apparatus for 1 minute; One milliliter of the culture was added to 10 mL of Giolitti–Cantoni broth (Merck, Germany) and incubated at 37°C for 48 hours. One loop full of the culture with black color was then streaked onto Baird Paker Agar (Merck, Germany) and incubated at 37°C for 48 hours. The resulting presumptive *S. aureus* colonies were tested to biochemical screening using a coagulase test. For detection of *Salmonella* spp. Each 25g food sample was diluted with 225 mL of sterile Buffered Peptone Water (Merck, Germany) and pummeled in a Stomacher apparatus for 2 minutes; the mixture was then incubated for 18 hours at 37°C. One milliliter of the culture was added to 10 mL of Rappaport-Vassiliadis soy peptone broth (Merck, Germany) and incubated at 42, 5°C for 18 hours. One loop full of the culture was then streaked onto Xylose lysine desoxycholate agar (Merck, Germany) and incubated at 37°C for 24 hours. The resulting presumptive *Salmonella* colonies were tested to biochemical screening and serological confirmation using *Salmonella* polyvalent O, O1 antisera (Becton–Dickinson, USA). For detection of *L. monocytogenes*, 25 g of the food samples were mixed with 225 mL of sterile Fraser Broth Listeria enrichment broth (Merck, Germany) and pummeled in a Stomacher for 1 minute, followed by incubation for 48 hours at 30°C. One loopful of the culture broth was streaked onto Chromogenic *Listeria* agar with selective supplement (Oxoid, Hampshire, UK) and incubated at 37°C for 48 hours. Presumptive colonies were streaked onto horse blood agar and TSA plates and incubated at 35°C for 48 hours. The resulting presumptive *Listeria* colonies were submitted for biochemical screening (oxidase test, catalase test, and Gram staining). For detection of *V. cholera*, the composited sample of 25 g and to added 225 ml of enrichment medium alkaline saline peptone water (Merck, Germany)and pummeled in a Stomacher for 1 minute, followed by incubation for 24 hours at 41.5°C. One loopful of the culture was then streaked onto Thiosulfate Citrate Bile and Sucrose agar (Merck, Germany) and incubated at 37°C for 24 hours. The resulting presumptive *V. cholerae* colonies were submitted for biochemical screening and serological confirmation.

DNA isolation

DNA extraction was performed by boiling 1 ml of each overnight culture pre-enrichment samples or culture in all the strains was boiled for 10 min in an Isotemp heat block (Fisher Scientific, Pittsburgh, PA) and centrifuged at 12,000 × g for 2 min. The supernatants were saved for posterior use as DNA template for all PCR reactions at − 20°C.

Oligonucleotides primers and multiplex PCR assays

The oligonucleotide primers used in this study are shown in Table 2. For multiplex PCR analysis, two primers pairs were used: stx, which is specific primers to various *stx1* and *stx2* gene for *E. coli*O157:H7described by Yamasaki S, et al (1996) [28]; the primer inv A, specific for *Salmonella* spp. described by designed by Chiu et al [40]; the primer iap, specific for *L. monocytogenes* described by designed by Manzan et al [41]; and the primer ctxA, specific for *V. cholera* described by designed by Shirai et al [42]. The rest of target-specific primers in this work were designed to amplify for the *nuc*gene of *S.aureus* and 16S rRNA gene sequence of bacterial with a amplification product of 112 bp and 720 bp, respectively. The primers were selected from a complete sequence from *S. aureus* strain DQ399678 and 16S rRNA gene sequences for most bacteria and archaea are available on public databases from the Gen Bank database. Primers were designed using the software Fast PCR [43]. The specificity of pair of primers was evaluated by nucleotide similarity searched with the BLAST algorithm at the NCBI website (http: www.ncbi.nlm. nih.gov). Additionally, the PCR amplification was evaluated with DNA samples from the bacterial species listed in Table 2.

The 16S rRNA gene was also targeted as an internal control of the presence of amplifiable bacterial DNA. The single PCR was performed a total volume of 50 µl using the Veriti96-Well Thermal Cycler (Applied Biosystems, Foster City, CA)in reaction mixtures (Promega) containing 0.5 µM each primer, 200 µM each dNTP, 3 mM $MgCl_2$, 1.5 U Taq DNA polymerase, 1x PCR

Table2. Primer pairs and its characteristics employed for the multiplex PCR

Organisms	Forward primers (5' – 3')	Reverse primer (5'-3')	Target gene / primer	Amplicon size (bp)	Reference
Staphylococcus aureus	AATTACATAAAGAACCTGCGACT	GCACTTGCTTCAGGACCATATT	*nuc*	112	This study
Escherichia coli O157:H7	GAGCGAAATAATTTATATGTG	TGATGATGGCAATTCAGTAT	*stx*	518	[28]
Salmonella spp.	ACAGTGCTCGTTTACGACCTGAAT	AGACGACTGGTACTGATCGATAAT	*invA*	244	[40]
Listeria monocytogenes	GGGCTTTATCCATAAAATA	TTGGAAGAACCTTGATTA	*iap*	453	[41]
Vibrio cholerae	CTCAGACGGGATTTGTTAGGCACG	TCTATCTCTGTAGCCCCTATTACG	*ctxA*	301	[42]
Bacterial DNA	AGAGTTTGATCATGG CTCAGG	GGACTACCAGGGTATCTAATT	16S rRNA	720	This study

buffer and 2 l template. A negative control containing the same reaction mixture except the DNA template was included in every experiment. While the multiplex PCR conditions were the same as the single PCR assay except for the concentration of primers and incorporated Betaine into each reaction. The optimized concentrations of the six primer pairs in the multiplex were 0.135μM for stxF/ R; 1μM for nuc-F/ R; 1.2μM for invA-F/ R; 1.65μM iap-F/ R; 0.125 μM for ctxA-F/ R, and 0.8 μM for 16S -F/ R. Since PCR additives, such as dimethyl sulfoxide, glycerol, bovine serum albumin, or betaine, have been reported to be of benefit in multiplex PCR. The aim is to enhance amplification, adding with 2.5 M concentrations of Betaine were incorporated in all multiplex method.

The PCR program was carried out at 95°C for 5 min, followed denaturing by 35 cycles of 94°C for 1 min, 57. 5°C for 1 min, and 72°C for 1 min, and a final 5 min of 72°C for extension. PCR products were electrophoresed in 1% agarose at 100 V for 50 min followed by staining with ethidium bromide (0.5 g/ mL) then visualized under ultraviolet light, and the results were recorded by photography using an ultraviolet trans illuminator (Gel Doc XR system, Bio-Rad).

Specificity of the multiplex PCR assay

To assess the specificity of the multiplex PCR assay, cultures of 7 target pathogens and 10 non-target bacterial strains (Table 1) were prepared.The purity of the genomic DNA was assessed by determining the A260/ A280 ratio using a spectrophotometer(BioMate 3; Thermo spectronic, Rochester, NY). The multiplex PCR was performed individually for each of DNA samples of 17 strains in multiplex primer system using the experimental conditions described above. The inclusivity and exclusivity were calculated according to the MicroVal protocol [44]. Inclusivity is the ability of the PCR method to detect the target analyze from a wide range of strains. Inclusivity is defined as the percentage of target DNA samples that gave a correct positive signal. Exclusivity is defined as the percentage of non-target DNA samples that gave a correct negative signal.

Multiplex PCR evaluation in spiked food samples

In order to validate the multiplex PCR assay in food, the three Vietnam of food samples matrices including vegetables, seafood products and raw meat fork that were purchased from a local super market and were immediately transported in insulated coolers at 4°C to the laboratory for inoculation and analysis on the same day. Before inoculating experiments, the food samples were carefully tested for the presence of E. coli O157:H7, S. aureus, Salmonella spp. L. monocytogenes, and V. cholera using the conventional culture method as described above for each of once. Make sure none of these pathogens could be detected by culture in these food samples. The five food borne pathogens were cultured in TSB at 37°C for 18 h prior to decimal dilution in a sterile saline solution (0.85% NaCl) in order to obtain levels of inoculation (colony-forming units [CFU]/g). One mL in each dilution was spread on TSB agar to determine bacterial counts and the equal concentration also used to prepare bacterium representing to spiked 25g food samples. Each 25g food samples

were inoculated with 1 mL of each level of inoculation, and then placed with 225 mL of sterile SEB medium. A nature sample with non-inoculated was employed as a negative control. The mixture was homogenized in a Stomacher apparatus for 1 minute. After each 12-h, 18-h and 24-hincubation times, a 1mL aliquot was collected from each sample in each period of time incubation and DNA was extracted as described.

Results and Discussion

Identification of food-borne pathogens

Rapid and simultaneous detection of multiple pathogenic bacteria in foods is of great importance to ensure food safety. In this study, we developed and evaluated a multiplex PCR for simultaneous detection of the five food borne pathogens including E. coli O157:H7, S. aureus, Salmonella spp, L. monocytogenes and V. cholera in a single reaction. In order to ensure specificity and sensitivity and toavoid cross-reactions, primer pair selection is critical in the multiplex PCR assay for the simultaneous detection of five foodborne pathogens. All six of primers (and one is 16S rRNA) that were erected, designed and analysed for the simultaneous detection of five food borne pathogens using the online available software as described by Kalendar et al. [43]. The table 2 shows all information of the primer 16S-F/ R for amplification of a 720-bp sequence from the 16S rRNA gene of bacterial DNA, stx-F/ R for amplification of a 518-bp sequence from the stx gene of E. coli O157:H7,iap-F/ R for amplification of a 453-bp sequence from the iap gene of L. monocytogenes, ctxA-F/ Rfor amplification of a 301-bp sequence from the ctxA geneof V. cholerae, inv-F/ R for amplification of a 244-bp sequence from the invA gene of Salmonella spp., and nuc-F/ R for amplification of a 112-bpsequence from the nuc gene of S. aureus. Furthermore, successful multiplexing of multiplex PCR assay requires careful experimental design and optimization of reaction conditions. To achieve accurate template quantification in a multiplex PCR assay, each reaction must efficiently amplify a single product, and amplification efficiency must be independent of template concentration and the amplification of other templates. The annealing temperature of a multiplex PCR assay is one of the most critical parameters for reaction specificity. We tested a range of temperatures above and below the calculated Tm of the primers. Fortunately, based on the yield of PCR products for the seven target genes, the results showed an optimal multiplex annealing temperature of 59, 5°C. As can be seen, the results showed that these six primer pairs in the multiplex PCR assay worked well independently, and could distinguish the five pathogens from each other with high specificity, and PCR amplification was obtained the size of the PCR product followed by identification in terms of the expected size. Five food-borne pathogens, including E. coli O157:H7 (NLU), S. aureus ATCC6538, S. enterica ATCC 14028, L. monocytogenes ATCC15313, and V. Cholera TCC 17802were detected simultaneously through the multiplex PCR assay using multiplex primer set and the conditions as described above. As the Table 3 and Fig. 1 show the multiplex PCR assay was successfully developed to simultaneously identify the five foodborne pathogens based on the generation of the expected PCR fragments of 112 bp, 244 bp, 301 bp, 453 bp and 518 bp for *S. aureus* ATCC6538, *Salmonella enterica* ATCC 14028, *V. cholera* ATCC 17802, *L. monocytogenes* ATCC15313and *E. coli* O157: H7(NLU), respectively. In addition, no PCR product corresponding with target microorganism was detected in negative control using the multiplex primer. In addition, PCR products corresponding with the positive-control 16S rRNA gene (720 bp) were detected from pure cultures of seven pathogens.

Specificity of the multiplex PCR Assay

The specificity of the multiplex PCR conducted with the seven target strains and 10 non-target bacterial strains. All three *E. coli* O157:H7 strains *E. coli* O157:H7(NIHE), *E. coli* O157:H7(HCMUS), *E. coli* O157:H7(NLU), *S. aureus* ATCC6538, *S. enterica* ATCC 14028, *L. monocytogenes* ATCC15313, and *V. cholera* ATCC 17802strains were positive in the multiplex PCR assay and all non-target bacterial including *E. coli* ATCC 11775, *E. coli* ATCC 25922, *E. coli*(I1), *E. coli*(I2), *E. coli*(I3), *E. coli*(I4), *E. coli* (I5), *C. perfringens* ATCC13124, *B. cereus*ATCC11778 and *S. sonnei* ATCC 9290were negative in the assay, whereas 16S rRNA was amplified as expected. No mispriming or non-specific amplification was observed. Expectedly, the size of each pathogen amplicon was obtained only from the target foodborne pathogens, resulted in 100% inclusivity and 100% exclusivity. Even of target food borne pathogens and ten non-target pathogens were used to evaluate and verify the specificity of primers in this study, each primer pair by the multiplex PCR on DNA templates (Table 4). These results demonstrated that our multiplex PCR assay could be used to identify each of these five foodborne pathogens.

Evaluation of the multiplex PCR assay with spiked food samples

In order to assess the detection sensitivity of the multiplex PCR assay for its application to food samples, three of kind of food samples (vegetables, seafood products and raw meat fork) inoculated with *E. coli* O157:H7(NLU), *S. aureus* ATCC6538, *S. enterica* ATCC 14028, *L. monocytogenes* ATCC15313, and *V. cholera* ATCC 17802with seven level of the number of viable cells (0, 10^0, 10^1, 10^2, 10^3, 10^4 and 10^5 CFU/ ml) were employed and a

Table 4. Specificity test for the multiplex PCR assay; a minus (-) indicates the absence of a band and a plus (+) indicates the presence of a band

Species	Genes					
	nuc	invA.	ctxA	iap	stx	16S rRNA
E. coli O157:H7 (NIHE)	-	-	-	-	+	+
E. coli O157:H7 (HCMUS)	-	-	-	-	+	+
E. coli O157:H7 (NLU)	-	-	-	-	+	+
*S. aureus*ATCC6538	+	-	-	-	-	+
*Salmonella entericа*ATCC14028	-	+	-	-	-	+
*L.monocytogenes*ATCC15313	-	-	-	+	-	+
*V. cholera*ATCC17802	-	-	+	-	-	+
E. coli ATCC 11775	-	-	-	-	-	+
E. coli ATCC 25922	-	-	-	-	-	+
E. coli(I1)	-	-	-	-	-	+
E. coli(I2)	-	-	-	-	-	+
E. coli(I3)	-	-	-	-	-	+
E. coli(I4)	-	-	-	-	-	+
E. coli(I5)	-	-	-	-	-	+
*C. perfringens*ATCC13124	-	-	-	-	-	+
*B. cereus*ATCC11778	-	-	-	-	-	+
S. sonnei ATCC 9290	-	-	-	-	-	+

nature sample of each categories was included as negative control after carefully tested using the conventional culture method as described above for the target pathogenic bacteria. Additionally, in order to archive the sensitivity and reproducibility of the multiplex PCR assay, the artificially inoculated and non-inoculated of these categories food samples that were incubated for 12, 18 and 24 h in SEB enrichment medium. As can be seen from Table 5 shows that the multiplex PCR assay was able to correctly identify the presence of the five foodborne pathogens at all different inoculated in the lowest concentration of 10 CFU/ mL in each category of the samples after enrichment for 12 hours in SEB medium (Figure 2). Comparatively, our multiplex PCR assay developed was similar or more sensitive with the same the lowest level of 10 CFU/ mL when compared with Kim et al. [26].

Furthermore, in the more recently reported [45] multiplex PCR assays, Lee et al. (2014) reported a multiplex PCR for simultaneous detection of *E. coli* O157:H7, *B. cereus*, *V. parahaemolyticus*, *Salmonella* spp. *L. monocytogenes*, and *S. aureus* in various Korean ready-to-eat foods. The multiplex PCR assay developed by Lee et al. (2007) also allowed for simultaneous detection at concentrations of 10^0 CFU/ mL of the pathogenic bacteria, after only 24 h of incubation time. The multiplex PCR assay established in this study could similar the incubation time when compared with Lee et al. (2007). It could also detect the five foodborne pathogens with the lowest level of 10 CFU/ mL after 12 h of enrichment. Consequently, a 12-h enrichment period is

Table 3. Evaluation of the specificity of all PCR primers using various pathogenic bacterial

Strain	Source	Genes/ Primers					
		stx	nuc	invA	iap	ctxA	16S rRNA
E. coli O157:H7	NIHE	+	-	-	-	-	+
E. coli O157:H7	HCMUS	+	-	-	-	-	+
E. coli O157:H7	NLU	+	-	-	-	-	+
S. aureus	ATCC6538	-	+	-	-	-	+
Salmonella spp.	ATCC14028	-	-	+	-	-	+
L. monocytogenes	ATCC15313	-	-	-	+	-	+
V. cholerae	ATCC17802	-	-	-	-	+	+

Table 5. Multiplex PCR results of five pathogens from artificially inoculated three food samples matrices

Pathogens	Incubation time (h)	CFU/ml	Multiplex PCR results detection in food samples		
			Vegetables	Seafood products	Raw meat fork
E. coli O157:H7 (stx)	12	0	-	-	-
		10^0	-	-	-
		10^1	+	+	+
		10^2	+	+	+
		10^3	+	+	+
		10^4	+	+	+
		10^5	+	+	+
	18	0	-	-	-
		10^0	-	-	-
		10^1	+	+	+
		10^2	+	+	+
		10^3	+	+	+
		10^4	+	+	+
		10^5	+	+	+
	24	0	-	-	-
		10^0	-	-	-
		10^1	+	+	+
		10^2	+	+	+
		10^3	+	+	+
		10^4	+	+	+
		10^5	+	+	+
S. aureus (nuc)	12	0	-	-	-
		10^0	-	-	-
		10^1	+	+	+
		10^2	+	+	+
		10^3	+	+	+
		10^4	+	+	+
		10^5	+	+	+
	18	0	-	-	-
		10^0	-	-	-
		10^1	+	+	+
		10^2	+	+	+
		10^3	+	+	+
		10^4	+	+	+
		10^5	+	+	+
	24	0	-	-	-
		10^0	-	-	-
		10^1	+	+	+
		10^2	+	+	+
		10^3	+	+	+
		10^4	+	+	+
		10^5	+	+	+
Salmonella spp. (invA)	12	0	-	-	-
		10^0	-	-	-
		10^1	+	+	+
		10^2	+	+	+
		10^3	+	+	+
		10^4	+	+	+
		10^5	+	+	+
	18	0	-	-	-
		10^0	-	-	-
		10^1	+	+	+
		10^2	+	+	+
		10^3	+	+	+
		10^4	+	+	+
		10^5	+	+	+
	24	0	-	-	-
		10^0	-	-	-
		10^1	+	+	+
		10^2	+	+	+
		10^3	+	+	+
		10^4	+	+	+
		10^5	+	+	+
L. monocytogenes (iap)	12	0	-	-	-
		10^0	-	-	-
		10^1	+	+	+
		10^2	+	+	+
		10^3	+	+	+
		10^4	+	+	+
		10^5	+	+	+
	18	0	-	-	-
		10^0	-	-	-
		10^1	+	+	+
		10^2	+	+	+
		10^3	+	+	+
		10^4	+	+	+
		10^5	+	+	+
	24	0	-	-	-
		10^0	-	-	-
		10^1	+	+	+
		10^2	+	+	+
		10^3	+	+	+
		10^4	+	+	+
		10^5	+	+	+
V. cholera (ctxA)	12	0	-	-	-
		10^0	-	-	-
		10^1	+	+	+
		10^2	+	+	+
		10^3	+	+	+
		10^4	+	+	+
		10^5	+	+	+
	18	0	-	-	-
		10^0	-	-	-
		10^1	+	+	+
		10^2	+	+	+
		10^3	+	+	+
		10^4	+	+	+
		10^5	+	+	+
	24	0	-	-	-
		10^0	-	-	-
		10^1	+	+	+
		10^2	+	+	+
		10^3	+	+	+
		10^4	+	+	+
		10^5	+	+	+

required in this multiplex PCR assay for detecting food samples contaminated with a low level of foodborne pathogens.

For the final analysis of multiplex PCR assay, we chose five of the representative bacteriaus ually associated with food-borne illnesses: *E. coli O157:H7, S. aureus, Salmonella spp, L. monocytogenes* and *V. cholera*. The multiplex PCR method is capable of detecting these five pathogens in approximately 16 hr (12 hr for enrichment, 1 hr for DNA extraction, 2 hr for PCR amplification, 45 min for capillary electrophoretic separation, and 15 min for interpretation). This is, by far, faster than 4 to 7 days to complete for each pathogen using a conventional detection method, which relies primarily on direct plating methods and biochemical tests.

Conclusion

In conclusion, the multiplex PCR assay for simultaneous detection of the five foodborne pathogens in food including *E. coli O157:H7, S. aureus, Salmonella spp, L. monocytogenes* and *V. cholera* was successfully developed and validated. It was able to sufficient in specifically and simultaneously detecting as few as 10 CFU/ mL of the five pathogens in artificially inoculated food samples after enrichment for 12 h. Finally, each 25-g sample was mixed with 225mL of SEB medium and incubated at 37°C for 12-h. Then, each1mL of the culture broth was subjected to the multiplex PCR assay as the schematic representation of detection procedure is presented in Figure 3. The assay is reliable, rapid, specific, and robust. Therefore, it can be another tool for the investigation of microbial contamination in raw food and food products, and will also be useful for identifying the sources of food borne out break

Acknowledgment

This work was supported by the Vietnam Education Foundation for funding (VEF Fellowship to V.V.G.)

Authors' contributions

Vo Van Giau collected the samples, performed the experiments, analyzed and interpreted the data and wrote the manuscript. The other authors read, reviewed and provided feedback on the final manuscript.

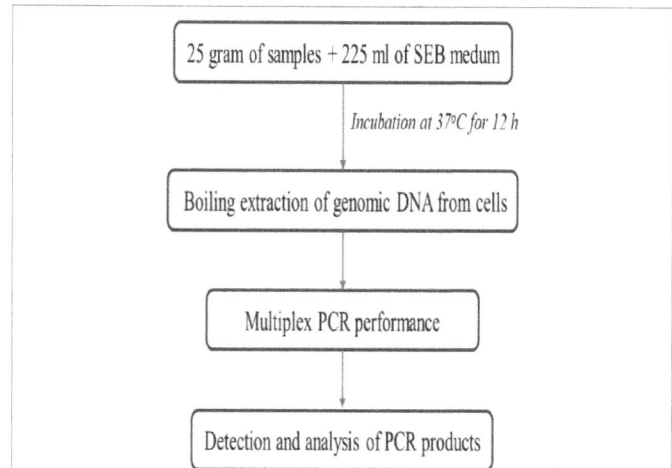

Figure 3: The scheme of multiplex PCR assay for simultaneous detection of *E. coli* O157:H7, *S. aureus*, *Salmonella* spp., *L. monocytogenes* and *V. cholera*.

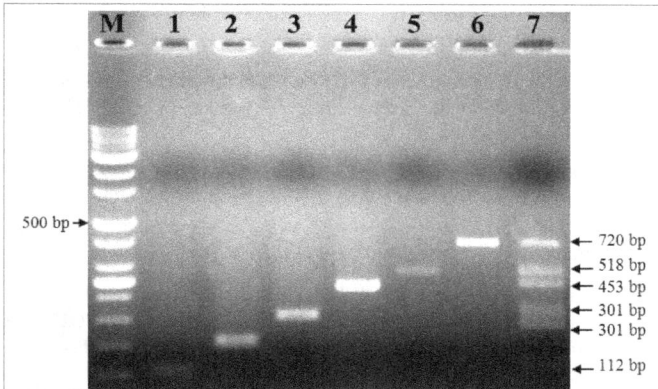

Figure 1: Multiplex PCR reaction applied tosingle and multiple pathogen detection.M, 100-bp DNAladder; lane 1, *S. aureus*ATCC6538; lane 2, *Salmonella enterica*ATCC14028; lane 3, *V. cholera*ATCC17802; lane 4, *L. monocytogenes*ATCC15313; lane 5, *E. coli* O157:H7NLU; lane 6, 16S rRNA; and lane 7,the five-pathogen mixture.

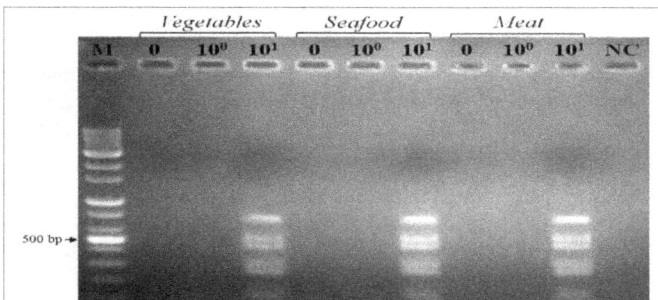

Figure 2: The results of the multiplex PCR assay in three categories of spiked food samples inoculated with different concentrations (showing 0, 10^0 and 10^1 colony-forming units mL^{-1}only) of five pathogens mixture after 12-h enrichment. M, 100 bp DNA ladder; N, negative control

Reference

1. Chao G, Zhou X, Jiao X, Qian X, Xu L. Prevalence and antimicrobial resistance of foodborne pathogens isolated from food products in China. Foodborne pathogens and disease. (2007);4(3):277–284. doi:10.1089/fpd.2007.0088.

2. Law JW, Ab Mutalib NS, Chan KG, Lee LH. Rapid methods for the detection of foodborne bacterial pathogens: principles, applications, advantages and limitations. Frontiers in microbiology. (2014);5:770. doi.org/10.3389/fmicb.2014.00770.

3. Karmali MA. Infection by Shiga toxin-producing *Escherichia coli:* an overview. Mol Biotechnol. (2004);26(2):117-22. DOI: 10.1385/MB:26:2:117.

4. Coffey B, Rivas L, Duffy G, Coffey A, Ross RP, McAuliffe, O. Assessment of *Escherichia coli* O157:H7-specific bacteriophages e11/2 and e4/1c in model broth and hide environments. Int J Food Microbiol. (2011):147(3):188-94. doi.org/10.1016/j.ijfoodmicro.2011.04.001.

5. Melton-Celsa A, Mohawk K, Teel L, O'Brien A. Pathogenesis of Shiga-toxin producing *Escherichia coli.* Curr Top Microbiol Immunol. (2012):357:67-103. doi.org/10.1007/82_2011_176.

6. Gobius KS, Higgs GM, Desmarchelier PM. Presence of activatable Shiga toxin genotype (stx(2d)) in Shiga toxigenic *Escherichia coli* from livestock sources. J Clin Microbiol. (2003):41(8):3777-83. doi.org/10.1128/JCM.41.8.3777-3783.2003.

7. Baron F, Cochet MF, Pellerin JL, Ben Zakour N, Lebon A, Navarro A, Proudy I, et al. Development of a PCR test to differentiate between *Staphylococcus aureus* and *Staphylococcus intermedius*. J Food Prot. 2004;67(10):2302-5.

8. Brakstad OG, Maeland JA. Generation and characterization of monoclonal antibodies against *Staphylococcus aureus*. APMIS. 1989;97(2):166-74.

9. B M Madison, V S Baselski. Rapid identification of *Staphylococcus aureus* in blood cultures by thermo nuclease testing. J Clin Microbiol. 1983;18(3): 722–724.

10. Becker K, von Eiff C, Keller B, Brück M, Etienne J, Peters G. Thermo nuclease gene as a target for specific identification of *Staphylococcus intermedius* isolates: Use of a PCR-DNA enzyme immunoassay. Diagn Microbiol Infect Dis. 2005;51(4):237-44. doi.org/10.1016/j.diagmicrobio.2004.11.010.

11. Wladimir Padilha da Silva; Jorge Adolfo Silva; Márcia Raquel Pegoraro de Macedo; Márcia Ribeiro de Araújo; Márcia Magalhães Mata, et al. Identification of *Staphylococcus aureus*, *S. intermedius*, and *S. hyicus* by PCR amplification ofcoa and nuc genes. Brazilian Journal of Microbiology. (2003);34:125-127. doi.org/10.1590/S1517-83822003000500043.

12. Crump JA, Mintz ED. Global trends in typhoid and paratyphoid Fever. Clin Infect Dis. 2010;50(2):241-6. doi: 10.1086/649541.

13. Seonghan Kim, Jonathan G. Frye, Jinxin Hu, Paula J. Fedorka-Cray, Romesh Gautom, David S. Boyle. Multiplex PCR-based method for identification of common clinical serotypes of *Salmonella enterica* subsp. enterica. J Clin Microbiol. 2006; 44(10): 3608–3615. doi: 10.1128/JCM.00701-06.

14. Galan, JE, Pace J, Hayman MJ. Involvement of the epidermal growth factor receptor in the invasion of cultured mammalian cells by *Salmonella typhimurium*. Nature. 1992;357:588–589. doi.org/10.1038/357588a0.

15. Galan, J..E., R. Curtiss. Distribution of the invA, -B, -C, and -D genes of *Salmonella Typhimurium* among other *Salmonella* serovars: invA mutants of *Salmonella Typhi* are deficient for entry into mammalian cells. *Infection and Immunity*.(1991):59:2901–2908.

16. Chen S, Wang F, Beaulieu JC, Stein RE, Ge B. Rapid detection of viable *Salmonellae* in produce by coupling propidium monoazide with loop-mediated isothermal amplification Appl Environ Microbiol. 2011;77(12):4008-16. doi: 10.1128/AEM.00354-11.

17. Krascsenicsová K, Piknová L, Kaclíková E, Kuchta T. Detection of *Salmonella enterica* in food using two-step enrichment and real-time polymerase chain reaction. Lett Appl Microbiol. 2008;46(4):483-7. doi: 10.1111/j.1472-765X.2008.02342.x.

18. Wang L, Shi L, Alam MJ, Geng Y, Li L. Specific and rapid detection of foodborne *Salmonella* by loop-mediated isothermal amplification method.*Food Research International*. 2008;41(1):69–74. doi.org/10.1016/j.foodres.2007.09.005.

19. Modzelewska-Kapituła M, Maj-Sobotka K. The microbial safety of ready-to-eat raw and cooked sausages in Poland: *Listeria monocytogenes* and *Salmonella* spp. occurrence. Food Control. 2014;36(1):212–216. doi.org/10.1016/j.foodcont.2013.08.035.

20. Ryu J, Park SH, Yeom YS, Shrivastav A, Lee SH, Kim YR, et al. Simultaneous detection of *Listeria* species isolated from meat processed foods using multiplex PCR. Food Control. 2013;32(2):659–664. doi.org/10.1016/j.foodcont.2013.01.048.

21. Cabanes D, Dehoux P, Dussurget O, Frangeul L, Cossart P. Surface proteins and the pathogenic potential of *Listeria monocytogenes*. Trends Microbiol. 2002;10(5):238-45.

22. Farber JM, Peterkin PI. *Listeria monocytogenes*, a food-borne pathogen. Microbiol Rev. 1991;55(3):476-511.

23. Todd ECD, Notermans S. Surveillance of listeriosis and its causative pathogen, *Listeria monocytogens*. *Food control*. 2011;22(9):1484-1490. doi.org/10.1016/j.foodcont.2010.07.021.

24. Pina M. Fratamico, Arun K. Bhunia, James L. Smith (Eds.). Foodborne Pathogens: Microbiology and Molecular Biology. Emerg Infect Dis. 2006;12(12): 2003. doi: 10.3201/eid1212.061077.

25. Motarjemi Y, Käferstein F. Food safety, Hazard Analysis and Critical Control Point and the increase in foodborne diseases: a paradox? Food Control. 1999;10(4-5):325–333. doi.org/10.1016/S0956-7135(99)00008-0.

26. Kim JS, Lee GG, Park JS, Jung YH, Kwak HS, Kim SB, et al. A novel multiplex PCR assay for rapid and simultaneous detection of five pathogenic bacteria: *Escherichia coli* O157:H7, Salmonella, Staphylococcus aureus, *Listeria monocytogenes* and *Vibrio* parahaemolyticus. J Food Prot. 2007;70(7):1656–1662.

27. Mortensen JE, Ventrola C, Hanna S, Walter A. Comparison of time-motion analysis of conventional stool culture and the BD MAXTM Enteric Bacterial Panel (EBP). BMC Clinical Pathology. 2015;15:9. doi.org/10.1186/s12907-015-0010-8.

28. Yamasaki S, Lin Z, Shirai H, Terai A, Oku Y, Ito H, et al. Typing of verotoxins by DNA colony hybridization with poly- and oligo-nucleotide probes, a bead-enzyme-linked immunosorbent assay, and polymerase chain reaction. Microbiol Immunol. 1996;40(5):345-52.

29. Cabrera-Garcı́a ME, Vázquez-Salinas C, Quiňones-Ramı́re, EI. Serologic and molecular characterization of Vibrio parahaemolyticus strains isolated from seawater and fish products of the Gulf of Mexico. Appl Environ Microbiol. 2004; 70(11): 6401–6406. doi: 10.1128/AEM.70.11.6401-6406.2004.

30. Jofre' A, Martin B, Garriga M, Hugas M, Pla M, Rodr'iguezLa'zaro D, et al. Simultaneous detection of Listeria monocytogenes and Salmonella by multiplex PCR in cooked ham. Food Microbiol. 2005;22(1):109–115. doi.org/10.1016/j.fm.2004.04.009.

31. Park SH, Kim HJ, Kim JH, Kim TW, Kim HY. Simultaneous detection and identification of Bacillus cereus group bacteria using multiplex PCR. J Microbiol Biotechnol. 2007;17(7):1177– 1182.

32. Ramesh A, Padmapriya BP, Chrashekar A, Varadaraj MC. Application of a convenient DNA extraction method and multiplex PCR for direct detection of *Staphylococcus aureus* and Yersinia enterocolitica in milk samples. Mol Cell Probes. 2002;16(4):307–314.

33. Sharma VK, Dean-Nystrom EA, Casey TA. Semi automated flurogenic PCR assay (TaqMan) for rapid detection of *Escherichia coli* O157:H7 and other Shiga toxigenic *E. coli*. Mol Cell Probes. 1999;13:291–302.

34. Fratamico PM, Bagi LK, Pepe T. A multiplex PCR for rapid detection and identification of Escherichia coli O157:H7 in foods and bovine feces. J Food Prot. 2000;63(8):1032-7.

35. Xu YG, Cui LC, Tian CY, Li SL, Cao JJ, Liu ZM, Zhang GC. A multiplex polymerase chain reaction coupled with highperformance liquid chromatography assay for simultaneous detection of six foodborne pathogens. Food Control. (2012);25(2):778–783. doi.org/10.1016/j.foodcont.2011.12.014.

36. Germini A, Masola A, Carnevali P, Marchelli R. Simultaneous detection of Escherichia coli O157:H7, *Salmonella spp.*, and *Listeria monocytogenes* by multiplex PCR. Food Control. (2009);20(8):733–738. doi.org/10.1016/j.foodcont.2008.09.010.

37. Vo Van Giau, Thuy Trang Nguyen, Thi Kim Oanh Nguyen, Thi Thuy Hang Le, Tien Dung Nguyen. A novel multiplex PCR method for the detection of virulence-associated genes of *Escherichia coli* O157:H7 in food. 3 Biotech. (2016);6(1):5. doi: 10.1007/s13205-015-0319-0.

38. Nguyen TT, Van Giau V, Vo TK. Multiplex PCR for simultaneous identification of E. coli O157:H7, Salmonella spp. and L. monocytogenes in food. 3 Biotech. 2016; 6(2): 205. doi: 10.1007/s13205-016-0523-6.

39. Kobayashi H, Kubota J, Fujihara K, Honjoh K, Ilo M, Fujiki N, et al.

Simultaneous enrichment of *Salmonella spp, Escherichia coli* O157:H7, *Vibrio parahaemolyticus, Staphylococcus aureus, Bacillus cereus* and *Listeria monocytogenes* by single broth and screening of the pathogens by multiplex real-time PCR. Food Sci Technol Res. 2009;15:427–438. doi.org/10.3136/fstr.15.427.

40. C H Chiu, J T Ou. Rapid Identification of Salmonella Serovars in Feces by Specific Detection of Virulence Genes, invA and spvC, by an Enrichment Broth Culture-Multiplex PCR Combination Assay. J Clin Microbiol. 1996 Oct; 34(10): 2619–2622.

41. Manzano M, Cocolin L, Ferroni P, Cantoni C, Comi G. A simple and fast protocol to detect PCR *Listeria monocytogenes* from Meat. J Sci Food Agric. 1997;74(1):25-30.

42. H Shirai, M Nishibuchi, T Ramamurthy, S K Bhattacharya, S C Pal, Y Takeda. Polymerase chain reaction for detection of the cholera toxin operon of *Vibrio cholera*. J. Clin. Microbiol. 1991;29(11):2517-2521.

43. Kalendar R, Lee D, Schulman AH. FastPCR software for PCR primer and probe design and repeat search. *Genes Genom Genomics*. (2009);3:1–14.

44. Anonymous. Microbiology of food and animal feeding stuffs-protocol for the validation of alternative methods (EN ISO 16140). European Committee for Standardization, Paris, France(2002).

45. Lee N, Kwon KY, Oh SK, Chang HJ, Chun HS, Choi SW. A multiplex PCR assay for simultaneous detection of *Escherichia coli* O157:H7, *Bacillus cereus, Vibrio parahaemolyticus, Salmonella* spp., *Listeria monocytogenes*, and *Staphylococcus aureus* in Korean ready-to-eat food. Foodborne Pathog Dis. 2014;11(7):574-80. doi: 10.1089/fpd.2013.1638.

6

E-Waste: Metal Pollution Threat or Metal Resource?

Shailesh R. Dave*, Monal B. Shah and Devayani R. Tipre

Department of Microbiology and Biotechnology, University School of Sciences, Gujarat University, Ahmedabad 380 009, Gujarat, India

***Corresponding author:** Shailesh R. Dave, UGC Emeritus Fellow, University School of Sciences, Gujarat university, Ahmedabad, Pin code - 380009
E-mail: shaileshrdave@yahoo.co.in*

Abstract

The decreasing costs and increasing availability of electronic equipments like mobiles, televisions, computers and their accessories with advanced technology and the fast rate at which the outdated units are changed, has given rise to a new stream of waste known as Electronic waste (E-waste). 'E-waste' is one of the rapidly growing problems of the present world. The article provides a concise overview of the current scenario of global and national E-waste generation, environmental and health hazards, existing legal networks as well as organizations working on this issue and current technologies for E-waste treatment namely pyrometallurgy, hydrometallurgy and bio hydrometallurgy. It further confers why researchers have shown more interest in bio hydrometallurgical techniques for recovery of metals from E-waste as compared to the conventional methods. If a sustainable technique of metal extraction from E-waste is developed, it would help to conserve the depleting high grade metallic ores, provide the extracted metals to industries and conserve the biotic and abiotic components of the ecosystem from the hazards of E-waste. Moreover, the purity of base and precious metals in E-wastes is about ten times higher than the ores hence, if potential E-waste recycling methods are developed and implemented, it would change the problem of pollution into a profitable metal resource.

Keywords: Autotrophs; Bio hydrometallurgy; Bioleaching; Biosorption; Cyanogenic microbes; Electronic waste; Global problem; Printed circuit board

Introduction

Stone Age cultures began around 10,000 BC, giving way to Copper Age in 3500 BC and Bronze Age in 2500 BC. Cultures continued to grow in complexity and technical advancement through the Iron Age (1200 BC), eventually giving rise to present day "Electronic Age". It is also known as the 'Digital Age" [1]. The increased necessities of mankind in Modern Era have led to a myriad of inventions from a simple calculator to a highly improvised super computer [2]. It has reduced the reliance on human labour, simplified the day-to-day activities and has made the life more comfortable. It has helped in shaping the future of the people, business and nation. In the early 1970, the occurrence of first battery powered calculator marked the beginning of new era in technology. Calculators were not the end of the technological evolutionary process [1]. The tremendous consumer demand drove manufacturers to improve the integrated circuit process with greater performance, better production yields and lower prices which gave rise to personal computers, mobile-phones, laptops and several other electronic equipments [3]. The whole world remains connected through various sophisticated electronic devices like a palmtop, mobile-phone or laptop which has also made the world's populace severely addicted to the technology [4]. But the murky side of this technological roar is the accumulation of the electronic wastes (e-waste).

Electronic Waste: Classification, Composition and Hazards

E-waste classification

Although, electronics have immensely helped in the nation's development, education, trade and commerce, it has also given mankind a reason to worry [5]. The active use of electronic equipments has led to a remarkable increase in their production and at the same time it has given rise to the problem of discarding them [6]. The massive generation of EEE due to the escalating consumer demands has created a new channel of waste, known as e-waste. Each and every electronic equipment has a pre-determined life-span after which they become obsolete. These obsolete electronic equipments are now known as Waste Electrical and Electronic Equipments (WEEE) or more commonly Electronic waste (e-waste) [7,8].

EMPA (Eidgenossische Materialprufungs-und Forschungs Anstalt) Swiss federal laboratories have classified WEEE into two categories viz. Electrical waste and Electronic waste collectively forming Waste Electrical and Electronic Equipments [9]. Approximately 50% of the WEEE consists of electrical wastes, which include household appliances like refrigerators, washing machines, dryers, air-conditioners, vacuum cleaners, coffee machines, toasters, irons, etc. The remaining 50% of WEEE consists of electronic wastes, which includes monitors, televisions and other electronic appliances like computers, telephones, faxes, printers, Video Compaq Disk (VCD) players, radios, other electronic tools like medical instruments, drilling and sewing machines, etc. The EEE are classified into three categories viz. white, brown and grey goods on the basis of their types and utility [10]. The typically large electrical goods and heavy consumer equipments which are mostly painted with white enamel are called white goods. It includes refrigerator, stoves, washing machines, etc. The relatively light consumer equipments as well as the IT and telecommunication equipments such as television, computers, mobile-phones, radio-sets, printers, etc. are called brown goods. Whereas; branded goods sold outside the authorized territory by unauthorized dealers at a price lower than the manufacturing territory are called grey goods [11]. They are presented in market as the resale of new products through channels unintended by the original manufacturers, which mostly includes automatic dispensers for money, cold drinks, sewing machines, video-games, etc. Categories of waste electrical and electronic equipments along with their classification as per the WEEE Directive are listed in Table 1 [12].

Chemical and physical composition of E-waste

On the basis of chemical composition, the WEEE consists of various metals, metalloids, precious metals, halogenated compounds and radioactive elements. Metals and metalloids include Aluminium, Arsenic, Antimony, Barium, Beryllium, Cadmium, Chromium, Copper, Europium, Lead, Lithium, Iron, Manganese, Mercury, Nickel, Selenium, Silica, Tin, Yttrium, Zinc, etc. Precious metals include Gold, Indium, Silver, Palladium, Platinum, etc. Halogenated compounds like Polychlorinated Biphenyls (PCB), Tetrabromobisphenol (TBBA), Polybrominated Biphenyls (PBB), Polybrominated Diphenyl Ethers (PBDE), Chlorofluorocarbon (CFC), Polyvinylchloride (PVC) are present. Radioactive metal like Americium is found in the electronic scrap. The percent composition of the above mentioned types can be represented as 60% metallic content, 30% sophisticated blends of plastics and 10% of dangerous pollutants [13,14].

On the basis of physical composition, the harmful substances found in large quantities include Cathode Ray Tubes (CRT), Printed Circuit Boards (PCB), epoxy resins, Polyvinyl Chlorides (PVC), thermosetting plastics, fibreglass, lead glass, concrete, ceramics, rubber and plywood [15,16]. The composition of e-waste is very diverse and differs in products belonging to different categories. It contains more than 1000 different substances, which are hazardous upon disposal. Many materials and wastes are currently traded internationally, but e-waste has drawn particular attention from government officials,

NGOs, researchers, and practitioners at both the domestic and international levels [6].

E-Waste hazards and its effect on human health

Fast technological progress, industrial growth in the developed and developing countries and product obsolescence have led to the rise in all types of WEEE. The human health and environmental impacts have become rising concerns due to rapidly growing e-waste and the complex and hazardous mixture of materials and metals contained in them. The amount of E-waste was estimated to be around 6 million tonnes in the European Union (EU) and the growth rate of E-waste is expected to rise by 3–5% per year [17]. Presently, the e-waste is one of the fastest growing waste streams in the world [18]. The e-wastes from developed countries enter into developing countries, easily, in the name of free trade or donation. They are transported largely through waterways or roadways, further complicating the problems associated with waste management. With the presence of deadly chemicals and toxic substances present in the electronic devices, disposal of e-waste is becoming an environmental and health nightmare [19,20]. It is reported that only 15-20% of the total 50 million tonnes of global e-wastes are recycled properly. If not, they are exported to the developing or under-developed countries where there are no standard methods of WEEE treatment [18]. Thus it is an alarming threat, especially for the workers who are employed to separate the e-waste components for obtaining valuable metals [21]. Some of the e-waste toxins and their effects on human health are depicted in Table 2 [22,23]. The ill-effects are mainly due to improper dismantling and treatment methods adapted generally and the workers involved for the treatment of e-waste in particular. These workers are mostly unaware of hazardous health effects due to toxic substances in e-wastes and moreover, they are not provided with the protective attires, gloves, glasses and the masks [24,25]. The disassembling and separation of certain parts of e-wastes with the bare hands can cause skin and respiratory disorders [26] whereas; using an open flame to separate and

Table 1: Categories of WEEE with their examples and classification of goods (WEEE Directive 2012/19/EU [12].

No	Categories	Electrical and electronic equipments	Groups
1.	Large household equipments	Refrigerator, washing machines, microwaves, electric radiators, stoves, large medical utility equipments, etc.	White goods
2.	Small household equipments	Vacuum cleaners, iron, toasters, clocks and watches, etc.	Brown goods
3.	Information Technology and telecommunication equipments	Printers, type-writers, telephones, calculators, facsimile, computers, mobile-phones, etc.	Brown goods
4.	Consumer equipments	Radio, television, video-cameras, musical instruments, recorders, etc.	Brown goods
5.	Lightning equipments	Fluorescent lamps, high intensity discharge lamps, sodium lamps, etc.	Brown goods
6.	Medical devices	Radiotherapy equipment, laboratory equipments for in-vitro diagnosis, dialysis machine, pulmonary ventilators, etc.	White or Brown goods
7.	Monitoring and control instruments	Smoke detector, thermostats, weighing and measuring appliances, etc.	Brown goods
8.	Automatic dispensers	Automatic dispensers for hot drinks and cold drinks, automatic dispenser for money	Grey Goods
9.	Electrical and electronic tools	Drills, sewing machines, tools for riveting, nailing or screwing, etc.	Grey Goods
10.	Toys, leisure and sports equipments	Video-games, electric trains or car racing toy set, etc.	Grey Goods

Table 2: E-waste sources, constituents and affected body parts. Source Pant, et al. [113], Monika and Kishore, [22], Padiyar [23].

No.	Sources	Constituents	Hazards and health effects
1.	Solder in printed circuit boards, glass panels and gaskets in computer monitors	Lead	Damage to kidney, central nervous system, circulatory systems and adverse effect on brain development of children
2.	Chip resistors and semi-conductors	Cadmium	Toxic irreversible effects on human health, damage to kidney, liver and nervous system
3.	Relays, Switches and printed circuit boards	Mercury	Chronic damage to the brain, respiratory and skin disorders
4.	Galvanized steel plates and decorator or hardener for steel housing	Chromium	Causes bronchitis and other respiratory disorders
5.	Cabling and computer housing	Plastics and PVC	Burning produces dioxin that causes reproductive and developmental problems
6.	Electronic scrap and circuit boards	Brominated flame retardants	Disorder of Endocrine system
7.	Front panels of CRT's	Barium, phosphorus and heavy metals	Causes muscular problems, damage to heart, liver and spleen
8.	Copper wires and PCB	Copper	Stomach cramps, nausea, liver damage or Wilsons's disease
9.	Lithium ion batteries	Lithium	Lithium can pass into breast milk and may harm a nursing baby, inhalation causes lung edema
10.	Nickel-cadmium rechargeable batteries	Nickel	Dermatitis, Asthma
11.	Motherboard	Beryllium	Lung cancer, inhalation causes chronic beryllium disease or Beryllicosis

Table 3: Top ten states and cities of India leading in E-waste generation during the year 2009. Source Vats and Singh. [57], Guidelines of E-waste [15].

No.	States	E-waste (metric tonnes/year)	Cities	E-waste (metric tonnes/year)
1.	Maharashtra	20,271	Mumbai	11,017
2	Tamil Nadu	13,486	Delhi	9729
3.	Andhra Pradesh	12,780	Bangalore	4648
4.	West Bengal	10,059	Chennai	4132
5.	Uttar Pradesh	10,381	Kolkata	4025
6.	Delhi	9,729	Ahmedabad	3287
7.	Karnataka	9,119	Hyderabad	2833
8.	Gujarat	8,994	Pune	2584
9.	Madhya Pradesh	7,800	Surat	1836
10.	Punjab	6,958	Nagpur	1769

dismantle the e-waste component can lead to the exposure of volatilized contaminants and may lead to the chronic respiratory disorders to the workers who are working in the place and the nearby vicinity [27]. It was observed that majority of the e-waste contaminants are spread through air which is the main pathway for their entry into the humans through inhalation, ingestion or skin absorption [28].

Environmental hazards due to E-wastes

Every year, large quantities of WEEE are discarded by several countries, most of which either end up in landfill or incineration or unauthorized recycling yards [20] which cause severe air, water and soil pollution. Solid waste management, which is already a huge task, is becoming more complicated by the incursion of e-waste, particularly computer waste. The CRT and personal computer constitute the first and second largest component of the e-waste stream, respectively [29]. The heavy metals, halogenated compounds and radioactive element present

in e-waste affect the environment during their inappropriate recycling processes. While burning the e-waste in the open environment, polycyclic aromatic compounds and dioxins are produced which are hundred times more toxic than domestic waste burning [30]. It was reported by Bertram et al. [31] that, approximately 5000 tonnes of Cu released into the environment, globally, was due to the e-wastes disposal.

If the combustion retardants like PBDEs from e-wastes are released into the environment, they are bio accumulated in living organisms due to their lipophilic nature [32] whereas; release of CFCs from the e-waste dumping site would eventually destroy the ozone layer [33]. The Toxicity Characteristic Leaching Procedure (TCLP) reported that the leachate generated from the e-wastes dumping sites has proved to be fatal for aquatic organisms and has made the water resources non-potable. The non-governmental organization such as Greenpeace reported the secret flow of tonnes of e-waste to the countries like China, India, Pakistan, Vietnam, Philippines, Malaysia, Nigeria, Ghana, etc. where strict environmental set of laws do not exist [34].

Statistics of E-Waste Generation

Global scenario

The US Environmental Protection Agency (EPA) estimated that the world generates 20-50 million tonnes (MT) of E-wastes each year [35] and it is expected to rise to 65-72 MT by the year 2017 [36,37]. The United Nation's Solving the E-waste Problem (UN StEP) Initiative forecasted that by the year 2017, the mass of the e-waste generated will be equal to eleven times of The Great Pyramid of Giza and 200 times of The Empire State Building of New York [38]. At present, majority of e-wastes are generated in Europe, the United States and Australasia. Whereas, in the next ten years, China, Eastern Europe and Latin America will become major e-waste producers in the world [39]. In the year 2005, the U.S. was leading the way with 3.0 MT of e-waste per year followed by China with 2.3 MT and India not far behind 0.3 MT per year [40]. The present report by Un StEP Initiative states that in the year 2012, the highest amount of e-waste was generated by European Union (EU) followed by, U.S., China, India, Japan, Russia, Brazil and Mexico. Whereas, a year later in 2013, China generated 11.1 MT of e-wastes which was higher than the U.S. with 10.0 MT [38].

In the year 2014, the U.S. and China produced the highest amount of e-waste weighing 10.0-11.0 MT in the world followed by Japan, Germany and India [41]. The top three Asian Nations with the highest e-waste generation were China, Japan and India which produced 16.0 MT corresponding to 3.7 kg per inhabitant. The lowest amount of e-waste generating continent was Africa with 1.9 MT of total e-waste in the year 2015 [41].

The inspection of 18 European sea-ports in 2005 revealed that 47% of e-waste intended to export, was illegal [42]. In UK alone, at least 23,000 tonnes of undeclared 'Grey market' electronic waste was illegally shipped to the Far East, India, Africa and China [18]. In the 1990s, governments in the European Union, Japan and some places of the United States had set up an e-waste recycling systems but they could not manage the increasing e-wastes they had generated. So they began exporting the problem to the developing countries where laws to protect workers and the environment are inadequate or not compulsory and moreover, these countries were unaware with the hazardous nature of e-waste [43]. Additionally, it was also cheaper to recycle e-waste in developing countries, for e.g.; the cost of recycling of computer monitors in the China is ten times cheaper than the U.S [44]. In the United States, it was estimated that 50-80% of the e-waste collected every year for recycling is being exported in this way [40]. This practice was legal because the US had not ratified the Basel Convention [45]. At the other end of the world, the average annual e-waste production in Greece for the period 2003-2006 came up to approximately 1,70,000 tonnes, representing 3.8% of the total amount of domestic solid waste [14]. The demand for electronic waste in the Asian countries began to grow when it was found that scrap yards could extract valuable metals such as copper, iron, aluminium, silicon, nickel, silver and gold during the recycling processes [46,47]. The e-waste dumping grounds in Asia are China, India and Pakistan. China is facing a dual problem of E-waste treatment both from domestic generation and illegal trans boundary movement [48]. The amounts of E-waste in China are increasing at the rate of 5-10% annually. A small town Guiyu situated in Hong Kong city of China is the largest e-waste collector in the world [48]. It was affirmed that the e-waste collected at Guiyu mainly came from the US, Canada, Japan and South Korea [41]. Guiyu is followed by Accra in Ghana and Lagos port in Nigeria (Africa) in the import of e-wastes [18]. A study on e-wastes dumping showed that an average of 5 million second-hand computers weighing 60,000 tonnes entered the country through Lagos port, out of which nearly 30,000 tonnes were non-functional or irreparable [21]. The Basel Action Network (BAN) estimated that about 45% of the imports in Africa were from the European Union, 45% from the U.S. and the remaining 10% from Japan, Korea, Finland, Germany, Norway, Netherlands, Italy and Singapore [29]. Other e-waste dumping places include Karachi and Islamabad in Pakistan and Delhi in India [1].

Indian scenario

In the last twenty years, the information and communication sector in India has revolutionized life of one and all, creating a huge effect on electronic industries and leading to a phenomenal rise in every nation [20]. In India, e-waste scenario has undergone a drastic change. After China imposed a ban on the import of e-waste in 2002, India has emerged as one of the largest dumping grounds for the developed countries of the world [49-51]. In India, it has been observed that the electronic items are stored unattended because of lack of knowledge about their management. Such e-wastes lie in houses, offices, warehouses etc. Many a times, these wastes are mixed with household wastes, which are finally disposed of in landfills or incinerated to reduce the bulk [52]. Moreover, there are no strict environmental laws or specific guidelines for e-waste disposal in India [45]. The obsolescence rate of e-wastes in India has been estimated to be 1,46,180 tonnes for the year 2005, which reached 8,00,000 tonnes in the year 2012 [15,51] and is expected to reach to more than 12,00,000 tonnes by the year 2020 [53-55]. During the year 2005, about 80% of the e-waste generated in the United States was exported to China, India and Pakistan [20] out of which, only 3% of total e-waste generated was recycled properly in India while the remaining e-waste was handled by the workers with bare hands, under unhygienic conditions without masks in order to earn their livelihood [56]. Ever since then, over 1 million poor people in India are involved in the manual recycling operations. Most of the people working in this recycling sector are the urban poor with very low literacy levels and hence very little awareness regarding the hazards of e-waste toxins [1]. Several laws were framed but these laws failed to stop this informal recycling. Sixty-five cities in India generate more than 60% of the total e-waste generated in the country [57]. The top ten states generating 70% of the total e-waste in India are Maharashtra, Tamil Nadu, Andhra Pradesh, Uttar Pradesh, West Bengal, Delhi, Karnataka, Gujarat, Madhya Pradesh and Punjab [3] and among the top ten cities generating e-waste, Mumbai ranks first followed by Delhi, Bengaluru, Chennai, Kolkata, Ahmedabad, Hyderabad, Pune, Surat and Nagpur [15,57]. The data of e-waste generated in metric tonnes by the top ten states and cities in India are depicted in Table 3. The state of art recyclers have been set up in India for recovery of metals from e-waste [1]. Central Pollution Control Board (CPCB) has participated in registering 23 formal recycling unit in India for e-waste [57] whereas; the informal recycling units include Mandoli industrial area, Kantinagar, Ibrahimpur, Brijgang and Shastri park area of East Delhi as well as workshops in Gaziabad and Zafarabad in Delhi which are well-known e-waste dismantling areas of India [58]. These areas are considered as the danger zone due to the grave health faced by the workers. In addition to Delhi, another rising concern area is the e-waste dismantling unit in Mumbai [59]. According to the study conducted by the NGO Toxics Link, some remote areas of Mumbai faces critical health and environmental risk posed by a huge 19,000 tonnes of e-waste produced here apart from a good amount of same being imported illegally [47].

India has become the fifth biggest producer of the e-waste in the world which generated 1.7 million tonnes of e-waste in 2014 [41]. It also stated that in the year 2014, the total e-waste generated in India contained approximately 16,500, 1900, 300 and 100-200 tonnes of iron, copper, gold and mixture of silver, aluminium and palladium which was equivalent to 52 billion USD (United States Dollar). However, none of the existing environmental laws has taken strict action on the inappropriate e-waste recycling or its hazardous nature [54]. In India, there are only two authorized functional e-waste dismantling facilities; one in Chennai and second in Bangalore. These facilities are M/s. Trishiraya Recycling Facilities, Chennai and M/s E-Parisara, Bangalore [60]. A report by United Nations predicted that by the year 2020, e-waste from old computers would jump by 400% in China whereas; as high as 500% in India as compared to 2007 e-waste records [41]. Additionally, amount of e-waste from discarded mobile phones would be about seven times higher in China and 18 times higher in India by the year 2020 as compared to the 2007 records [36,61]. The escalating e-waste and its hazardous content would definitely constitute a toxic mine in India and the world however, if managed with extreme care and appropriate technology can convert e-wastes to valuable urban mine as it has a large potential reservoir of recyclable materials [62].

A survey by CPCB accounted Ahmedabad and Surat amongst the highest e-waste generators in Gujarat (CPCB, 2011). A case by Toxic Link reported, import of 30 metric tonnes of e-waste in Ahmedabad in a single month [63,64]. Due to the ever increasing information and communication technology sector, Gujarat generates quite a large amount of e-waste annually [15,57]. But unfortunately only 5% of the total e-waste generated in the state reaches the recycling sites [56,65]. The rest is sold to the informal and local markets where the workers dismantle the computers, mobiles, television in an unsystematic manner to extract out the valuable and precious metals unknowingly, damaging their own health and posing a threat to the environment [20]. Looking to the up-front problem faced by many regions across the state and nation, Gujarat Pollution Control Board (GPCB) has authorized seven private e-waste management projects in the state. The e-waste recyclers are M/s. E-Process House in Valsad, M/s. E-coli Waste Management Pvt. Ltd. in Sabarkantha, M/s.

ECS Environment Ltd. in Ahmedabad, M/s. Pruthvi E-recycler Pvt. Ltd. and M/s. Green care E-recycle Company in Rajkot, M/s. Earth e-waste management Pvt. Ltd. in Surat and M/s. Gujarat refilling centre in Vadodara. These recycling units have received no-objection certificate for treatment of e-waste and their registration is valid till the year 2019 [15]. These projects absolutely follow the norms laid by Central Pollution Control Board of India and Ministry of Environment and Forests (MOEF). To implement the project, government has approved Special Purpose Vehicles (SPV) in the name of Gujarat e-Nirmal Ltd. and these vehicles help the recycling units to collect e-waste from door-to-door in closed vans. A report from Sulaimani, [66] states that, the collected e-waste by the recycling plants is then separated into functional and non-functional parts [66]. The functional parts are renovated and sold. Non-functional parts are dismantled and shredded into small sizes. These shredded parts are then differentiated on the basis of recyclable and reusable components like metals, non-metals, glass, plastic, etc. The hazardous components are carefully separated from the recyclable material during e-waste processing and then disposed off according to the guidelines laid by CPCB. The Guidelines for Environmentally Sound Management of E-waste published by CPCB provide the approach and methodology for environmentally sound management of e-waste. The set-up of e-recycling units in Gujarat has created a wave in bringing about an improvement in e-waste management, operational treatment plans, and protective protocols for workers and awareness amongst the public [56].

Organizations/Networks Working On E-Wastes Issues and Hazards

International networks

- The Consumer Electronics Association (CEA) (GreenerGadgets.org)

- The Basel Action Network (BAN.org)

- Silicon Valley Toxics Coalition (SVTC.org)

- The World Reuse, Repair and Recycling Association (wr3a.org)

- Texas Campaign for the Environment (texasenvironment.org)

- U.S. Environmental Protection Agency (US EPA)

- Greenpeace association (greenpeace.org)

- European Union (EU) directives such as WEEE (Waste Electrical and Electronic Equipment) and ROHS (Restrictions of Hazardous Substances)

- Solving the E-waste problem – Initiative (Step-Initiative) (www.step-initiative.org)

National networks

- The E-waste Guide, India (www.ewaste.in)

- Solid Waste Association of India (NSWAI) (www.nswai.com)

- Toxics Link (www.toxicslink.org)

Other networks are WEEE Forum, Umicore (www.umicore. com), Clean India, Indian Environmental Society, India Habitat Centre, Microbial Biotechnology Area of Tata Energy Research Institute. These networks prevent all forms of toxic trade, monitor and control e-waste management systems, settle e-waste recycling and disposal standards, thereby strengthening co-operation and harmonization of global e-waste related activities [10,63,67]. Switzerland was the first country in the world where an official e-waste management system was established and operated [7]. The legislation regarding e-waste management was introduced for the first time in 1998 through ORDEA Law (Ordinance on the Return, The Taking Back and the Disposal of Electrical and Electronic Appliances) [68]. Two different e-waste recycling systems were established in the country viz. SWICO Recycling Guarantee (The Swiss Association for Information, Communication and Organizational Technology) which manages the brown goods and S.EN.S (Stiftung Entsorgung Schweiz) system which manages the white goods [7,14].

Printed circuit board as an essential part of electrical and electronic equipments

E-waste differs physically and chemically from the domestic and industrial wastes. It contains both, valuable and dangerous components which require special attention and recycling practices to avoid adverse environmental impact [69]. The concentration of selected metals in earth-crust, ores and solid waste compared to the metal content of e-scrap and e-waste PCB is listed in Table 4. Whereas; the detailed metal content data of mobile phone, computer, television, LongXiang (LX) and tube-light PCB is presented in Table 5 [70]. All the PCBs showed presence of 10 to 12 metals. Copper content range from 6.4 to as high as 36.0% lead 1.2 to 15.5% and silver is 0.02 to 0.05%. The data presented in these tables indicates that e-waste is a rich source of base and precious metals as compared to other metal resources [71-73] which can be exploited to compensate the diminishing high grade ores and can also meet up the increasing demands of heavy metals in the industries [51,74]. According to the report by Li et al. [75] and Erust et al. [2], PCB is the basic component of almost all EEE present right from simple calculators and transistors to largest super computers. Thus the

rise in e-waste generation, PCB accumulation has also increased creating a danger zone due to the presence of high metallic content and hazardous materials in the PCB.

According to solid waste management professionals, the PCB waste generation, their trans-boundary movement and disposal are the new issues of concern which is equivalent to the existing global environmental problems like acid rain, ozone depletion and global warming [29]. The PCBs are the platform upon which microelectronic components such as semiconductor chips and capacitors are mounted. The PCB is a base which provides the electrical interconnections between components [76]. The compositions of PCB are metals, polymers and ceramics however; it varies depending on the type of the electronic devices [77,55]. The PCBs show metal content of around 28%, plastics 19%, bromine 4%, glass and ceramics 49%. Precious metals like gold, platinum and silver are also present and they constitute around 0.3-0.4%. In PCB, the average metal contents detected are Cu 12.6%, Zn 5.6%, Pb 3.1%, Ni 2.4%, Al 1.4%, Fe 1.2%, Ag 0.003% and Au 0.0014% [25]. Besides these, inorganic elements like isocyanates and phosgenes from polyurethanes, acrylic and phenolic resins, epoxides and phenols from microchips are also found in the PCB [40]. Due to its complex composition, PCB recycling requires a multidisciplinary approach to separate valuable metals, fibres and plastic fractions and thereby reduce the environmental pollution [5, 26].

The e-waste recycling generally starts from the disassembling stage in which reusable and toxic parts are separated. Thereafter, the PCBs are treated using physical recycling process [78]. The physical recycling process involves a preliminary step where, size reduction of the PCB e-waste is performed, followed by a step in which metallic and non-metallic fractions are separated and collected for further treatments [79]. Physical process for separating the metallic and non-metallic fraction of e-waste includes shape separation, magnetic separation, electric conductivity-based separation, density-based separation and corona electrostatic separation [73,80]. At the end of the physical process, non-metals are finally separated from the metallic fractions. The obtained metallic fractions can be treated by pyrometallurgical, hydrometallurgical or by biotechnological methods.

Table 4: Concentration of selected metals in E-scrap and PCB compared to earth-crust, ore, fly ash and slag. Li, et al. [25], Krebs, et al. [71].

Metals	Metal concentration (%) Various metal resources					
	Fly ash	Slag	Earth crust	Ore	E-scrap	E-waste PCB
Cu	0.09	0.16	0.007	0.2	8.0 - 26	12.6
Ni	0.014	0.014	0.008	1.5	0.5 - 2.0	2.4
Pb	0.8	0.06	0.0016	4.0	1.0 - 3.15	3.1
Zn	2.7	0.19	0.008	4.0	2.6	5.6
Sn	0.6	0.05	0.004	1.0	2.3	3.5
Precious metals (Ag, Au, Pd)	ND	ND	ND	BDL	0.01-0.33	0.002-0.003

*ND: Not Detected, BDL: Below Detectable Limit, PCB: Printed Circuit Board

Table 5: Metal content of various e-wastes PCB studied (Shah et al. (2014))[70].

Metals	Solubilized metals in various e-waste PCB (mg.g^{-1})				
	Mobile-phone	Computer	Television	Long Xiang (LX) PCB	Tube-light
Cu	360.00	300.00	118.25	64.25	167.75
Zn	7.96	37.00	19.27	1.23	22.70
Ni	8.55	3.84	13.00	0.62	1.48
Al	6.66	45.93	56.27	14.28	53.22
Pb	12.07	136.5	154.80	133.70	80.50
Fe	10.50	60.12	64.00	9.46	69.75
As	4.34	7.82	5.33	5.52	1.65
Cr	0.59	1.61	1.10	0.93	1.21
Au	0.10	0.14	ND	ND	ND
Ag	0.28	0.23	0.50	0.22	0.22
Pd	0.64	0.27	0.37	0.27	0.68
Cd, Co, K, Na, Se	BDL	BDL	BDL	BDL	BDL

*ND: Not Detected; BDL: Below Detectable Limit

Current Technologies for the Recycling of E-Waste and PCB

State-of-the-art e-waste recycling technology

Due to the hazardous nature of the land filling, incineration and open-burning technique for e-waste treatment [81-84], new recycling technologies with reduced environmental pollution and dealing with the metal recovery from e-waste came into existence. These technologies working on a large scale and treating tonnes of e-waste are known as the state-of-the-art technologies. The technology for e-waste recycling comprises of three steps. The first step is known as the detoxification step in which the critical components like lead glass from CRT and chlorofluorocarbon (CFC) gases from refrigerator, light bulbs and batteries are removed in order to prevent contamination of these toxic substances during the downstream processing [85]. In the second step known as shredding or mechanical processing, the recyclable materials like metals, plastics and glass are separated by crushing, grinding, magnetic and eddy current separation and air separation. Whereas, the hazardous materials and emissions like PBDE, BFR, flue gas, furans and dioxins are filtered and treated to minimize the environmental impact [86]. The third step of the state-of-art recycling is the refining step. In this step, the separated recyclable fractions of metals, plastics and glass, obtained from the second step are refined and conditioned to be sold as secondary raw materials. The residues of the refining process are disposed of at the final disposal site as per the norms set up by CPCB [3].

Drawbacks of State-of-the-art E-waste recycling technologies

- The residual particulate matter in the form of ash is prone to spread in the vicinity and can be dangerous when inhaled.

- Vaporization of the toxic dimethylene mercury can cause respiratory disorders.

- Susceptible to uncontrolled fires which can release toxic fumes.

Pyrometallurgical and hydrometallurgical technologies

The pyrometallurgy consists of the treatment of e-wastes at a very high temperature of around 600-800°C in the pre-combustion step and 900-1200°C in the post combustion step [73] to bring about physical and chemical transformations and recover the metals. In this process, the crushed scraps are burned in a smelter or a furnace to remove plastics in one fraction and the refractory matter in another. Thereafter, these refractory matter form a slag phase with the metal oxides [73,87] from which the metals are recovered. A typical Umicore Isa Smelt process involves two steps for metal extraction. The first step is the recovery of precious metals, known as Precious Metal Operations (PMO) is done by Isa Smelt furnace [73,88]. It includes incineration, smelting in a plasma-arc or blast furnace, drossing, sintering, melting and reactions in a gas phase at a high temperature [89]. Plastic and other organic substances are also removed from the feed in this step. The smelter separates the precious metals from the base metals at the end of the first step. The second step involves the recovery of base metals which is known as Base metal operation (BMO) where the metals are mostly concentrated in a lead slag. It is further treated with lead refining to separate different base metals [90,91]. Consequently, pure metals are recovered in a special metal refinery. Pyrometallurgical treatment has now become a traditional method for recovery of precious metals from e-wastes in the past twenty years [91].

Comparing with the pyrometallurgical processing, hydrometallurgical method is more reliable and easily controlled [5,92]. It involves the use of aqueous solutions containing a lixiviant (acid, cyanide, halide) which is brought in contact with a material containing a valuable metal. After the extraction of metals, they are recovered by using precipitation, chemical reduction, cementation and solvent extraction or ion exchange processes [51]. Various researches have been carried out and few hydrometallurgical methods have been patented [93]. It is a vast field and involves variable methods for recovery of precious and base metals. Few studies have been carried out with acid leaching [93,94], cyanide leaching [95], halide leaching [96,97], thiourea leaching [98] and thiosulphate leaching [99]. These methods mostly have their starting point with sulphuric acid leaching in which the base metals are recovered followed by the chloride, cyanide and halide leaching sequentially to recover silver, gold and palladium, respectively [100]. A patent by Zhou et al. [94], described a technique for recovery of metals from e-waste. The scrap is first heated at 400-500°C for 8-12 h to remove the plastic. In the second step, the crude metal residue is treated with HCl or H_2SO_4 at 90°C to dissolve base metals like Cu, Zn, Ni, Al, etc. and the third step involves the use of dilute HNO_3 with a solid liquid ratio of 1:2 at 60°C to dissolve Ag. Finally, the last step involves the recovery of gold and palladium by using HCl and $NaClO_3$. Various other techniques and recycling methods have been developed which involves series of reactions for metal extraction but it is not the main aim, hence the detail of each are not included.

Drawbacks of pyrometallurgical and hydrometallurgical techniques

- The pyrometallurgical technique generates atmospheric pollution through the release of toxic gases like dioxins and furans and they are very expensive due to high energy consumption in the mechanical process for separation of metals in one fraction and polymers and ceramics in another fraction [101,102].

- Moreover, precious metal recovery is not so efficient by pyrometallurgical methods [73,103].

- Hydrometallurgical processes generate high volume of acidic water due to the use of concentrated acids which create a problem for discarding it [70]. Moreover, it requires different chemical lixiviants for separation of base and precious metals which increases the number of steps for metal recovery.

- Both the processes are highly dependent on investment and regarded as uneconomical ways to extract metals from e-waste [3,55].

Bio hydrometallurgical technology

It was observed that pyrometallurgy was the predominant method during the year 1970s and mid-1980s for metal recovery, which was taken over by the hydrometallurgy during mid-1980s and 2005, and presently it has been overtaken by the bio hydrometallurgy. In the last decade, recovery of metals by bio hydrometallurgy is one of the most promising and revolutionary biotechnology [55,104,105]. However, pyrometallurgy and hydrometallurgy are still being applied for metal extraction from e-wastes and the threat to the environmental pollution due to these methods persists. Bio hydrometallurgical technique exploits microbiological processes for recovery of base and precious metal ions. It is soon-to-be a major technology break-through for the materials and mineral industry. Moreover, great interest is shown by the international and national companies for adopting the bio hydrometallurgical techniques due to its eco-friendly nature. It mainly includes two individual processes namely 'Bioleaching' and 'Bio sorption'. Over the past 40 years many researchers have investigated the application of biotechnology in mining industries [106]. Bioleaching has been applied commercially for extracting copper, zinc, nickel, cobalt and precious metals from low-grade ores and tailings for many years and now even extended to recover metals from concentrates and fly ash [72]. In bioleaching processes, solubilization of metals is based on the interactions between metals and microorganisms. This technique allows metal recycling by processes similar to that in the natural biogeochemical cycles and is therefore eco-friendly and competent technique [3,107].

Sand and colleagues [108,109] worked on mechanisms for bacterial leaching of metal sulphide ores via thiosulphate and polysulphate pathways and also described bioleaching mechanism involving Fe^{3+} ions. From the reports by Sand et al. [109] and Ehrlich, [110]. It was inferred that the direct leaching mechanism involved the enzymatic oxidation of the sulphur moiety present in the heavy metal sulphides. Whereas; in indirect mechanism, there is a non-enzymatic metal sulphide oxidation by Fe^{3+} iron in combination with an enzymatic re-oxidation of the resulting Fe^{2+} iron. Studies carried out by Rawlings, [111] and Rohwerder et al. [112] justifies indirect mechanism as the relevant and better approach as compared to direct mechanism. The indirect mechanism can be divided into two sub-types viz. the "contact" and "non-contact" mechanisms. In the contact mechanism, the microbial cells attach to the surface of sulphide mineral due to the presence of the exopolymeric layer whereas, in non-contact mechanism, plank tonic cells oxidize the Fe^{2+} iron in the medium and regenerate the Fe^{3+} iron resulting from the Fe^{2+} iron due to bioleaching, which react chemically with metal sulphides afterwards. Studies carried out with ores using various sulphur and iron-oxidizing microbes as well as heterotrophic fungi showed 50-90% of various metal extractions under different experimental conditions [113]. The bioleaching experiments carried out by Tipre and Dave, [114] for metal extraction from polymetallic concentrate showed 80.0-88.0% of Cu and Zn extraction in a shake flask and 5 L laboratory stirred tank reactor under optimized conditions. The iron and sulphur oxidizers were employed in the latter study and it exhibited direct mechanism whereas; indirect process carried out by Patel et al. [115] showed 80.0-81.0% of Cu and Zn extraction from polymetallic concentrate in comparatively shorter time period. Though these processes have been successfully applied for the leaching of metals from ores and concentrates, data pertaining to their application for the extraction of metals from e-wastes are still scanty. The ability of microorganisms to leach and mobilize metals from e-wastes comprises of three principles viz. redox reactions, formation of organic acids, inorganic acids and metal leaching agents.

Bioleaching of metals by chemolithoautotrophs

Chiefly acidophilic group of bacteria plays an important role in bioleaching of base and precious metals [113]. Formation of sulphuric acid by *Acidithiobacillus thiooxidans* and Fe^{3+} iron by *Leptospirillium* spp. help in dissolving the metallic fraction from e-wastes [116,117]. Metal extraction from e-wastes was carried out using various acidophiles like *Acidithiobacillus ferrooxidans*, *Acidithiobacillus thiooxidans*, *Leptospirillium ferriphilum*, *Sulfolobus thermosulfidooxidans*, etc. which resulted in efficient metal extraction [2]. Studies by Brandl et al. [71], showed more than 90% of the Cu, Zn, Ni, Al solubilization by *Acidithiobacillus thiooxidans* at scrap concentration of 5-10 g/L. Choi et al. [118] reported that the iron oxidation rates of *A. ferrooxidans* did not change much with the addition of c-PCB to the medium. An interesting finding was that the concentration of leached copper in the solution increased with the addition of 1 g/L of citric acid as chelating agent. The latter studies showed 37-40% of Cu leaching without adding citric acid and 81-83% of Cu leaching with the addition of citric acid. Precious metals like Au and Ag were extracted using *Acidithiobacillus* sp, *Leptospirillium* sp. and *Acidiphilium*sp. [119]. But, better results for Au leaching were obtained using cyanogenic bacterial strains [113]. It was also observed that the mixed cultures of *A. thiooxidans* and *A. ferrooxidans* were more efficient than each taken separately for the extraction of metals like Cu, Ni, Zn and Pb from PCB scrap [120]. To improve copper recovery from PCB, a multiple point PCB addition was adopted in which the PCB was added in instalments to prevent the toxic effects of e-waste on the viable cells [121]. Brandl et al. [71] and Yang et al. [122], proposed the effect of single step and two-step bioleaching process and the latter was found to be more effective for substantial metal extraction from e-waste. In two-step process, the bacterially generated lixiviant plays the main role in metal extraction and the microbe-metal interaction is not much significant. Moreover, these processes employ the independent lixiviant generation and helps in optimizing each parameter thereby maximizing the productivity [123]. Comparisons of metals extractions from computer and mobile PCB by chemical and biological means showed more than 90% copper, zinc and nickel extractions [55,103]. Table 6 [124-131] enlists some of the chemolithoautotrophs employed in extraction of metals from e-wastes.

Bioleaching of metals by cyanogenic microbes

Cyanide is formed by a variety of cyanogenic bacteria viz. *Chromo bacterium violaceum*, *Pseudomonas aeruginosa*, *Pseudomonas fluorescens*, *Bacillus megaterium*and fungi viz. *Maramiusoreades*, *Clitocybe* sp., *Polysporus*sp. etc. many of them belonging to soil micro flora [132]. The microbially produced cyanide helps in bioleaching of metals like Cu, Zn, Fe, Ni, Pt, Au and Ag. It is believed that cyanide formation has a benefit for the organism by inhibiting the growth of competing microorganisms growing with the cyanogenic microbes in the soil. Glycine is the precursor for cyanide synthesis which is formed by oxidative decarboxylation reaction [133,134]. Cyanide occurs in solution in two forms viz., the cyanide anion (CN^-) and hydrogen cyanide (HCN). At pH 7, cyanide is mostly present as HCN, according to its pKa of 9.3. Both the forms have their roles in extraction of metals at pH 7-10 [134,135]. The formation HCN by an organism was first described in mesophilic bacterium *Chromo bacterium violaceum* by Sneath in 1953 and in the basidiomycete *Marasmius* in 1871 by Losecke [135,136]. In the presence of CN^- or HCN, many metals and metalloids such as Ti, V, Cr, Mn, Fe, Co, Ni, Cu, Zn, Mo, Cd, Pd, Ag, Pt, Au, Hg, Po, and U form well-defined metal-cyanide complexes, which show a very good water solubility and exhibit high chemical stability [137,138]. The cyanogenic microbes are capable to form water soluble metal cyanides with different efficiencies [139]. Gold was microbially solubilized as dicyanoaurate $[Au(CN)_2^-]$ and nickel was solubilized as tetracyanonickelate $[Ni(CN)_4^{2-}]$ [138]. Pham and Ting [140] extracted Au from e-waste by *Chromo bacterium violaceum* and *Pseudomonas fluorescens* after the biooxidation of e-waste by *A. ferrooxidans* which specifically removes Cu leaving Au residues behind. The HCN forming *Pseudomonas aeruginosa* was

Table 6: List of microorganisms used for extraction of metals from E-waste.

No.	Microorganisms used in the study	Metals extracted	% metal recovery	References
1.	*Acidiphilium acidophilum*	Ni, Zn	40 – 86	Hudec, et al. [124]
2.	*Acidithiobacillus ferrooxidans, Leptospirillium ferrooxidans, Acidithiobacillus thiooxidans*	Cu, Ni, Zn	89 – 98	Groudev, et al. [125]; Liang, et al. [120]; Choi, et al. [118]; Bas, et al. [126]
3.	*Aspergillus Niger, Penicillium simplicissimum*	Cu, Sn, Al, Ni, Pb, Zn	65 – 95	Brandl, et al. [72]
4.	*Aspergillus Niger, Acidithiobacillus thiooxidans*	Cu	82	Saidan and Valix, [127]
5.	*Desulfovibrio desulphuricans*	Au, Pb	68, 95	Creamer, et al. [128]
6.	*Chromobacterium violaceum, Pseudomonas fluorescens, Bacillus megaterium*	Ni, Au	43.5, 14.9, respectively	Faramarzi, et al. [138]
7.	*Chromobacterium violaceum* and *Pseudomonas aeruginosa* mixed culture	Cu, Zn, Fe, Au, Ag	8 – 83	Pradhan and Kumar, [141]
8.	*Pseudomonas plecoglossicida, Pseudomonas fluorescens, Chromobacerium violaceum*	Au, Ag, Pt	5 - 68.5	Brandl, et al. [129]
9.	*Sulfobacillus thermosulfidooxidans, Thermoplasma acidiphilum*	Cu, Ni, Zn, Pb, Sn, Al, Fe, Ag	74 – 89	Ilyas, et al. [130], Ilyas, et al. [131]
10.	*Leptospirillum ferriphilum* dominated consortium	Cu, Zn, Ni	86-99.8%	Shah, et al. [70] Shah, et al. [55]

applied in the bioleaching process for the first time by Pradhan and Kumar, [141] for mobilization of metals from electronic waste. The studies showed that mixture of *C. violaceum* and *P. aeruginosa* exhibited higher metal mobilization as compared to *C. violaceum* alone. The mixture of *C. violaceum* and *P. aeruginosa* exhibited more than 83, 73, 49, 13 and 8% of total Cu, Au, Zn, Fe, and Ag solubilization, respectively as compared to 79, 69, 46, 9 and 7% of Cu, Au, Zn, Fe and Ag solubilization, respectively with *C. violaceum* alone at a scrap concentration of 10 g/L. The cyanogenic microbes play a main role in precious metal recovery which is a shortcoming when iron-oxidizers are used [142].

Bioleaching of metals by organic acid producing fungi

Heterotrophic fungi like *Aspergillus Niger*, *Penicilliumsimplicissimum*, *Penicilliumbilaiae*, *Saccharomyces cerevisiae*, *Yarrowialipolytica*, etc. have been reported to grow in presence of electronic scrap. These fungi produce organic acids like citric acid, tartaric acid and oxalic acid which act as complexing agents and help in extraction of metals like Cu, Cd, Sn, Al, Ni, Pb, Zn etc. The studies carried out by Brandl et al. [71] showed feasibility for using *A. niger* and *P.simplicissimum* to leach metals from e-wastes. The findings showed that the one-step leaching experiments with fungi adversely affected its growth and resulted in only 25-55% metal solubilization. On the other hand, when the fungal biomass was separated from medium after maximum production of organic acid (two-step), it resulted in 65% of Cu and Sn extraction and 95% of Al, Ni, Pb and Zn extraction. The two-step leaching process was demonstrated using a gluconic acid producing fungi *A. Niger*. Ren etal. [143] found that *A. Niger* can be utilized to extract Cd and Al from metal containing wastes. Bosshard et al. [144] and Brombacher et al. [145] demonstrated the application of fungal bioleaching for metal extraction from fly ash. The reports of fungal bioleaching with e-wastes are very limited due to the heterogeneity of the e-wastes and its toxicity on developing fungal mycelium. Table 6 enlists involvement of different microbes for the recovery of various metals present in e-waste.

The metals extracted by bioleaching are further recovered by bio sorption methods. It involves the use of biomass of bacteria, fungi, algae, microbial proteins and chitosan (deacetylated derivative of chitin) etc. as adsorbents. Studies carried out by Dave et al. [146], showed *Eichhornia* sp. biomass played an important role in copper sorption from metal containing waste. As high as 85% of copper was removed from waste in 24 h contact time at *pH* 5. Parameswari et al. [147] reported around 86-95% sorption of heavy metals like Cr (VI) and Ni by *Azotobacter chroococcum*, *Bacillus* sp. and *P. fluorescens* within 72 h at 35°C temperature. Ilhan et al. [148] investigated effects of pH, temperature and initial concentration of metal ions on the bio sorption capacity by *Staphylococcus saprophyticus*. The optimum pH values for chromium, lead and copper was found to be 2.0, 4.5 and 3.5, respectively and the maximum adsorption for Cr^{3+}, Pb^{2+} and Cu^{2+} was observed at initial concentrations of 193.66 mg/L, 100 mg/L and 105 mg/L, respectively which resulted in 46, 100 and 43% bio sorption of Cr, Pb and Cu. This infers that pH, initial metal concentration and bio sorbent capacity have a great influence on bio sorption studies.

Savitha et al. [149] demonstrated manganese bio sorption from e-waste by *Helminthosporium solani*, *Aspergillus Niger*, *Fusarium oxysporum* and *Cladosporiumcladosporoides*, amongst which *H. solani* showed the best results. The maximum adsorption was found to be 97% at pH 7. Various other bio sorbents used for metal sorption are *Streptomyces erythraeus*, *Spirulina plantensis*, *Desulfovibriodesulfuricans*, *Bacillus subtilis*, *Neurosporacrassa*, *Rhizopusarrhizus*, *Chlorella vulgaris* which belong to either bacteria, fungi or algae whereas, animal bio sorbents include hen eggshell membrane, ovalbumin, lysozyme, bovine serum albumin, etc. [113].

Compared with other existing methods, biohydrometallurgy offers a number of advantages including low operating costs, eco-friendly nature, minimization of the volume of chemicals and biological sludge to be handled and high efficiency in detoxifying effluents. Moreover, this technology is well accepted by industry as it goes along with the current need of maintaining nature's harmony. It has become a widely accepted option for the clean-up of contaminated sites and aquifers [113]. However, few more developments in this field are required for its application in metal extraction from e-wastes on large scale and thus solving the problem in an environmentally friendly manner.

Conclusions

In the recent decades, the world has witnessed a technological development fuelled by continuous demand for latest gadgets and devices. In no case, this revolution will decline, thus making electronic waste; a major global issue. Management of E-waste is thus a daunting task and need to be tackled efficiently. Recycling of E-waste for the extraction of base and precious metals are of real concern not only due to the presence of high concentration of these metals but also its hazardous nature. Pyrometallurgy and hydrometallurgy have their own limitations in recycling of metals from E-waste. Bio hydrometallurgical recovery of metals from E-waste appears to be an attractive treatment technique, which can result in more than 90% of metal extraction. This would lead to an acceptable solution of some of the metallic pollutants and would provide value-added metal recovery thereby helping to compensate the gap between the demand and supply of metals. Looking to the brighter side of the current issue, E-waste can be a resource of base and precious metals if appropriate recycling technologies are applied. However, there is a need to scale-up the process to advance commercial application of bio hydrometallurgy for remediation and profitable recovery of metals.

Acknowledgements

The authors are thankful to Gujarat State Biotechnology Mission (GSBTM), Financial Assistance Program (FAP-10), Gujarat, India, for the project Grant (No. GSBTM/MD/PROJECTS/ FAP/451/2010-11) and research scholarship to Monal B. Shah.

References

1. Agnihotri V. In: E-waste in India. Research Unit in Larrdis, Rajya Sabha secretariat, Rajya Sabha, Delhi, India. 2011; pp. 1-127.

2. Ceren Erüst, Ata Akcil, Chandra Sekhar Gahan, Aysenur Tuncuk and Haci Deveci. Biohydrometallugy of secondary metal resources: a potential alternative approach for metal recovery. J. Chem.Tech. Biotechnol. 2013; 88(12): 2115-32. DOI: 10.1002/jctb.4164.

3. Sharma Pramila, Fulekar M.H. and Pathak Bhawana. E-waste: A challenge for tomorrow. Res. J. Recent Sci. 2012; 1(3): 86-93.

4. Hirschhorn J. Technology addiction in the electronic age: worldwide progress or servitude? In: Global Research, USA. 2013; p. 1.

5. Sohaili. J, Muniyandi, S, and Suhaila, S.M. A review on printed circuit board recycling technology. J. Emer. Trends. Eng. Appl. Sci. 2012; 3 (2): 12-8.

6. Atsushi Terazono, Shinsuke Murakami, Naoya Abe, Bulent Inanc, Yuichi Moriguchi, Shin-ichi Sakai, et.al. Current status and research on E-waste issues in Asia. J. Mater. Cycles Waste Manag, 2006: 8(1): 1-12.

7. Deepali Sinha-Khetriwala, Philipp Kraeuchib, Markus Schwaninger. A comparison of electronic waste recycling in Switzerland and India. Environ. Imp. Assess. Rev. 2005; 25(5): 492-504. Doi:10.1016/j.eiar.2005.04.006.

8. M. Khurrum, S. Bhutta, Adnan Omar, Xiaozhe Yang. Electronic waste: A growing concern. Econom. Res. Int. 2011; 2011(2011): Doi:10.1155/2011/474230.

9. UNEP, Basel convention on the control of trans boundary movements of hazardous wastes and their disposal. United Nations Environment Programme. 2009.

10. Saini A, Taneja A. Managing E-waste in India - A review. Int. J. Appl. Eng. Res. 2012; 7(11): 1-6.

11. Yilmax A. Comparison of heavy metals of Grey mullet (M.Cephalus L) and Sea bream (S.Aurata L.) caught in Iskenderun bay (Turkey). Turk. J. Vet. Anim. Sci. 2005; 29(2): 257-62.

12. WEEE Directive. Directive 2012/19/EU of European Parliament and of the council of 4th July 2012 on waste electrical and electronic equipments (WEEE). In: EUR-Lex. 2012; (34): pp. 38-71.

13. Rolf Widmera, Heidi Oswald-Krapf a, Deepali Sinha-Khetriwalb, Max Schnellmannc, Heinz Bo¨ nia. Global perspective on e-waste. Environ. Impact Assess. Rev. 2005; 25(2005): 436-58. Doi:10.1016/j.eiar.2005.04.001

14. Gaidajis G, Angelakoglou K, Aktsoglou D. E-waste: Environmental problems and current management. J. Eng. Sci. Technol. Rev. 2010; 3(1): 193-9.

15. Guidelines for Environmentally Sound Management of E-waste, In: MoEF letter No.23-23/2007-HSMD, Ministry of Environment and Forests and Central Pollution Control Board, Delhi, India. 2008; pp.1-93.

16. Priyadarshini S. A survey on electronic waste management in Coimbatore. Int. J. Eng. Sci. Technol. 2011; 3(3): 2099-104.

17. EC. In: Report from the commission to the council and the European parliament on the implementation of community waste legislation. 2000.

18. UNEP. Call for global action on e-waste, United Nations Environment Programme. 2006

19. Kurian J. Electronic waste management in India: issues and strategies. In: Proceedings Sardinia.11th International Waste Management and Landfill Symposium 1-5 October. Cagliari, Sardinia, Italy. 2007.

20. Saoji A. E-waste management: an emerging environmental and health issue in India. Natl. J. Med. Res. 2012; 2(1): 107-10.

21. Schmidt CW. Unfair trade: E-waste in Africa. Environ Health Perspect. 2006; 114(4): 232-5.

22. Monika and Jugal Kishore. E-waste management: as a challenge to public health in India, Ind. J. Comm. Med. Indian J Community Med. 2010; 35(3): 382–5. Doi: 10.4103/0970-0218.69251.

23. Padiyar N. Nickel allergy-is it a cause of concern in everyday dental practice. Int. J. Contemp. Dentist. 2011; 2(1), 80-1.

24. LaDou J, Lovegrove S. Export of electronics equipment waste. Int J Occup Environ Health. 2008; 14(1): 1-10.

25. Li Y, Xu X, Liu J, Wu K, Gu C, Shao G, et. al. The hazard of chromium exposure to neonates in Guiyu of China. Sci Total Environ. 2008; 403(1-3): 99-104. Doi: 10.1016/j.scitotenv.2008.05.033.

26. He W, Li G, Ma X, Wang H, Huang J, Xu M, et.al. WEEE recovery strategies and the WEEE treatment status in China. J. Hazard. Mater.2006; 136(3): 502-12.

27. Manomaivibool P. Extended producer responsibility in a non-OECD context: the management of waste electrical and electronic equipment in India. Res. Conserv. Recycl. 2009; 53(3): 136-44. DOI:10.1016/j.resconrec.2008.10.003.

28. Priyadarshini S. A survey on electronic waste management in Coimbatore. Int. J. Eng. Sci. Technol. 2011; 3(3): 2099-104.

29. Nnorom IC, Osibanjo O. Electronic waste (e-waste): material flows and management practices in Nigeria, Waste Manag. 2008; 28(8): 1472-9.

30. Brian K. Gullets, William P. Linak, Abderrahmane Touati, Shirley J.Wasson, Staci Gatica, Charles J. King et al. Characterization of air emissions and residual ash from open burning of electronic wastes during simulated rudimentary recycling operations. J. Mater. Cycl. Waste Manag. 2007; 9(1): 69-79.

31. Bertram M, Graedel T, Rechberger H, and Spatari S. The contemporary European copper cycle: waste management subsystem, Ecol. Econ. 2002; 42(1-2): 43-57. DOI: 10.1016/S0921-8009(02)00100-3.

32. Deng WJ, Zheng JS, Bi XH, Fu JM, Wong MH. Distribution of PBDEs in air particles from an electronic waste recycling site compared with Guangzhou and Hong Kong, South China. Environ Int. 2007; 33(8): 1063-9.

33. Scheutz C, Mosbaek H, Kjeldsen P. Attenuation of methane and volatile organic compounds in landfill soil covers. J Environ Qual. 2004; 33(1): 61-71.

34. Cobbing M. Toxic Tech: Not in our backyard. Uncovering the hidden flows of e-waste. Report from Greenpeace International, Amsterdam. 2008.

35. Burke M. The gadget scrap heap. Chem. World UK. 2007; 4(6): 45-8.

36. Lewis T. Worlds e-waste to grow 33% by 2017. Live Science, Washington, US, IST. 2013.

37. Wang B, Kuehr F and Huisman J. In: The global e-waste monitor-2014-Quantities, flows and resources. United Nations University, IAS –SCYCLE, Bonn, Germany. 2015; pp. 1-80.

38. Ecowatch. Worldwide e-waste to reach 65 million tonnes by 2017, Yale Environment 360. Ohio, U.S. 2013.

39. Robinson BH. E-waste: An assessment of global production and environmental impacts. Sci. Sci Total Environ. 2009; 408(2): 183-91. Doi: 10.1016/j.scitotenv.2009.09.044.

40. Huang K, Guo J, Xu Z. Recycling of waste printed circuit boards: A review of current technologies and treatment status in China. J Hazard

Mater. 2009; 164(2-3): 399-408. Doi: 10.1016/j.jhazmat.2008.08.051

41. UN report. India fifth biggest generator of e-waste in 2014: NDTV, All India, 2015.

42. Hischier R, Wager P and Gauglhofer J. Does WEEE recycling makes sense from an environmental prospective? The environmental impacts of the Swiss take-back and recycling systems for waste electrical and electronic equipment (WEEE). Environ. Impact Assess. Rev. 2005; 25, 525-39.

43. Barba-Gutierrez Y, Adenso-Diaz B and Hopp M. An analysis of some environmental consequences of European electrical and electronic waste regulation. Res. Conserv. Recycl. 2008; 52(3): 481-95.

44. Greenpeace International. Where does E-waste go? 2008.

45. BAN and SVTC. Exporting Harm: The High-Tech Thrashing of Asia, Basel Action Network and Silicon Valley Toxics Coalition. 2002.

46. CPCB. Implementation of e-waste rules. In: Central pollution control boards in association with Ministry of Environment and Forests. Delhi, India. 2011; pp. 1-55.

47. Krishnan S. Challenges in managing E-waste in India, In: Disaster, Risk and Vulnerability Conference March 12-14, Kerala, India. 2011; 1.

48. Puckett J and Smith T. Exporting harm: The High-Tech Trashing of Asia. The Basel Action Network, Silicon Valley Toxics Coalition, Seattle. 2002.

49. Yu J, Welford R, Hills P. Industry responses to EU WEEE and ROHS directives: Perspectives from China. Corpor. Soc. Res. Environ. Manag. 2006; 13(5): 286-99. DOI: 10.1002/csr.131.

50. Nnorom C, Osibanjo O, Okechukwu K, Nkwachukwu O, Chukwuma R. Evaluation of heavy metal release from the disposal of waste computer monitors at an open dump. Int. J. Env. Sci. Dev. 2010; 1(3): 227-33.

51. Jha M, Kumari A, Choubey, P. Lee, Kumar V, and Jeong J, et al. Leaching of lead from solder material of waste printed circuit boards (PCBs), Hydrometallurgy. 2012; 121-4, 28–34. DOI: 10.1016/j. hydromet.2012.04.010.

52. Khattar V, Kaur J, Chaturvedi A, and Arora R. E-Waste Assessment in India: Specific focus", New Delhi, India. 2007.

53. Alam T. Gurgaon E-waste likely to grow 300% by 2020. In: The Economic Times, India. 4; 2011, 12.24 pm IST. 2011.

54. Basu, M. New e-waste management plan lucrative for states, The Pioneer, New Delhi, India. 2010.

55. Shah MB, Tipre DR, Purohit MS, Dave SR. Development of two-step bioleaching process for enhanced biorecovery of Cu-Zn-Ni from computer printed circuits boards. J Biosci Bioeng. 2015; 120(2): 167-73. Doi: 10.1016/j.jbiosc.2014.12.013.

56. Gupta R, Sangita Kaur V. Electronic waste: a case study. Res. J. Chem. Sci. 2011; 1(9), 49-56.

57. Vats M and Singh S. Status of e-waste in India –A review. Int. J. Inno. Res. Sci. Eng. Technol. 2014; 3(10): 16917-31.

58. Bridgen K, Labunska I, Santillo D and Allsopp M. Recycling of electronic wastes in China and India: workplace and environmental contamination. Greenpeace Research Laboratories Technical Note .2005; 9, 1-74.

59. Pratap A. Interview on e-waste in India and the Basel Ban. The Pioneer, New Delhi, India. 2009.

60. Borthakur A, Singh P. Electronic waste in India: problems and policies. Int. J. Environ. Sci. 2012; 3(1): 353-62.

61. Ahmed S, Panwar R and Sharma A. Forecasting e-waste amounts in India. Int. J. Eng. Res. Gen. Sci. 2014; 2(6): 324-41.

62. Malone D. Worldwide e-waste constitutes a valuable 'urban mine'. UN under Secretary General, NDTV, India. 2015.

63. Violet N. Pinto. E-waste hazard: the impending challenge. Indian J Occup Environ Med. 2008; 12(2): 65–70. Doi: 10.4103/0019-5278.43263.

64. ToxicsLink. Scrapping the hi-tech myth: Computer waste in India. ToxicsLink, Mumbai, India. 2003.

65. Sinha T. Downside of digital revolution. Toxics Link. 2008.

66. Kamran Sulaimani, Kamran Sulaimani. Status of e-waste in Gujarat. 2010.

67. Babu R, Parande A and Basha A. Electrical and electronic waste: A global environmental problem. Waste Manag. Res. 2007; 25(4): 307-10. Doi: 10.1177/0734242X07076941.

68. Fishbein B. End-of-life management of electronics abroad. Waste in the wireless world: the challenge of cell-phones. INFORM Inc. New York. 2002; pp.1-13.

69. Bleiwas D and Kelly T. In: Obsolete computers, "Gold Mines", or High-Tech Trash? Resource recovery by recycling [R], United States Geological Survey. 2001; p. 7.

70. Shah MB, Tipre DR, Dave SR. Chemical and biological processes for multimetal extraction from waste printed circuit boards of computers and mobile-phones. Waste Manag Res. 2014; 32(11): 1134-41. Doi: 10.1177/0734242X14550021.

71. Krebs W, Brombacher C, Bosshard P, Bachofen R, and Brandl H. Microbial recovery of metals from solids. FEMS Microbiol. Rev. 1997; 20(3-4): 605-17. DOI: 10.1111/j.1574-6976.1997.tb00341.x.

72. H. Brandl, R. Bosshard, M. Wegmann. Computer-munching microbes: metal leaching from electronic scrap by bacteria and fungi. Hydrometallurgy. 2001; 59(2001): 319-26.

73. Cui J, Zhang L. Metallurgical recovery of metals from electronic waste: a review. J Hazard Mater. 2008; 158(2-3): 228-56. Doi: 10.1016/j. jhazmat.2008.02.001

74. Abhilash, Pandey B, and Natarajan K. Microbial extraction of uranium from ores. In: Microbiology for Minerals, Metals, Materials and the Environment. CRC press, Boca Raton, FL. 2015; pp. 59-98.

75. Li J, Lu H, Guo J, Xu Z, Zhou Y. Recycle technology for recovering resources and products from waste printed circuit boards. Environ Sci Technol. 2007; 41(6): 1995-2000.

76. Duan H, Hou K, Li J, Zhu X. Examining the technology acceptance for dismantling of waste printed circuit boards in light of recycling and environmental concerns. J Environ Manage. 2011; 92(3): 392-9. Doi: 10.1016/j.jenvman.2010.10.057.

77. Guo J, Rao Q, Xu Z. Application of glass-nonmetals of waste printed circuit boards to produce phenolic moulding compound. J Hazard Mater. 2008; 153(1-2): 728-34.

78. Askiner Gungor, Surendra M. Gupta. Disassembly sequence planning for products with defective parts in product recovery. Comput. Ind. Eng. 1998; 35(1-2): 161-4. Doi: 10.1016/S0360-8352(98)00047-3.

79. Bernardes A, Bohlinger I, Milbrandt H, Rodrigues D, and Wuth W. Recycling of printed circuit boards by melting with oxidizing/reducing top blowing process. In: TMS Annual Meeting, Orlando, Florida.1997; 363-75.

80. Cui J, Forssberg E. Mechanical recycling of waste electrical and

electronic equipment: a review. J. Hazard Mater. 2003; 99(3): 243-63.

81. Funcke W and Hemminghaus H. PXDF/D in flue gas from an incinerator charging wastes containing Cl and Br and a statistical description of the resulting PXDF/D combustion profiles. Organo. Compd.1997; 31, 93-8.

82. Stewart E and Lemieux P. Emissions from the incineration of electronics industry waste. Proc. Electronics Environ. IEEE International Symposium. 2003; 19-22: 271-5. Doi: 10.1109/ISEE.2003.1208088.

83. Osako M, Kim YJ, Sakai S. Leaching of brominated flame retardants in leachates from landfills in Japan. Chemosphere. 2004; 57(10): 1571-9.

84. Townsend G, Musson S, Jang, Y and Chung I. Leaching of hazardous chemicals from discarded electronic devices. Florida center for solid and hazardous waste management, Gainesville FL. 2004.

85. Peters-Michaud N, Katers J and Barry J. Occupational risks associated with electronics demanufacturing and CRT glass processing operations and the impact of mitigation activities on employee health and safety. Proc. Electronics Environ. IEEE International Symposium. 2003; 19-22, 323-328. DOI: 10.1109/ISEE.2003.1208098.

86. Aucott M, McLinden M, Winka M. Release of mercury from broken fluorescent bulbs. J Air Waste Manag Assoc. 2003; 53(2): 143-51.

87. Jadhav U and Hocheng H. A review of recovery of metals from industrial waste. J. Achiev. Mater Manufact. Eng. 2012; 54(2): 159-67.

88. Brusselaers J. An eco-efficient solution for plastics-metals-mixtures from electronic waste: the integrated metals smelter. In: 5th Identiplast 2005. The Biennial conference on the recycling and recovery of plastics identifying the opportunities for plastics recovery, Brussels, Belgium.

89. Hoffmann J. Recovering precious metals from electronic scrap. Jom-J. Miner. Met. Mater. Soc. 1992; 44(7): 43-48. DOI: 10.1007/BF03222275.

90. Heukelem A. Eco efficient optimization of pre-processing and metal smelting. In: Electronics goes green. Berlin, Germany. 2004; pp. 657–62.

91. Hageluken C. Recycling of electronic scrap at umicore's integrated metals smelter and refinery, World. Metallur. Erzmetall. 2006; 59(3): 152–61.

92. Ogata T, Nakano Y. Mechanisms of gold recovery from aqueous solutions using a novel tannin gel adsorbent synthesized from natural condensed tannin. Water Res. 2005; 39(18): 4281-6.

93. Kogan V. Process for the recovery of precious metals from electronic scrap by hydrometallurgical technique, International Patent, WO/2006/013568 (C22B 11/00), W.I.P. Organization. 2006.

94. Zhou P, Zheng Z, and Tie J. Technological process for extracting gold, silver and palladium from electronic industry waste. Chinese Patent, CN1603432A (C22B 11/00). 2005.

95. Dorin R and Woods R. Determination of leaching rates of precious metals by electrochemical techniques. J. Appl. Electro chem. 1991; 21(5): 419. Doi: 10.1007/BF01024578.

96. Pangum L and Browner R. Pressure chloride leaching of a refractory gold ore. Miner. Eng. 1996; 9(5): 547-56. DOI: 10.1016/0892-6875(96)00042-8.

97. S. Çolak B. Dönmez F. Sevim. Study on recovery of gold from decopperized anode slime. Chem Eng. Technol. 2001; 24(1): 91-5. DOI: 10.1002/1521-4125(200101)24:1<91::AID-CEAT91>3.0.CO;2-A.

98. Ubaldini S Fornari P and Massidda R. Innovative thiourea gold leaching process. Hydrometallurgy. 1998; 48(1), 113-24. Doi:10.1016/S0304-386X(97)00076-5.

99. Jeffrey M and Brunt S. The quantification of thiosulphate and polythionates in gold leach solutions and on anion exchange resins. Hydrometallurgy. 2007; 89(1): 52-60. DOI: 10.1016/j.hydromet.2007.05.004

100. Quinet P, Proost A and Lierde V. Recovery of precious metals from electronic scrap by hydrometallurgical processing routes. Miner. Metall. Process. 2005; 22(1): 17-22.

101. Menad N and Bjorkman B and Allain E. Combustion of plastics contained in electric and electronic scrap. Res. Conserv. Recycl. 1998; 24: 65-85.

102. Owens CV Jr, Lambright C, Bobseine K, Ryan B, Gray LE Jr, Gullett BK, et al. Identification of estrogenic compounds emitted from combustion of computer printed circuit boards in electronic waste. Environ Sci Technol. 2007; 41(24): 8506-11.

103. Ilyas S. Bioleaching of metals from ores and electronic scrap. Ph.D. thesis, University of Agriculture, Faisalabad, Pakistan. 2010; DOI: 10.1016/j.hydromet.2007.04.007.

104. Sum E. The recovery of metals rom electronic scrap. Jom-J. Miner. Met. Mater. Soc. 1991; 43(4): 53-61. DOI: 10.1007/BF03220549.

105. Mabbett AN, Sanyahumbi D, Yong P, Macaskie LE. Biorecovered precious metals from industrial wastes: Single-step conversion of a mixed metal liquid waste to a bioinorganic catalyst with environmental application. Environ Sci Technol. 2006; 40(3): 1015-21.

106. Morin D, Lips a, Pinches T, Huisman J, Frias C, Norberg A, et al. BioMinE-Integrated project for the development of biotechnology for metal bearing materials in Europe. Hydrometallurgy. 2006; 83(1): 69-76. DOI: 10.1016/j.hydromet.2006.03.047.

107. Bosecker K. Bioleaching: metal solubilization by microorganisms. FEMS Microbiol. Rev. 1997; 20(3-4): 591-604. DOI: 10.1111/j.1574-6976.1997.tb00340.x.

108. Sand W, Gehrke T and Hallmann R, Schippers A. Sulfur chemistry, biofilm and the (in) direct attack mechanism-a critical evaluation of bacterial leaching. Appl. Microbiol. Biotechnol. 1995; 43(6): 961-966. DOI: 10.1007/BF00166909.

109. Wolfgang Sanda, Tilman Gehrkea, Peter-Georg Jozsaa, Axel Schippersb. (Bio) chemistry of bacterial leaching – direct vs. indirect bioleaching. Hydrometallurgy. 2001; 59(2-3), 159-75. Doi: 10.1016/S0304-386X(00)00180-8.

110. Ehrlich H. Beginnings of rational bioleaching and highlights in the development of bio hydrometallurgy: A brief history. Eur. J. Miner. Process. Environ. Prot. 2003; 4(2): 102-12.

111. Rawlings DE. Heavy metal mining using microbes. Annu Rev Microbiol. 2002; 56: 65-91.

112. Rohwerder T, Gehrke T, Kinzler K, Sand W. Bioleaching review part A: progress in bioleaching: fundamental and mechanisms of bacterial metal sulfide oxidation. Appl Microbiol Biotechnol. 2003; 63(3): 239-48.

113. Pant D, Joshi D, Upreti MK, Kotnala RK. Chemical and biological extraction of metals present in E-waste: a hybrid technology. Waste Manag. 2012; 32(5): 979-90. Doi: 10.1016/j.wasman.2011.12.002.

114. Tipre D and Dave S. Bioleaching processes for Cu-Pb-Zn bulk concentrate at high pulp density. Hydrometallurgy. 2004; 75(1-4): 37-43. DOI: 10.1016/j.hydromet.2004.06.002.

115. Patel B. Biotechnology of metal extraction and ferric regeneration for GMDC concentrate, Ph.D. thesis, Gujarat University, Gujarat, India. 2012.

116. Willner J. Leaching of selected heavy metals from electronic waste in the presence of the Acidithiobacillus ferrooxidans bacteria. J. Achiev. Mater. Manufac. Eng. 2012; 55(2): 860-3.

117. Saidan M, Brown B and Valix M. Leaching of electronic waste using biometabolised acids. Chinese J. Chem. Eng. 2012; 20(3): 530-4.

118. Choi M, Cho K. Kim, DS and Kim, DJ. Microbial recovery of copper from printed circuit boards of waste computer by Acidithiobacillus ferrooxidans. J. Env. Sci. Health, Part A- Toxic/Hazard. Subs. Environ. Eng. 2012; 39 (11): 2973-82. DOI:10.1081/LESA-200034763.

119. Olson G. Microbial oxidation of gold ores and gold bioleaching. FEMS Microbiol. Lett. 2006; 119(1-2): 1-6. DOI: 10.1111/j.1574-6968.1994.tb06858.x.

120. Liang G, Mo Y and Zhou Q. Novel strategies of bioleaching metals from printed circuit boards (PCBs) in mixed cultivation of two acidophiles. 2010; 47(7). Doi: 10.1016/j.enzmictec.2010.08.002.

121. Liang G, Tang J, Liu W, Zhou Q. Optimizing mixed culture of two acidophiles to improve copper recovery from printed circuit boards (PCBs). J Hazard Mater. 2013; 250-251: 238-45. Doi: 10.1016/j.jhazmat.2013.01.077.

122. Yang T, Xu Z, Wen J and Yang L. Factors influencing bioleaching of copper from waste printed circuit boards by Acidithiobacillus ferrooxidans. Hydrometallurgy. 2009; 97, 29-32.

123. Mishra D and Rhee Y. Current research trends of microbiological leaching for metal recovery from industrial waste. In: Current Research, technology and education topics in applied microbiology and microbial biotechnology. A. Mendez-Vilas edi. Korea. 2010; pp. 1289-95.

124. Hudec R, Sodhi M and Arora D. Biorecovery of metals from electronic waste. In: 7th Latin American and Carribean Conference for Engineering and Technology, 2-5 June. San Cristobal, Venezula. 2009.

125. Groudev S, Spasova I, Nicolova M, Georgiev P and Angelov A. Biological and chemical leaching of non-ferrous and precious metals from electronic scrap. Ann. Rev. Geology Geophys. 2007; 50(1): 191-4.

126. Bas A, Deveci H, Yazici E. Bioleaching of copper from low grade scrap TV circuit boards using mesophilic bacteria. Hydrometallurgy. 2013; 138: 65-70. DOI: 10.1016/j.hydromet.2013.06.015.

127. Saidan M, Valix M. Bioleaching of copper from electronic waste using Aspergillusniger and Acidithiobacillus. In: Chemeca. 2011; 18-21.

128. Creamer NJ, Baxter-Plant VS, Henderson J, Potter M, Macaskie LE. Palladium and gold removal and recovery from precious metal solutions and electronic scrap leachates by Desulfovibriodesulfuricans. Biotechnol Lett. 2006; 28(18): 1475-84.

129. Brandl H, Rehm H and Reed G. Microbial leaching of metals-Biotechnology. In: Rehm H and Reed G. (Eds.) Second Edi, Wiley online, Weinheim. 2008; pp. 191-224. DOI: 10.1002/9783527620999.ch8k.

130. Ilyas S, Anwar M, Niazi S and Ghauri A. Bioleaching of metals from electronic scrap by moderately thermophilic acidophilic bacteria, Hydrometallurgy. 2007; 88(1-4): 180-8. DOI: 10.1016/j.hydromet.2007.04.007.

131. Ilyas S, Ruan C, Bhatti H, Ghauri M and Anwar M. Column bioleaching of metals from electronic scrap. Hydrometallurgy. 2009; 101(3-4): 135-40. DOI: 10.1016/j.hydromet.2009.12.007.

132. Knowles CJ, Bunch AW. Microbial cyanide metabolism. Adv Microb Physiol. 1986; 27: 73-111.

133. Laville J, Blumer C, Schroetter C, Gaia V, Defago G, Keel C, et al. Characterization of the hcn ABC gene cluster encoding hydrogen cyanide synthase and anaerobic regulation by ANR in the strictly aerobic biocontrol agent Pseudomonas fluorescens CHAO. J. Bacteriol. 1998; 180(12): 3187-96.

134. Blumer C, Haas D. Mechanism, regulation and ecological role of bacterial cyanide biosynthesis. Arch Microbiol. 2000; 173(3): 170-7.

135. Michaels R, Corpe WA. Cyanide formation by Chromo bacterium violaceium. J Bacteriol. 1965; 89: 106-12.

136. Bach E. On hydrocyanic acid formation in mushrooms. Physiol. Plantar.1948; 1: 387-89.

137. R A Askeland and S M Morrison. Cyanide production by Pseudomonas aeruginosa. Appl Environ Microbiol. 1983; 45(6): 1802–7.

138. Faramarzi MA, Stagars M, Pensini E, Krebs W, Brandl H. Metal solubilization from metal containing solid materials by cyanogenic Chromobacterium violaceum. J Biotechnol. 2004; 113(1-3): 321-6.

139. Oh CJ, Lee SO, Yang HS, Ha TJ, Kim MJ. Selective leaching of valuable metals from waste printed circuit boards. J Air Waste Manag Assoc. 2003; 53(7): 897-902.

140. Pham V and Ting Y. Gold bioleaching of electronic waste by cyanogenic bacteria and its enhancement with biooxidation. Adv. Mater. Res. Vols. 2009; 71-73: 661-4. Doi: 10.4028/www.scientific.net/AMR.71-73.661.

141. Pradhan JK, Kumar S. Metals bioleaching from electronic waste by Chromobacterium violaceum and Pseudomonas sp. Waste Manag Res. 2012; 30(11): 1151-9. Doi: 10.1177/0734242X12437565.

142. Bhagat M, Burgess J, Antunes M, Whiteley C and Duncan J. Precipitation of mixed metal residues from wastewater utilising biogenic sulphide. Minerals Engineering. 2004; 17: 925-32. DOI: 10.1016/j.mineng.2004.02.006.

143. Ren WX, Li PJ, Geng Y, Li XJ. Biological leaching of heavy metals from a contaminated soil by Aspergillus Niger. J Hazard Mater. 2009; 167(1-3): 164-9. Doi: 10.1016/j.jhazmat.2008.12.104.

144. Bosshard P, Bachofen R and Brandl H. Metal leaching of fly ash from municipal waste incineration by Aspergillus Niger. 1996; 30(10): 3066-070. DOI: 10.1021/es960151v.

145. Christoph Brombacher, Reinhard Bachofen and Helmut Brandl. Development of a laboratory scale leaching plant for metal extraction from fly ash by Thiobacillus strains. Appl Environ Microbiol. 1998; 64(4): 1237–41.

146. Dave S, Damani M, Tipre D. Copper biosorption and bioprecipitation by Eichhornia spp. and sulphate reducing bacteria. Adv. Mater. Res. 2009; 71: 561-564.

147. Parameswari E, Lakshmanan and Thilagavathi T. Biosorption of chromium (VI) and nickel (II) by bacterial isolates from an aqueous solution. Electronic J. Env. Agri. Food Chem. 2009; 8(3): 150-156.

148. Ilhan S, Nourbaksh M, Kilicarslan S and Ozdag H. Removal of chromium, lead and copper ions from industrial waste waters by Staphylococcus saprophyticus. Turk. Electronic J. Biotechnol. 2004; 2: 50-57.

149. Savitha J, Sahana N and Praveen V. Metal biosorption by Helminthosporium solani – a simple microbiological technique to remove metal from E-waste. Curr. Sci. 2010; 98(7): 903-4.

7

An overview on the application of genus *Chlorella* in biotechnological processes

Cristiano José de Andrade[1]*and Lidiane Maria de Andrade[1]

[1]Chemical Engineering Department of Polytechnic School of the University of São Paulo

*Corresponding author: Dr. Andrade C. J, Chemical Engineering Department of Polytechnic School of the University of São Paulo, São Paulo, SP, Brazil.
E-mail: eng.crisja@gmail.com

Abstract

Chlorella is a genus of green algae (single-cell) that shows spherical shape. The cultivation of microalgae, such as *Chlorella*, is promising due to technical easiness of this type of bioprocesses. The production of microalgae at industrial scale started in the early 1960s. The microalgae cultivation is aligned to the green chemistry concept. Thus, there is a strong trend that this type of bioprocess will be applied all over the world. The aim of this study was review some characteristics of genus *Chlorella*, and mainly the potential industrial applications of genus *Chlorella* and their biocompounds. Microalgae are able of two types of trophy: autotrophy and heterotrophy. The photoautotrophy organisms are used when the CO_2 fixation is the mainstream appeal, whereas heterotrophy and mixtrophy organisms can be mainly use for the production of microalgae biomass. Some of the main applications of genus *Chlorella* are: productions of biofuels (biodiesel, biomethane and biohydrogen), cosmetics (skin care), supplementary foods (polyunsaturated fatty acids), pigments (carotenoids and chlorophyll) and wastewater treatments (reduction of chemical oxygen demand and bioremediation).

Keywords: *Chlorella;* Biofuels; Supplementary Foods; Wastewater Treatments.

Introduction

Chlorella is a genus of green algae (single-cell) that shows spherical shape ≈ 2 to 10 μm (diameter) (Figure 1). Compared to others microalgae, *Chlorella* species present higher photosynthetic efficiency. In addition, it was predicted that 10,000 tons of proteins per year could be produced by 20 people staff (4-square kilometer) - *Chlorella* farm [1]. The production of microalgae at industrial scale started in the early 1960s in Japan, in which *Chlorella* species were applied as food additive. Then, in 1980s the microalgae cultivation spread out all over the world (e.g. USA, India, Israel and Australia) [2].

Microalgae are able of two types of trophy (nourishment) (i) autotrophy (phototrophy) and (ii) heterotrophy (phagotrophy). Autotrophy organisms absorb light in order to reduce CO_2 (to obtain energy). Photoautotrophy organisms require only inorganic minerals, in which a photoautotrophy obligate cannot grow in dark.

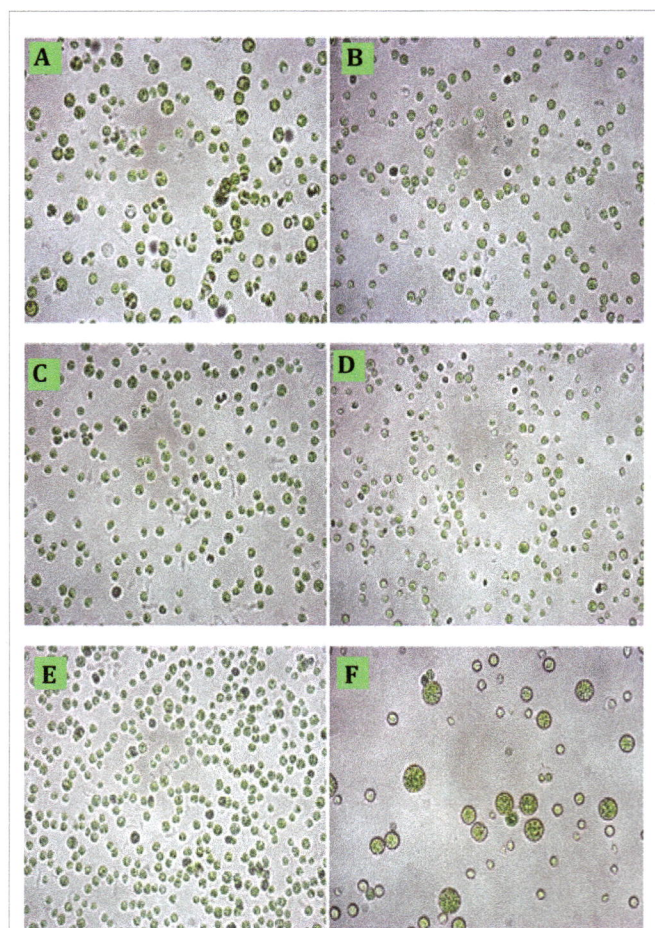

Figure 1: *Chlorella desiccata* (Cde), *Chlorella kessleri* (Cke), *Chlorella luteoviridis* (Clu), *Chlorella prototothecoides* (Cpr), *Chlorella sorokiniana* and *Chlorella vulgaris* (Cv); respectively A, B, C, D, E and F (Source: Authors).

On the other hand, heterotrophic organisms obtain energy from organic compounds. Photoheterotrophic organisms use light for energy, nevertheless they cannot use CO_2 as sole carbon source, that is, they take organic compounds from environment to complete the carbon requirement (Table 1) [3].

Table 1: Trophic possibilities for the microalgae

	Inorganic carbon	Organic carbon	Light
Autotrophy	✓		✓
Heterotrophy†		✓	
Photoautotrophic††	✓		✓
Photoheterotrophic		✓	✓
Mixotrophic	✓	✓	✓
Auxotrophy*	✓	✓	

Last but not least, mixtrophy is other interesting type of microalgae's metabolism. Mixotrophy is the cultivation in which organic carbon and CO_2 are simultaneously assimilated (a mixture between autrotophy and heterotrophy). This metabolic pathway gives these microorganisms advantages over photosynthetic mode of cultivation. This metabolism is approximately the sum of specific growth rates of cells under heterotrophic and photoautrophic. In addition, mixotrophy growth allowed the cultivation in the dark (higher productivity) and higher cell concentration, for instance the maximum specific growth rate for *Chlorella vulgaris* 0.11 h^{-1} (photoautotrophy mode) and 0.098 h^{-1} (heterotrophic mode) and 0.198 h^{-1} (mixotrophic) [4-5].

Regarding culture media for the cultivation of microalgae, they are very simple (Table 2). In this sense, even similar one another, some compositions are widely used for the cultivation of genus *Chlorella*, for instance, Guillard's WC [6-7]; Bold [7-8].

Table 2: Culture media composition of culture media for microalgae cultivation

WC

$NaNO_3$ (g/L)	$CaCl_2 \cdot 2H_2O$ (g/L)	$MgSO_4 \cdot 7H_2O$ (g/L)	$NaHCO_3$ (g/L)	$Na_2SiO_3 \cdot 9H_2O$ (g/L)	K_2HPO_2 (g/L)	$NaEDTA \cdot 2H_2O$ (g/L)	$FeCl_3 \cdot 6H_2O$ (g/L)	$CuSO_4 \cdot 5H_2O$ (g/L)	$ZnSO_4 \cdot 7H_2O$ (g/L)	$CoCl_2 \cdot 6H_2O$ (g/L)	$MnCl_2 \cdot 4H_2O$ (g/L)	$Na_2MoO_4 \cdot 2H_2O$ (g/L)	$MgSO_4$ (g/L)	Thiamine-HCl (μg/L)	Biotin (g/L)	Cyanocobalamin (g/L)
0.085	0.037	0.037	0.012	0.028	0.008	0.04	0.04	0.010	0.022	0.01	0.180	0.006	0.019	0.297	0.005	0.005

Bold

$NaNO_3$ (g/L)	$CaCl_2.2H_2O$ (g/L)	$MgSO_4.7H_2O$ (g/L)	K_2HPO_4 (g/L)	KH_2PO_4 (g/L)	NaCl (g/L)	NaEDTA (g/L)	$FeSO_4 \cdot 7H_2O$ (g/L)	H_3BO_3 (g/L)	$ZnSO_4.7H_2O$ (g/L)	$MnCl_2.4H_2O$ (g/L)	MoO_3 (g/L)	$CuSO_4.5H_2O$ (g/L)	$Co(NO_3)_2.6H_2O$ (g/L)
0.25	0.026	0.075	0.075	0.175	0.025	0.0049	0.0049	0.0115	0.0088	0.00144	0.00071	0.00157	0.00048

Sorokin and Krauss [8]

KNO_3 (g/L)	KH_2PO_4 (g/L)	$MgSO_4.7H_2O$ (g/L)	$CaCl_2.2H_2O_2$ (g/L)	$FeSO_4.7H_2O$ (g/L)	EDTA (g/L)	H_3BO_4 (μg/L)	$MnCl_2.4H_2O$ (μg/L)	$ZnSO4.7H_2O$ (μg/L)	$CuSO4.5H_2O$ (μg/L)	$Co(NO3)2.6H2O$ (μg/L)	MoO_3 (μg/L)
1.25	1.25	1.0	0.04	0.05	0.5	114	14	88	16	5	7

Therefore, the culture media described in Table 2 can be used for photoautotrophy organisms, in which the 2 (CO_2) fixation is the mainstream appeal. In addition, the phototrophic production is the most effective in terms of net energy balance. Nevertheless, this bioprocess show higher variation and lower productivity - when compared with heterotrophic production [2]. In this regard, since atmospheric 2 (CO_2) does not supply enough carbon to achieve high rates of autotrophic microalgae production – (the diffusion of atmosphere 2 (CO_2) → aqueous phase ≈ 10 g/m.d); the use of bicarbonate-carbonate buffer (medium) can be useful because it provides 2 (CO_2) for photosynthesis as detailed below:

$$2HCO_3^- \leftrightarrow CO_3^{2-} + H_2O + CO_2$$

$$HCO_3^- \leftrightarrow CO_2 + OH^-$$

$$CO_3^{2-} + H_2O \leftrightarrow 2OH^-$$

Obviously, the pH of culture medium tends to become alkali, in which at high microalgae density, it reaches pH as high as 11 [3].

On the other hand, heterotrophy and mixtrophy organisms can be, mainly use for the production of microalgae biomass. In addition, compared to photoautotrophy system, the mixotrophy cultivation mode shows lower production cost due to the higher biomass and lipid productivity and the possibility in use low-cost culture media such as industrial wastes (culture medium is ≈ 80% of the total production cost) [5].

The cultivation of microalgae, either photoautotrophy or heterotrophy modes, plays already an important role in biobased economy (very aligned to green chemistry concept). It is worth noting that in 2050 the world population is estimated to reach 9 billion people, that is, the demand for commodities will increase exponentially, in which the sustainable production (food and energy). Microalgae are not only one the most promising waste converters and recyclers, but can be efficiently cultivated in places that are inhospitable for agriculture, which can provide proteins and lipids (food) or raw material for bioplastic [9].

Some of the main applications of *Chlorella* are described in more details below, such as productions of biofuels, cosmetics, supplementary foods, pigments, by wastewater treatments.

Biofuels

It is well-known that global climate change has increased due to the green house gas emissions from fossil fuels. Thus, alternative sources of energy need to be investigated and explored. Biofuel is a fuel that is produced by biological processes, for instance bioalcohols (ethanol, propanol, butanol), biodiesel, green diesel, biogas (biomethane, biohydrogen), etc.

Regarding microalgae-based biofuels, mainly biodiesel and biogas have been investigated. The lipids from microalgae can be extracted and then esterified with alcohol, which produce biodiesel, whereas biogas (biomethane and biohydrogen) is

Biodiesel

Biodiesel is a clean alternative fuel source that could replace fossil fuels. Usually, biodiesel is produced from raw oleaginous such as soybean and sunflower, which leads to issues in terms of deforestation, world hunger and land pollution. Thus, other sources of lipids should be investigated. In this sense, microalgae lipids are one of the most promising feedstock to produce biodiesel, since they do not compete with food crops and show high content of lipids (up to 75 wt%). However, the current high cost of biodiesel from microalgae makes its production at industrial scale infeasible [11-12]. Other advantage on the use of microalgae lipids to biodiesel production is the mixotroph metabolism of some microalgae (auto- and heterotrophy metabolisms), for instance glucose can be used by *Chlorella* protothecoides. In particular, the heterotrophy cultivation is cheaper, easier, feasible in colder climates and allows the use of agro-industrial wastes as substrates [12-13].

Veillette et al. [12] detailed the esterification of microalgae free fatty acids using Amberlyst-15 as a catalyst, in which a conversion higher as 84% was reached [12].

Biogas

As already mentioned the production of biogas from microalgae occurs by the anaerobic digestion of microalgae biomass by anaerobic bacteria. The anaerobic digestion encompasses 4 general steps (i) hydrolysis, fermentation, acetogenesis and methanogenesis. The composition of biogas is CH_4 (55–75%) and CO_2 (25–45%) [10].

Jankowska et al. [10] compiled the biogas yields from microalgae, for instance, *C. kessleri* (0.335 L biogas/g.VS (65% CH_4) (0.218 L CH_4/gVS)); *C. vulgaris* (0.337 L CH_4/g.VS); *C. vulgaris* (0.180 L CH_4/g.CODin); *C. vulgaris* (0.156 L CH_4/g.COD); *C. vulgaris* ((0.364 LN biogas/g.VS) (62.6% CH_4) (0.228 LN CH_4/g. VS)); *C. vulgaris* ((0.366 L biogas/g.VS) (62.5% CH_4) (0.229 L CH_4/g.VS)); *C. vulgaris* (0.139 L CH_4/g.COD in) [10].

The biogas yield is highly affected by the specie of microalgae, type of pretreatment, presence of inhibitors of hydrogenesis or methanogenesis, organic loading, retention time, temperature, pH, substrate, etc. In this sense, as described by Choi et al. compared to others microorganisms, the cell walls of microalgae are more recalcitrant. Thus, the pretreatment (acid + thermal) of *C. vulgaris* was needed to increase the hydrolysis with consequent enhancement on the H_2 production [10, 14]

Cosmetics

Components of microalgae, typically *C. vulgaris* specie, are often used in cosmetics. One of the most interesting approaches on the applications of microalgae is in the cosmetic formulations. Microalgae have sun protection skills due to the presence of chlorophyll-a in its composition (light absorption) [15]. In addition, *Chlorella* extract is also used by the skin care industry, since some compounds from *Chlorella* extract have anti-aging, refreshing, regenerant, emollient and anti-irritant activities [15].

Microalgae of *Chlorella* genus can produce metabolites, such as sporopollenin and mycosporine-like amino acids, to protect themselves from ultra violet (UV) radiation (Table 3) [16].

Table 3: UV-Screening compounds produced by *Chlorella*

UV Screening Compound	Specie
Sporopollenin	*Chlorella fusca*
Mycosporine-Like Amino Acids	*Chlorella minutissima* *Chlorella sorokiniana*

In this sense, currently, there are some cosmetics microalgae-based commercially available. The world's first facial moisturizer, Sun *Chlorella* Cream® which is produced by Sun *Chlorella* Japanese Company, is based in *C. pyrenoidosa* extract. This facial moisturizer promotes the skin hydration and also aid the skin cell renewal. Other example is the Dermochlorella that is produced by the ProTec Ingredia French Company. Dermochlorella is produced from *C. vulgaris* extract, which shows firming, restructuring and eye contour effects besides it stimulates the synthesis of collagen. In addition, it decreases the morphology of stretch marks and reduces vascular imperfections.

Supplementary Food

Microalgae, in particular, genus *Chlorella,* can synthesize essential nutritional compounds. Microalgae composition is up to 50-70% protein (essential amino acids), 30% lipids (polyunsaturated fatty acids), up to 8-14% carotene and a fairly high concentration of vitamins B1, B2, B3, B6, B12, E, K, D, among others. Thus, *Chlorella* extract can be used as supplementary food as described below starch.

Starch

Microalgae under a wide range of conditions accumulate starch. Palacios et al. described an interesting system that integrated *Azospirillum brasilense* and *Chlorella sorokiniana. A. brasilense* is microalgae growth-promoting bacterium by mainly indole-3-acetic acid.

Human Nutrition

Microalgae have been used by humans for thousands years and the commercial large-scale of *Chlorella* genus started in the 1960's by the Japanese company Nohon *Chlorella*. In this sense, the human consumption of microalgae biomass is restricted to very few species, in which *Chlorella, Spirulina* and *Dunaliella* are the main genus produced at industrial scale. Microalgae biomass is usually marketed (human nutrition) as tablet or powder, in the health food market [2, 19-20].

Chlorella microalgae genus have positive impact on the health of humans due to their nutritional content (nutraceutical – proteins, lipids, pigments, carbohydrates), as shown in Table 4 (polyunsaturated fatty acids). In addition, *Chlorella* extract shows effects against renal failure and growth promotion of intestinal

Lactobacillus (probiotic) [2]. In others interesting approaches, Ebrahimi-Mameghani et al. [21] studied the *C. vulgaris* supplementation on glucose homeostasis, insulin resistance and inflammatory biomarkers in patients with nonalcoholic fatty liver disease [21]. The authors indicated that 1,200 mg of *C. vulgaris* supplementation can effects on weight loss, serum glucose and enhanced the inflammatory biomarkers as well as liver function in non-alcoholic fatty liver disease patients, whereas, Cherng et al. [22]. proved that *Chlorella* contains a peptide known as *Chlorella*-11 (Val-Glu-Cys-Tyr-Gly-Pro-Asn-Arg-Pro-Gln-Phe) that showed activity against inflammation caused by lipopolysaccharides from Gram-negative bacteria [21-22].

Carbohydrates are also found in *Chlorella* extract as starch, sugar, glucose and other polysaccharides [17]. Fatty acids from microalgae have encompasses omega families such as oleic acid, linoleic acid, linolenic acid and especially arachidonic acid (AA), eicosapentaenoic acid (EPA) and docosahexaenoic acid (DHA), which ones feature high added value as shown in Table 4.

Table 4: Fatty acids from microalgae.

Acids	Carbon Atoms	Chemical Formula	Omega Family
Palmitoleic	16	$C_{16}H_{30}O_2$	ω7
Oleic	18	$C_{18}H_{34}O_2$	ω9
Linoleic	18	$C_{18}H_{32}O_2$	ω6
α-Linolenic	18	$C_{18}H_{30}O_2$	ω3
γ-Linolenic	18	$C_{18}H_{30}O_2$	ω6
Homo γ-Linolenic	20	$C_{20}H_{34}O_2$	ω6
AA	20	$C_{20}H_{32}O_2$	ω6
EPA	20	$C_{20}H_{30}O_2$	ω3
DHA	22	$C_{22}H_{32}O_2$	ω3

Arachidonic acid from microalgae can be used as supplementary food when there is deficiency in linoleic acid or some difficulty to convert linoleic acid to Arachidonic acid [23-24].

"Regarding muscle growth, arachidonic acid repairs and promotes the growth of skeletal muscle tissue [25]. In addition, arachidonic acid is the most abundant fatty acid (20%) in the brain. Thus, the entire neurological health depends on the level of arachidonic acid [26-29]. Low content of arachidonic acid in the brain can contribute to diseases such as Alzheimer's and Bipolar disorder. Even more, arachidonic acid as supplementary food can has been shown to increase lean body mass, strength, resistance and present anti-inflammatory properties".

Eicosapentaenoic Acid (EPA)

In microalgae EPA is the precursor for prostaglandin-3, thromboxane-3, and leukotriene-5 group. EPA has the ability to reduce inflammation, decrease depression and suicidal behavior, schizophrenia and improves the chemotherapy response.

Docosahexaenoic Acid (DHA)

Docosahexaenoic acid is an omega-3 fatty acid, that is, a primary structural component of the human brain, eye, cerebral cortex, skin, retina and heart health. Among the applications of DHA are: infant formulations, products for pregnant and nursing women, food and beverage products, dietary supplements, immune modulating effects and capacity to inhibit growth of human colon carcinoma cells, more than other omega-3.

Companies that produce omega fatty acids, mainly EPA and DHA, from microalgae, are shown in Table 5.

Table 5: Companies that market omega-3 from *Chlorella*.

Company	Location	Started
Live Fuels	USA	2006
Aurora Algae	USA	2006
Martek Biosciences	USA	2007
Blue Biotech International GmBH	Germany	2000
Photonz Corporation	New Zealand	2002
Ingrepro BV	The Netherlands	2001

The possibilities of human nutrition with microalgae are very wide. Microalgae for human nutrition can be incorporated into candies, beverages, snacks, pastas and juices formulations, in which the commercial applications are dominated by some genus including *Chlorella* [19]. Companies that market *Chlorella* for human nutrition are shown in Table 6.

Table 6: Companies that market *Chlorella* for human nutrition.

Company	Location	Started	Production (tons/year)
Sun *Chlorella* Corporation	Japan	1969	NF
Yaeyama Shokusan Co Ltd.	Japan	1975	420
Maypro Industries Inc.	USA	1977	NF
Taiwan *Chlorella* Manufacturing Co ltd	Taiwan	1964	400
Far East Microalgae Ind Co., Ltd	Taiwan	1976	1000
Roquette Klötze GmbH & Co. KG	Germany	1995	130-150
Lotus Organics	Kazakhstan	2007	NF
Martek Biosciences Corporation	USA	2007	NF

*NF = Not Found

Animal Feed

Chlorella is one of the most frequently microorganisms used as animal nutrition which is directly consumed by larval (brief period), mollusks, penaeid shrimp or indirectly for live prey fed to small fish [2, 30]. In this sense, specific microalgae are feasible to be used as animal feed supplements, for instance, the supplementation animal nutrition by *Chlorella*, *Scenedesmus* and *Spirulina* has enhanced the immune response, improved fertility, better weight control, healthier skin and a lustrous coat [2].

Table 7: Composition of some species of *Chlorella* genus microalgae (% dried biomass) [31-33].

Specie	Protein	Carbohydrate	Lipid	Reference
Chlorella vulgaris	51-58	12-17	14-22	[17]
Chlorella calcitrans			14.6-16.4/39.8	
Chlorella emerson ii			25.0-63.0	
Chlorella protothecoides			14.6-57.8	
Chlorella pyrenoidosa	57	26	2.0	[34-35]
Chlorella sorokiniana			19.0-22.0	
Chlorella sp.			10.0-48.0	

Pigments

Microalgae pigments, carotenoids and chlorophyll, are often used by industries, such as food, nutraceutical, pharmaceutical, aquaculture, and cosmetic industry; as well by clinical/research laboratories (label for antibodies and receptors) [36].

Carotenoids

Carotenoids are pigments that have been draw attention due to their potential health benefits, in which microalgae are a natural source of carotenoids. They have a common C40 backbone structure of isoprene units; they are lypholilic and usually presented color such as red, orange or yellow. The carotenoids pigments can be divided in two groups: carotenes and xanthophylls [37].

Carotenes

The composition of carotenes contains only hydrocarbons

and the common carotenes found in microalgae are lycopene and ß-carotene (Figure 2).

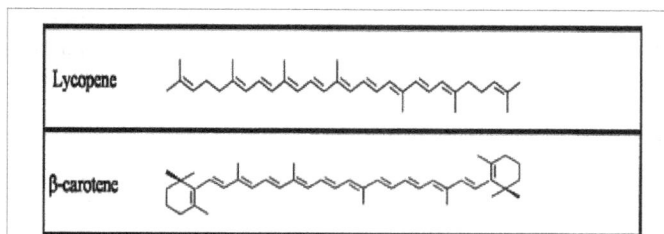

Figure 2: Chemical structure of carotenes from microalgae.

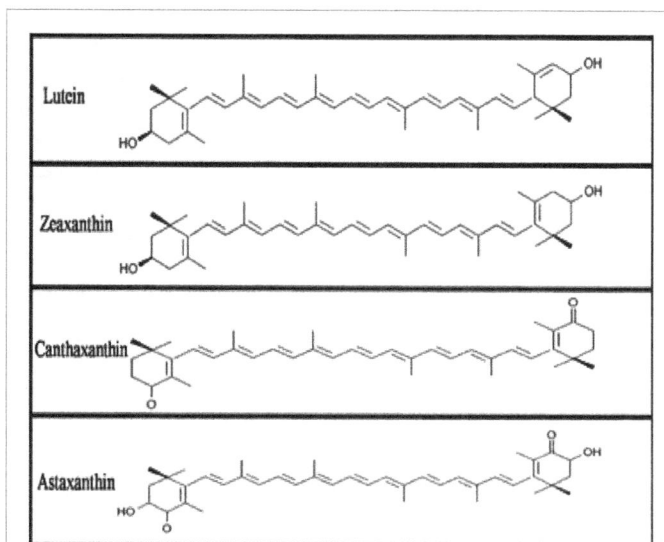

Figure 3: Chemical structure of xanthophylls from microalgae.

Xanthophylls

Xanthophylls are oxygenated derivatives of carotenes. The common carotenes found in microalgae are lutein, zeaxanthin, canthaxantin and astaxanthin and their chemical structures are shown in Figure 3.

The main carotenoids synthesize by *Chlorella* microalgae genus are mainly astaxanthin, ß-carotene, lutein, lycopen and canthaxantin. In general, carotenoids have an anti-oxidant property. Therefore they protect the cells from reactive radicals, prevent lipid oxidation, promote the stability and functionality of the photosynthetic machinery of cells [3].

Chlorophyll

Chlorella is also known as 'Emerald food' due to its high content (7% of biomass) of chlorophyll a [38]. Chlorophyll is a pigment present in microalgae that produce carbohydrates from carbon dioxide and water light through energy absorption (photosynthesis process – Calvin cycle). The chlorophylls present in *Chlorella* microalgae are: (i) chlorophyll-a - responsible for oxygenic photosynthesis, and (ii) chlorophyll-b - absorb energy to aid in photosynthesis process.

Chlorophyll can be recovered from microalgae biomass by organic solvent extraction. Chlorophyll has antioxidant and antimutagenic properties and can be used as additive in pharmaceutical, cosmetic products, and also as a natural food pigment. The Table 8 summarizes the health benefits by pigments produced by microalgae.

Table 8: Potential health human benefits by pigments produced by microalgae (Adapted from Gong and Bassi, [37]).

	Class	Pigment	Health Benefits	Reference
Carotenoids	Carotenes	Lycopene	Anti-cancer	[39]
			Cardiovascular health	
		ß-carotene	Anti-oxidant	[40]
			Prevents liber fibrosis	
	Xanthophylls	Lutein	Prevents cataract and age-related	[41]
			Cardiovascular health	[42]
		Canthaxantin	Anti-oxidant	[43]
			Creates tan color	
		Astaxanthin	Strong anti-oxidant	[44]
			Anti-cancer	[45]
			Cardiovascular health	[46]

Wastewater Treatment

Generally, there are two subsequent treatments (i) for the sedimentation of materials; (ii) to oxidize the organic materials. Then, the wastewaters are disposal to aquatic environment [53]. Aquaculture systems - as wastewater treatment and recycling - have been draw attention, mainly due to their capacity to simultaneously solve environmental and sanitary issues. In addition, these processes can be economically feasible [54]. In this sense, aquaculture systems, in particular those that apply algae take advantages of oxygen production, which favors heterotrophic bacteria.

Hammouda et al. [54] tested microbial consortium comprised by *Chlorella* and *Scenedesmus* for the treatment of wastewater in both batch and continuous modes. The authors described a progressive and high reduction of chemical oxygen demand; 89% and 91.7%, in batch and continuous modes, respectively [54].

Murwanashyaka et al. [53] detailed a study that green microalgae *Chlorella sorokiniana* FACHB-275 was cultivated under both light and lightless conditions. The cultivation aimed to remediate wastewater under heterotrophic conditions. Preliminary results showed high tolerance from nitrogenous and phosphorous compounds. The authors described the relation between initial nutrient content and removal efficiency. The highest removal efficiency reached 99% (123.6 mg N/L and 26.8 mg P/L) [53].

Chlorella sp. was already tested to remediate aquaculture wastewater (fish farm) aerated with boiler flue gas wastewater. Thus, the authors proved that is feasible simultaneously reduce CO_2 emission and produce microalgae biomass [54-55].

Bioremediation

Microalgae efficiently absorb heavy metals, for instance *Chlorella vulgaris* absorbed Pb^{+2}, in which the highest adsorption rate was 15.4 mg/g.min [56-57].

Hammouda et al. [54] tested microbial consortium comprised by *Chlorella* and *Scenedesmus* for the treatment of wastewater. The system proved to be efficient on simultaneous removal of Fe (from 0.99 to 0.02 mg/L), Ni (from 0.661 to 0.15 mg/L) and Cr (from 0.4 to 0.09 mg/L) [54].

Perspective, Advantages and Drawbacks

The microalgae cultivation is an eco-friendly and renewable process, which is aligned to the green chemistry concept, in particular when the microalgae cultivation is integrated to flue gases (source of carbon). Thus, there is a strong trend that this type of bioprocess will be applied all over the world. One of the main advantages of microalgae cultivation, over other bioprocesses, is its versatility, for instance lipids (biodiesel); lipids (polyunsaturated fatty acids – human nutrition and animal feed), pigments, proteins (peptide known as Chlorella-11), among others [2,36].

Among the advantages of microalgae cultivation are: (i) seasonality - microalgae are cultivable throughout the year, which gives the microalgae cultivation advantages over all oilseed crops; (ii) microalgae cultivation is a submerged bioprocess, nevertheless, it needs less water than agricultural land crop; (iii) microalgae cultivation does not compete with production of food, since brackish water on non-arable land can be used; (iv) microalgae have fast growth (≈ 3.5 h generation time) and high oil concentration (20-50% - dry weight of biomass); (iv) it is an environmentally friendly process (green chemistry concept), since occurs the biofixation of CO_2 (1 kg of dry microalgae biomass utilize ≈ 1.83 kg of CO_2); (v) the nutrients can be obtained from wastewater, in particular nitrogen and phosphorus ($\downarrow$ production cost and waste treatment), (vi) simultaneous production of valuable products (e.g. proteins and lipids) [2].

Although there are several potential microalgae applications, in particular those using genus *Chlorella*, there is still room for improvement, for instance, the production efficiency must be increased three times and production costs must be reduced ten times, whereas Brennan and Owende, pointed out the main challenges on microalgae biofuel technology, including (i) to achieve high photosynthetic efficiency on continuous production mode; (ii) the cultivation using single species of microalgae (not susceptible to contamination); (iii) few industrial scale plants are current in operation (lack of knowledge), (iv) integration between flue gases and microalgae cultivation [2,9].

Conclusion

The microalgae cultivation is one of the most promising bioprocesses due to its technical easiness and versatility. Very likely, the microalgae cultivation will be applied all over the world. Among the potential applications of genus *Chlorella*, the production of biofuels, in particular biodiesel, the supplementation of foods (polyunsaturated fatty acids) and wastewater treatments (reduction of chemical oxygen demand) are the most feasible ones.

References

1. Zuñiga C, Li C-T, Huelsman T, Levering J, Zielinski DC, McConnell BO, et al. Genome-scale metabolic model for the green alga *Chlorella vulgaris* UTEX 395 accurately predicts phenotypes under autotrophic, heterotrophic, and mixotrophic growth conditions. Plant Physiol. 2016;172:589-602. doi.org/10.1104/pp.16.00593

2. Brennan L, Owende P. Biofuels from microalgae - A review of technologies for production, processing, and extractions of biofuels and co-products. Renew Sust Energ Rev. 2010;14:557-577. doi.org/10.1016/j.rser.2009.10.009

3. Grobbelaar JU. Algal nutrition. Mineral nutrition. Richmond A, editor. Handbook of microalgal culture: biotechnology and applied phycology. Oxford:Blackwell Science;2004.

4. Lee Y-K. Algal nutrition. Heterotrophic carbon nutrition. Richmond A, editor. Handbook of microalgal culture: biotechnology and applied

phycology. Oxford:Blackwell Science;2004.

5. Abreu AP, Fernandes B, Vicente A A, Teixeira J, Dragone G. Mixotrophic cultivation of *Chlorella vulgaris* using industrial dairy waste as organic carbon source. Bioresource Technol. 2012;118:61-66. doi. org/10.1016/j.biortech.2012.05.055

6. Guillard RRL. Culture of phytoplankton for feeding marine invertebrates. Smith WL and Chantey MH, editors. Culture of marine invertebrate animals. Plenum Publishers, New York;1975.

7. Andersen R. Algal culturing technique. San Diego: Elservier;2005.

8. Sorokin C, Krauss RW. The effect of light intensity on the growth rates of green algae. Plant Physiol.1958;33:109-113.

9. Wolkers H, Barbosa M, Kleinegris D, Bosma R, Wijffels RH. Large-scale sustainable cultivation of microalgae for the production of bulk commodities. 2011 Available at:

www.groenegrondstoffen.nl/downloads/Boekjes/12Microalgae_UK.pdf

10. Jankowska E, Sahu AK, Oleskowicz-Popiel P. Biogas from microalgae: Review on microalgae's cultivation, harvesting and pretreatment for anaerobic digestion. Renew Sust Energ Rev. 2016.75:692-709. doi. org/10.1016/j.rser.2016.11.045

11. Karimi M. Exergy-based optimization of direct conversion of microalgae biomass to biodiesel. J Clean Prod. 2017;141: 50-55. doi. org/10.1016/j.jclepro.2016.09.032

12. Veillette M, Giroir-Fendler A, Faucheux N, Heitz M. Esterification of free fatty acids with methanol to biodiesel using heterogeneous catalysts: From model acid oil to microalgae lipids. Chem Eng J. 2017; 308:101-109. doi.org/10.1016/j.cej.2016.07.061.

13. EL-Sheekh MM, Bedaiwy MY, Osman ME. Ismail MM. Mixotrophic and heterotrophic growth of some microalgae using extract of fungal-treated wheat bran. Int J Recycl Org Waste Agricult. 2012;1:12. doi:10.1186/2251-7715-1-12

14. Choi J-M, Han S-K, Kim J-T, Lee C-Y. Optimization of combined (acid + thermal) pretreatment for enhanced dark fermentative H2 production from *Chlorella vulgaris* using response surface methodology. Int Biodeter Biodegr. 2016;108: 191-197. doi.org/10.1016/j.ibiod.2015.06.013

15. Stolz, Patrick, Obermayer, Barbara. Manufacturing microalgae for skin care. Cosmet Toiletries. 2005;120(3):99.

16. Priyadarshani I, Rath B. Commercial and industrial applications of micro alae - A review. J. Algal Biomass Utln. 2012;3(4):89-100.

17. Becker W. Microalgae in human and animal nutrition. Richmond A. editor, Handbook of microalgal culture. Oxford: Blackwell Science;2004.

18. Palacios OA, Choix FJ, Bashan Y, De-Bashan LE. Influence of tryptophan and indole-3-acetic acid on starch accumulation in the synthetic mutualistic *Chlorella sorokiniana - Azospirillum brasilense* system under heterotrophic conditions. Res Microbiol. 2016;167(5): 367-379. doi. org/10.1016/j.resmic.2016.02.005

19. Spalaore P, Cassan-Joannis C, Duran E, Isambert A. Commercial Applications of Microalgae. J Biosci Bioeng. 2006;101(2):87-96. doi. org/10.1263/jbb.101.87

20. Iwamoto H. Industrial production of microalgal cell-mass and second-ary products - major industrial species - *Chlorella*. Richmond A. editor, Handbook of microalgal culture. Oxford: Blackwell Science;2004.

21. Ebrahimi-Mameghani M, Sadeghi Z, Farhangi MA, Vaghef-Mehrabany E, Aliashrafi, S. Glucose homeostasis, insulin resistance and in flammatory biomarkers in patients with non-alcoholic fatty liver disease: Beneficial effects of supplementation with microalgae *Chlorella vulgaris*: A double-blind placebo-controlled randomized clinical trial. Clin Nutr. 2016. doi.org/10.1016/j.clnu.2016.07.004

22. Cherng JY, Liu CC, Shen CR, Lin HH, Shih MF. Beneficial effects of *Chlorella*-ll peptide on blocking LPS-induced macrophage activation and alleviating thermal injury-induced inflammation in rats. I J Immunopath Ph. 2010;23:811-820. doi.org/ 10.1177/039463201002300316

23. MacDonald ML, Rogers QR, Morris JG. Nutrition of the domestic cat, a mammalian carnivore. Annu Rev Nutr. 1984;4: 521-562. doi:10.1146/annurev.nu.04.070184.002513

24. Rivers JP, Sinclair AJ, Craqford MA. Inability of the cat to desaturate essential fatty acids. Nature. 1975;258:171-173. doi:10.1038/258171a0

25. Trappe TA, Liu SZ. Effects of prostaglandins and COX-inhibiting drugs on skeletal muscle adaptations to exercise. J Appl Physiol. 2013;115(6):909-19. doi:10.1152/japplphysiol.00061.2013

26. Crawford MA and Sinclair, AJ. Nutritional influences in the evolution of mammalian brain. In: Lipids, malnutrition & the developing brain. Ciba Found Symp.1971:267-292

27. Rapoport SI. Arachidonic acid and the brain. J Nutr. 2008;138(12):2515-20.

28. Ormes Jacob. Effects of arachidonic acid supplementation on skeletal muscle mass, strength, and power. NSCA ePoster Gallery. National Strength and Conditioning Association. 2014. Available at: https://forum.bodybuilding.nl/topics/effects-of-arachidonic-acid-supplementation-on-skeletal-muscle-mass-strength-and-power.370627/

29. Harris WS, Mozaffarian D, Rimm E, Kris-Etherton P, Rudel LL, Appel LJ, et al. Omega-6 fatty acids and risk for cardiovascular disease: a science advisory from the American Heart Association Nutrition Subcommittee of the Council on Nutrition, Physical Activity, and Metabolism; Council on Cardiovascular Nursing; and Council on Epidemiology and Prevention. Circulation. 2009;119(6):902–907. doi.org/10.1161/CIRCULATIONAHA.108.191627

30. Muller-Feuga A. The role of microalgae in aquaculture: situation and trends. J Applied Phycol. 2000;12(3-5):527-534. doi. org/10.1023/A:1008106304417

31. Yu X, Zhao P, He C, Li J, Tang X, Zhou J, et al. Isolation of a novel strain of Monoraphidium sp. and characterization of its potential application as biodiesel feedstock. Bioresource Technol. 2013;121:256-262. doi. org/10.1016/j.biortech.2012.07.002

32. Hu G, Fan Y, Zhang L, Yuan C, Wang J, Li W, et al. Enhanced lipid productivity and photosynthesis efficiency in a *Desmodesmus sp.* mutant indued by heavy carbon ions. 2013;8:e60700. doi.org/10.1371/journal.pone.0060700

33. Chisti Y. Biodiesel from microalgae. Biotechnol Adv. 2007;25:294-306. doi.org/ 10.1016/j.biotechadv.2007.02.001

34. Um BH and Kim YS. Review: A chance for Korea to advance algal-biodiesel technology. J Ind Eng Chem. 2009;15:1-7. doi.org/10.1016/j.jiec.2008.08.002

35. Sydney EB, Sturm W, Carvalho JC, Thomaz-Soccol V, Larroche C, Pandey A, Soccol CR. Potential carbon dioxide fixation by industrially important microalgae. Bioresource Technol. 2010;101: 5892-5896. doi.org/10.1016/j.biortech.2010.02.088

36. Begum H, Yusoff FM, Banerjee S, Khatoon H, Shariff M. Availability and utilization of pigments from microalgae. Crit Rev Food Sci Nutr. 2016;56(13):2209-2222. doi.org/ 10.1080/10408398.2013.764841

37. Gong M, Bassi A. Carotenoids from microalgae: A review of recent developments. Biotechnol Adv. 2016;34(8):1396-1412. doi.org/10.1016/j.biotechadv.2016.10.005

38. Bewicke D, Potter B. Chlorella: The emerald food. Ronin Publishing: Berckley- CA,2009.

39. Viuda-Martos M, Sanchez-Zapata E, Sayas-Barberá E, Sendra E, Pérez-Álvarez JA, Fernández-López J. Tomato and tomato byproducts. Human health benefits of lycopene and its application to meat products: a review. Crit Rev Food Sci Nutr. 2014; 54(8): 1032–1049. doi.org/10.1080/10408398.2011.623799

40. Virtamo J, Taylor PR, Kontto J, Männistö S, Utriainen M, Weinstein SJ, Huttunen J, Albanes D. Effects of α-tocopherol and β-carotene supplementation on cancer incidence and mortality: 18-year postintervention follow-up of the alpha-tocopherol, beta-carotene cancer prevention study. Int J Cancer. 2014;135(1):178–185. doi.org/10.1002/ijc.28641

41. Manayi A, Abdollahi M, Raman T, Nabavi SF, Habtemariam S, Daglia M, et al. Lutein and cataract: from bench to bedside. Crit Rev Biotechnol. 2016;36(5):829. doi.org/10.3109/07388551.2015.1049510

42. Vijayapadma V, Ramyaa P, Pavithra D, Krishnasamy R. Protective effect of lutein against benzo(a)pyrene-induced oxidative stress in human erythrocytes. Toxicol Ind Health. 2014;30(3):284-293. doi.org/10.1177/0748233712457439

43. Zhang W, Wang J, Wang J, Liu T. Attached cultivation of *Haematococcus pluvialis* for astaxanthin production. Bioresour Technol. 2014; 158: 329-335. doi.org/10.1016/j.biortech.2014.02.044

44. Fasano E, Serini S, Mondella N, Trombino S, Celleno L, Lanza P, et al. Antioxidant and anti-inflammatory effects of selected two human immortalized keratinocyte lines. Biomed Res Int. 2014:2014; 1–11. doi.org/10.1155/2014/327452

45. Li J, Zhu D, Niu J, Shen S, Wang G. An economic assessment of astaxanthin production by large scale cultivation of *Haematococcus pluvialis*. Biotechnol Adv. 2011;29(6):568-574. doi.org/10.1016/j.biotechadv.2011.04.001

46. Park JS, Chyun JH, Kim YK, Line LL, Chew BP. Astaxanthin decreased oxidative stress and inflammation and enhanced immune response in humans. Nutr Metab. 2010;7:18. doi.org/10.1186/1743-7075-7-18

47. Stenblom EL, Montelius C, Östbring K, Håkansson M, Nilsson S, Rehfeld JF, Erlanson-Albertsson C. Supplementation by thylakoids to a high carbohydrate meal decreases feelings of hunger, elevates CCK levels and prevents postprandial hypoglycaemia in overweight women. Appetite. 2013;68:118-123. doi.org/10.1016/j.appet.2013.04.022

48. Jubert C, Mata J, Bench G, Dashwood R, Pereira C, Tracewell W, et al. Effects of chlorophyll and chlorophyllin on low-dose aflatoxin B(1) pharmacokinetics in human volunteers. Cancer Prev Res (Phila). 2009;2:1015-22. doi.org/10.1158/1940-6207.CAPR-09-0099

49. Shaughnessy DT, Gangarosa LM, Schliebe B, Umbach DM, Xu Z, MacIntosh B, et al. Inhibition of fried meat-induced colorectal DNA damage and altered systemic genotoxicity in humans by crucifera, chlorophyllin, and yogurt. PLoS One. 2011;6(4):e18707. doi.org/10.1371/journal.pone.0018707

50. Zhang YL, Guan L, Zhou PH, Mao LJ, Zhao ZM, Li SQ, et al. The protective effect of chlorophyllin against oxidative damage and its mechanism. Zhonghua Nei Ke Za Zhi. 2012;51(6):466-470.

51. Maekawa LE, Lamping R, Marcacci S, Maekawa MY, Nassri MRG, Koga-Ito CY. Antimicrobial activity of chlorophyll-based solution on *Candida albicans* and *Enterococcus faecalis*. Revista Sul-brasileira de Odontologia. 2007;4:36-40.

52. Miret S, Tascioglu S, van der Burg M, Frenken L, Klaffke W. In vitro bioavailability of iron from the heme analogue sodium iron chlorophyllin. J Agric Food Chem. 2010;58(2):1327-1332. doi.org/10.1021/jf903177q

53. Murwanashyaka T, Shen L, Ndayambaje JD, Wang Y, He N, Lu Y. Kinetic and transcriptional exploration of *Chlorella sorokiniana* in heterotrophic cultivation for nutrients removal from wastewaters. Algal Res. 2016. doi.org/10.1016/j.algal.2016.08.002

54. Hammouda O, Gaber A, Abdel-Raouf N. Microalgae and wastewater treatment. Ecotox Environ Safe, 1995;31:205-210. doi.org/10.1006/eesa.1995.1064

55. Kuo C-M, Jian J-F, Lin T-H, Chang Y-B, Wan X-H, Lai J-T, et al. Simultaneous microalgal biomass production and CO_2 fixation by cultivating *Chlorella* sp. GD with aquaculture wastewater and boiler flue gas. Bioresource Technol. 2016;221:241-250. doi.org/10.1016/j.biortech.2016.09.014

56. Aksu Z, Sag Y, Kutsal T. A comparative study of the adsorption of chromium(V1) ions to *C. vulgaris* and *Z. ramigera*. Environ Technol Lett. 1990;11:3340. doi.org/10.1080/09593339009384836

57. Shah M. Microbial Community Structure of Activated Sludge As Investigated With DGGE. J Adv Res Biotech. 2016;1(1): 7

8

Socio-Commercial Agri-Biotech Model for Rural Development in India by Combining Livestock and Organic Farming Practices

Abhishek Cukkemane*

Bijasu Agri Research Laboratory LLP, Kondhwa, Pune-411048, Maharashtra, India

*Corresponding author: *Abhishek Cukkemane, Bijasu Agri Research Laboratory LLP, Kondhwa, Pune-411048, Maharashtra, India,*
E-mail: director@bijasu.org

Abstract

India experienced major success in agriculture productivity, post-green revolution, due to use of fertilizers. Around the same time, operation flood made India from a milk-deficient to highest producer by introducing exotic breeds. These approaches resulted in food and milk security but at a cost of decrease in soil health and loss in animal farming especially indigenous breeds of cattle. Since few years, educated medium land holding farmers have adopted organic farming practises. But, the supply of high quality organic manures remains a bottle neck in many sectors. It is therefore necessary to highlight the importance of combining organic- and cattle farming to the rural folk by introducing socio-commercial models based on modern biotechnological processes that can convert animal wastes into high quality manures on a mass-scale that can be efficiently marketed in the rural areas. Such practise will make small and medium land holding farmers adopt cattle farms, which will in turn be an epicenter for rural development by constant supply of manures to the farmers at a reasonable price. Therefore, we briefly highlight the impact of farming revolutions and the problems faced by farmers in India. Our analysis of the total cattle waste in India and the area of arable land, we highlight the possibility of organic farming across the nation. By utilizing readily available raw materialist is possible to accommodate socio-commercial model that can result in revenue generation by combining organic- and animal- farming practises for agronomic and rural development.

Keywords : Socio-corporate; Agronomic; Organic farming; Cattle rearing; Rural development; Environmental friendly

Introduction

Soil is one of the most important hosts which influence various biogeochemical and microbial processes. Traditionally, maintenance of soil health in Indian farming was based on animal wastes from farm especially cattle to replenish organic content. Reformation of Indian agriculture under green revolution in the late 1960s [1] and white revolution that began in 1970 [2,3] has raised food and dairy production, which aided millions in India and made it self-sufficient and food secure. But usage of chemical fertilizers and pesticides has led to decrease in organic content of soil taking a heavy toll on the environment and other useful flora-fauna [4]. This resulted in poor soil health, less crop yield and overall economic loss to farmers. Therefore, it is pertinent to develop faster and efficient Organic Manures (OM) by improving traditional approaches of organic farming by incorporating latest biotechnological that can compete and curtail the use of their synthetic counterparts.

For example, organic fertilizer like Kunapanjala [5], has been successfully used in the state of Sikkim, India, which was officially declared an "organic state" previous year. Yet in many other parts of India, unprocessed animal waste such as cow dung is directly applied to the field which maintains soil health but does not provide yield comparable to synthetic fertilizers. Therefore it is important to produce OM, for modern farmers that have a short processing time, efficient and high yield providing qualities such as Panchagavya and Jeevamrut [6-8] that has met with decent success in West and South Indian states, which are prepared by fermenting cow dung and urine. Recently, we performed metabolomic analysis and have identified the various biochemical's that make these OM efficient for agriculture purposes [6]. Although very efficient, these OM are used by a small percentage of the farming communities' i.e. 650,000 hectares [9] in a total arable land of 160 million hectares [10]. This is because of several short- comings such as: preparation time (10-30 days); manual mixing (laborious and expensive); undesirable results and off-odors (if not mixed properly) causing farmers to abandon organic farming practises and reverting to synthetic fertilizers that are readily available and easy -to-use.

Hence, the need of the hour is to develop technology and strategy for preparation of OM which will overcome all the above mentioned short-coming sand increase farmer's revenue. To incorporate all these benefits it is necessary to develop model wherein, cattle farms have an electrically driven fermenter, installed to produces the above mentioned OM in 7-10 days. Such an agronomic model should incorporate various social and cultural values for successful rural development. In this review, we consider various factors in rural India and the prospect of socio-commercial system that can benefit the rural farming society at large.

Cattle Farming and Biodiversity- More the Merrier

Cattle rearing hold a very significant religious and cultural value in Indian farming society. It is revered as an esteemed practise but two major agricultural revolutions have negatively impacted it.

Impact of green revolution: Before the green revolution and introduction of chemical fertilizers, animal farming was an important mainstay of agricultural activities, which maintained the organic content of soil. This is chiefly expressed as the C:N ratio, such as soil condition, the nature of applied organic manure and the crops that were cultivated. Farmers resorted to adding animal dung and mixed/inter cropping to maintain a steady balance of C:N ratio of 20:1 as suggested by the Indian council of agricultural research [11]. But now, with the extensive use and ease in the availability of synthetic fertilizers, farmers abandoned their sizeable live stocks, which indirectly resulted in gradual decline of organic content of soil over many decades.

Impact of operation flood: Operation flood helped to improve the scenario of dairy farming in India but caused gradual decrease in the rearing of indigenous cattle. Interestingly, the Indian breeds are resistant to drought; consume less water and feed, which result in lower yield of milk. Typically, an average Indian cow provides milk ranging from 5 to 15 L/day, though some well-maintained breeds can provide up to 20 L. But, dwarfs in comparison with the exotic varieties like Jersey and Holstein-Friesian, which produces 20 L/ day on average. To increase milk capacity, the administration encouraged farmers to rear the exotic species of cows which caused major loss in the biodiversity and population of indigenous cattle breeds [3]. As of now, it is estimated that India has 190 million cows out of which only 33 million account for the 37 indigenous breeds [12,13]. In the last 16 years, India has experienced 4 periods of drought. These episodes caused farmers to rear indigenous breeds of cows because of their drought tolerance quality and moreover the success story of Brazilian farms rearing high milk-yielding Indian cattle varieties encouraged them to follow their footsteps.

Is It Possible to Cultivate The Entire Nation Solely by Organic Farming?

India has a total of 328.7 million hectares of land and is the second largest country after the United States of America in total arable land. Estimates suggest that 160 million is arable, but only 95 million is utilized [10]. It highlights a major agricultural concern of under-utilization of nearly 33.33% of the arable land. This raises a very important question, i.e., is it possible to apply the organic farming technology to the entire arable land in India?

Let us perform a simple calculation considering that a cattle produces 10Kg of cow dung, therefore 37 million cattle will produce 370 million Kg of dung daily. Assuming soil of 1% organic carbon (OC), this is usually enriched in the top soil 0-10 cm (0-0.1 m; also called humus layer) with a bulk density of 1.4 g/cm^3. One will require adding 10,000 x 0.1 x 1.4 = 1.44 tonnes Kg of organic carbon per hectare (10,000m^2). Generally, cattle waste contains ~ 15% OC [14]. Therefore, one needs to add 93.24Kg of cow dung

to get a soil of 1% OC content; for 160 million hectares, a total 14,918 million (~ 15 billion) Kg of dung will be required. By only using cow dung from indigenous varieties, soil application is possible every 40 days assuming 100% utility of the organic matter. But if we were consider the entire cattle population of 190 million then the application rate would be weekly. Cow cattle are the third largest milch animal across India, which would imply that the total animal waste generated should suffice for regular application on all arable plots.

But what are The Problems Faced by the Farmers in Using Agricultural Wastes?

Animal dung represents complex undigested material that is broken down into simple biochemical and metabolites gradually over a period of 3-6 months, whereas chemical fertilizers are ready to use. Furthermore, it may contain harmful pathogens that may prove to be life threatening [15,16].

Apart from this, several policy making bodies have advocated usage of chemicals for over 5 decades due to which the current generation of farmers have lost traditional farming knowledge and practise are not willing to take up organic farming because it is labor intensive and cumbersome. Lastly, poor education and awareness of farmers have played a major share. For instance, China adopted and practised biogas technology [17], whereas the rest of the world moved towards green revolution. With this, China not only benefited from harvested of gas-energy but also used OC rich slurry, which is processed manure devoid of pathogens, for farming purposes along with judicial use of chemical fertilizers.

Socio-Economic Model

The socio-economic situation amongst Indian farmers is rather heterogeneous. More than 60% are small land holders (1-3 acres of land), 19% fall into the medium land holders (4-9 acres) and 7% are large farmers (< 10 acres). Landholding size and governmental initiatives are important factor determining motivation and agriculture as a good [18,19] source of income. In many cases, middle and large scale farmers have other sources of revenue [19,20] apart from farming. The larger farmers prefer any farming technology and inputs that increase their crop yield, but it is the middle scale educated farmers and to some extent small scale farmers who practise organic farming. Small land holding farmers on the other hand are heavily dependent [19,20] on the outcome of their yield and hence are not willing to take up organic farming practises. Therefore, it is important to develop socio-commercial models that will encourage them to adopt organic farming practises, reduce their spending and improve their livelihood.

A very important criterion in rural development is self-sustenance, which is reflected by the agricultural output and index above poverty line. The cost of chemical farming is directly related to petroleum costs, which has increased considerably since 1960s. Recently, with fluctuating oil prices, low outputs due to poor soil health and insufficient animal farming; we need to develop cost-effective alternatives and newer commercial models. It is therefore essential to integrate livestock management and cropping together for sustainable and profitable farming, especially for poor farmers in rain-fed regions [21, 22].

One such socio-commercial model is suggested for mass-production of OM described above [6-8] in a cooperative manner among farmers taking into account readily available raw materials (Table 1). The OM that requires fermentation periods of 14-30 days by manually mixing of the contents can be converted by installing electrically driven fermenter. This reduces manual labour and preparation time to 4-7 days, resulting in increased production of up to 4 and 2.5-5 folds in time and volume, respectively.

In order to highlight, the profitability of such a model system, we first calculated the costs associated with requirements (Table 1) and the revenue generation scheme (Table 2) based on current pricing. We assumed a small cattle farm of 30 animals that produces 10Kg of dung and 2L of urine per day by an animal. In such a small setup it is possible to generate up to 270,000INR (~ 4,000USD) per month. Assuming high spending of 2,000USD for maintenance, one is left with remaining 2,000USD. In India, cattle farms are highly revered due to religious sentiments amongst majority of Indians. Such a socio-commercial model (Figure 1) will lead to revenue generation and permit rural development programmes that include farmers meet, education and women empowerment.

On similar lines, one such project is Venu-Madhuri (VM) in Kolhapur district, India, where 5 different villages pool their resources. In this socio-commercial model, farming and women empowerment is encouraged via cattle farming. Cattle dung- and urine is procured and managed to produce a variety of commercial products such as soaps, incense sticks, ayurvedic medicine etc., along with energy generation and organic manure preparation. Unlike the idea mentioned above for mass-production and revenue generation, VM focuses on "Production by masses rather than mass-production".

Perspective- Good Practise Leads To Good Results

We have emphasized on the importance of cattle and organic farming as a profitable alternative to curtail use of chemical fertilizers. But our environment is already contaminated with various hazardous chemicals, out of which usage of pesticide, Dichlorodiphenyltrichloroethane (DDT) and their effects has been well documented [23]. In one of our studies, we noticed contamination of OM with chemical pesticides, most likely via cattle fodder. Farms that have been rigorously using organic farming practises since the last 14 years also had low level of chemical pesticides [6], this highlights the potential problems of obtaining residue free farming practise altogether. Regardless, in order to have a clean green initiative, one will have to start from base line, i.e. organic cultivation and feeding of fodder crops in chemical free environment, and thus obtaining organic-chemical free cow dung and urine (Figure 2). Such practises, where livestock management and agriculture can result in improving economic situation of poor farmers [21,22], should be encouraged. Moreover, when farming communities come together in the form of cooperative societies to maintain livestock and produce OM in

bulk; it can result in rural development and dramatically curtail use of synthetic fertilizers.

Table 1: Requirements for preparation and costing (in parentheses) of 500 L organic manures. INR is the abbreviation for Indian Rupees.

	Panchagavya (Pc)	**Jeevamrut (Ja)**
Cow Dung	125 Kg (750 INR)	75 Kg (450 INR)
Cow Urine	125 L (1875 INR)	75 L (1,125 INR)
Gram flour	--	12.5 Kg (1,500 INR)
Jaggery	--	12.5 Kg (750 INR)
Cow milk	60 L (2400 INR)	--
Cow curd	60 Kg (6000 INR)	--
Cow cream	1 Kg (340 INR)	--
Water	129 L	325 L
Production cost	= 11,365 INR/ 500L (19.45 INR/ L)	= 3,450 INR/ 500L (7.65 INR/ L)

Table 2: Revenue generation per month from raw materials available in Cattle farm with 30 cows

Quantity of Organic Manure in a single run	500 L
Quantity of Pc production in 1 month	= 1,500 L
Quantity of Ja production in 1 month	= 2,000 L
Selling price of 1 L (Pc/ Ja)	180/ 120 INR
Grant Total(Pc/ Ja)	**466,080 INR/ 313,400 INR**

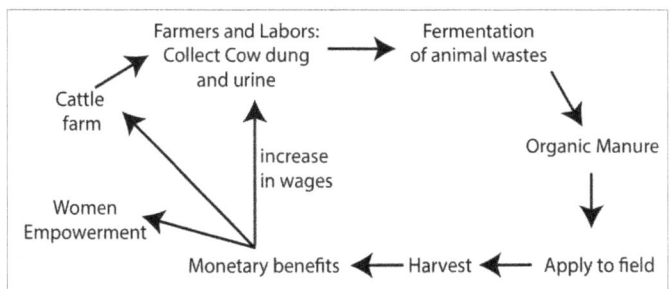

Figure 1: Socio-commercial model depicting the various channels for organic manure production via animal wastes and its scope in rural development.

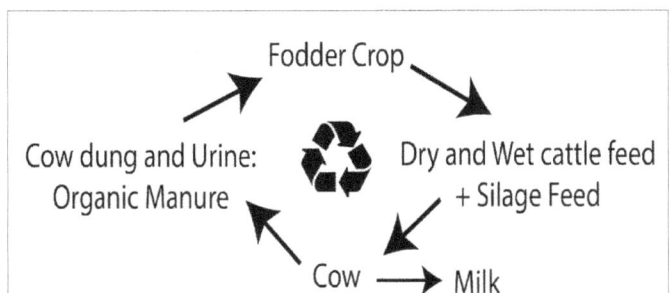

Figure 2: Organic green circle highlighting the process of farming in a closed circuit, thereby eliminating/reducing the scope of contamination by chemical fertilizers and pesticides.

Acknowledgement

I would like to thank Vijay Chemicals and Neeti Developers for support and suggestions; Rahul Deshpande from VenuMadhuri; and Dr. Nivedita Cukkemane for critically reading the manuscript.

References

1. Pingali PL. Green revolution: impacts, limits, and the path ahead. Proc Natl Acad Sci U S A. 2012;109(31): 12302-12308. doi: 10.1073/pnas.0912953109.

2. Bellur VV, Singh SP, Radharao C, Rajeswararao C. The white revolution - How Amul brought milk to India. Long Range Planning. 1990;23(6): 71-79. doi:10.1016/0024-6301(90)90104-C.

3. Gautam, Dalal RS, Pathak V. Indian dairy sector: Time to revisit operation flood. Livestock Science. 2010;127(2-3):164-175. doi.org/10.1016/j.livsci.2009.09.010.

4. Wang, ZH, Li S X, Malhi S. Effects of fertilization and other agronomic measures on nutritional quality of crops. Journal of the Science of Food and Agriculture. 2008;88:7-23.

5. Nene YL. Potential of Some Methods Described in Vrikshayurvedas in Crop Yield Increase and Disease Management. Asian Agri-History. 2012;16(1):45-54.

6. Ukale DU, Rohit VB, Santosh KU, Nivedita C, Abhishek AC. Metabolic analysis of liquid formulations of organic manures and its influence on growth and yield of *Solanum lycopersicum L.* (tomato) crop in field. Biocatalysis and Agricultural Biotechnology. 2016;8:50-54.

7. Ali MN, Ghatak S, Ragul T.Biochemical analysis of Panchagavya and Sanjibani and their effect in crop yield and soil health. Journal of Crop and Weed. 2011;7:84-86.

8. Gore N, Sreenivasan MN. Influence of liquid organic manures on growth, nutrient content and yield of tomato (Lycopersicon esculentum Mill.) in the sterilized soil. Karnataka Journal of Agriculture Sciences, 2011;24:153-157.

9. Helga W, Lernoud J. The world of organic agriculture - stastics and emerging trends 2016, ed. Research Institute of organic agriculture (FiBL), IFOAM Bonn. 2016.

10. World-bank [cited 2016] http://data.worldbank.org/indicator/AG.LND.AGRI.ZS. 2016.

11. ICAR, Handbook of Indian Agriculture. 6 ed. 2006;New Delhi: Indian Council of Agricultural Research.

12. National Dairy Development Board, [cited 2012]http://www.nddb.org/information/stats/pop.

13. Press Bureau India, Government of India. Ministry of Agriculture [cited 2014]http://pib.nic.in/newsite/PrintRelease.aspx?relid=109280 .

14. Roy S, Kashem AM. Effects of Organic Manures in Changes of Some Soil Properties at Different Incubation Periods. Open Journal of Soil Science. 2014;4(3):81-86. doi:10.4236/ojss.2014.43011.

15. King LA, Nogareda F, Weill FX, Mariani-Kurkdjian P, Loukiadis E, Gault G, et al. Outbreak of Shiga toxin-producing *Escherichia coli* O104:H4 associated with organic fenugreek sprouts, France, June 2011. Clin Infect Dis. 2012;54(11):1588-94. doi: 10.1093/cid/cis255.

16. Burger R. EHEC O104:H4 in Germany 2011: Large outbreak of bloody diarrhea and haemolytic uraemic syndrome by shiga-toxin producing *E. coli* via contaminated food. 2012: Institute of Medicine (US). Improving Food Safety through a One Health Approach: Workshop Summary. Washington (DC): National Academies Press (US).

17. Buren AV. A Chinese Biogas Manual, ed. A.v. Buren. Practical Action Publishing. 1979.

18. Raghav S, Sen C. Socio-economic status of farmers and their perception about technology: a case study. EPRA International Journal of Economics and Business Review. 2014;2(3):7-13.

19. Singh TP, Kumbhar V, Kumari S. Study of socioeconomic status of farmers in drought prone regions of Maharashtra, India- a case study. International Journal of Current Research. 2016;8(6):33304-33306.

20. Ananthnag K, Ali MMK, Kumar VHM. A study on socio-economic status of farmers practicing on organic farming in eastern dry zone of Karnataka. Online Journal of Bioscience and Informatics. 2014;1(2):75-84.

21. McDermott JJ, Staal SJ, Freeman HA, Herrero M, Van de Steeg JA. Sustaining intensification of smallholder livestock systems in the tropics. Livestock Science. 2010;130(1-3):95-109.

22. Tarawali S, Herrero M, Katrien D, Elaine G, Michael B. Pathways for sustainable development of mixed crop livestock systems: Taking a livestock and pro-poor approach. Livestock Science. 2011;139(1-2):11-21.

23. Henk van den Berg. Global Status of DDT and Its Alternatives for Use in Vector Control to Prevent Disease. Environ Health Perspect. 2009;117(11):1656–1663. doi: 10.1289/ehp.0900785.

9

Polymorphisms of *Shank3* gene in Chinese Han Children with Autism Spectrum Disorders (ASD)

Mei Jun Wang[1,3], Fei Gui[1], Xuan Jing[1], Fu Sheng Huang[1], Weipeng Wang[2], Haiqing Xu[2], Hong Yan Chai[1], Chunzi Liang[1] and Jian Cheng Tu[1*]

[1]Department of Clinical Laboratory & Center for Gene Diagnosis, Zhongnan Hospital of Wuhan University, Wuhan, china

[2]Women and Children's hospital of Hubei province, Wuhan China

[3]Department of Clinical laboratory, Shiyan People's Hospital, Affiliated to Hubei Medical College, Shiyan China

*Corresponding author: Jian Cheng Tu, Department of Clinical Laboratory & Center for Gene Diagnosis, Zhongnan Hospital of Wuhan University, Wuhan, china. E-mail: jianchengtu@whu.edu.cn

Abstract

This study investigated the co-relationship between single nucleotide polymorphisms (SNPs) variations of *SHANK3* (SH3 and multiple ankyrin repeat domains 3) gene and childhood autism in Chinese Han populations. high-resolution method (HRM) were used to examine 11 polymorphisms of *SHANK3* in 100 patients with ASD compared to 100 health controls. Risk estimates were expressed as odds ratio (OR) and 95% confidence interval (CI). Overall there were significant differences in the allele distributions of rs9616816, rs6010061 and rs756638 (for rs9616816, OR=1.918, 95%CI=1.287-2.858, P=0.001; for rs6010061, OR=0.517, 95%CI=0.291-0.919, P=0.023; for rs756638, OR=0.506, 95%CI=0.296-0.866, P=0.012; respectively). Logistic regression analyses confirmed that these three SNPs were directly linked with significant risk of ASD. Our data indicate that *SHANK3* gene polymorphisms may play a vital role in ASD genetic susceptibility in Chinese Han populations.

Introduction

The autism spectrum disorders (ASD), first reported in the early 1940sby psychiatristsis a severe neuropsychiatric disorder syndrome [1,2]. ASD is characterized by impairments of sociality and communication, lacking of reciprocal social interaction or responsiveness and repetitive or stereotyped behaviors (citation), the Onset of ASD mainly occurs in early childhood [3]. Despite the etiology of autism are largely unknown, it is widely accepted that autism is a multifactorial disease with a complex interaction of genetic and environmental factors [1,5,6]. In the past decades, though promising progress has been reported, ASD is still an urgent challenge for the public health [7].

Research proved that cause of ASD may reside in abnormalities at the synapse [8,9]. The *SHANK3* gene, encodes a synaptic scaffolding protein [10]. In human beings, *SHANK3* is expressed preferentially in cerebral cortex and cerebellum [11,12]. With its multiple protein interaction domains, this molecule directly or indirectly connects with neurotransmitter receptors and cytoskeleton proteins [13,14]. It also participates in the formation, maturation and enlargement of dendritic spines and is essential for the formation of functional synapse [10,15,16]. More importantly, a number of studies have been performed on *SHANK3* polymorphisms with autism risk in different populations. However, the results were still controversial. Only one study of polymorphisms in *SHANK3* found significant difference between ASD cases and controls [17], while the others were negative [18,19,20].

The genetic variations of *SHANK3* of ASD in Chinese Han population has been reported by Qin and Shao, however, it's not sufficient to demonstrate the co-relation between *SHANK3* gene with other SNPs for the lack of diversity of the investigated SNPs. Therefore, further research about the co-relation between genetic variations of *SHANK3* with ASD in Chinese Han populations is necessary and important. So far 21586 SNPs (according to NCBI) of *SHANK3* gene has been reported. Accordingly, we selected 11 tag SNPs of *SHANK3* gene in 100 cases and 100 controls to ascertain the association between this SNP and ASD susceptibility in Chinese Han children to gain a better understanding of the way it exerts its effect on ASD.

Material and method

Participants

A total of 100 ASD patients (76 males and 24 females, 2-12 years) and 100 healthy controls (69 males and 31 females, 2-12 years) were enrolled from Zhongnan Hospital of Wuhan

University, Hubei Provincial Maternal and Child Health Hospital. All patients met DSM-IV diagnostic criteria for ASD. Written informed consent was obtained from each participant, and the study protocol was approved by the ethics committees of Zhongnan Hospital of Wuhan University.

Genotyping

Genomic DNA was extracted from the blood using a TIANamp Blood DNA Kit (TIANGEN, BeiJing, China). Eleven SNPs were selected from the HapMap HCB database with the criteria used in our SNP selection procedure [a minor allele frequency over 0.1 and tag SNPs with an r^2 value above 0.8] to examine the association between *SHANK3* and ASD (Table 1). SNPs were genotyped by high-resolution melting of small amplicons on LightScanner 96 instrument (Idaho Technology, USA). Primer details and product lengths are shown in table S1. About 5% of the samples were randomly selected using direct PCR sequencing (Life Technologies Corporation, Shanghai, China) and the concordance was 100%.

Table 1: Characteristics of SNPs in shank3 gene cluster

SNP	Position[1]	Minor allele	Major allele	MAF[2]
rs2301584	51171497	A	G	0.223
rs2341011	51139635	T	C	0.321
rs41281537	51171667	A	G	0.058
rs5770820	51150473	A	G	0.253
rs5770992	51146139	G	A	0.108
rs6010061	51151724	C	T	0.417
rs6010065	51158017	C	G	0.490
rs756638	51171693	A	G	0.299
rs8137951	51165664	A	G	0.374
rs9616816	51123505	A	G	0.362
rs9616915	51117580	C	T	0.358

[1] Position in basepairs was derived from dbSNP Build 137. Based on NCBI Human Genome Build 37.3 (November, 2014) of chromosome 22
[2] MAF, minor allele frequency

Table S1. Amplification primers utilized in the genotype

SNP		Primers (5'→3')	product length (bp)	Tm (℃)
rs9616816	Forward	GCTCTCAGCATGGAAAGA	57	54.2
	Reverse	TCCCATCACTGTTGTTTT		
rs6010061	Forward	GGAGTTTTCTCTCCATTCATATCTT	60	55
	Reverse	CTTAAGCACCATACTCC		
rs756638	Forward	TGTGTCTGTCCCTCATACC	102	54.5
	Reverse	CATGTGGTCCAGGCTGA		
rs6010065	Forward	TGGTACTTCTGCGTCGG	89	59
	Reverse	GCCAGTACAGGGCTCC		
rs2301584	Forward	GTTCCGCTTCACCTCCTT	69	57
	Reverse	GCCTCAGGACTGGAGCA		
rs41281537	Forward	GCTCAGTTGCCTGCTTG	86	58
	Reverse	CCGGTATGAGGGACAGA		
rs2341011	Forward	TCCGCTTCACCTCCTTT	67	56.8
	Reverse	GCCTCAGGACTGGAGCA		
rs5770992	Forward	TGGTCAGAATTTTCAC	50	45
	Reverse	TTATCTACATGGGGTT		
rs5770820	Forward	CTCTAGGGAGCAGGGAGAC	112	55
	Reverse	GACCAGCAGAAAGAAGCAA		
rs9616915	Forward	TCTCCACGACCACGC	52	63
	Reverse	CTCCTGCCAGCCATT		
rs8137951	Forward	ATGTCATACATACTATTTTTGCATT	55	53.6
	Reverse	TAGCACAAAGCCAGGAA		

Statistical methods

Hardy-Weinberg Equilibrium and allele frequency distributions were analyzed by the chi-square test (SPSS, version 18.0). Each genotype of examined polymorphisms was assessed by logistic regression analyses under the additive (major homozygotes versus heterozygotes versus minor homozygotes), dominant (major homozygotes versus heterozygotes plus minor homozygotes) and recessive (major homozygotes plus heterozygotes versus minor homozygotes) models of inheritance after adjusting for sex and age, respectively (SPSS, version 18.0). Linkage disequilibrium (LD) analysis of SNPs and the haplotype association were analyzed using Haploview 4.2 and SHEsis software. P values less than 0.05 were considered statistically significant.

Results

Genotype and Allele Frequencies

No significant deviation from HWE was observed for all tested SNPs in the control groups (P > 0.05). As shown in table 2, logistic regression analysis revealed that rs9616816 was associated with ASD in both additive model [OR = 1.791, 95% CI (1.217-2.635), p = 0.003] and recessive model [OR = 2.569, 95% CI (1.382-4.774), p = 0.003], rs6010061 and rs756638 were also associated with ASD in both additive model [OR = 0.550, 95% CI(0.320-0.946), p = 0.031; OR = 0.510, 95% CI(0.296-0.878), p = 0.015] and dominant model [OR = 0.472, 95% CI (0.242-0.922), p = 0.028; OR = 0.461, 95% CI (0.248- 0.858), p = 0.014].

Table 2: Risk estimate based on the distributions of genotype and allele frequency.

SNP	genotype	Control (n = 100)	Case (n =100)	Allele OR(95% CI), P value[1]	Additive OR(95% CI), P value[2]	Dominant OR(95% CI), P value[2]	Recessive OR(95% CI), P value[2]
Rs9616816	GG	31	19				
	GA	45	37				
	AA	24	44	1.918 (1.287-2.858), **0.001**	1.791 (1.217-2.635), 0.003	1.927(0.994-3.737), 0.052	2.569(1.382-4.774), 0.003
rs6010061	TT	69	82				
	CT	25	15				
	CC	6	3	0.517 (0.291-0.919),**0.023**	0.550 (0.320-0.946), 0.031	0.472 (0.242-0.922), 0.028	0.436 (0.104-1.818), 0.254
rs756638	GG	61	77				
	GA	34	21				
	AA	5	2	0.506 (0.296-0.866),**0.012**	0.510 (0.296-0.878), 0.015	0.461 (0.248-0.858), 0.014	0.386 (0.072-2.064), 0.266
rs6010065	GG	28	37				
	GC	43	39				
	CC	29	24	0.755 (0.509-1.119), 0.161	0.788 (0.543-1.142), 0.208	0.674 (0.368-1.233), 0.200	0.764 (0.403-1.452), 0.412
rs2301584	GG	70	71				
	GA	26	27				
	AA	4	2	0.896 (0.526-1.524), 0.684	0.920 (0.542-1.562), 0.759	0.992 (0.536-1.834), 0.979	0.468 (0.083-2.637), 0.389
rs41281537	GG	79	73				
	GA	18	26				
	AA	3	1	1.194 (0.666-2.141), 0.552	1.157 (0.647-2.068), 0.623	1.355 (0.703-2.612), 0.365	0.304 (0.031-3.002), 0.308
rs2341011	CC	49	40				
	CT	41	47				
	TT	10	13	1.310 (0.864-1.987), 0.204	1.309 (0.861-1.988), 0.208	1.457 (0.829-2.560), 0.191	1.325 (0.548-3.205), 0.532
rs5770992	AA	60	65				
	AG	32	28				

	GG	8	7	1.189 (0.769-1.838), 0.437	0.859 (0.552-1.338), 0.502	0.805 (0.453-1.433), 0.462	0.871 (0.302-2.512), 0.799
rs5770820	GG	23	33				
	GA	44	34				
	AA	33	33	0.818 (0.552-1.212), 0.317	0.829 (0.578-1.188), 0.307	0.568 (0.301-1.073), 0.082	0.991 (0.548-1.791), 0.975
rs9616915	TT	87	87				
	CT	12	11				
	CC	1	2	1.077 (0.506-2.295), 0.847	1.027 (0.501-2.104), 0.942	0.940 (0.407-2.169), 0.885	2.174 (0.187-25.238), 0.535
rs8137951	GG	55	58				
	GA	36	32				
	AA	9	10	0.950 (0.609-1.481), 0.821	0.955 (0.627-1.453), 0.828	0.886 (0.505-1.554), 0.672	1.116 (0.430-2.895), 0.822

The minor C allele of rs6010061 was associated with a lower risk of ASD [OR = 0.517, 95% CI (0.291-0.919), p = 0.023] and the minor A allele of rs756638 was associated with a lower risk of ASD [OR = 0.506, 95% CI (0.296-0.866), p = 0.012], while carriers of the rs9616816 A allele were associated with a higher risk of ASD [OR = 1.918, 95% CI (1.287-2.858), p = 0.001]. Other SNPs, like rs6010065, rs2301584, rs41281537, rs2341011, rs5770992, rs5770820, rs9616915, and rs8137951, did not show any association with ASD.

Linkage disequilibrium and haplotype analysis

The Haploview 4.2 software was used for linkage disequilibrium analysis, and none of SNPs of this section are closely linked (D' and r2 > 0.85). The LD block of 11 SNPs were constructed (Figure 1). Table 3 presents the haplotype frequencies (≥ 3%) of three positive polymorphisms in patients and controls. The most common haplotype A-T-G, G-C-G, G-T-A was assigned as the reference group in risk estimates. But the haplotype ATG may increase the risk of autism (OR: 2.702, 95%:1.769~4.126, p < 0.01). Details are listed in table S1.

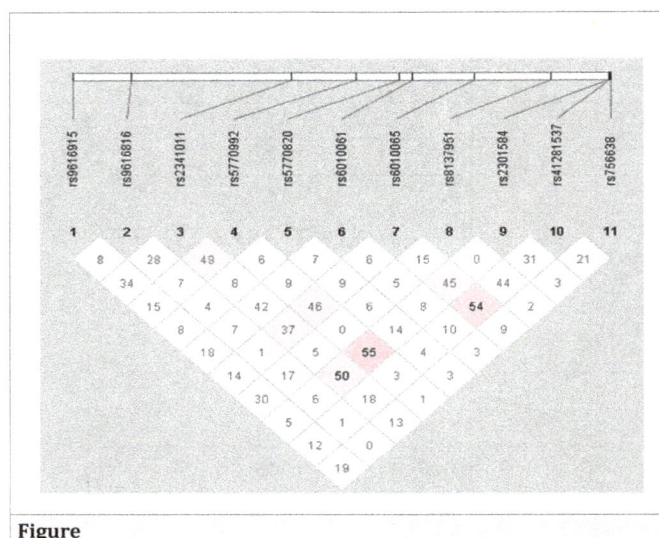

Figure

Table 3: Haplotype analysis in the control and the autism group

Haplotypes	Cases (Freq.)	Controls (Freq.)	$\chi2$	P	Odds ratio (95%CI)
A C A	2.90(0.015)	2.70(0.013))	/	/	/
A C G*	12.17(0.061)	16.97(0.085)	0.875	0.35	0.695 [0.324~1.494]
A T A*	12.52(0.063)	20.88(0.104)	2.326	0.13	0.570 [0.274~1.183]
A T G*	97.41(0.487)	52.45(0.262)	21.676	3.3E-06	2.702 [1.769~4.126]
G C A	2.47(0.012)	3.43(0.017)	/	/	/
G C G*	3.45(0.017)	13.90(0.070)	6.626	0.01	0.234 [0.071~0.773]
G T A*	7.11(0.036)	17.00(0.085	4.365	0.04	0.395 [0.161~0.970]
G T G*	61.96(0.310)	72.67(0.363)	1.367	0.24	0.779 [0.512~1.184]

Discussions

In this case-control study, we investigated the relationship between the 11 tag SNPs in *SHANK3* gene and the risk of ASD in the Chinese population. Among the 11 SNPs, three SNPs (rs9616816, rs756638, rs6010061) were found to be significantly associated with the risk of ASD in Chinese Han populations.

Since the first report of *SHANK3* mutations in ASD was published by Moessner et al. in 2007, several studies have been investigated the relationship between *SHANK3* gene polymorphisms and ASD in different populations. Qin found none of the five SNPs was significant evidence (P < 0.05) for preferential transmission of an allele by FBAT in all samples [18]; Sykes's data suggested that *SHANK3* deletions may be limited to lower functioning individuals with autism[19]; Chien's research revealed that the 5 tag SNPs (rs2341011, rs5770992, rs5770820, rs6010065, and rs2301584) were not significant statistically [19]. However, there were few positive results of association between *SHANK3* polymorphisms and ASD. Shao's study of rs9616915 polymorphisms in *SHANK3* found significant difference between ASD cases and controls in Chinese Han population [17], while the others reported that *SHANK3* might not represent a major susceptibility gene for ASD. The inconsistency with these pioneer works is possibly due to the difference of the sample size, individual genetic background, research design and environmental factors. However, the results were not fully consistent with previous reports. In the present study, we found the rs9616816, rs756638 and rs6010061 polymorphisms in the *SHANK3* gene has a statistically significant association with ASD susceptibility and may affect the subject susceptibility toward autism in the Chinese Han population. We established genotyping methods of 11 SNPs in the *SHANK3* gene cluster by high-resolution melting and successfully found both the rs9616816 and rs6010061 were associated with ASD risk. The protective role of rs9616816 A allele against the risk of ASD suggested *SHANK3* a possible candidate gene involved in the pathogenesis of ASD. Furthermore, Analysed with three models for genotype distributions, the association between *SHANK3* SNPs (rs9616816, rs756638, rs6010061) and ASD remained significant after performing statistical adjustments for age and sex. This result supported previous reports that *SHANK3* gene is a susceptible predictor of ASD risk factors [21,22,23,24,25].

Till now, little information is known about the role of *SHANK3* gene in the diverse pathological processes to ASD children. *SHANK3* gene, encodes a protein of the postsynaptic density of excitatory synapses, had been shown to bind to neuroligin, which, form a complex at glutamatergic synapses. In humans, *SHANK3* was found expressed predominantly in cerebral cortex and cerebellum [4,12, 26]. Durand and coworkers then identified two alterations in *SHANK3* in subjects with an ASD, one is a de novo insertion of a G nucleotide in exon 21 of SHANK3, which leads to a frame-shift and presumed loss of function; the other was found in an unrelated family with a de novo deletion of terminal 22q13, with the breakpoint in intron 8 of *SHANK3* [27]. Genetic and functional data implicate *SHANK3* as a potential genic cause of ASD, which lead us to seek to further assess the involved polymorphisms and associated phenotypic outcomes. Recent studies indicate that autism is a disease of polygenic inheritance. Analysis of polymorphisms in the *SHANK3* gene allows to effectively screen for autism risk [17]. Most previously reported studies narrowed on the mutations region. In our study, we covered the whole region of the *SHANK3* gene, 11 tag SNPs, spreading in coding regions, 5'- and 3'- UTR regions, were selected and studied in our cohort. Positive SNPs, found only in intron (rs9616816 and rs6010061) and 3' untranslated region (3' UTR) (rs756638) of *SHANK3* gene, indicated potential mechanism on the affection of the expression of *SHANK3* by binding with transcription factors or micro RNA, could be altering the interaction of Shank with miRNAs.

The correlation between genotype and phenotype is very complicated, both genetic and environmental infaectors have significant effect on ASD. The further investigation is that positive SNPs how to regulate the expression of *Shank3* gene by micRNA to reveal the role *shank3* gene on the pathogenesis of ASD.

Conclusions

Our study supports that *SHANK3* be a critical gene for the etiology of autism in Han Chinese population and three SNPs of *SHANK3* gene potentially function as risky factor for ASD upon further validation and functional studies.

Acknowledgements

This work was supported by the National Natural Science Foundation of China and National Institutes of Health of United States of America (grant no.812111103), the Foundation of Hubei Provincial Population and Family Planning Commission (grant no.JS-2013002) and the National Natural Science Foundation of China (grant no.81472033). We express our deep appreciation to participants for their contribution to this study and all members of Center for Gene Diagnosis who collected and processed data for this project.

References

1. Abrahams BS, Geschwind DH. Advances in autism genetics:on the threshold of a new neurobiology. Nat Rev Genet. 2008;9(5):341-55. doi: 10.1038/nrg2346

2. H A. Die. "Autistischen Psychopathen" im Kindesalter. European Archives of Psychiatry and Clinical Neuroscience. 1944; 117: 76-136.

3. Lamb JA, Moore J, Bailey A, Monaco AP. Autism: recent molecular genetic advances. Hum Mol Genet. 2000;9(6):861-868.

4. Baron MK, Boeckers TM, Vaida B, Faham S, Gingery M, Sawaya MR. An architectural framework that may lie at the core of the postsynaptic density. Science. 2006;311(5760):531-535. doi: 10.1126/science.1118995

5. Veenstra-VanderWeele J, Cook EH Jr. Molecular genetics of autism spectrum disorder. Mol Psychiatry. 2004;9(9): 819-832. doi: 10.1038/sj.mp.4001505

6. Roberts JL, Hovanes K, Dasouki M, Manzardo AM, Butler MG. Chromosomal microarray analysis of consecutive individuals with autism spectrum disorders or learning disability presenting for genetic services. Gene. 2014; 535(1):70-8. doi: 10.1016/j.gene.2013.10.020

7. Newschaffer CJ, Curran LK. Autism: an emerging public health problem. Public Health Rep. 2003; 118(5):393-399. doi:10.1093/phr/118.5.393

8. Garber K. Neuroscience. Autism's cause may reside in abnormalities at the synapse. Science. 2007;317(5835):190-191. doi: 10.1126/science.317.5835.190

9. O'Conno rEC, Bariselli S, Bellone C. Synaptic basis of social dysfunction: a focus on postsynaptic proteins linking group-I mGluRs with AMPARs and NMDARs. Eur J Neurosci. 2014;39(7), 1114-1129. doi: 10.1111/ejn.12510

10. Tu JC, Xiao B, Naisbitt S, Yuan JP, Petralia RS, Brakeman P. Coupling of mGluR/Homer and PSD-95 complexes by the Shank family of postsynaptic density proteins. Neuron. 1999;23(3):583-592.

11. Lim S, Naisbitt S, Yoon J, Hwang JI, Suh PG, Sheng M, et al. Characterization of the Shank family of synaptic proteins. Multiple genes, alternative splicing, and differential expression in brain and development. J Biol Chem. 1999;274(41):29510-29518.

12. Bonaglia MC, Giorda R, Borgatti R, Felisari G, Gagliardi C, Selicorni A. Disruption of the ProSAP2 gene in a t(12;22)(q24.1;q13.3) is associated with the 22q13.3 deletion syndrome. Am J Hum Genet. 2001;69(2):261-268. doi:10.1086/321293

13. Ehlers MD. Synapse structure: glutamate receptors connected by the shanks. Curr Biol. 1999;9(22), R848-850.

14. Boeckers TM, Bockmann J, Kreutz MR, Gundelfinger ED. ProSAP/Shank proteins - a family of higher order organizing molecules of the postsynaptic density with an emerging role in human neurological disease. J Neurochem. 2002;81(5):903-910.

15. Roussignol G, Ango F, Romorini S, Tu JC, Sala C, Worley PF, et al. Shank expression is sufficient to induce functional dendritic spine synapses in aspiny neurons. J Neurosci. 2005;25(14):3560-3570.doi: 10.1523/JNEUROSCI.4354-04.2005

16. Durand CM, Betancur C, Boeckers TM, Bockmann J, Chaste P, Fauchereau F, et al. Mutations in the gene encoding the synaptic scaffolding protein SHANK3 are associated with autism spectrum disorders. Nat Genet. 2007;39(1):25-27. doi: 10.1038/ng1933

17. Shao S, Xu S, Yang J, Zhang T, He Z, Sun Z,et al. A commonly carried genetic variant, rs9616915, in SHANK3 gene is associated with a reduced risk of autism spectrum disorder: replication in a Chinese population. Mol Biol Rep. 2014;41(3): 1591-1595. doi: 10.1007/s11033-013-3005-5

18. Qin J, Jia M, Wang L, Lu T, Ruan Y, Liu z. Association study of SHANK3 gene polymorphisms with autism in Chinese Han population. BMC Med Genet. 2009;10: 61. doi: 10.1186/1471-2350-10-61

19. Sykes NH, Toma C, Wilson N, Volpi EV, Sousa I, Pagnamenta AT. Copy number variation and association analysis of SHANK3 as a candidate gene for autism in the IMGSAC collection. Eur J Hum Genet. 2009;17(10): 1347-1353. doi:10.1038/ejhg.2009.47

20. Chien YL, Wu YY, ChiuYN, Liu SK, Tsai WC, Lin PI, et al. Association study of the CNS patterning genes and autism in Han Chinese in Taiwan. Prog Neuropsychopharmacol Biol Psychiatry. 2011;35(6):1512-1517. doi: 10.1016/j.pnpbp.2011.04.010

21. Gauthier J, Spiegelman D, Piton A, Lafreniere RG, Laurent S, St-Onge J. Novel de novo SHANK3 mutation in autistic patients. Am J Med Genet B Neuropsychiatr Genet. 2009;150B(3):421-424. doi: 10.1002/ajmg.b.30822

22. Kolevzon A, Cai G, Soorya L, Takahashi N, Grodberg D, Kajiwara Y. Analysis of a purported SHANK3 mutation in a boy with autism: clinical impact of rare variant research in neurodevelopmental disabilities. Brain Res. 2011;1380:98-105. doi: 10.1016/j.brainres.2010.11.005

23. Soorya, Kolevzon A, Zweifach J, Lim T, Dobry Y, Schwartz L. Prospective investigation of autism and genotype-phenotype correlations in 22q13 deletion syndrome and SHANK3 deficiency. Mol Autism. 2013; 4(1):18. doi: 10.1186/2040-2392-4-18

24. Uchino S, Waga C. SHANK3 as an autism spectrum disorder-associated gene. Brain Dev. 2013; 35(2):106-10. doi:10.1016/j.braindev.2012.05.013

25. Leblond CS, Nava C, Polge A, Gauthier J, Huguet G, Lumbroso S, et al. Meta-analysis of SHANK Mutations in Autism Spectrum Disorders: a gradient of severity in cognitive impairments. PLoS Genet. 2014;10(9):e1004580. doi: 10.1371/journal.pgen.1004580

26. Meyer G, Varoqueaux F, Neeb A, Oschlies M, Brose N. The complexity of PDZ domain-mediated interactions at glutamatergic synapses: a case study on neuroligin. Neuropharmacology. 2004; 47(5), 724-733. doi: 10.1016/j.neuropharm.2004.06.023

27. Durand CM, Perroy J, Loll F, Perrais D, Fagni L, Bourgeron T, et al. SHANK3 mutations identified in autism lead to modification of dendritic spine morphology via an actin-dependent mechanism. Mol Psychiatry. 2012;17(1):71-84. doi: 10.1038/mp.2011.57

28. Veenstra-VanderWeele J, Cook EH Jr. Molecular genetics of autism spectrum disorder. Mol Psychiatry. 2004;9(9): 819-832. doi: 10.1038/sj.mp.4001505

10

Shift from Gel based to Gel Free Proteomics to Unlock Unknown Regulatory Network in Plants

Sajad Majeed Zargar[1]*, Nancy Gupta[2], Rakeeb A Mir[3], Vandna Rai[4]

[1]Centre for Plant Biotechnology, Division of Biotechnology, Sher-e-Kashmir University of Agricultural Sciences & Technology of Kashmir, Shalimar, Srinagar, J&K, India
[2]School Biotechnology, Sher-e-Kashmir University of Agricultural Sciences & Technology of Jammu, Chatha, Jammu, J&K, India
[3]School of Bio resources & Biotechnology, BGSB University, Rajouri, J&K, India
[4]NRCPB, New Delhi, India

*Corresponding author: Sajad Majeed Zargar, Centre for Plant Biotechnology, Division of Biotechnology, Sher-e-Kashmir University of Agricultural Sciences & Technology of Kashmir, Shalimar, Srinagar, J&K, India, E-mail: smzargar@gmail.com

Abstract

Proteomics has emerged as a vital tool to identify the novel proteins and explore the cellular dynamics by employing highly proficient and innovative techniques introduced in the past few years. The expedition began with the emergence of gel-based 2DE approaches for evaluation of diverse proteins and differential protein analysis. Later, advancements in the field lead to the advent of gel-free approaches that render more accuracy and reliable results. In spite of introduction of high-throughput analysis offered by new techniques, the overall beneficence of gel-based approaches cannot be ignored. The combined effort of all these approaches can generate astonishing results. Protein analysis under specific conditions attempts to deduce the expression pathways and thus will help to expand our knowledge pertaining to metabolic and regulatory routes. Further, understanding the dynamics of these biomolecules and metabolites involved in the pathways can facilitate their desired manipulation for altering the gene expression. This will assist in clinical purposes for development of drugs and in systems biology to get the broad picture of metabolic processes and their regulation. This review highlights the trend with which progress in plant proteomics technology has been made right from gel based to gel free strategies with discussion of basic principles and procedures involved in each technique. Advances in this field can lead to precise interpretation of several biological processes.

Keywords: 2DE; DIGE; iTRAQ; ICAT; SILAC; plant proteomics; quantitative proteomics

Introduction

Proteomics refers to the high throughput systematic analysis of protein expression and function with the aid of protein biochemistry, mass spectrometry and bioinformatics' tools. Most of the biological researches aim to decipher the metabolic regulatory pathways to gain deeper insight into various cellular processes and thus contributing to our understanding of biological systems. Although, there are ongoing efforts to gather information contained in the human genome sequence through genomics and transcriptomics approaches; elucidating the dynamic changes in proteins will provide a better picture of cellular genetics and its regulation. In this context, quantitative proteomics has been emerged recently as a fascinating platform to explore the hidden pathways and reveal fundamentals of biological systems. With the advent of the most feasible and spectacular techniques, quantitative proteomics is directed to exemplify the identification and quantification of diverse proteins that represent a rich source of biological information. Quantitative proteomics employs various techniques for separation of proteins based on their physicochemical properties (molecular mass, pH, charge etc) followed by their recognition and quantification by mass spectrometry methods utilizing data systems and software's. The major steps involved in these analytical processes are i) cell sampling ii) extraction, isolation and solubilisation of proteins iii) separation through electrophoresis (isotachophoresis, zonal, isoelectric focussing, off gel) and chromatography (ion exchange chromatography, 2- dimensional liquid chromatography, reverse phase) iv) quantification using gel based and gel free approaches and v) identification using mass spectrometry approaches. The chief principles and procedure of aforementioned steps are briefly illustrated in (a,b,c,d,e). For detailed understanding of MS protein identification principles, readers are advised to visit http://www.spectroscopynow.com/ ; http://www.ionsource.com/; http://www.asms.org/whatisms/index.html. Many advances have been made so far in the field of quantitative proteomics, however there is no technique that demonstrates the actual expression of these biological entities (proteins) in one go as in case of genomics (microarray analysis, genome sequencing). Proteomics studies require the collective efforts of different fields of sciences (biochemistry, physics, computer sciences and statistics) that collaborate to analyse structural and functional aspects of proteins. In this review, we have highlighted various efficient and reliable quantification methods that differ in many ways in terms of progression, sensitivity, robustness, accuracy and quality of data obtained. Quantitative mass spectrometry approaches can be mainly categorized into gel based (2- dimensional electrophoresis) and gel-free approaches. Later are further divided to label-based (chemical, enzymatic and metabolic tagging) and label-free (data independent and data dependent) methods. Each technique has its pros and cons and rather than replacement, these methods complement each other. The choice of methodology adopted for analysis depends on the biological sample taken under consideration and the information required. In order to make successful systematic and quantitative profiling, we require good quality separation and quantitative techniques accompanied with protein/ nucleotide databases that promise high accuracy to search peptide masses with the assistance of powerful automated bioinformatic tools and softwares. Here, we have summarized the basic strategies involved in the aforementioned techniques that expedite the functional analysis of proteins on global scale and thus assisting the conventional molecular biology methods.

Need for proteomics studies in plants

Genes are considered to be the major repositories of biological information required for the processing of molecular mechanisms inside the cell. However, the information stored in the gene is in coded form and needs to be expressed and translated to proteins for performing vital functions of life. The genes are transcribed to mRNA in coded form of nucleotides that are further translated and decoded to amino acid sequence of proteins. Thus, for proper investigation of cellular processes, studies can be conducted at genomics (genes), transcriptomics (mRNA) and proteomics (protein) level. With the emergence of sequencing technique, it is possible to deduce the DNA sequence in short time but focus has now been shifted to predict the innumerable functions which are performed by the expression components of these genes i.e., Proteins. The study of cell at genomic and transcriptomic level provides a rough estimate of the expression and genome annotation. Thus study of proteins is important as these are the ultimate bio molecules that complement the structure and function of living systems. The concept of proteome was given by Marc Wilkins and his associates to describe the protein complement of the genome [1]. The large scale study of diverse proteins and precise measurement of their expression by utilizing powerful analytical techniques is referred to as "Proteomics" [2]. The wide acceptance of this recently emerged technology is attributed to advancements in the methodologies that attempt to accomplish accurate identification and quantification of proteins. The expression of genes can be detected by mRNA profiling but it neither reflects the correct amount of proteins nor their regulatory status [3]. Transcripts are highly unstable and are translated differently under varying environmental conditions producing different proteins. There is a dire need to analyse multiple protein products arising from single gene to figure out expression pathways thoroughly. Various post-translational modifications occur that affect the structure, localization and function of proteins and presents challenges in front of analytical tools to detect multiple forms of proteins [4]. A gene expressed under different set of environmental conditions generates separate transcripts which are further modified chemically and

expressed in various ways to produce several forms of proteins. Thus, relying completely on genomics tools is not a good option. Expression studies should be carried out at proteome level to attain the complete knowledge of ongoing cellular processes. The first proteomic work in plants was published on evaluation of proteins in *Arabidopsis thaliana* using 2-dimensional gel electrophoresis [5]. The proteomic studies have made excellent progress in the past few years, right from the sub-cellular proteome expression analysis of various organelles [6-11]. And understanding the developmental process by spot comparison [12-14]. To deciphering the proteins involved in stress physiology [15-24] and plant pathogen interactions [25-28]. The detailed inspection of several proteins synthesized by various intriguing metabolic processes in relation to the external environment will help to assign functions to orphan genes and understanding their regulation ultimately serving the purpose. As proteins are directly involved in the molecular networks, accurate changes can be measured in response to alternating surroundings. Table 1 represents the key studies carried out on proteomic analysis of plants using advanced gel based and gel free methodologies. Thus, proteomics approaches can be considered as the essential key to unlock mysteries of life and its processes finally governing the efficient manipulation of pathways for improvement of crops.

Proteomics Methodologies: Progress Till Date

Gel based methods

Gel based protein separation and quantification is performed by integrating simple analytical methods of protein biochemistry with high throughput mass spectrometry analysis. The first report in this context was published about three decades ago by O'Farrel for analysis of complex proteins [68]. Since then, tremendous progress has been made in the field of protein studies due to continued development of methodologies in terms of precision and accuracy. This technique allows global analysis of thousands of protein isoforms expressing under specific set of conditions [4]. Gel based approach mainly employs 2D-PAGE (Two-dimensional polyacrylamide gel electrophoresis) and 2D-DIGE (Two dimensional differential gel electrophoresis) that efficiently separates proteins on the basis of their physicochemical properties, differentially analyses the spots obtained and finally identifies the components using mass spectrometry. Figure 2 (a, b) represents basic steps involved in gel based proteomics approach. In spite of huge advancements in the field of proteomics, these techniques have not been replaced and are routinely used for quantification of proteins.

2D-PAGE (Two Dimensional Polyacrylamide Gel Electrophoresis): Electrophoresis technique (separation of charged molecules under the influence of electric current) has been extensively utilized for the separation of biomolecules [69] based on their specific characteristics. Two dimensional approach attempts to separate proteins depending on two parameters i.e., pH and molecular mass. Former employs IEF (iso-electric focusing) technique and later utilizes polyacrylamide gels in electrolytic medium subjected to current under the influence of electric field. The second dimension represents better resolution and precision

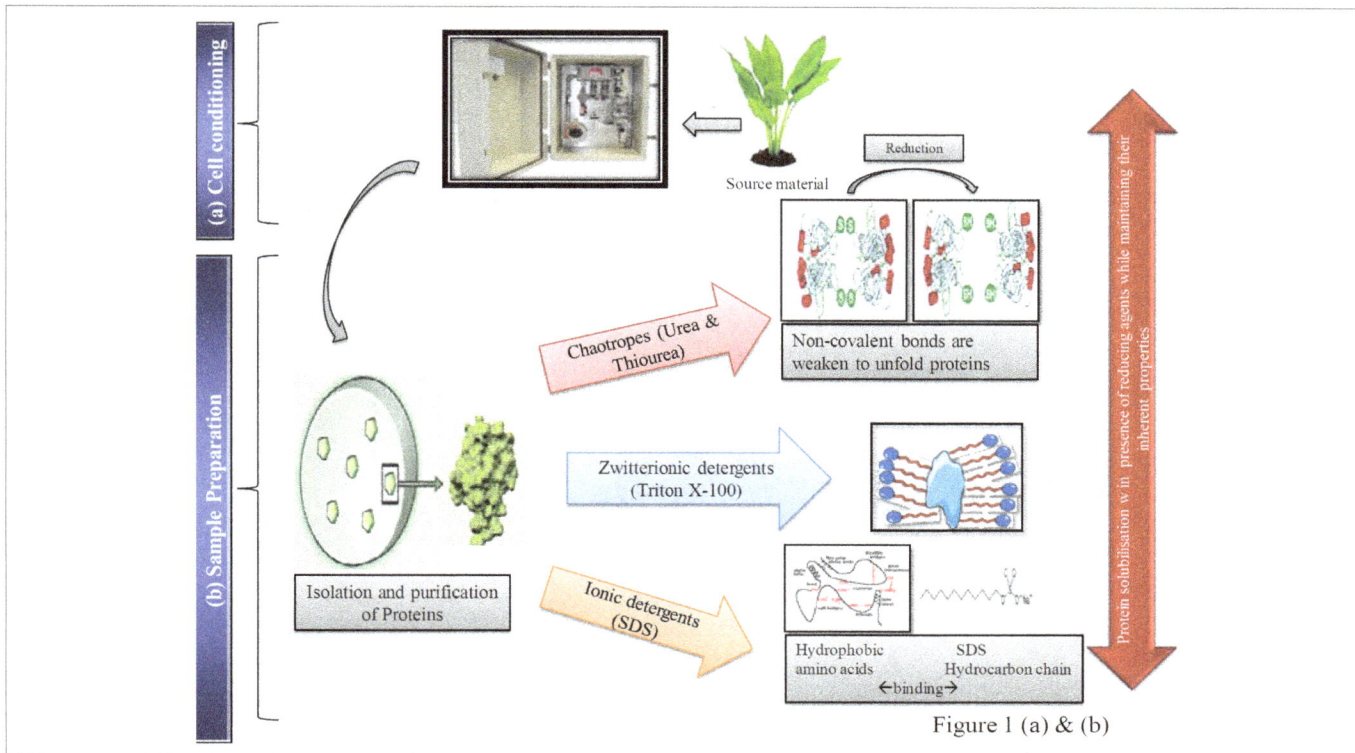

Figure 1 (a) & (b)

Figure 1ab: Major steps involved in proteomic analysis. (a) Cell conditioning; (b) Sample preparation.

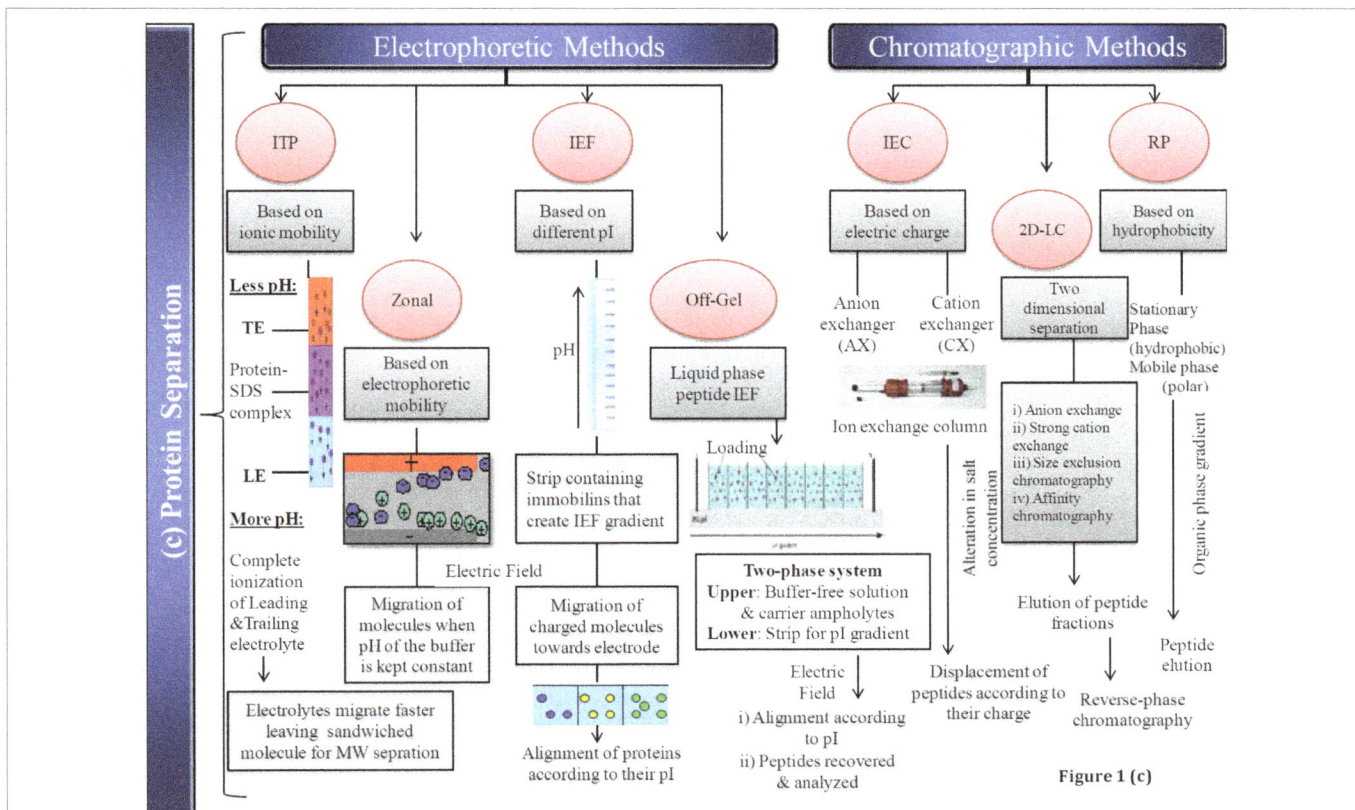

Figure 1 (c)

Figure 1c: Protein Separation.

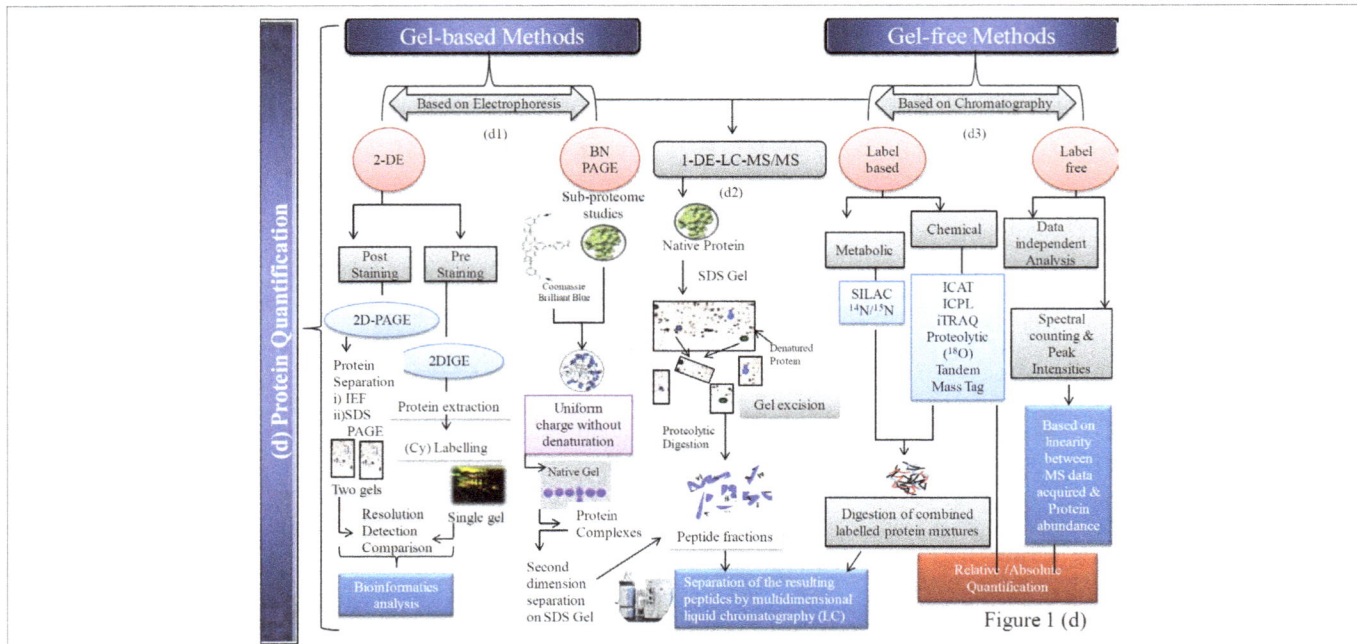

Figure 1d: Protein Quantification.

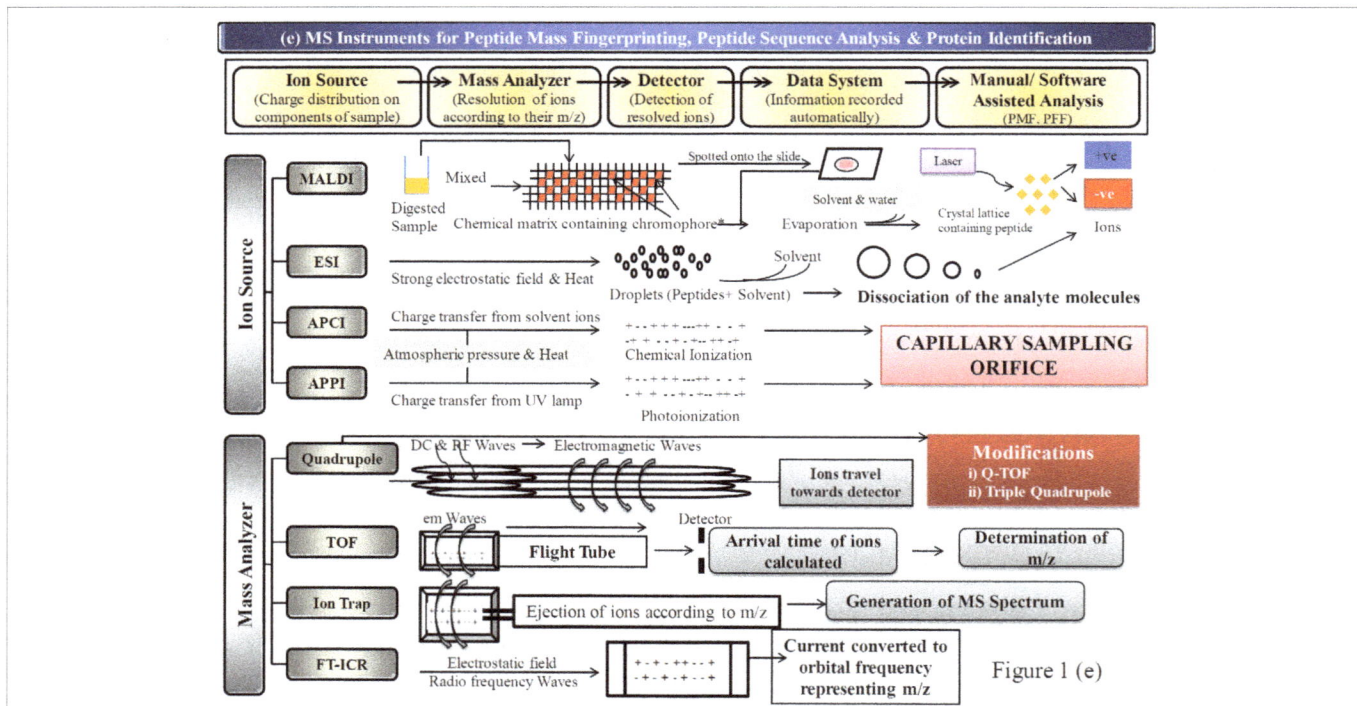

Figure 1e: Protein Identification (MS Instruments for Peptide Mass Fingerprinting, Peptide Sequence Analysis & Protein Identification).
SDS: Sodium-dodecyl Sulfate; ITP: Isotachophoresis; TE: Trailing Electrolyte; LE: Leading Electrolyte; MW: Molecular Weight; IEF: Isoelectric Focusing; pI: Iso-Electric Point; IEC: Ion Exchange Chromatography; 2D-LC: Two Dimensional Liquid Chromatography; RP: Reverse Phase; 2DE: Two Dimensional Electrophoresis; 2D-PAGE: Two Dimensional Polyacrylamide Gel Electrophoresis; 2D-DIGE: Two Dimensional Difference in Gel Electrophoresis; (Cy): Cyanine Dye; BN-PAGE: Blue Native Polyacrylamide Gel Electrophoresis; SILAC: Stable Isotope Labelling by Amino Acids in Cell Culture; ICAT: Isotope-Coded Affinity Tags; ICPL: Isotope Coded Protein Labeling; iTRAQ: Isobaric Tags for Relative and Absolute Quantification; MS: Mass Spectrometry; PMF: Peptide Mass Fingerprinting; MALDI: Matrix Assisted Laser Desorption Ionization; ESI: Electro Spray Ionization; APCI: Atmospheric Pressure Chemical Ionization, APPI: Atmospheric Pressure Photo Ionization; DC: Direct Current; RF: Radiofrequency; TOF: Time-of- Flight; Em Waves: Electromagnetic Waves; m/z: Mass to Charge Ratio; FT-ICR: Fourier Transform-Ion Cyclotron Resonance

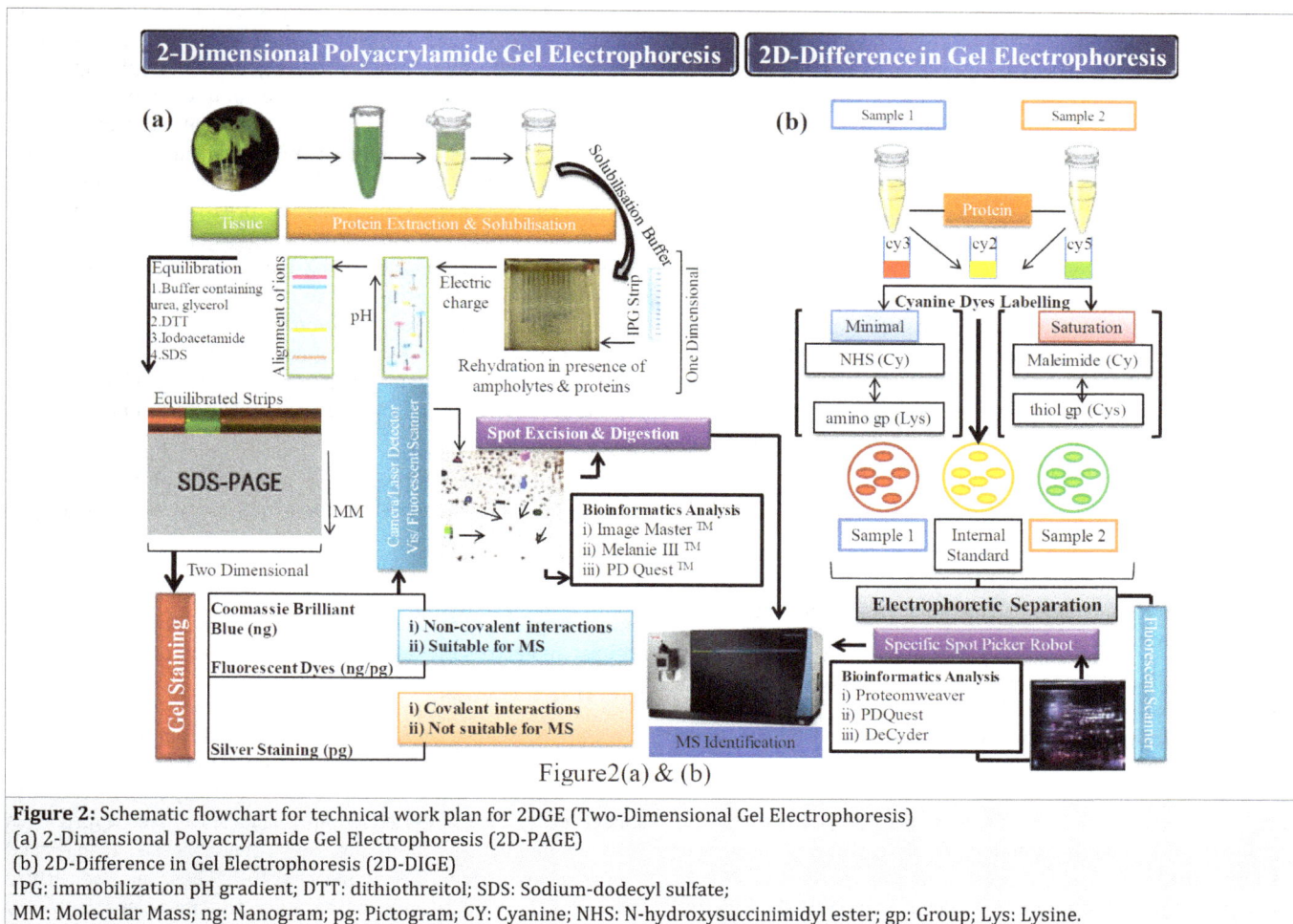

Figure 2: Schematic flowchart for technical work plan for 2DGE (Two-Dimensional Gel Electrophoresis)
(a) 2-Dimensional Polyacrylamide Gel Electrophoresis (2D-PAGE)
(b) 2D-Difference in Gel Electrophoresis (2D-DIGE)
IPG: immobilization pH gradient; DTT: dithiothreitol; SDS: Sodium-dodecyl sulfate;
MM: Molecular Mass; ng: Nanogram; pg: Pictogram; CY: Cyanine; NHS: N-hydroxysuccinimidyl ester; gp: Group; Lys: Lysine.

to analyse complex proteins in the sample. First of all, the sample (tissue/cell) is collected and stored under suitable conditions and used to extract proteins free from impurities. Then proteins are solubilised using ionic/non-ionic detergents and reducing agents (DTT) and finally separated. Iso-electric focusing requires IPG strips (Immobilized pH gradient strips) that contain acrylamide polymerized with bis-acrylamide and immobilins covalently bind in gel. These immobilins were used to create stable pH gradient [70] inside the gel. These strips are dipped in the solubilisation buffer containing extracted proteins of the sample. Both positively and negatively charged proteins (amphoteric) have specific pI (iso-electric point) at which the charge of the protein is neutralized. Different amphoteric molecules get focused on the gel corresponding to their respective pI when subjected under electric field. The strips are further treated with SDS and chemicals like DTT, iodoacetamide for equilibration of their chemical properties and thus leaving proteins to be analysed only on the basis of their molecular masses. The hydrocarbon tail in the SDS (sodium-dodecyl sulphate) binds to the hydrophobic amino acids through non-covalent interactions and gets distributed over entire proteins imparting negative charge all over and causes denaturation [71]. The proteins then travel from porous polyacrylamide gel (PAGE) according to their molecular weight under the influence

of electric current. Different pH of buffer systems present more accurate resolution due to isotachophoresis in which proteins are aligned on the stacking gel (less pH) and further separated onto resolving gel (high pH) thus giving a fair start to all proteins. The buffer systems possess leading and trailing ions whose electrophoretic mobility is affected by the external pH. Some of the most commonly used buffer systems are Tris-glycine, Bis-Tris, Tris-acetate and Tris-Tricine. Next step after resolution is the detection of the proteins which can be achieved by in-gel staining methods. The most commonly used chemicals for staining are coomassie brilliant blue, silver nitrate and fluorescent dyes that differ in terms of sensitivity, utility and suitability for MS analysis. Coomassie brilliant blue and fluorescent dyes bind non-covalently to the proteins and can be separated easily, thus suitable for MS analysis but silver staining involves covalent linking to the proteins thus not recommended for MS analysis in spite of its high sensitivity than other dyes [72]. Fluorescent dyes such as Deep Purple, SYPRO Ruby, SYPRO Red, SYPRO Orange, RuBPS, FlamingoTM, Krypton TM, ASCQ Ru, ProQdiamond, ProQemerald etc present very high sensitivity and accurate determination of protein isoforms in the sample due to their efficient compatibility with MS [73]. Stained gels are visualized using scanners/detectors and differential spots are taken into consideration for further analysis (depicted in Figure 2a). The abundance of proteins is roughly assessed by the intensity of spots. However, for absolute detection, these spots are excised from the gel and subjected to enzymatic digestion producing variable peptides that undergo ionization in MS and analysed based on their different mass to charge ratios. The speed with which these peptides move towards detector is critically observed and fed to data systems. The data is then used to search matches in the databases (protein/nucleotide) for peptide mass fingerprinting using powerful automated softwares for identification of proteins [74]. This technique ensures the utilization of cost-effective technologies for adequate resolution of complex proteins and can be used along with the advanced tandem MS approaches as the first fractionation step. However, it renders errors due to greater number of steps involved and low proteome coverage due to insolubility of highly acidic or basic, hydrophobic proteins. The detection of spot intensity permits inaccuracy due to resolution of diverse proteins in the same location and thus prevents the low abundant proteins to be detected [4,75-77]. In addition to this, different experimental treatment of samples and gel-to-gel variation also contribute to the poor reproducibility of 2D-PAGE. These drawbacks are overcome by 2D-DIGE technique that allows resolution on single gel rather than separate gels.

2D-DIGE (Two Dimensional Differential Gel Electrophoresis) This modified gel based technique offers few advantages over 2D-PAGE by minimizing the errors caused by gel–to-gel variation and need for analysis of more than one gel thereby reducing manual error. The protein samples are pre-stained by using cyanine-based fluorescent dyes. The NHS (N-hydroxysuccinimidyl) ester group and maleimide derivatives of these dyes react with the amino and thiol groups of the protein respectively. The labelling is performed in two ways- minimal labelling and saturation labelling [78]. Former deals with the labelling of N'-terminal of lysine residues by amide linkage to NHS ester group (required for maintaining the multiple charges on the surface of protein thereby preventing in solubilisation) and later facilitates the binding of thiol groups of cysteine residues to the maleimide derivatives of dyes (recommended for low abundant proteins due to high sensitivity) leading to comparatively wide proteome coverage. The cyanine based dyes should be tagged such that they might not influence the mobility of proteins when subjected to electrophoresis [79]. The labelling of different samples with resolvable fluorescent cyanine based dyes allows differential expression studies. The protein samples to be quantified are labelled with Cy3 and Cy5 dyes that impart different colours when visualized by fluorescence scanner and thus depict the amount of protein within the sample by measurement of spot intensity (Illustrated in Figure 2b). These labelled proteins along with the internal standard Cy2 (representing presence of both the samples) is used for normalization of the ratios of intensities retrieved from different samples paving way for accurate quantification of proteins. The labelled protein samples are mixed and subjected to electrophoretic separation in a single gel thereby eliminating the need to analyse more gels and reducing the experimental error. The stained images are captured, scanned, digitalized and the fluorescence intensities of variable samples are analysed and

the data is fed to efficient softwares such as DeCyder, Proteom weaver PDQuest and Progenesis [70] for comparison of spot intensities and useful information is generated depicting to the abundance of specific proteins. DIGE technique is more appealing as compared to 2D-PAGE in terms of sensitivity, reproducibility, reliability, accuracy, automation, and more suitability for more diverse proteins & MS analysis. The gel based techniques have been applied for differential protein expression studies in plants [17,80-84]. And few research findings emphasized the equivalent need of gel based techniques to accomplish the task of protein annotation [85-88]. Regardless of so many advantages of 2D-DIGE, it is unable to beat some of the immanent drawbacks of gel based approaches like narrow-coverage of proteins due to tagging of lysine and cysteine amino acids only, insolubility of some of the membrane proteins and sample preparation variation.

Advancements in the field of proteomics have led to the emergence of gel-free proteomics approach that addresses the issues such as reproducibility, low-proteome coverage, quality of data obtained that are observed in case of gel-based methods. Gel-free methods are mainly dependent on LC-MS/MS technique and instead of examining one spot at a time, it takes into consideration all the peptides generating from the proteins proving to be more robust and extremely informative high-throughput strategy.

Gel free methods

Gel free methods can be considered as a direct consequence of the numerous innovative developments in the past two decades. These methods eliminate few of the major experimental errors observed in case of gel based approaches. Gel free approaches can be mainly classified into two categories i.e., label-based approaches and label free approaches [89]. These methods utilize the proficient liquid chromatography and mass spectrometry tools to generate the quantification data thereby promising efficient protein studies. The gel free strategy is quite simple and engages few steps in which proteins are isolated, digested (labelled/non-labelled), eluted on liquid chromatography and detected to be analyzed by mass spectrometry approaches for absolute or relative quantification. Wide methodologies pertaining to label-based and label-free approaches are discussed in the following sections.

Label based methods

Label-based approaches utilize specialized isotope/isobar tags having specific groups that label proteins and peptides chemically/metabolically or enzymatically [90]. The steps undertaken for label-based quantification are depicted in Figure 3 a & b. The separation of these labelled peptides on LC and further analyses by highly sensitive MS technique yields highly informative data for retrieving precise results by comparing the relative abundances of heavy and light samples. Numerous experiments have been conducted using label based strategies to understand the mechanisms of various developmental stages, stress physiology by differential expression studies [26,28,91-95]. The label based approaches assure automated high-throughput quantitative proteome analysis of unknown proteins with reliable automated and multiplexing abilities. Proteins/

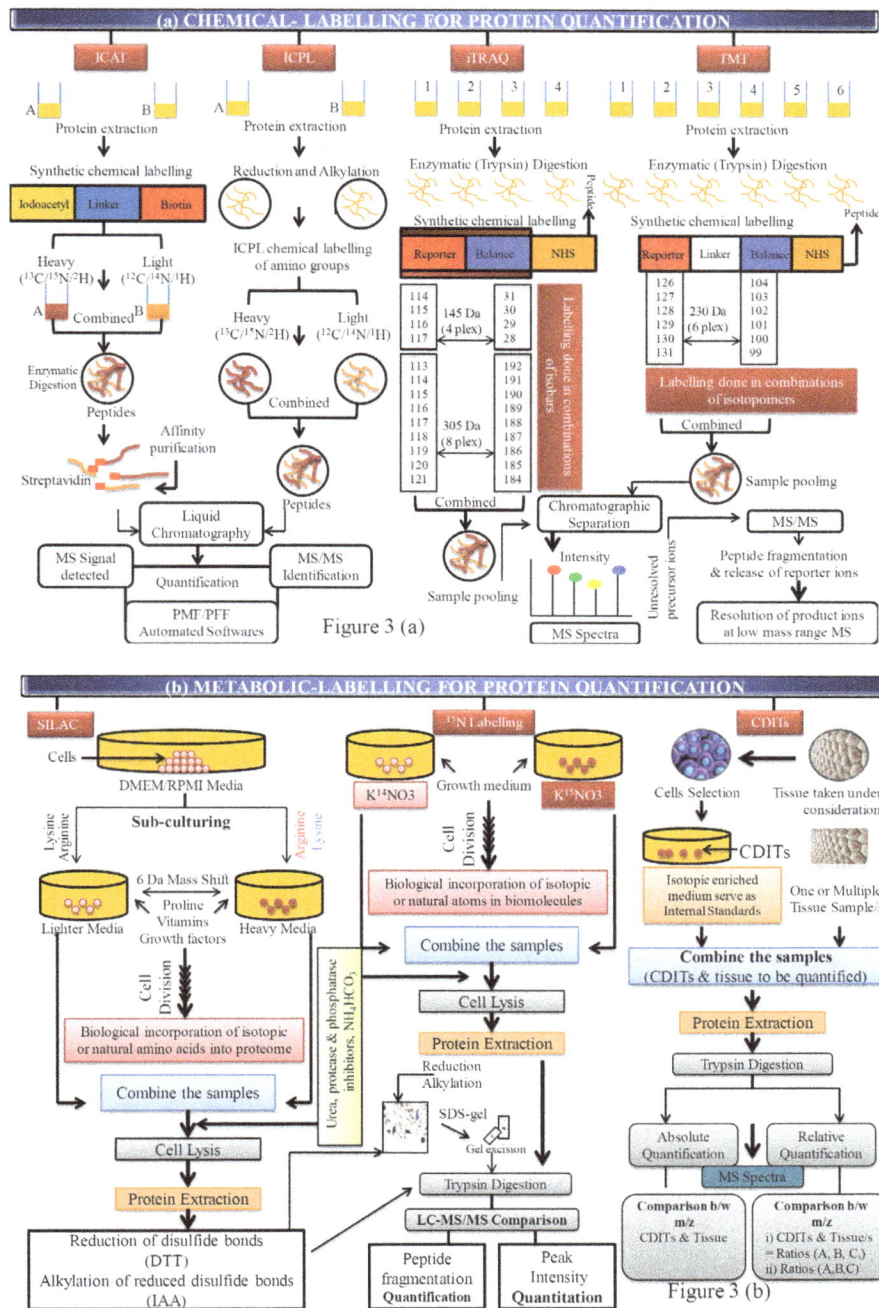

Figure 3: Outline of technical work plan for label-based proteomics

(a) Chemical Labelling for Protein Quantification

(b) Metabolic Labelling for Protein Quantification

ICAT: Isotope-Coded Affinity Tags; tag consists of three functional elements i.e. iodoacetyl group (yellow) binds to thiol-specific groups, linker (blue) introduces mass shifts and biotin (red) used for reducing complexity by affinity purification.

ICPL: Isotope Coded Protein Labelling; modified version of ICAT that permits multiplexing and labelling of almost all peptides.

iTRAQ: Isobaric Tags for Relative and Absolute Quantification; tag consists of reporter (red) that introduces mass differences, balance group (blue) that maintains the similar weight of all reporter group in tags and NHS group (yellow) that binds specifically to peptides.

TMT: Tandem Mass Tag; differs from iTRAQ with respect to the presence of additional linker group (white) and isobars used.

SILAC: Stable Isotope Labelling by Amino Acids in Cell Culture; DMEM: Dulbecco's Modified Eagle Medium, RPMI Medium: Roswell Park Memorial Institute Medium, DTT: dithiothrietol, IAA: indole acetic acid

15N Labelling: Differs from SILAC in terms of incorporation of labelled elements through inorganic chemicals instead of labelled amino acids.

CDITs: Culture-Derived Isotope Tags; can be used for absolute quantification as well.

peptides are labelled either chemically (using isotopes or isobars) or metabolically to introduce mass shifts for distinguishing the relative intensities of the proteins in the samples.

Chemical labelling: The method involves utilization of chemically synthesized tags that incorporate variable isotopes and isobars which introduce mass difference within the labelled proteins (ICAT) and peptides (ICPL, iTRAQ, TMT) for their differential expression studies based on the abundance of peptides detected by their peak intensities (based on m/z) and MS/MS fragmentation. This strategy offers accuracy and simultaneous comparison of more than two samples. It has been further sub-divided into isotopic (ICAT & ICPL) and isobaric (iTRAQ & TMT) labelling.

Isotope-Coded Affinity Tags (ICAT): The first *in vitro* method that permits tagging of proteins and peptides of all types of biological samples using stable isotopes was developed by Gygi and his associates in 1999 [96]. This technique employs ICAT reagents that consist of mainly iodo-acetamide group or N-ethymaleimide, a spacer or linker arm and biotin. Iodo-acetamide group/N-ethymaleimide is highly specific chemical reactive group that alkylates thiol group of cysteine residues in the protein sample. A spacer or linker arm is meant for introduction of mass shift by incorporation of different isotopes in different samples. Isotopes used for this purpose are proton (H)/deuterium (D), $^{12}C/^{13}C$, $^{15}N/^{16}N$ that introduce mass difference of upto 8Da due to presence of these elements in the labelled or unlabelled amino acid residues producing light and heavy tags. Biotin group assists purification of labelled peptides by affinity chromatography in biotin-avidin/streptavidin systems and thus captures all the cysteine containing peptides from the mixture. The strategy employs isolation of protein from two samples followed by synthetic chemical labelling using ICAT reagents; one with heavy and other with light isotopes. The labelled samples are then pooled, enzymatically digested to peptides and subjected to affinity purification by biotin moiety to reduce sample complexity. The next step involves the removal of biotin as it decreases the resolution efficiency of mass spectrometry [96] using acid-cleavable (ALICE/ introduction of disulphide bond in linker) or photo-cleavable linkers (UV light) [97,98]. The peptides are then allowed to resolve on liquid chromatographic separations and further analysed by tandem mass spectrometric techniques. Relative peak intensities obtained in MS spectra directly correlates to the abundance of respective peptides in the sample whereas tandem MS allows peptide mass fingerprinting by detection of product ions generated during peptide fragmentation which leads to identification of proteins. The technique offers accuracy as samples are similarly treated by protease preventing experimental variations and reduced sample complexity due to tagging of only cysteine residues but is also associated with loss of information leading to lower proteome coverage. To overcome these limitations, other tags have been developed that allow multiplexing as well.

Isotope-Coded Protein Labelling (ICPL): The ICPL strategy, referred to as modified version of ICAT approach was developed by Schmidt and co-workers that solved major shortcomings of

confined sequence coverage and low throughput [99]. It utilizes amine-reactive N-nicotinoyloxy-succinimide tags that cause derivatization of free amino-terminal groups and ε-amino groups of lysine residues and introduce mass difference of 4 Da and ~6 Da in case of H/D and $^{12}C_6/^{13}C_6$ labelling respectively. The first step involves extraction, reduction and alkylation of protein samples to ensure uniform labelling of all free amino terminals of lysine residues. The heavy and light labelled protein samples are then pooled and subjected to enzymatic digestion. Since lysine residues are labelled, treatment with trypsin will generate longer fragments, thus Glu-C endoproteinases in addition to trypsin are used for digestion to obtain shorter fragments. ICPL tags are hydrophilic in nature and help to maintain the intrinsic characteristics of peptides that allow efficient quantification. The isotopes used in ICPL tags can be used in varied combinations for multiplexing (triplex and quadruplex) and differentiates the sample by even 2 Da. This method permits labelling of lysine containing peptides that are found abundant as compared to cysteine residues in most of the proteins and eliminates the confined sequence coverage of proteome to some extent but not completely as lysine is also absent in some of the proteins. To overcome this issue, post-digest ICPL approach was developed that allows labelling of peptides after enzymatic digestion. All the free N-terminals of peptides are labelled uniformly and thus the results obtained are non-biased and considered to be more accurate [100]. The digested labelled peptides are separated on liquid chromatography and finally analysed on MS for quantification and identification of proteins. ICPL strategy is observed to be very suitable for efficient MS/MS fragmentation and detection of peak intensities and also offers the analysis of post-translational modifications and isoforms [101]. Despite of advantages over ICAT method, post digest ICPL involves labelling after digestion which can lead to manual error and the tags can interfere in the mobility of peptides when subjected to liquid chromatography [102]. In addition to isotopes, isobars have been employed for the quantification purpose that can introduce mass difference of even 1 Da and efficient multiplexing (iTRAQ & TMT).

Isobaric Tags for Relative and Absolute Quantification (iTRAQ): ITRAQ strategy has been utilized efficiently to explore the diverse molecular mechanisms occurring in plants. The strategy involves isobaric labelling of peptides; introduced by Ross and his associates in 2004 [103]. The isobaric tags bind covalently to the N-terminal of peptides; introduce a mass shift of even 1Da and thus paves way for multiplexing. The tag is comprised of three main functional elements i) Reporter group that introduces mass shifts ii) Balance group that is required to maintain the overall mass of isobaric tag iii) NHS group which specifically binds to the peptide. Chemically, reporter group is N-methylpiperizine which provides a mass shift range; balance group is mainly carbonyl group whose mass is adjusted according to the reporter group so that all tags have same mass for combined reporter and balance group and NHS ester group is amine reactive. The isobars are employed in varied combinations leading to efficient comparison of two, four and eight samples simultaneously which is otherwise not possible in case of ICAT strategy. Reporter groups have a mass range varying from 114-

117 Da and 113-121 Da that are compensated by balance group having a range from 28-31 Da and 184-192 Da in 4-plex (mass tag = 145 Da) and 8-plex (mass tag = 305 Da) respectively. Thus, the overall mass of reporter and balance group remains constant (as shown in Figure 3 a). The strategy includes labelling of enzymatically digested protein samples in which NHS-group binds covalently to all the N-termini of peptides equally. The labelled peptides are then further analysed by LC-MS/MS. All the labelled peptides are resolved by chromatographic separation and detected by MS to generate spectra. However, the peptides remain unresolved as second round fragmentation of peptides is necessary for producing product ions associated with release of reporter ions. For this purpose high quality MS with triple quadrupole is used. The ions entering are fragmented by less collision energies to produce precursor ions which cannot be distinguished and presented as a single peak. These precursor ions are again introduced under the influence of high collision energies to give product ions that are detected and resolved to generate spectra for simultaneous protein quantification and identification when retrieved information is searched against available protein and nucleotide databases. Although iTRAQ strategy has been utilized extensively in the past years due to its high accuracy and multiplexing abilities, it is associated with the need of high-throughput data acquisition system and modified versions of MS that can read slight mass differences.

Tandem mass tag: This technique shares the same principle as iTRAQ strategy with slight variation in the chemical structure of the tag used for labelling. $^{13}C/^{15}N$ isotopomers are mainly used in varying proportion to create mass difference. The tag is comprised of reporter region (creates mass difference), linker region (conjugates reporter to balance group and is easily cleavable), balance group (maintains constant mass) and protein reactive group (binds to amine/cysteine/carbonyl) [104]. The reporter group mass varies from 126-131 Da which is maintained to a constant mass of 230 Da by balance group having mass ranging from 99-104 Da which leads to 6-plex analysis. Similar to iTRAQ, this technique involves efficient uniform labelling of peptides (amine-) and cysteine residues (cys TMT) after enzymatic digestion of proteins. The reporter ions are released at the time of peptide fragmentation in MS/MS, produce spectra that is recorded and finally the abundance of peptides is interpreted that allows protein identification and relative quantification.

Although chemical labelling presents accuracy in determination of peptide abundance and overcomes in-gel experimental variation, it requires careful sample preparation methods and highly efficient mass spectrometric analysis which can discriminate peptides varying by even 1 Da mass. Multiplex versions are undoubtedly presenting proficient comparison but are associated with complicated data analysis due to its inability to select peptides after one round of fragmentation. The major drawback of these techniques is that pooling of samples is done just before LC-MS/MS analysis which creates space for experimental biasness and inaccuracy. To overcome the limitations of *in vitro* techniques, metabolic labelling came in limelight that incorporates the tags in the samples from the very beginning.

Enzymatic labelling: This method employs substitution of natural (^{16}O) and isotopic oxygen (^{18}O) in the carboxyl groups of amino acid residues. The isolated proteins are digested by proteases that target serine, lysine and arginine residues in presence of heavy water ($H_2^{18}O$) and light water ($H_2^{16}O$). Later, Hcl was used to serve as catalyst for labelling of carboxyl terminal residues along with water ($H_2^{18}O/ H_2^{16}O$) referred to as acid mediated oxygen substitution [105]. During the enzymatic digestion, amide bond is broken and one isotopic oxygen atom is substituted in carboxyl group. The cleaved peptide undergoes one more substitution in place of second oxygen of carboxyl group in presence of enzyme creating three possibilities for mass differences i.e., $^{16}O/^{16}O$ (0Da), $^{16}O/^{18}O$ (2Da), $^{18}O/^{18}O$ (4Da) when compared to unlabelled peptides. The mass differences are detected to depict the relative abundance of peptides by comparing their ionic intensities. This method allows efficient incorporation of tags but permits side reactions that interferes with accurate data analysis. However, to inhibit undesirable reactions, suitable buffers and esterification is performed methanol and deuterated methanol [106].

Metabolic labelling: Metabolic labelling strategy employs biological incorporation of isotopic amino acids and elements through cell culture in plants and dietary food in animals. It surpasses the major drawback of *in vitro* labelling and eliminates experimental error to a great extent. After the intake of labelled amino acids inside the body, the cells are rapidly multiplied and undergo vast array of cellular processes that ensures efficient incorporation of isotopes. Metabolic labelling can be carried out by either isotopic essential amino acids (arginine, lysine, leucine, and tyrosine) or isotopic elements (^{13}C, ^{15}N, ^2H, ^{18}O). Despite of so many advantages, it lacks applicability for all biological samples and is somewhat tedious and expensive.

Stable Isotopic Labelling of Amino Acids in Cell Culture (SILAC): SILAC is a simple *in vivo* technique that was first developed in 2002 [107]. Later on, the method was efficiently introduced in eukaryotic organisms as well. Isotopic lysine ($C_6H_{14}N_2O_2$) and arginine ($C_6H_{14}N_4O_2$) amino acid tags are mainly used in this strategy. Looking at the chemical structure, it can be estimated that isotopic and normal amino acid will have a mass difference of 6Da ($^{12}C/^{13}C$), 2 Da ($^{14}N/^{15}N$) and likewise varied combinations of isotopic amino acids will lead to simultaneous high-throughput analysis of more samples. First of all, a cell culture medium is prepared and cells whose proteome is to be analysed are grown. The medium is divided into two sections, one is provided with labelled amino acids and other with unlabelled amino acids (as shown in Figure 3b). The cells are allowed to divide for four to five generations after subculturing in presence of similar growth regulators and conditions to confirm the unbiased incorporation of amino acid residues. To validate the efficient labelling of cells with heavy isotopes, mass spectrometric analysis is performed. The cells from both the culture medium are then pooled and proteins are extracted, treated with reducing agents and alkylated using iodoacetamide. The reduced proteins are then treated with trypsin for digestion to peptides. This can also be performed after separation of reduced proteins on SDS-PAGE. The mixture of peptides is co-eluted in liquid chromatography (reverse phase or strong ion exchange chromatography). The separated fractions are then further analysed by mass spectrometric techniques. This technique ensures less chance of biasness and handling errors with 100% incorporation of tags if grown for sufficient time [108]. SILAC strategy can provide absolute quantification of proteins provided there is no undesirable metabolic conversion of labelled amino acids to other by products (eg arginine is converted to proline). Except the labelled amino acids, all other amino acids present in sample should be non-isotopic and present in sufficient amount to eliminate the probability of side reactions. The technique has been modified to identify post-translational modifications, especially methylation by using heavy-methyl SILAC approach [109]. The technique has certain limitations associated to suitability of biological material in question and laborious steps involving highly proficient tools and expensive synthesises of tags.

^{15}N labelling: The first metabolic labelling study was performed using isotopically enriched media containing ^{15}N and is observed to be appropriate mainly for prokaryotes. The technique utilizes same strategy as SILAC with the difference of incorporating isotopic elements instead of amino acids. The labelled and unlabelled elemental nitrogen is provided to the growing cells by introducing inorganic salts in the culture medium (as shown in figure 3b). The cells are grown separately in the medium containing all the essential components required for growth. One sample is provided with the isotopic labelled element (heavy ^{15}N) and other is grown in presence of natural Nitrogen element (light ^{14}N). Multiple division of cells is allowed for few generations and then samples are mixed, protein is extracted, reduced, alkylated and digested by trypsin. The labelled and unlabelled peptides are then eluted on chromatographic separation and finally analysed by MS to generate ion chromatograms depicting intensities that are directly proportional to their relative abundances. The technique facilitates incorporation of ~98% tags but lacks in precision due to variation in number of nitrogen in peptides and also peptide sequences. Thus it makes the analysis and interpretation complicated.

Culture-Derived Isotope Tags (CDITs): This technique provides absolute and relative quantification of proteomes by using labelled internal standards. This strategy has been derived from SILAC technique as it is also associated with in vivo introduction of isotopes in cultured cell [110]. The cells are chosen from the tissue (taken under consideration for protein quantification). These cells are grown in suitable culture medium in presence of isotopes, thus referred to as culture-derived isotope tags (CDITs). The CDITs are mixed with the tissue sample which is to be analysed. The strategy further allows combined extraction of proteins followed by reduction and digestion to peptides. Now, the peptides from labelled (CDITs) and unlabelled (tissue sample) are analysed by mass spectrometry. Ion chromatograms are generated that depict m/z ratio of labelled and unlabelled peptides, isotopic distribution of peptides is observed and thus the quantity of the peptide in question is estimated. The calculated ratio of sample (m/z of same sequence of labelled peptides serving as reference and unlabelled peptides of tissue) depicts the absolute abundance of respective peptide. In case of more than one sample, same CDITs are added to all the different samples serving as internal standard. Protein from different samples ($T_1/T_2/T_3/T_4$ + CDITs) is extracted and digested separately. Mass to charge ratio for all the samples having internal standards are calculated which shows isotopic distribution between CDITs and tissue sample. The number of ratios calculated is equal to the number of samples addressed and these calculated ratios are finally compared to estimate the relative abundance of peptides (as depicted in Figure 3b).

To date, a lot of experiments have been conducted and published for biological studies in plants using label-based approaches due to wide applicability and accuracy. However, this method involves many steps and the samples that can be analysed are limited. Moreover, handling errors could lead to incomplete incorporation of tags and give way to side reactions. These constraints have been taken away by more advanced label-free approaches that allow direct analysis on LC and comparison on the basis of mass spectrometric data.

Label Free Methods

Label-free approaches evolving rapidly are considered to the most robust and accurate technique that provides higher dynamic range of quantification. In this strategy, the isolated protein are enzymatically digested and subjected to high resolution chromatographic separation. The eluted peptides are then transferred to MS where m/z ratios are analyzed and chromatogram depicting signal intensities are retrieved. It has been validated that the signal intensities of peptides provide a direct measure for its abundance in the sample [111,112]. The peptides are further fragmented by triple quadrupole mass analyzer where high and low collision energies are responsible to collect information for precursor and product ions; referred to as Tandem MS (MS/MS). MS/MS data provides correct information regarding the identity of proteins by comparing peptide masses obtained from analysis to that in the nucleotide/protein databases through powerful dedicated softwares. This reveals the identity of unknown proteins as well. The first software to be used for peptide matching was Sequest introduced in 1994 [113]. The most important advantage of label-free approaches is that we can compare as many protein samples as possible by comparing the signal peak intensities and MS spectra after individual separation and detection of all samples on LC-MS/MS. (Figure 4 a & b) demonstrates various global and targeted approaches for relative/absolute quantification of proteins.

Relative quantification

This approach is based on data dependent MS and MS/MS based label-free analysis. MS data is retrieved mainly in the form of MS spectra and signal intensities. These criteria's are assessed to interpret the identification and relative abundance of peptides in the sample. This approach has been broadly categorized in to two methods i.e., spectral intensities measurements and spectral

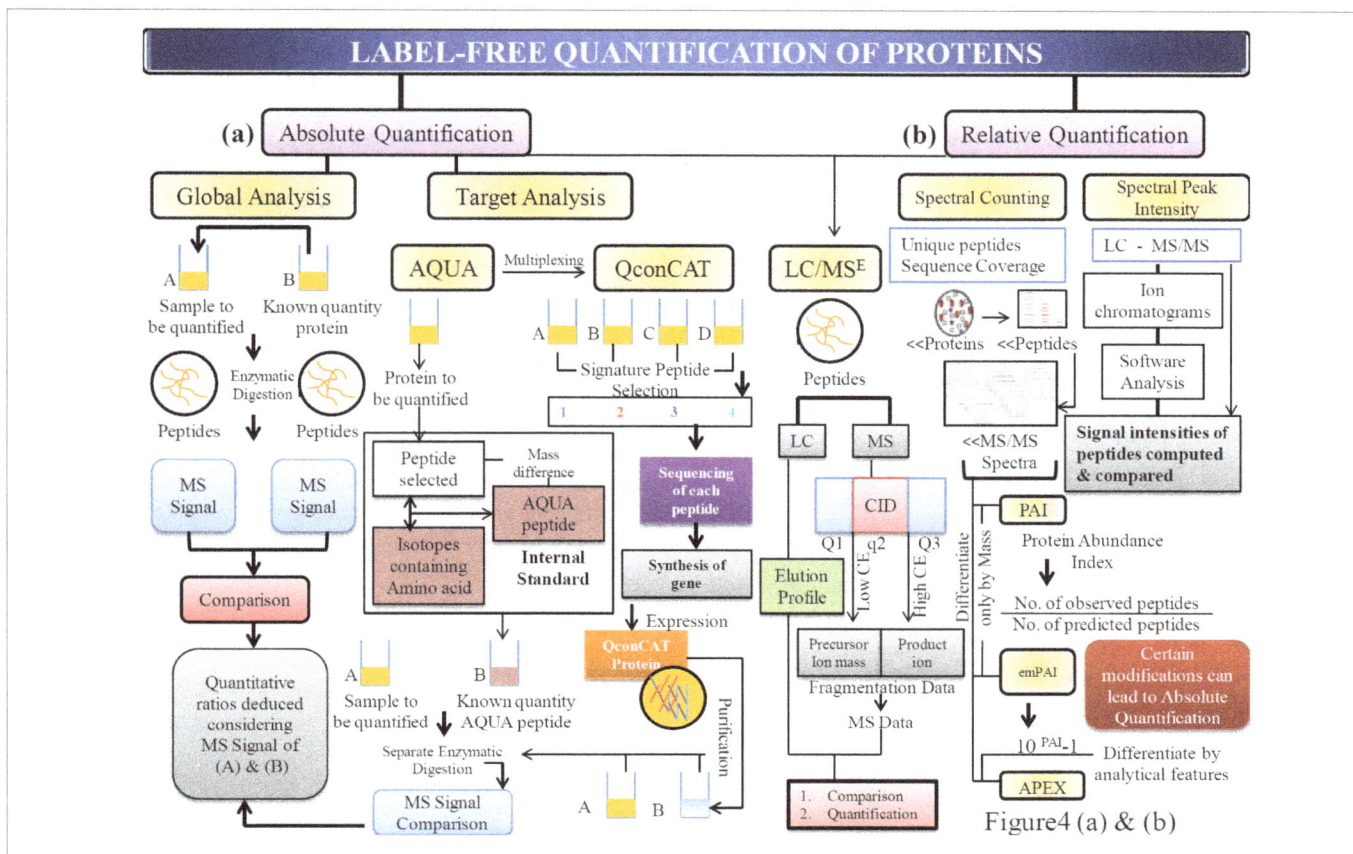

Figure 4: Major approaches and steps involved in label-free proteomics
(a) Absolute Quantification
(b) Relative Quantification
AQUA: Absolute Quantification of Proteins; QconCAT: Quantification by Concatenated Signature Peptides Coded Affinity Tags; LC/MSE : Liquid Chromatography Mass Spectrometry and E (superscripted) Stands for Elevated Energy; CID: Collision Induced Dissociation; Q1, q2, Q3 stands for Quadrupoles 1, 2 & 3 respectively, Low CE: Low Collision Energy; High CE: High Collision Energy; emPAI: Exponentially Modified Protein Abundance Index; APEX: Absolute Protein Expression; LC-MS/MS: Liquid Chromatography Tandem Mass Spectrometry.

counting [114]. Certain modifications in these techniques can lead to absolute measurement of protein (APEX).

Spectral intensities measurements

This technique employs the detection of ion intensities to estimate the quantity of protein in the sample. The first step involves preparation of sample in which the protein to be analysed is isolated individually from the source and subjected to high resolution liquid chromatographic separation; ultra-HPLC is the method of choice for this purpose. All the peptides under consideration are eluted by presenting a specific m/z ratio and specific retention time [115,116]. Using dedicated software's, ion chromatograms are generated that are automatically aligned. Now, the same peptide present in the samples to be quantified possess a particular m/z ratio and retention time. On the basis of these observed properties, peptide matching is performed either to that with other sample or reference in databases. When analysed on MS, peptides face collision energy under the influence of voltages inside the quadrupole and produce precursor ions. The total ion current of signal is computed and

calculated for each peptide. On the basis of m/z ratio, retention time (elution profile) and signal intensities, area under curve (AUC) is calculated using automated software's. AUC is directly proportional to the quantity of peptide and used to compare the abundance of peptides in different samples. In short, the strategy involves identification of peptides on the basis of m/z ratio and retention time when compared to reference and quantification of identified peptides by computing area under curve depicting signal intensity. For accurate measurements, one should be aware of basic elution profile characteristics of the peptide; this leads to bypass of some of the low abundant or unrecognized peptides. Calculation of number of spectra provides more dynamic range in this regard.

Spectral counting

Spectral counting can be regarded as a quantitative approach based on the rationale that the number of MS/MS spectra of peptides is directly proportional to the quantity of protein from which they are obtained. Recent study has reported the correlation of spectral count to that of fold change observed and that the number of spectra is correlated to the peptide count [117]. This strategy is mainly based on MS/MS analysis for simultaneous identification and quantification of proteins in contrast to peak intensity measurements that relies on MS analysis. In this approach, sample preparation and digestion steps are undertaken to produce tryptic peptides. These peptides are subjected on liquid chromatography and further analysis is performed by tandem mass spectrometry. The peptides are eluted independent of their physical characteristics but the analysis yields different retention time and chromatographic peaks data which distinguishes the spectral count of each peptide without the need to align chromatograms. The more the quantity of proteins, the higher will be the number of MS/MS spectra of the related peptides. Again, large proteins produce greater number of tryptic peptides and thus more spectral count. The spectral or peptide count from the samples to be analysed are compared and relative abundance of proteins are determined. Certain methods have been introduced to measure the peptide count i.e., i) protein abundance index [118], ii) exponentially modified protein abundance index [110]. And iii) absolute protein expression [119]. Protein Abundance Index (PAI) is calculated by computing the ratio of the number of peptides observed in the sample within the given specific range of MS to the expected peptides that can be observed after trypsin digestion of analysed protein. This method has been modified by its normalization which distinguishes the sequence on the basis of their characteristics presenting a better correlation for assessment of protein quantity. The method is referred to as the exponential form of PAI represented as emPAI (exponentially modified protein abundance index) and represented as $10^{PAI}-1$ [70]. This is based on the concept that larger proteins have more peptides and thus show a direct correlation to the protein content in the sample. The major drawback of these methods is the neglected intrinsic characteristics of proteins that are surpassed by certain advancements. Another modified version of emPAI is Absolute Protein Expression (APEX) which takes into consideration the physicochemical properties of proteins and analytical features of MS analysis. In this strategy, computational models are used to predict the possibility of tryptic peptides that can be produced within the given dynamic range of MS. This estimate of peptides is then compared to the observed peptides that are actually detected. It deduces a correct order of magnitude to depict the influence of intrinsic features of protein. From the estimate, a correction factor for each molecule of a particular protein is included in the analysis, called as O_i value [120]. In this way, the likelihood of the occurrence of a peptide of particular protein is estimated and gives absolute protein concentration per cell.

Absolute quantification

This strategy is the direct consequence of various modifications and rapid growth in the field of proteomics. The techniques offer accurate measurements of proteins present in the sample by taking into account the reference or internal standards. In this regard, we can classify the methods in two major categories i.e., Global Analysis and Target Analysis.

Global Analysis: In the former approach, a known quantity of protein is added to the sample which is to be quantified. Separate enzymatic digestion is done for known protein and its MS signal intensity is observed and recorded. Then the sample to be analyzed (containing known amount of protein) is separately digested and analysed on MS. The mass spectrometric data obtained for both the known protein and the sample is then compared. The same peptides of reference and test sample produce similar elution profile. The ratio of the signal intensity is thus used to deduce the absolute abundance of the peptides in the sample.

Target Analysis: As described in the former approach, all the peptides of the corresponding protein are considered for analysis. But in target approach, only a few selective peptides are quantified by utilizing isotopic labelling strategy. The most previous approach employed for targeted protein analysis was ELISA but it requires specific antibodies and complete information of the proteins to be analyzed and thus not suitable for analysis of noval proteins. Recently, AQUA strategy has been introduced that employs standard peptide for absolute measurement of particular peptides present in the protein. For reference, targeted peptide which is to be analysed is selected from the protein and labelled using isotopes containing amino acids. This labelled peptide is also referred to as AQUA peptide. The reference peptide is added to the protein sample from which peptide abundance is to be measured. The sample is enzymatically digested and analysed on mass spectrometric tools. The signal intensities appear as duplets for a specific peptide as there is slight mass difference between the AQUA peptide (isotopically labelled) and natural peptide present in sample. The relative intensities of both the peptides are observed and recorded. The calculation of ratio of AQUA peptide to that of sample peptide depicts the absolute quantification of protein.

Another strategy that has been derived from AQUA for simultaneous analysis of more than one sample is QconCAT technique. This multiplex strategy involves the selection of unique peptide from each sample of protein to be analysed. These peptides are then sequenced and a chimeric gene is synthesized using computational tools [73]. The expression of chimeric gene leads to the production of protein representing signature peptide of each protein and is referred to as concatenated protein (QconCAT). The isotopic labelling of QconCAT protein serves as internal standard [121]. It is mixed with different samples of protein to be quantified. Again, enzymatic digestion and analysis on MS tools produces signals that are recorded and used to compare and derive ratios for measurement of absolute peptide abundance.

LC/MS^E approach is based on data independent acquisition and is amenable for absolute and relative quantification. In this technique, the digested peptides are eluted on high resolution chromatographic system and subjected to advanced high mass resolution MS that utilizes triple quadrupole/TOF. The instrument is provided with electromagnetic waves and alternating high and low collision energies. As the product enters the system, gets fragmented in presence of low collision energies to generate precursor ions. However, selection is not done at this level for further splitting and thus termed as data independent analysis. The precursor ions are again fragmented in presence of high collision energies generated by inert gases to yield product ions. The record of retention time, m/z ratio and signal intensities assist in the grouping of generated ions to their respective source. Product ions and precursor ions have similar elution profile and generate signatures, called as Exact Mass Retention Time (EMRT) signatures [122]. The intensity of these ions when analysed on MS, generates particular EMRT signatures for specific peptides and thus can be correlated to the abundance of their corresponding peptides.

Label-free approaches are considered to be the most advanced innovative emerging techniques that attempt to generate highly informative data for identification of novel proteins. The multiplexing technology eliminates the need for laborious sample processes, isotopic labelling and the number of steps involved but requires highly efficient and advanced versions of proteomic tools thereby leading to cost issues. However, it has contributed excessively in the recent years towards identification of proteins and understanding of biological mechanisms.

All the proteomic strategies are complementary to each other associated with several advantages and drawbacks. The integration of all these techniques will be useful to get a clear picture of wide mechanisms taking place in living systems. In this regard, few of the recently published works on plant proteomics have been depicted in Table 1. Incremental improvements in these strategies will lead to uncover orphan genes and gain in-depth knowledge of protein dynamics.

Summary

Table 1: Recent key studies for identification of novel proteins and differential expression studies in plants under specific conditions using proteomic approaches.

S.No.	Plant	Technique	Biological Study	Reference
1.	*Hordeum vulgare* (Barley)	2DE	Differential proteomics for identification of proteins associated with grain quality	Finnie, et al. [14]
2.	*Magnolia seaboldii*	2DE	Comparative protein profiling at seed germination stage	Lu, et al. [29]
3.	*Hordeum vulgare* (Barley)	2DE	Study of proteins produced in response to biotic stress (Fusarium infection) during grain development	Trumper, et al. [30]
4.	Petunia	2DE	Differential expression studies of proteins associated with anthocyanin	Prinsi, et al. [31]
5.	*Prunus persica* (Peach)	2DE	Mesocarp and leaf proteome study for understanding chilling stress response	Almeida, et al. [32]
6.	*Zea mays* (Maize)	2DE	Proteome analysis of mid-rib determining the size of leaf angle	Wang, et al. [33]
7.	*Glycine max* (Soybean)	2DE	Analysis of proteins associated with seed filling	Hadjuch, et al. [34]
8.	*Citrus sinensis* & *Citrus grandis*	2DE	Differential proteomics to demonstrate boron toxicity in two species differing in boron tolerance	Sang, et al. [35]
9.	*Brachypodium distachyon*	2DE-MALDI/TOF	Leaf and root proteome analysis in response to H_2O_2 stress	Bian, et al. [36]
10.	*Pisum sativum* (Pea)	2D-DIGE	Identification of proteins produced in response to *Orobanche crenata*	Castellejo, et al. [37]
11.	*Hordeum vulgare* Barley	ICAT	Detection of thioredoxin target disulfide in proteins released from aleurone layer	Hagglund, et al. [92]
12.	*Solanum lycopersicon* (Tomato)	TMT	Detection of redox proteins responsive to biotic stress (*P.syringae*)	Parker, et al. [38]
13.	*Ricinus communis* (Castor bean)	ICPL & iTRAQ	Analysis of proteins involved in development of endosperm	Nogueria, et al. [93]
14.	*Vitis vinefera* Grape	iTRAQ	Analysis of mesocarp and endocarp proteins synthesized in response to pathogen	Melo-Braga, et al. [26]
15.	*Oryza sativa* (Rice)	iTRAQ	Analysis of cold-responsive proteins	Neilson, et al. [94]
16.	*Triticum aestivum* (Wheat)	iTRAQ	Study of protein responses to biotic stresses (powdery mildew)	Fu, et al. [24]
17.	*Camellina sinensis* (Tea plant)	iTRAQ	Differential protein studies concerned to chlorophyll content and abnormal chloroplast development	Wang, et al. [39]
18.	*Nicotiana tabacum* (Tobacco)	iTRAQ	Understanding TMV resistance mechanism by differential protein analysis of susceptible and resistant strains	Wang, et al. [40]
19.	*Glycine max* (Soybean)	iTRAQ	Proteins and pathways associated with male sterility were revealed	Li, et al. [41]
20.	*Cucumis sativus* (Cucumber)	iTRAQ	Identification of proteins produced in phloem sap in response to salt stress	Fan, et al. [42]
21.	Arabidopsis	iTRAQ-OFFGEL	Detection of proteins required for regulation of iron homeostasis and transport mechanisms	Zargar, et al. [21]

22.	*Medicago trunculata*	iTRAQ-OFFGEL	Analysis of microsomal proteins	Abdallah, et al. [70]
23.	Arabidopsis	iTRAQ-OFFGEL	Proteomic investigation of shoot microsomal proteins to reveal the impact of Fe deficiency on photosynthesis	Zargar, et al. [130]
24.	*Crotolaria juncea*	^{15}N	Production of ^{15}N labeled green manure	Ambrosano, et al. [43]
25.	Sugarcane	^{15}N	^{15}N-labeled nitrogen from green manure and ammonium sulphate utilization by the sugarcane ratoon.	Ambrosano, et al. [44]
26.	*Chlamydomonas reinhardii*	SILAC	Protein analysis in response to salt stress	Mastrobuoni, et al. [95]
27.	*Quercus ilex* (Holm-oak)	2DE & LC-MS/MS	Differential profiling of storage and stress/defense protein by acorn seed analysis	Galvan, et al. [45]
28.	Arabidopsis	2DE & LC-MS	Proteome analysis in response to the smoke-derived growth regulator karrikin	Baldrianova, et al. [46]
29.	*Hevea brasiliensis* (Rubber tree)	2DE, Western Blot & LC-MS	Study of protein regulation under biotic stress (fungal infection)	Havanapan, et al. [47]
30.	*Brassica napus*	2DE & LC-MS/MS	Proteome studies from plastid of developing embryo and leaves	Demartini, et al. [48]
31.	*Triticum aestivum* (Wheat)	2DE & LC-MS/MS	Aleurone layer proteome analysis at different stages of grain development	Nadaud, et al. [49]
32.	*Nicotiana tabacum* (Tobacco)	BN -PAGE & LC-MS/MS	Exploration of BY-2 protein complexes	Remmerie, et al. [50]
33.	*Glycine max* (Soybean)	2DIGE & gel-free shotgun approach	Analysis of inter and intracellular proteins of calli developed from hypocotyls	Miemyk, et al. [51]
34.	*Oryza sativa* (Rice)	Immune affinity purification & LC-MS/MS	Detection of acetylation motifs, lysine acetylated proteins and their localization	Xiong, et al. [52]
35.	Arabidopsis	iTRAQ & LC-MS	Identification of proteins involved in glucosinolate metabolism	Mostafa, et al. [53]
36.	*Jatropha curcas*	Histologica and Transmission electron microscopy analysis	Exploration of structural changes associated with the plastid to gerontoplast transition	Shah, et al. [54]
37.	*Hevea brasiliensis* (Rubber tree)	iTRAQ & LC-MS/MS	Identification of ethylene-/jasmonate responsive proteins to understand defense mechanism	Dai, et al. [55]
38.	*Beta vulgaris* (Sugar Beet)	iTRAQ & 2DLC-MS/MS	Membrane proteome analysis for salt dtress	Li, et al. [56]
39.	*Musa acuminate* (Banana)	Gel-free	Study on the plasma membrane proteome of Banana	Vertommen, et al. [57]
40.	*Fragaria ananassa* (Strawberry)	LC-MS	Proteomic investigation for flavonoid and anthocyanin biosynthesis at different ripening stages	Song, et al. [58]
41.	*Oryza sativa* (Rice)	Label free quantitative MS	Analysis of proteins synthesized during heat stress	Timabud, et al. [59]
42.	*Glycine max* (Soybean)	Label free quantitative MS	Annotation of proteins synthesized in response to jasmonic acid and salicylic acid under flooding stress	Kamal, et al. [60]
43.	*Brassica napus*	Label-free quantitative MS	Analysis of plasma membrane proteins in response to phosphorous deficiency	Chen, et al. [61]
44.	*Solanum lycopersicon* (Tomato)	Gel-LC-Orbitrap-MS	Membrane proteome analysis pertaining to development in male gametophyte	Paul, et al. [62]
45.	*Nelumbo nucifera* (Lotus)	Label-free Shotgun approach	Analysis of proteins involved in cellular dedifferentiation and callus formation	Liu, et al. [63]
46.	*Glycine max* (Soybean)	Label-free quantitative proteomics	Protein analysis under abiotic stresses (flood and drought)	Wang, et al. [39]
47.	*Saccharum officinarum* (Sugarcane)	Shotgun associated with nano ESI-HDMS technology	Identification of proteins associated with somatic embryogenesis development for protection of cells	Reis, et al. [64]
48.	*Araucaria angustifolia*	Label free quantitative proteomics	Protein analysis of embryogenic cell cultures	Santos, et al. [65]
49.	*Momordica charantia* (Bitter Melon)	LC-MS/MS	Seed proteome analysis to identify angiotensin-1 converting enzyme inhibitory peptides.	Priyanto, et al. [66]
50.	Arabidopsis	LC-MS	Proteomic analysis of cytosolic ribosomal proteins	Hummel, et al. [67]

The curiosity to unravel the concealed mysteries of life and various processes involved in its sustenance inspired the researchers to understand the basic regulatory pathways of biological systems. In this regard, tremendous efforts have been made in the field of genomics, transcriptomics, proteomics, metabolomics and ionomics. We have confined this article to proteomics approach only. Initially, protein profiling and protein-protein interactions were performed to study the diverse proteins. Then the focus was shifted to proteome analysis by employing 2DE in the late 20th century. Later, with advancements in technologies, gel-free proteomics strategies came into limelight that attempts to enumerate the functional dimension of proteins by employing more reliable and proficient tools for precise and accurate results. To date, huge numbers of entries have been made concerned to study of insects, worms, animals, human diseases studies [123-126]. These approaches have found wide applicability in the field of medicines for discovery of biomarkers and early disease diagnosis [127]. And evolutionary ecology studies [128]. As well. However, the field of plant proteomics is still in its infancy and needs to be explored due to presence of vast diverse proteins and metabolites. As mentioned above, there is no single strategy that can be considered self-sufficient to evaluate all the proteins. The detection of novel proteins and their dynamics requires combined efforts of modern technologies. The choice of a particular proteomic strategy depends on the biological question under consideration. Amendment in the employed scientific technologies, tools and databases is bound to render extremely informative facts concerned to metabolism. The sub-cellular proteome studies have been carried out in the past years [129]. Recently, the concept of proteome atlas has been introduced that emphasizes the need to integrate specific proteome studies to create a big picture of diverse proteins involved in metabolic and regulatory pathways [130]. The proteome analysis of different organelles at different developmental stages under varied environmental conditions will help to elucidate the metabolic regulations producing diverse proteins and metabolites. In-depth knowledge of these metabolic routes will expand our knowledge and assist in manipulation of key elements to produce desired results and development of superior genotypes having agronomic merit with biotic and abiotic stress tolerance. Thus, we can conclude that the incremental refinement in proteomic technologies and ongoing efforts of the researchers will facilitate unveiling the metabolic networks for deducing unexplored mechanisms and creating wealth of information.

References

1. Wilkins MR, Sanchez JC, Gooley AA, Appel RD, Humphery-Smith I, Hochstrasser DF et al.: Progress with proteome projects: why all proteins expressed by a genome should be identified and how to do it. Biotechnol Genet Eng Rev. 1996;13:19-50.

2. Tyers M, Mann M. From genomics to proteomics. Nature. 2003;422(6928):193-7.

3. Liebler DC. Introduction to Proteomics: Tools for the New Biology. Totowa, New Jersey Humana Press. Inc. 2002.

4. Chevalier F. Highlights on the capacities of "Gel-based" proteomics.

Proteome Sci. 2010;8:23. Doi: 10.1186/1477-5956-8-23.

5. Kamo M, Kawakami T, Miyatake N, Tsugita A: Separation and characterization of Arabidopsis thaliana proteins by two-dimensional gel electrophoresis. Electrophoresis. 1995;16(3):423-30.

6. Santoni V, Rouquié D, Doumas P, Mansion M, Boutry M, Degand H, et al. Use of a proteome strategy for tagging proteins present at the plasma membrane. Plant J. 1998;16(5):633-41.

7. Peltier JB, Friso G, Kalume DE, Roepstorff P, Nilsson F, Adamska I, et al. Proteomics of the chloroplast: systematic identification and targeting analysis of lumenal and peripheral thylakoid proteins. Plant Cell. 2000;12(3):319-41.

8. Kruft V, Eubel H, Jänsch L, Werhahn W, Braun HP. Proteomic approach to identify novel mitochondrial proteins in Arabidopsis. Plant Physiol. 2001;127(4):1694-710.

9. Chivasa S, Ndimba BK, Simon WJ, Robertson D, Yu XL, Knox JP, et al. Proteomic analysis of the Arabidopsis thaliana cell wall. Electrophoresis. 2002;23(11):1754-65.

10. Bae MS, Cho EJ, Choi EY, Park OK. Analysis of the Arabidopsis nuclear proteome and its response to cold stress. Plant J. 2003;36(5):652-63.

11. Borner GH, Lilley KS, Stevens TJ, Dupree P. Identification of glycosylphosphatidylinositol-anchored proteins in Arabidopsis. A proteomic and genomic analysis. Plant Physiol. 2003;132(2):568-77.

12. Porubleva L, Vander Velden K, Kothari S, Oliver DJ, Chitnis PR. The proteome of maize leaves: use of gene sequences and expressed sequence tag data for identification of proteins with peptide mass fingerprints. Electrophoresis. 2001;22(9):1724-38.

13. Gallardo K, Job C, Groot SP, Puype M, Demol H, Vandekerckhove J, et al. Proteomics of Arabidopsis Seed Germination. A Comparative Study of Wild-Type and Gibberellin-Deficient Seeds. Plant Physiol. 2002;129(2):823-37.

14. Finnie C, Melchior S, Roepstorff P, Svensson B. Proteome Analysis of Grain Filling and Seed Maturation in Barley. Plant Physiol. 2002;129(3):1308-19.

15. Bhushan D, Jaiswal DK, Ray D, Basu D, Datta A, Chakraborty S, et al. Dehydration-responsive reversible and irreversible changes in the extracellular matrix: Comparative proteomics of chickpea genotypes with contrasting tolerance. J Proteome Res. 2011;10(4):2027-46. Doi: 10.1021/pr200010f.

16. Staudinger C, Mehmeti V, Turetschek R, Lyon D, Egelhofer V, Wienkoop S. Possible role of nutritional priming for early salt and drought stress responses in Medicago truncatula. Front Plant Sci. 2012;3:285. Doi: 10.3389/fpls.2012.00285.

17. Zadražnik T, Hollung K, Egge-Jacobsen W, Meglič V, Šuštar-Vozlič J, et al. Differential proteomic analysis of drought stress response in leaves of common bean (Phaseolus vulgaris L.). J Proteomics. 2013;78:254-72. Doi: 10.1016/j.jprot.2012.09.021.

18. Muneer S, Hakeem KR, Mohamed R, Lee JH. Cadmium toxicity induced alterations in the root proteome of green gram in contrasting response towards iron supplement. Int J Mol Sci. 2014;15(4):6343-55. Doi: 10.3390/ijms15046343.

19. Irar S, González EM, Arrese-Igor C, Marino D. A proteomic approach reveals new actors of nodule response to drought in split-root grown pea plants. Physiol Plant. 2014;152(4):634-45. Doi: 10.1111/ppl.12214.

20. Swigonska S, Weidner S. Proteomic analysis of response to long-

term continuous stress in roots of germinating soybean seeds. J Plant Physiol. 2013;170(5):470-9. Doi: 10.1016/j.jplph.2012.11.020.

21. Zargar SM, Kurata R, Inaba S, Fukao Y. Unravelling the iron deficiency responsive proteome in Arabidopsis shoot by iTRAQ-OFF GEL approach. Plant Signal Behav. 2013;8(10): Doi: 10.4161/psb.26892.

22. Sato D, Akashi H, Sugimoto M, Tomita M, Soga T. Metabolomic profiling of there sponse of susceptible and resistant soybean strains to foxglove aphid, Aulacorthum solani. Kaltenbach. J Chromatogr B Analyt Technol Biomed Life Sci. 2013;925:95-103. Doi: 10.1016/j.jchromb.2013.02.036.

23. Watson BS, Bedair MF, Urbanczyk-Wochniak E, Huhman DV, Yang DS, Allen SN, et al. Integrated metabolomics and transcriptomics reveal enhanced specialized metabolism in Medicago truncatula root Border Cells. Plant Physiol. 2015;167(4):1699-716. Doi: 10.1104/pp.114.253054.

24. Fu Y, Zhang H, Mandal SN, Wang C, Chen C, Ji W, et al. Quantitative proteomics reveals the central changes of wheat in response to powdery mildew. J Proteomics. 2016;130:108-19. Doi: 10.1016/j.jprot.2015.09.006.

25. Kaffarnik FA, Jones AM, Rathjen JP, Peck SC. Effector proteins of the bacterial pathogen Pseudomonas syringae alter the extracellular proteome of the host plant, Arabidopsis thaliana. Mol Cell Proteomics. 2009;8(1):145-56. Doi: 10.1074/mcp.M800043-MCP200.

26. Melo-Braga MN, Verano-Braga T, León IR, Antonacci D, Nogueira FC, Thelen JJ, et al. Modulation of protein phosphorylation, glycosylation and acetylation in grape (Vitis vinifera) mesocarp and exocarp due to Lobesia botrana infection. Mol Cell Proteomics. 2012;11(10):945-56.

27. Marsh E, Alvarez S, Hicks LM, Barbazuk WB, Qiu W, Kovacs L, et al. Changes in protein abundance during powdery mildew infection of leaf tissues of Cabernet Sauvignon grapevine (Vitis vinifera L.). Proteomics. 2010;10(10):2057-64. Doi: 10.1002/pmic.200900712.

28. Fan J, Chen C, Yu Q, Brlansky RH, Li ZG, Gmitter FG Jr et al. Comparative iTRAQ proteome and transcriptome analyses of sweet orange infected by, 'Candidatus Liberibacter asiaticus'. Physiol Plant. 2011;143(3):235-45. Doi: 10.1111/j.1399-3054.2011.01502.x.

29. Lu XJ, Zhang XL, Mei M, Liu GL, Ma BB. Proteomic analysis of Magnolia sieboldii K. Koch seed germination. J Proteomics. 2016;133:76-85. Doi: 10.1016/j.jprot.2015.12.005.

30. Trümper C, Paffenholz K, Smit I, Kössler P, Karlovsky P, Braun HP, et al. Identification of regulated proteins in naked barley grains (Hordeum vulgare nudum) after Fusarium graminearum infection at different grain ripening stages. J Proteomics. 2016;133:86-92. Doi: 10.1016/j.jprot.2015.11.015.

31. Prinsi B, Negri AS, Quattrocchio FM, Koes RE, Espen L.Proteomics of red and white corolla limbs in petunia reveals a novel function of the anthocyanin regulator ANTHOCYANIN1 in determining flower longevity. J Proteomics. 2016;131:38-47. doi: 10.1016/j.jprot.2015.10.008.

32. Almeida AM, Urra C, Moraga C, Jego M, Flores A, Meisel L, et al. Proteomic analysis of a segregant population reveals candidate proteins linked to mealiness in peach. J Proteomics. 2016;131:71-81. doi: 10.1016/j.jprot.2015.10.011.

33. Wang N, Cao D, Gong F, Ku L, Chen Y, Wang W, et al. Differences in properties and proteomes of the midribs contribute to the size of the leaf angle in two near-isogenic maize lines. J Proteomics. 2015;128:113-22. Doi: 10.1016/j.jprot.2015.07.027.

34. Hajduch M, Ganapathy A, Stein JW, Thelen JJ. A systematic proteomic study of seed filling in soybean: establishment of high-resolution two-dimensional reference maps, expression profiles, and an interactive proteome database. Plant Physiol. 2005;137(4):1397-419.

35. Sang W, Huang ZR, Qi YP, Yang LT, Guo P, Chen LS, et al. An investigation of boron-toxicity in leaves of two citrus species differing in boron-tolerance using comparative proteomics. J Proteomics. 2015;123:128-46. Doi: 10.1016/j.jprot.2015.04.007.

36. Bian YW, Lv DW, Cheng ZW, Gu AQ, Cao H, Yan YM, et al. Integrative proteome analysis of Brachypodium distachyon roots and leaves reveals a synergetic responsive network under H_2O_2 stress. J Proteomics. 2015;128:388-402. Doi: 10.1016/j.jprot.2015.08.020.

37. Castillejo MÁ, Fernández-Aparicio M, Rubiales D. Proteomic analysis by two dimensional in gel electrophoresis (2D DIGE) of the early response of Pisum sativum to Orobanche crenata. J Exp Bot. 2012;63(1):107-19. Doi: 10.1093/jxb/err246.

38. Parker J, Zhu N, Zhu M, Chen S. Profiling thiol redox proteome using isotope tagging mass pectrometry. J Vis Exp. 2012;(61). Pii: 3766. Doi: 10.3791/3766.

39. Wang J, Wang XR, Zhou Q, Yang JM, Guo HX, Yang LJ, et al. iTRAQ protein profile analysis provides integrated insight into mechanism of tolerance to TMV in Tobacco (Nicotiana tabacum). J Proteomics. 2016;132:21-30. Doi: 10.1016/j.jprot.2015.11.009.

40. Wang L, Cao H, Chen C, Yue C, Hao X, Yang Y, et al. Complementary transcriptomic and proteomic analyses of a chlorophyll-deficient tea plant cultivar reveal multiple metabolic pathway changes. J Proteomics. 2016;130:160-9. Doi: 10.1016/j.jprot.2015.08.019.

41. Li J, Ding X, Han S, He T, Zhang H, Yang L, et al. Differential proteomics analysis to identify proteins and pathways associated with male sterility of soybean using iTRAQ-based strategy. J Proteomics. 2016;138:72-82. Doi: 10.1016/j.jprot.2016.02.017.

42. Fan H, Xu Y, Du C, Wu X. Phloem sap proteome studied by iTRAQ provides integrated insight into salinity response mechanisms in cucumber plants. J Proteomics. 2015;125:54-67. doi: 10.1016/j.jprot.2015.05.001.

43. Ambrosano EJ, Trivelin PCO, Cantarella H, et al. Nitrogen-15 labelling of Crotolaria juncea green manure. Scientia Agricola. 2003;60(1):181-184.

44. Ambrasano EJ, Trivelin PCO, Cantarella H, Ambrasano GMB, Schammass EA, Muraoka T , Rossi F. ^{15}N-labeled nitrogen from green manure and ammonium sulphate utilization by the sugarcane ratoon. Scientia Agricola. 2011;68(3):361-368.

45. Valero Galván J, Valledor L, Navarro Cerrillo RM, Gil Pelegrín E, Jorrín-Novo JV. Studies of variability in Holm oak (Quercus ilex subsp. Ballota [Desf.] Samp.) Through acorn protein profile analysis. J Proteomics. 2011;74(8):1244-55. Doi: 10.1016/j.jprot.2011.05.003.

46. Baldrianová J, Černý M, Novák J, Jedelský PL, Divíšková E, Brzobohatý B, et al. Arabidopsis proteome responses to the smoke-derived growth regulator karrikin. J Proteomics. 2015;120:7-20. Doi: 10.1016/j.jprot.2015.02.011.

47. Havanapan PO, Bourchookarn A, Ketterman AJ, Krittanai C. Comparative proteome analysis of rubber latex serum from pathogenic fungi tolerant and susceptible rubber tree (Hevea brasiliensis). J Proteomics. 2016;131:82-92. Doi: 10.1016/j.jprot.2015.10.014.

48. Demartini DR, Jain R, Agrawal G, Thelen JJ. Proteomic comparison of plastids from developing embryos and leaves of Brassica napus. J Proteome Res. 2011;10(5):2226-37. doi: 10.1021/pr101047y.

49. Nadaud I, Tasleem-Tahir A, Chateigner-Boutin AL, Chambon C, Viala D, Branlard G et al. Proteome evolution of wheat (Triticum aestivum L.) aleurone layer at fifteen stages of grain development. J Proteomics. 2015;123:29-41. doi: 10.1016/j.jprot.2015.03.008.

50. Remmerie N, De Vijlder T, Valkenborg D, Laukens K, Smets K, Vreeken J, et al. Unravelling tobacco BY-2 protein complexes with BN PAGE/LC-MS/MS and clustering methods. J Proteomics. 2011;74(8):1201-17. Doi: 10.1016/j.jprot.2011.03.023.

51. Miernyk JA, Jett AA, Johnston ML. Analysis of soybean tissue culture protein dynamics using difference gel electrophoresis. J Proteomics. 2016;130:56-64. Doi: 10.1016/j.jprot.2015.08.023.

52. Xiong Y, Peng X, Cheng Z, Liu W, Wang GL. A comprehensive catalog of the lysine-acetylation targets in rice (Oryza sativa) based on proteomic analyses. J Proteomics. 2016;138:20-9. Doi: 10.1016/j.jprot.2016.01.019.

53. Mostafa I, Zhu N, Yoo MJ, Balmant KM, Misra BB, Dufresne C, et al. New nodes and edges in the glucosinolate molecular network revealed by proteomics and metabolomics of Arabidopsis myb28/29 and cyp79B2/B3 glucosinolate mutants. J Proteomics. 2016;138:1-19. Doi: 10.1016/j.jprot.2016.02.012.

54. Shah M, Soares EL, Lima ML, Pinheiro CB, Soares AA, Domont GB, et al. Deep proteome analysis of gerontoplasts from the inner integument of developing seeds of Jatropha curcas. J Proteomics. 2016. Pii: S1874-3919(16)30047-1. Doi: 10.1016/j.jprot.2016.02.025.

55. Dai L, Kang G, Nie Z, Li Y, Zeng R.Comparative proteomic analysis of latex from Hevea brasiliensis treated with Ethrel and methyl jasmonate using iTRAQ-coupled two-dimensional LC-MS/MS. J Proteomics. 2016;132:167-75. doi: 10.1016/j.jprot.2015.11.012.

56. Li H, Pan Y, Zhang Y, Wu C, Ma C, Yu B, et al. Salt stress response of membrane proteome of sugar beet monosomic addition line M14. J Proteomics. 2015;127(Pt A):18-33. Doi: 10.1016/j.jprot.2015.03.025.

57. Vertommen A, Møller AL, Cordewener JH, Swennen R, Panis B, Finnie C, et al. A workflow for peptide-based proteomics in a poorly sequenced plant: A case study on the plasma membrane proteome of banana. J Proteomics. 2011;74(8):1218-29. doi: 10.1016/j.jprot.2011.02.008.

58. Song J, Du L, Li L, Kalt W, Palmer LC, Fillmore S, et al. Quantitative changes in proteins responsible for flavonoid and anthocyanin biosynthesis in strawberry fruit at different ripening stages: A targeted quantitative proteomic investigation employing multiple reaction monitoring. J Proteomics. 2015;122:1-10. doi: 10.1016/j.jprot.2015.03.017.

59. Timabud T, Yin X, Pongdontri P, Komatsu S. Gel-free/label-free proteomic analysis of developing rice grains under heat stress. J Proteomics. 2016;133:1-19. Doi: 10.1016/j.jprot.2015.12.003.

60. Kamal AH, Komatsu S. Jasmonic acid induced protein response to biophoton emissions and flooding stress in soybean. J Proteomics. 2016;133:33-47. Doi: 10.1016/j.jprot.2015.12.004.

61. Chen S, Luo Y, Ding G, Xu F. Comparative analysis of Brassica napus plasma membrane proteins under phosphorus deficiency using label-free and MaxQuant-based proteomics approaches. J Proteomics. 2016;133:144-52. Doi: 10.1016/j.jprot.2015.12.020.

62. Paul P, Chaturvedi P, Selymesi M, Ghatak A, Mesihovic A, Scharf KD, et al. The membrane proteome of male gametophyte in Solanum lycopersicum. J Proteomics. 2016;131:48-60. Doi: 10.1016/j.jprot.2015.10.009.

63. Liu Y, Chaturvedi P, Fu J, Cai Q, Weckwerth W, Yang P, et al. Induction

64. Reis RS, Vale Ede M, Heringer AS, Santa-Catarina C, Silveira V. Putrescine induces somatic embryo development and proteomic changes in embryogenic callus of sugarcane. J Proteomics. 2016;130:170-9. Doi: 10.1016/j.jprot.2015.09.029.

65. dos Santos AL, Elbl P, Navarro BV, de Oliveira LF, Salvato F, Balbuena TS, et al. Quantitative proteomic analysis of Araucaria angustifolia (Bertol.) Kuntze cell lines with contrasting embryogenic potential. J Proteomics. 2016;130:180-9. Doi: 10.1016/j.jprot.2015.09.027.

66. Priyanto AD, Doerksen RJ, Chang CI, Sung WC, Widjanarko SB, Kusnadi J, et al. Screening, discovery, and characterization of angiotensin-1 converting enzyme inhibitory peptides derived from proteolytic hydrolysate of bitter melon seed proteins. J Proteomics. 2015;128:424-35. doi: 10.1016/j.jprot.2015.08.018.

67. Hummel M, Dobrenel T, Cordewener JJ, Davanture M, Meyer C, Smeekens SJ, et al. Proteomic LC-MS analysis of Arabidopsis cytosolic ribosomes: Identification of ribosomal protein paralogs and re-annotation of the ribosomal protein genes. J Proteomics. 2015;128:436-49. Doi: 10.1016/j.jprot.2015.07.004.

68. O'Farrell PH. High resolution two-dimensional electrophoresis of proteins. J Biol Chem. 1975;250(10):4007-21.

69. Smithies O. How it all began: A personal history of gel electrophoresis. Methods Mol Biol. 2012;869:1-21. Doi: 10.1007/978-1-61779-821-4_1.

70. Abdallah C, Dumas-Gaudot E, Renaut J, Sergeant K. Gel-Based and Gel-Free Quantitative Proteomics Approaches at a Glance. Int J Plant Genomics. 2012;2012:494572. doi: 10.1155/2012/494572.

71. Weber K, Osborn M. The reliability of molecular weight determinations by dodecyl sulfate-polyacrylamide gel electrophoresis. J Biol Chem. 1969;244(16):4406-12.

72. Switzer RC 3rd, Merril CR, Shifrin S. A highly sensitive silver stain for detecting proteins and peptides in polyacrylamide gels. Anal. Anal Biochem. 1979;98(1):231-7.

73. Deracinois B, Flahaut C, Duban-Deweer S, Karamanos Y. Comparative and Quantitative Global Proteomics Approaches: An Overview Proteomes. 2013;1:180-218. Doi: 10.3390/ proteomes 1030180.

74. Liebler DC. Introduction to Proteomics: Tools for the New Biology. Totowa, New Jersey Humana Press. Inc. 2002.

75. Zhou S, Bailey MJ, Dunn MJ, Preedy VR, Emery PW. A quantitative investigation into the losses of proteins at different stages of a two-dimensional gel electrophoresis procedure. Proteomics. 2005;5(11):2739-47.

76. Lilley KS, Razzaq A, Dupree P. Two-dimensional gel electrophoresis: recent advances in sample preparation, detection and quantitation. Curr Opin Chem Biol. 2002;6(1):46-50.

77. Gygi SP, Corthals GL, Zhang Y, Rochon Y, Aebersold R. Evaluation of two-dimensional gel electrophoresis- based proteome analysis technology. Proc Natl Acad Sci U S A. 2000;97(17):9390-5.

78. Borner GH, Lilley KS, Stevens TJ, Dupree P. Identification of glycosyl phosphatidylinositol-anchored proteins in Arabidopsis. A proteomic and genomic analysis. Plant Physiol. 2003;132(2):568-77.

79. Unlü M, Morgan ME, Minden JS. Difference gel electrophoresis: A single gel method for detecting changes in protein extracts. Electrophoresis. 1997;18(11):2071-7.

80. Khatoon A, Rehman S, Hiraga S, Makino T, Komatsu S. Organ-specific proteomics analysis for identification of response mechanism in soybean seedlings under flooding stress. J Proteomics. 2012;75(18):5706-23. Doi: 10.1016/j.jprot.2012.07.031.

81. Salavati A, Khatoon A, Nanjo Y, Komatsu S. Analysis of proteomic changes in roots of soybean seedlings during recovery after flooding. J Proteomics. 2012;75(3):878-93. Doi: 10.1016/j.jprot.2011.10.002.

82. Schenkluhn L, Hohnjec N, Niehaus K, Schmitz U, Colditz F. Differential gel electrophoresis (DIGE) to quantitatively monitor early symbiosis- and pathogenesis-induced changes of the Medicago truncatula root proteome. J Proteomics. 2010;73(4):753-68. doi: 10.1016/j.jprot.2009.10.009.

83. Sergeant K, Spiess N, Renaut J, Wilhelm E, Hausman JF. One dry summer: a leaf proteome study on the response of oak to drought exposure. J Proteomics. 2011;74(8):1385-95. Doi: 10.1016/j.jprot.2011.03.011.

84. Li T, Xu SL, Oses-Prieto JA, Putil S, Xu P, Wang RJ, et al. Proteomics analysis reveals post-translational mechanisms for cold-induced metabolic changes in Arabidopsis. Mol Plant. 2011;4(2):361-74. doi: 10.1093/mp/ssq078.

85. Fu C, Hu J, Liu T, Ago T, Sadoshima J, Li H. Quantitative Analysis of Redox-sensitive Proteome with DIGE and ICAT. J Proteome Res. 2008;7(9):3789-802. doi: 10.1021/pr800233r.

86. Braisted JC, Kuntumalla S, Vogel C, Marcotte EM, Rodrigues AR, Wang R, et al. The APEX Quantitative Proteomics Tool: Generating protein quantitation estimates from LC-MS/MS proteomics results. BMC Bioinformatics. 2008;9:529. Doi: 10.1186/1471-2105-9-529.

87. Charro N, Hood BL, Faria D, Pacheco P, Azevedo P, Lopes C, et al. Serum proteomics signature of cystic fibrosis patients: a complementary 2-DE and LC-MS/MS approach. J Proteomics. 2011;74(1):110-26. doi: 10.1016/j.jprot.2010.10.001.

88. Finamore F, Pieroni L, Ronci M, Marzano V, Mortera SL, Romano M, et al. Proteomics investigation of human platelets by shotgun nUPLC-MSE and 2DE experimental strategies: a comparative study. Blood Transfus. 2010;8 Suppl 3:s140-8. doi: 10.2450/2010.021S.

89. Syahir a, Usui K, Tomizaki K, Kajikawa K, Mihara H. Label and Label-free Detection Techniques for Protein Microarrays. Microarrays. 2015;4:228-244.

90. Bantscheff M, Lemeer S, Savitski MM, Kuster B. Quantitative mass spectrometry in proteomics: a critical review. Anal Bioanal Chem. 2012;404(4):939-65. Doi: 10.1007/s00216-012-6203-4.

91. Yang Y, Qiang X, Owsiany K, Zhang S, Thannhauser TW, Li L, et al. Evaluation of different multidimensional LC- MS/MS pipelines for isobaric tags for relative and absolute quantitation (iTRAQ)-based proteomic analysis of potato tubers in response to cold storage. J Proteome Res. 2011;10(10):4647-60. Doi: 10.1021/pr200455s.

92. Hägglund P, Bunkenborg J, Yang F, Harder LM, Finnie C, Svensson B, et al. Identification of thioredoxin target disulfides in proteins released from barley aleurone layers. J Proteomics. 2010;73(6):1133-6. doi: 10.1016/j.jprot.2010.01.007.

93. Nogueira FC, Palmisano G, Schwämmle V, Campos FA, Larsen MR, Domont GB, et al. Performance of isobaric and isotopic labeling in quantitative plant proteomics. J Proteome Res. 2012;11(5):3046-52. Doi: 10.1021/pr300192f.

94. Neilson KA, Mariani M, Haynes PA. Quantitative proteomic analysis of cold-responsive proteins in rice. Proteomics. 2011;11(9):1696-706. Doi: 10.1002/pmic.201000727.

95. Mastrobuoni G, Irgang S, Pietzke M, Assmus HE, Wenzel M, Schulze WX, et al. Proteome dynamics and early salt stress response of the photosynthetic organism Chlamydomonas reinhardtii. BMC Genomics. 2012;13:215.

96. Gygi SP, Rochon Y, Franza BR, Aebersold R. Correlation between protein and mRNA abundance in yeast. Mol Cell Biol. 1999;19(3):1720-30.

97. Qiu Y, Sousa EA, Hewick RM, Wang JH. Acid-Labile isotope-coded extractants: A class of reagents for quantitative mass spectrometric analysis of complex protein mixtures. Anal Chem. 2002;74(19):4969-79.

98. Bottari P, Aebersold R, Turecek F, Gelb MH. Design and synthesis of visible isotope-coded affinity tags for the absolute quantification of specific proteins in complex mixtures. Bioconjug Chem. 2004;15(2):380-8.

99. Schmidt A, Kellermann J, Lottspeich F. A novel strategy for quantitative proteomics using isotope-coded protein labels. Proteomics. 2005;5(1):4-15.

100. Leroy B, Rosier C, Erculisse V, Leys N, Mergeay M, Wattiez R, et al. Differential proteomic analysis using isotope- coded protein-labeling strategies: comparison, improve- ments and application to simulated microgravity effect on Cupriavidus metallidurans CH34. Proteomics. 2010;10(12):2281-91. doi: 10.1002/pmic.200900286.

101. Lottspeich F, Kellermann J. ICPL labeling strategies for proteome research. Method. Methods Mol Biol. 2011;753:55-64. Doi: 10.1007/978-1-61779-148-2_4.

102. Mann M, Kulak NA, Nagaraj N, Cox J. The coming age of complete, accurate, and ubiquitous proteomes. Mol Cell. 2013;49(4):583-90. Doi: 10.1016/j.molcel.2013.01.029.

103. Ross PL, Huang YN, Marchese JN, Williamson B, Parker K, Hattan S, et al. Multiplexed protein quantitation in Saccharomyces cerevisiae using amine-reactive isobaric tagging reagents. Mol Cell Proteomics. 2004;3(12):1154-69.

104. Dayon L, Hainard A, Licker V, Turck N, Kuhn K, Hochstrasser DF, et al. Relative quantification of proteins in human cerebrospinal fluids by MS/MS using 6-plex isobaric tags. Anal Chem. 2008;80(8):2921-31. doi: 10.1021/ac702422x.

105. Niles R, Witkowska HE, Allen S, Hall SC, Fisher SJ, Hardt M et al. Acid-Catalyzed oxygen-18 labeling of peptides. Anal Chem. 2009;81(7):2804-9. doi: 10.1021/ac802484d.

106. Goodlett DR, Keller A, Watts JD, Newitt R, Yi EC, Purvine S, et al. Differential stable isotope labeling of peptides for quantitation and de novo sequence derivation. Rapid Commun Mass Spectrom. 2001;15(14):1214-21.

107. Ong SE, Blagoev B, Kratchmarova I, Kristensen DB, Steen H, Pandey A, et al. Stable isotope labeling by amino acids in cell culture, SILAC, as a simple and accurate approach to expression proteomics. Mol Cell Proteomics. 2002;1(5):376-86.

108. Yates JR, Ruse CI, Nakorchevsky A. Proteomics by mass spectrometry: approaches, advances, and applications. Annu Rev Biomed Eng. 2009;11:49-79. doi: 10.1146/annurev-bioeng-061008-124934.

109. Ong SE, Mittler G, Mann M. Identifying and quantifying in vivo methylation sites by heavy methyl SILAC. Nat Methods. 2004;1(2):119-26.

110. Ishihama Y, Oda Y, Tabata T, Sato T, Nagasu T, Rappsilber J, et al. Exponentially modified protein abundance index (emPAI) for estimation of absolute protein amount in proteomics by the

number of sequenced peptides per protein. Mol Cell Proteomics. 2005;4(9):1265-72.

111. Chelius D, Bondarenko PV. Quantitative profiling of proteins in complex mixtures using liquid chromatography and mass spectrometry. J Proteome Res. 2002;1(4):317-23.

112. Liu H, Sadygov RG, Yates JR 3rd. A model for random sampling and estimation of relative protein abundance in shotgun proteomics. Anal Chem. 2004;76(14):4193-201.

113. Eng JK, McCormack AL, Yates JR. An approach to correlate Tandem Mass Spectral Data of peptides with amino acid sequences in a protein database. J Am Soc Mass Spectrom. 1994;5(11):976-89. doi: 10.1016/1044-0305(94)80016-2.

114. Wang M1, You J, Bemis KG, Tegeler TJ, Brown DP. Label-Free mass spectrometry-based protein quantification technologies in proteomic analysis. Brief Funct Genomic Proteomic. 200;7(5):329-39. doi: 10.1093/bfgp/eln031.

115. Strittmatter EF, Ferguson PL, Tang K, Smith RD. Proteome analyses using accurate mass and elution time peptide tags with capillary LC time-of-flight mass spectrometry. J Am Soc Mass Spectrom. 2003;14(9):980-91.

116. Zimmer JS, Monroe ME, Qian WJ, Smith RD. Advances in Proteomics Data Analysis and Display Using an Accurate Mass and Time Tag Approach. Mass Spectrom Rev. 2006;25(3):450-82.

117. Zhang B, VerBerkmoes NC, Langston MA, Uberbacher E, Hettich RL, Samatova NF et al. Detecting differential and correlated protein expression in label- free shotgun proteomics. J Proteome Res. 2006;5(11):2909-18.

118. Rappsilber J, Ryder U, Lamond AI, Mann M. Large-Scale proteomic analysis of the human spliceosome. Genome Res. 2002;12(8):1231-45.

119. Lu P, Vogel C, Wang R, Yao X, Marcotte EM. Absolute protein expression profiling estimates the relative contributions of transcriptional and translational regulation. Nat Biotechnol. 2007;25(1):117-24.

120. Braisted JC, Kuntumalla S, Vogel C, Marcotte EM, Rodrigues AR, Wang R, et al. The APEX Quantitative Proteomics Tool: Generating protein quantitation estimates from LC-MS/MS proteomics results. BMC Bioinformatics. 2008;9:529. doi: 10.1186/1471-2105-9-529.

121. Beynon RJ, Doherty MK, Pratt JM, Gaskell SJ. Multiplexed absolute quantification in proteomics using artificial QCAT proteins of concatenated signature peptides. Nat Methods. 2005;2(8):587-9.

122. Silva JC, Denny R, Dorschel CA, Gorenstein M, Kass IJ, Li GZ, et al. Quantitative proteomic analysis by accurate mass retention time pairs. Anal Chem. 2005;77(7):2187-200.

123. Cao X, Fu Z, Zhang M, Han Y, Han H, Han Q, et al. iTRAQ-based comparative proteomic analysis of excretory-secretory proteins of schistosomula and adult worms of *Schistosoma japonicum*. J Proteomics. 2016;138:30-9. doi: 10.1016/j.jprot.2016.02.015.

124. Brandi J, Pozza ED, Dando I, Biondani G, Robotti E, Jenkins R, et al. Secretome protein signature of human pancreatic cancer stem-like cells. J Proteomics. 2016;136:1-12. doi: 10.1016/j.jprot.2016.01.017.

125. Bruschi M, Santucci L, Ravera S, Candiano G, Bartolucci M, Calzia D, et al. Human urinary exosome proteome unveils its aerobic respiratory ability. J Proteomics. 2016;136:25-34. doi: 10.1016/j.jprot.2016.02.001.

126. Borges MH, Figueiredo SG, Leprevost FV, De Lima ME, Cordeiro Mdo N, Diniz MR, et al. Venomous extract protein profile of Brazilian tarantula Grammostola iheringi: searching for potential biotechnological applications. J Proteomics. 2016;136:35-47. doi: 10.1016/j.jprot.2016.01.013.

127. Khan A, Khan AU. Biomarker Discovery and Drug Development: A Proteomics Approach. J Proteomics Bio inform. 2012;5:3.

128. Baer B, Millar AH. Proteomics in evolutionary ecology. J Proteomics. 2016;135:4-11. doi: 10.1016/j.jprot.2015.09.031.

129. Rossignol M, Analysis of the plant proteome. Current Opinion in Biotechnology. 2001;12(2): 131-134.

130. Zargar SM, Kurata R, Inaba S, Oikawa A, Fukui R, Ogata Y, et al. Quantitative proteomics of Arabidopsis shoot microsomal proteins reveals a cross-talk between excess zinc and iron deficiency. Proteomics. 2015;15(7):1196-201. doi: 10.1002/pmic.201400467.

11

Impact of Abiotic Elicitors on *In vitro* Production of Plant Secondary Metabolites

Poornananda M. Naik and Jameel M. Al-Khayri*

Department of Agricultural Biotechnology, College of Agriculture and Food Sciences, King Faisal University, P.O. Box 420, Al-Hassa 31982, Saudi Arabia

***Corresponding author:** Jameel M. Al-Khayri, Department of Agricultural Biotechnology, College of Agriculture and Food Sciences, King Faisal University, P.O. Box 420, Al-Hassa 31982, Saudi Arabia, E-mail: jkhayri@kfu.edu.sa*

Abstract

A wide variety of secondary metabolites are synthesized from primary metabolites by plants which are used for the defense purpose. The secondary metabolites had a great scope in the pharmaceuticals, food additives, flavors, and industrial applications. The secondary metabolites are accumulated in the plant body due to stress. The production of plant secondary metabolites by cultivation of plants and chemical synthesis are important agronomic and industrial objectives. The chemical synthesis in most cases has not been economically feasible. The alternative promising option is in vitro culture, which represents a potential source of bioactive compounds, but very few cultures synthesize secondary metabolites in comparison to those produced in intact plants. Elicitor is the one of the stress agent that enhances the production of secondary metabolites in a particular tissue, organs and cells. Elicitors are classified into biotic and abiotic based on their nature. In recent years the use of elicitors in the plant tissue culture has opened a new path for the production of secondary metabolite compounds. Abiotic elicitors are of non-biological origin, includes metals, light, osmotic, drought, salinity, thermal and hormonal elicitors. Abiotic elicitors have different effects on the cellular processes in the plant system, such as growth, photosynthesis, carbon partitioning, carbohydrate and lipid metabolism, osmotic homeostasis, protein synthesis, and gene expression. The present review deals with the effects of different abiotic elicitors on the production of secondary metabolites from in vitro culture.

Keywords: Abiotic Elicitor; Callus; Cell Culture; Hairy Roots; Secondary Metabolites; Stress

Introduction

Plants are the complex organisms, it forms an important part of our everyday diet and their constituents and nutritional value have been intensively studied for decades. In addition to essential primary metabolites like carbohydrates, lipids and amino acids, higher plants are also able to synthesize a number of low molecular weight compounds called the secondary metabolites. Plant secondary metabolites are the diverse group of organic compounds that are produced by plants to facilitate interaction with the biotic and abiotic environment to establish the defense mechanism [1]. Plant secondary metabolites are unique sources of pharmaceuticals, food additives, flavors and industrially important biochemicals [2]. Plants will continuously produce the novel products as well as chemical models for new drugs in the coming centuries, because the chemistry of the majority of plant species is yet to be characterized. The advent of chemical analyses and the characterization of molecular structures have helped in precisely identifying these plants and correlating them with their activity under controlled experimentation. Despite advancements in synthetic chemistry, we still depend upon biological sources for a number of secondary metabolites including pharmaceuticals.

The plant, cell, tissue and organ culture techniques have emerged as an escapable tool with the possibilities of complimenting and supplementing the conventional method in plant breeding, plant improvement and biosynthetic pathways. Plant tissue culture plays a major role in conservation of germplasm, rapid clonal propagation, regeneration of genetically manipulated superior clones, production of secondary metabolites and ex vitro conservation of valuable phytodiversity [3,4]. Especially, plant cell and organ cultures are promising technologies to obtain plant-specific valuable metabolites [5]. Cell and organ cultures have a higher rate of metabolism than field grown plants because the initiation of cell and organ growth in culture leads to the rapid proliferation and to a condensed biosynthetic cycle [6]. Callus induction is necessary, as the first step, in many tissue culture experiments. Callus and cell suspension can be used for long-term cell cultures maintenance. Cell suspension culture systems could be used for large scale culturing of plant cells from which secondary metabolites could be extracted. The advantage of this method is that it can ultimately provide a continuous, reliable source of natural products. Due to the limited availability and complexity of chemical synthesis, plant cell culture becomes an alternative route for large-scale production of this desired compound [7].

Recent research in the in vitro culture systems, a wide variety of elicitors have been employed in order to modify cell metabolism. These modifications are designed to enhance the productivity of useful metabolites in the cultures of the plant cells/tissues. The cultivation period in particular, can be reduced by the application of elicitors, although maintaining high concentrations of product [6]. "Elicitor is a scientifically described term for stress factors that directly or indirectly triggers the inducible defense changes in a plant system that results in an activation of array of protection mechanisms, including induction or expansion of biosynthesis of fine chemicals which do have a major role in the adaptation of plants to the stressful environment" [8].

Classification of Elicitors

Elicitors can be divided into two types on the basis of nature, biotic and abiotic. Biotic elicitors are the substances of biological origin, which includes polysaccharides originated from plant cell walls (chitin, pectin, cellulose, etc.) and micro-organisms. Abiotic elicitors consist of the substances that are of non-biological origin and are grouped into physical, chemical and hormonal factor. The classification of abiotic elicitor is depicted in figure 1.

Abiotic Elicitors

Abiotic elicitors have wide range of effects on the plants and in the production of secondary metabolites. The use of abiotic elicitors in plant cell cultures has received less attention compared with the biotic elicitors [9]. Recent research works explained the functions of many key genes, proteins, metabolites and molecular networks involved in plant responses to heavy metals, light, drought, salinity, thermal, hormonal and other abiotic elicitors [10]. In this review, actions of some of these elicitors are discussed in response to the production of secondary metabolites from in vitro culture.

Effect on the Production of Secondary Metabolites

Chemical Elicitors: Metals are influenced to alter the production of secondary metabolites by changing the aspects of secondary metabolism [11]. Metals have become one of the main abiotic stress agents for living organisms because of their increasing use in the developing fields like industry, agrotechnics, high bioaccumulation and toxicity [12]. Metals like Ni, Ag, Fe and Co have been shown to elicit the production of secondary metabolites in a number of plants [1]. In the cell suspension culture of *Vitis vinifera* the cobalt at all three used concentrations (5, 25 and 50 µM), Ag and Cd at low concentration (5 µM) were most effective to stimulate the phenolic acid production, and also increasing the 3-O-glucosyl-resveratrol up to 1.6-fold of the control level after the 4 hours (h) of treatments [12].

In hairy root cultures of *Ambrosia artemisiifolia*, an eightfold increase of thiarubrine A production was obtained when the 16-day-old culture was challenged with 50 mg/L vanadyl sulfate ($VOSO_4$) for 72 h [13]. In an attempt to enhance betalaines production, the hairy roots were exposed to metal ions [14]. It was reported that Ca^{2+}, Ag^+ and Cd^{2+} could improve the production of tropane alkaloids, scopolamine and hyoscyamine, in hairy roots cultures of *Brugmansia candida* [15,16]. Many kinds of heavy metal were also used as elicitors to induce accumulations of bioactive compounds in *Salvia miltiorrhiza*, such as Co^{2+}, Ag^+, Cd^{2+}, Cu^{2+}, Ce^{3+}, La, Mn^{2+} and Zn^{2+} [17-20]. Among them, Ag^+ was considered as an effective elicitor for phenolic compound and tanshinone production in *S. miltiorrhiza* hairy roots and could improve rosmarinic acid, salvianolic acid B and tanshinones production. Silver nitrate ($AgNO_3$) stimulated the production of tanshinone in the root culture of *Perovskia abrotanoides* [21]. The yields of atropine content in the *Datura metel* hairy roots were increased by nanosilver as an elicitor, after 12, 24 and 48 h of the treatment [22].

Physical Elicitors: Ultrasound, light, osmotic stress, salinity, drought and thermal stress are some of the physical elicitors.

Ultrasound: The low-energy ultrasound (US) also act as an abiotic elicitor to induce plant defense mechanism and stimulate the secondary metabolite production in plants [23]. In addition, US can induce cell membrane permeabilization so as to enhance intracellular product release. This cell-permeabilizing effect may be complementary to the two-phase culture to accomplish product release from the cells and removal from the medium. Lin and Wu [24] reported that, the combination of US stimulation and in situ solvent extraction in a *Lithospermum erythrorhizon* cell culture led to 2 to 3-fold increase in the yield of shikonin. While, in *Taxus chinensis* 1.5 to 1.8-fold increase in taxol yield with 2 minutes (min) US treatment once or twice during a week-culture period was achieved [25]. In *Taxus baccata* cell culture the amount of taxol was increased by 3 times when treated with US [26], and ginsenoside saponins enhanced by 75% in *Panax ginseng* cell culture [27].

Light: The light is a physical factor which can affect the metabolite production in plants. Light can stimulate secondary

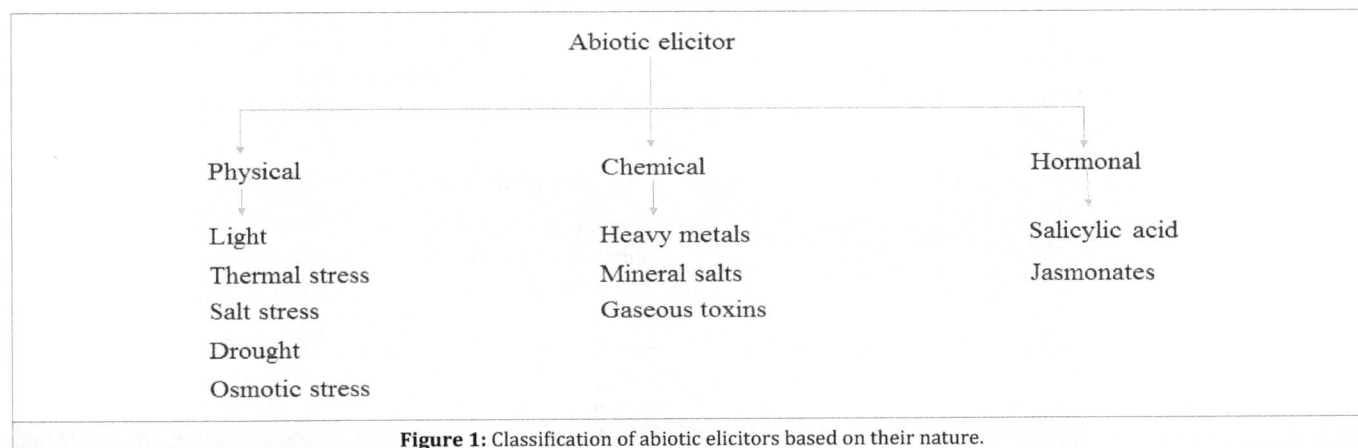

Abiotic elicitor

Physical
Light
Thermal stress
Salt stress
Drought
Osmotic stress

Chemical
Heavy metals
Mineral salts
Gaseous toxins

Hormonal
Salicylic acid
Jasmonates

Figure 1: Classification of abiotic elicitors based on their nature.

metabolites include gingerol and zingiberene production in *Zingiber officinale* callus culture [28]. Light plays a role in both growth and secondary metabolite production in the hairy roots. Sauerwein et al. [29] found that the alkaloid content of both normal and hairy roots of *Hyoscyamus albus* was greater in roots grown in the light compared to roots grown in the dark. The study of Yu et al. [30] in *P. ginseng* hairy roots also showed that the exposure of hairy roots to different light spectral ranges affected growth and metabolite biosynthesis. The light induced the growth as well as indole alkaloid production in the hairy root cultures of *Catharanthus roseus* [31]. The effect of light irradiation influenced artemisinin biosynthesis in hairy roots of *Artemisia annua* [32].

Ultraviolet (UV) light acts as an abiotic factor which stimulates the biosynthesis of secondary metabolites [33]. UV radiation is divided into three regions: UV-C (wavelengths below 280 nm), UV-B (280- 315 nm) and UV-A (315-400 nm). UV-C is the most damaging, but it is almost completely absorbed by the stratosphere. By contrast, UV-B radiation is only partially absorbed by the stratospheric ozone layer and UV-A is not at all absorbed. UV-B radiation was exclusively seen as a stress factor, UV-B trigger distinct changes in the plant's secondary metabolism resulting in an accumulation of phenolic compounds such as flavonoids and glucosinolates [33]. UV-B irradiation induced a rise of nitric oxide (NO) production, activities of nitric oxide synthase and phenylalanine ammonia lyase (leading to flavonoid synthesis), as well as flavonoid level in *Ginkgo biloba* callus [34]. Ramani and Jayabaskaran [35] reported enhanced catharanthine and vindoline production in suspension cultures of *C. roseus* by UV-B light. UV-B elicited an increase in the total terpenoid indole alkaloids (TIAs) concentrations in *C. roseus* hairy roots [36]. High doses of artificial UV-B radiation modified the antioxidant content by increasing the content of vitamin C and decreased the phenolic content of *in vitro* cultured *Turnera diffusa* plants [37]. Ku et al. [38] reported that synthesis of resveratrol and piceatannol were promoted by UV-C radiation in callus cultures of peanut. UV-C irradiation is an effective method to enhance stilbene production in grape calli of different genotypes [39]. UV-C together with methyl jasmonate (MeJA) or salicylic acid (SA) also used to enhance stilbene production in *V. vinifera* cell cultures [40].

Osmotic, Salt and Drought Stress: Osmotic stress is an important abiotic elicitor affecting plant growth, development, morphogenesis and the formation of secondary metabolites [41]. Sucrose is a typical osmotic stress agent used for the induction of water stress in plants that also serves as a vital carbon and energy source. The influence of osmotic stress enhanced the accumulation of capsaicin in cell suspension cultures of *Capsicum chinensis* [42]. It also enhanced the production of steviol glycosides content in both callus as well as suspension culture of *Stevia rebaudiana* [43].

Plants have developed complex mechanisms for adaptation to the osmotic, ionic and oxidative stresses that are induced by the salt stress. Exposure to salinity is known to induce or stimulate production of secondary plant products, such as phenols, terpenes

and alkaloids [44,45]. The salt stress decreased the anthocyanin level in the salt-sensitive species [46]. The effect of KCl and $CaCl_2$ induced stress on *in vitro* cultures of *Bacopa monnieri* enhanced the accumulation of medicinally important bacoside A content [47]. An improved synthesis of vinblastine and vincristine was observed in *C. roseus* embryogenic tissue culture by using sodium chloride (NaCl) as an elicitor [48]. In *Nitraria tangutorum* cell suspension the increased sitosterol content was observed at 250 mM NaCl treatment [49]. Small increase of the canavanine content in *Sutherlandia frutescens in vitro* shoot culture growing on 100 mM NaCl medium was detected, indicating that salinity stress was not a major limitation on cavanine production [50].

The drought is an important stress factor limiting the plant growth, reproductive development and finally survival. Drought stress tolerance is seen in all plants, but its extent varies from species to species. Plants which are exposed to drought stress frequently were affected the synthesis and accumulation of secondary metabolite contents [51]. A weak water deficit greatly increased the glycyrrhizic acid content in roots of *Glycyrrhiza uralensis* [52]. The drought stress was induced by polyethylene glycol (PEG), when it was treated to *in vitro* grown date palm callus, the proline content was increased gradually in response to increasing PEG-concentration. At higher concentration (30%) accumulation of proline started to decline as an indication of disturbance of physiological system [53]. PEG as a supplement had little to no effect on canavanine synthesis in *Sutherlandia frutescens in vitro* shoot culture [50].

Thermal Stress: Extreme temperature is an adverse environmental factor limiting growth and productivity of plants, it also hinders the plant growth by manipulating various metabolic processes including synthesis and degradation of primary metabolites [54]. Temperature range of 17–25°C is normally used for the induction of callus tissues and growth of cultured cells [6]. The *Melastoma malabathricum* cell cultures incubated at a lower temperature range (20 ± 2°C) grew better and had higher anthocyanin production than those grown at 26 ± 2°C and 29 ± 2°C [55]. Optimum temperature (25°C) maximizes the anthocyanin yield as demonstrated in cell cultures of *Perilla frutescens* [56] and strawberry [57]. Although temperature around 25°C is normally used for hairy root cultures, lowering the cultivation temperature (19.5°C) increased the proportion of linolenic acid and the total content of indole alkaloids in *C. roseus* hairy roots [58]. A 5C increase in temperature significantly increased the ginsenoside content in roots of *Panax quinquefolius* [59].

Hormonal Elicitors:

Salicylic Acid: Salicylic acid is the one of the important abiotic elicitor, which has the capability to induce the secondary metabolites from *in vitro* cultures. SA induced the stilbene production in the cell suspension of *V. vinifera* [40]. A high concentration of 200 µM SA was required to induce substantial quantities of gymnemic acid in the suspensions that reached a maximum after 48 h treatment. The SA induced response towards gymnemic acid accumulation resulted in a 4.9-fold increase in comparison to the control cultures [60]. In the cell culture of *S. miltiorrhiza*, the different concentrations of the SA were affected the accumulation of salvianolic acid B and of caffeic acid. Both phenolic acid accumulations were significantly increased at 8 and 96 h after the applications of 3.125–25 mg/L of SA, but were significantly less with 32-50 mg/L of SA. After the 96 h treatments with 3.125–25 mg/L of SA, the concentration of the phenolic acids decreased significantly compared to the amount 8 h after the treatments, but were still higher than that of the control [61]. SA with transgenic technology, highly enhanced the production of tanshinones in *S. miltiorrhiza* hairy roots [62]. Optimum production of withanolide A, withanone and withaferin A were reported in the elicited-hairy roots of *Withania somnifera* [63].

Jasmonates: Jasmonic acid has been proposed as key compounds of the signal transduction pathway involved in the elicitation of secondary metabolite biosynthesis which takes part in plant defense reactions [64]. The application of a two-stage culture system with a combined treatment of mannitol (2mM) and JA (40µM) resulted in the optimum accumulation of resveratrol in the callus biomass of *V. vinifera* [65]. JA and its more active derivative MeJA can trigger the production of a wide range of plant secondary metabolites such as rosmarinic acid, terpenoid indole alkaloid and plumbagin in various cell cultures [66-68]. JA and MeJA have been used as elicitors for stilbene biosynthesis in *V. vinifera* cell cultures [69,70]. The treatment of MeJA to *V. vinifera* cell cultures also promoted anthocyanin accumulation [71]. In the cell culture of *Andrographis paniculata,* the MeJA induced the optimum accumulation of andrographolide at 24 h compared with 48 and 72 h of treatments [72]. In the *V. vinifera* cell system, a rapid accumulation of trans-resveratrol was recorded with MeJA treatement, starting from 2 h and reaching its maximum value at 96 h [70].

In a study, JA elicitation is reported to enhance the production of plumbagin in hairy root culture of *Plumbago indica* [73]. JA and MeJA have been used as elicitors for stilbene biosynthesis in *Vitis rotundifolia* hairy root cultures [74]. MeJA with transgenic technology, highly enhanced the production of tanshinones in *S. miltiorrhiza* hairy roots [62]. In the hairy root culture of *W. somnifera*, MeJA elicited the production of withanolide A, withanone and withaferin A [63]. The root cultures of *Taverniera cuneifolia* treated with different concentrations of MeJA, the glycyrrhizic acid content increased gradually with an increase in MeJA (1–100 µM) concentration. Approximately 2.5-fold elevation in glycyrrhizic acid production was noticed in MeJA (100 µM) treated roots, when compare to the control. However, further increase in MeJA (1000 µM) concentration resulted in the decrease of glycyrrhizic acid production [75]. The MeJA enhanced the production of bacoside A, a valuable triterpenoid saponin having nootropic therapeutic activity in *in vitro* shoot cultures of *B. monnieri* [76].

Conclusion

The evolutionary process made plant to produce new bioactive compounds time to time. The plant produces a number of secondary metabolites in varying concentration in different parts of the tissue. For more than three decades an *in vitro* culture plays an important role in the production of secondary metabolites from the particular tissue, organ and cells. In the past, less attention was paid to abiotic elicitor, but now the use of abiotic elicitors has emerged as one of the most effective strategy for enhancing the productivity of different commercial secondary metabolites from in vitro cultures. Though, elicitation enhances secondary metabolism in in vitro culture of plant cells/organ, but the exact mechanism of elicitationis still not fully understood. Moreover, molecular and biosynthetic pathway of the secondary metabolites should be understood and extensive research must be carried out in this way to determine the optimum conditions for each specific medicinal plant to enhance the secondary metabolites.

References

1. Wang JW, Wu JY. Effective elicitors and process strategies for enhancement of secondary metabolite production in hairy root cultures. Adv Biochem Eng Biotechnol. 2013; 134: 55-89. doi: 10.1007/10_2013_183.

2. Murthy HN, Lee EJ, Paek KY. Production of secondary metabolites from cell and organ cultures: strategies and approaches for biomass improvement and metabolite accumulation. Plant Cell Tiss Org Cult. 2014; 118: 1-16.

3. Anis M, Husain MK, Faisal M, et al. In vitro approaches for plant regeneration and conservation of some medicinal plants. In: Kumar A, Sopory SK, editors. Recent advances in plant biotechnology and its application. New Delhi: IK International Pvt. Ltd; 2009. p. 397-410.

4. Anis M, Husain MK, Siddique, I, et al. Biotechnological approaches for the conservation of forestry species. In: Jenkins JA. editor. Forest decline: causes and impacts. USA: Nova Science Publishers Inc; 2011. p. 1-39.

5. Kehie M, Kumaria S, Tandon P, Ramchiary N. Biotechnological advances on in vitro capsaicinoids biosynthesis in capsicum:a review. Phytochem Rev.2015; 14(2): 189–201. doi: 10.1007/s11101-014-9344-6.

6. Rao SR, Ravishankar GA. Plant cell cultures: Chemical factories of secondary metabolites. Biotechnol Adv. 2002; 20(2): 101-53.

7. Savitha BC, Timmaraju R, Bhagyalaksami N, Ravishankar GA. Different biotic and abiotic elicitors influence betalain production in hairy root cultures of *Beta vulgaris* in shake flask and bioreactor. Process Biochem. 2006; 41: 50–60.

8. Goel MK, Mehrotra S, Kukreja AK. Elicitor-induced cellular and molecular events are responsible for productivity enhancement in hairy root cultures: an insight study. Appl Biochem Biotechnol. 2011; 165(5-6): 1342-55. doi: 10.1007/s12010-011-9351-7.

9. Radman R, Saez T, Bucke C, Keshavarz T. Elicitation of plants and microbial cell systems. Biotechnol Appl Biochem. 2003; 37(Pt 1): 91-102.

10. Rodziewicz P, Swarcewicz B, Chmielewska K, Wojakowska A, Stobiecki M. Influence of abiotic stresses on plant proteome and metabolome changes. Acta Physiol Plant. 2014; 36: 1–19.

11. Nasim SA1, Dhir B. Heavy metals alter the potency of medicinal plants. Rev Environ Contam Toxicol. 2010; 203: 139-49. doi: 10.1007/978-1-4419-1352-4_5.

12. Cai Z, Kastell A, Speiser C, SmetanskaI. Enhanced resveratrol production in *Vitis vinifera* cell suspension cultures by heavy metals without loss of cell viability. Appl Biochem Biotechnol. 2013 Sep;171(2):330-40. doi: 10.1007/s12010-013-0354-4.

13. Bhagwath SG, Hjortsø MA. Statistical analysis of elicitation strategies for thiarubrine A production in hairy root cultures of *Ambrosia artemisiifolia*. J Biotechnol. 2000; 80(2): 159–167.

14. Rudrappa T, Neelwarne B, Aswathanarayana RG. In situ and ex situ adsorption and recovery of betalains from hairy root cultures of *Beta vulgaris*. Biotechnol Progr. 2004; 20(3): 777-85.

15. Pitta-Alvarez SIP, Spollansky TC, Giulietti AM. The influence of different biotic and abiotic elicitors on the production and profile of tropane alkaloids in hairy root cultures of *Brugmansia candida*. Enzym Microb Technol. 2000; 26(2-4): 252–8.

16. Angelova Z, Georgiev S, Roos W. Elicitation of plants. Biotechnol Biotechnol Equip. 2006; 20(2): 72-83.

17. Yan Q, Shi M, Ng J, Wu JY. Elicitor-induced rosmarinic acid accumulation and secondary metabolism enzyme activities in *Salvia miltiorrhiza* hairy roots. Plant Sci. 2006; 170(4): 853–8.

18. Zhao JL, Zhou LG, Wu JY. Effects of biotic and abiotic elicitors on cell growth and tanshinone accumulation in *Salvia miltiorrhiza* cell cultures. Appl Microbiol Biotechnol. 2010; 87(1): 137-44. doi: 10.1007/s00253-010-2443-4.

19. Zhou J, Fang L, Wang X, Guo LP, Huang LQ. La dramaticaly enhances the accumulation of tanshinones in *Salvia miltiorrhiza* hairy root cultures. Earth Sci Res. 2013; 2(1): 187-192.

20. Shi M, Luo X, Ju G, Yu X, Hao X, Huang Q, et al. Increased accumulation of the cardio-cerebrovascular disease treatment drug tanshinone in *Salvia miltiorrhiza* hairy roots by the enzymes 3-hydroxy-3-methylglutaryl CoA reductase and 1-deoxy-D-xylulose 5-phosphate reductoisomerase. Funct Integr Genomics. 2014; 14(3): 603-15. doi: 10.1007/s10142-014-0385-0.

21. Arehzoo Z, Christina S, Florian G, Parvaneh A, Javad A, Seyed H, et al. Effects of some elicitors on tanshinone production in adventitious root cultures of *Perovskia abrotanoides* Karel. Ind Crops Prod. 2015; 67: 97–102.

22. Zahra S, Mehrnaz K, Gholamreza A, Mustafa G. Improvement of atropine production by different biotic and abiotic elicitors in hairy root cultures of *Datura metel*. Turk J Biol. 2015; 39: 111-118.

23. Yu M, Liu H, Shi A, Liu L, Wang Q. Preparation of resveratrol-enriched and poor allergic protein peanut sprout from ultrasound treated peanut seeds. Ultrason Sonochem. 2016; 28: 334-40. doi: 10.1016/j.ultsonch.2015.08.008.

24. Lin L, Wu J. Enhancement of shikonin production in single and two-phase suspension cultures of *Lithospermum erythrorhizon* cells using low-energy ultrasound. Biotechnol Bioeng. 2002; 78(1):81–8.

25. Wu J, Lin L. Enhancement of taxol production and release in *Taxus chinensis* cell cultures by ultrasound, methyl jasmonate and in situ solvent extraction. Appl Microbiol Biotechnol. 2003; 62(2-3):151–155.

26. Rezaei A, Ghanati F, Dehaghi MA. Stimulation of taxol production by combined salicylic acid elicitation and sonication in *Taxus baccata* cell culture. In: 2011 International Conference on Life Science and Technology IPCBEE; 2011. p. 193–7.

27. Lin L, Wu J, Ho KP, Qi S. Ultrasound-induced physiological effects and secondary metabolite (saponin) production in *Panax ginseng* cell cultures. Ultrasound Med Biol. 2001; 27(8): 1147–52.

28. Anasori P, Asghari G. Effects of light and differentiation on gingerol and zingiberene production in callus culture of *Zingiber officinale* Rosc. Res Pharm Sci. 2008; 3(1): 59-63.

29. Sauerwein M, Wink M, Shimomura K. Influence of light and phytohormones on alkaloid production in transformed root cultures of *Hyoscyamus albus*. J Plant Physiol. 1992; 140(2): 147–52.

30. Yu KW, Murthy HN, Hahn EJ, Paek KY. Ginsenoside production by hairy root cultures of *Panax ginseng*: influence of temperature and light quality. Biochem Eng J. 2005; 23(1): 53–56.

31. Bhadra R, Morgan JA, Shanks JV. Transient studies of light-adapted cultures of hairy roots of *Catharanthus roseus*: Growth and indole alkaloid accumulation. Biotechnol Bioeng. 1998; 60(6): 670–8.

32. Liu CZ, Guo C, Wang Y, Ouyang F. Effect of light irradiation on hairy root growth and artemisinin biosynthesis of *Artemisia annua* L. Process Biochem.2002; 38(4): 581-5.

33. Schreiner M, Mewis I, Neugart S, Zrenner R, Glaab J, Wiesner M, et al. UV-B Elicitation of secondary plant metabolites. In: Kneissl M, Rass J,editors. III-Nitride ultraviolet emitters. Springer series in materials science 227. Springer International Publishing, Switzerland: p. 387-414. doi: 10.1007/978-3-319-24100-5_14.

34. Hao G, Du X, Zhao F, Shi R, Wang J. Role of nitric oxide in UV-B-induced activation of PAL and stimulation of flavonoid biosynthesis in *Ginkgo biloba* callus. Plant Cell Tiss Org Cult. 2009; 97(2): 175–85.

35. Ramani S, Jayabaskaran C. Enhanced catharanthine and vindoline production in suspension cultures of *Catharanthus roseus* by ultraviolet-B light. J Mol Signal. 2008; 3: 9. doi: 10.1186/1750-2187-3-9.

36. Binder BY, Peebles CA, Shanks JV, San KY. The effects of UV-B stress on the production of terpenoid indole alkaloids in *Catharanthus roseus* hairy roots. Biotechnol Prog. 2009; 25(3): 861-5. doi: 10.1002/btpr.97.

37. Soriano-Melgar Lde A, Alcaraz-Meléndez L, Méndez-Rodríguez LC, Puente ME, Rivera-Cabrera F, Zenteno-Savín T. Antioxidant responses of damiana (*Turnera diffusa* Willd) to exposure to artificial ultraviolet (UV) radiation in an in vitro model; part II; UV-B radiation. Nutr Hosp. 2014; 29(5): 1116-22. doi: 10.3305/nh.2014.29.5.7092.

38. Ku KL, Chang PS, Cheng YC, Lien CY. Production of stilbenoids from the callus of *Arachis hypogaea*: a novel source of the anticancer compound piceatannol. J Agric Food Chem. 2005; 53(10): 3877-81.

39. Liu W, Chunyan Liu, Chunxiang Yang, Lijun Wang, Shaohua Li. Effect of grape genotype and tissue type on callus growth and production of resveratrols and their piceids after UV-C irradiation. Food Chem.2010; 122(3): 475-484.

40. Xu A, Zhan JC, Huang WD. Effects of ultraviolet C, methyl jasmonate and salicylic acid, alone or in combination, on stilbene biosynthesis in cell suspension cultures of *Vitis vinifera* L. cv. Cabernet Sauvignon. Plant Cell Tiss Org Cult. 2015; 122(1): 197–211.

41. Liu CZ, Cheng XY. Enhancement of phenylethanoid glycosides biosynthesis in cell cultures of *Cistanche deserticola* by osmotic stress. Plant Cell Rep. 2008; 27(2): 357-62.

42. Kehie M1, Kumaria S, Tandon P. Osmotic stress induced—capsaicin production in suspension cultures of *Capsicum chinense* Jacq.cv. Naga King Chili. Bioprocess Biosyst Eng. 2014; 37(6): 1055-63. doi: 10.1007/s00449-013-1076-2.

43. Gupta P, Sharma S, Saxena S.. Biomass yield and steviol glycoside

production in callus and suspension culture of *Stevia rebaudiana* treated with proline and polyethylene glycol. Appl Biochem Biotechnol. 2015; 176(3): 863-74. doi: 10.1007/s12010-015-1616-0.

44. Selmar D. Potential of salt and drought stress to increase pharmaceutical significant secondary compounds in plants. Agr Forest Res. 2008; 58: 139–144.

45. Haghighi Z, Karimi N, Modarresi M, Mollayi S. Enhancement of compatible solute and secondary metabolites production in *Plantago ovata* Forsk. by salinity stress. J Med Plants Res. 2012; 6(18): 3495–3500.

46. Daneshmand F, Arvin MJ, Kalantari KM. Physiological responses to NaCl stress in three wild species of potato in vitro. Acta Physiol Plant. 2010; 32(1): 91-101.

47. Ahire ML, Laxmi S, Walunj PR, Kavi Kishor PB, Nikam TD. Effect of potassium chloride and calcium chloride induced stress on in vitro cultures of *Bacopa monnieri* (L.) Pennell and accumulation of medicinally important bacoside A. J Plant Biochem Biotechnol. 2014; 23(4): 366–378. doi: 10.1007/s13562-013-0220-z.

48. Fatima S, Mujib A, Dipti T. NaCl amendment improves vinblastine and vincristine synthesis in *Catharanthus roseus*: a case of stress signalling as evidenced by antioxidant enzymes activities. Plant Cell Tiss Org Cult. 2015; 121(2): 445–458.

49. Ni J, Yang X, Zhu J, Liu Z, Ni Y, Wu H, et al. Salinity-induced metabolic profile changes in *Nitraria tangutorum* Bobr. suspension cells. Plant Cell Tiss Org Cult. 2015; 122(1): 239-48. doi: 10.1007/s11240-015-0744-0.

50. Colling J, Stander MA, Makunga NP. Nitrogen supply and abiotic stress influence canavanine synthesis and the productivity of in vitro regenerated *Sutherlandia frutescens* microshoots. J Plant Physiol. 2010; 167(17): 1521-4. doi: 10.1016/j.jplph.2010.05.018.

51. Al-Gabbiesh A, Kleinwächter M, Selmar D. Influencing the contents of secondary metabolites in spice and medicinal plants by deliberately applying drought stress during their cultivation. Jordan J Biol Sci. 2015; 8(1):1-10.

52. Li W, Hou J, Wang W, Tang X, Liu C, Xing D. Effect of water deficit on biomass production and accumulation of secondary metabolites in roots of *Glycyrrhiza uralensis*. Russ J Plant Physiol. 2011; 58(3): 538-542.

53. Al-Khayri JM, Ibraheem Y. In vitro selection of abiotic stress tolerant date palm (*Phoenix dactylifera* L.): A review. Emir J Food Agric. 2014; 26(11): 921-933.

54. Xu Y, Du H, Huang B. Identification of metabolites associated with superior heat tolerance in thermal bentgrass through metabolic profiling. CropSci. 2013; 53: 1626–1635.doi:10.2135/cropsci2013.01.0045.

55. Chan LK, Koay SS, Boey PL, Bhatt A. Effects of abiotic stress on biomass and anthocyanin production in cell cultures of Melastoma malabathricum. Biol Res. 2010; 43: 127-35. PMID: 21157639.

56. Zhong JJ, Yoshida T. Effects of temperature on cell growth and anthocyanin production in suspension cultures of *Perilla frutescen.* J Ferment Bioeng. 1993; 76(6): 530-1.

57. Zhang W, Seki M, Furusaki S. Effect of temperature and its shift on growth and anthocyanin production in suspension cultures of strawberry cells. Plant Sci. 1997; 127(2): 207-214.

58. Toivonen L, Laakso S, Rosenqvist H. The effect of temperature on hairy root cultures of *Catharanthus roseus*: growth, indole alkaloid accumulation and membrane lipid composition. Plant Cell Rep. 1992;

11(8): 395-9. doi: 10.1007/BF00234368.

59. Jochum GM, Mudge KW, Thomas RB. Elevated temperatures increase leaf senescence and root secondary metabolite concentration in the understory herb *Panax quinquefolius* (Araliaceae). Am J Bot. 2007; 94(5): 819-26. doi: 10.3732/ajb.94.5.819.

60. Chodisetti B, Rao K, Gandi S, GiriA. Gymnemic acid enhancement in the suspension cultures of *Gymnema sylvestre* by using the signaling molecules—methyl jasmonate and salicylic acid. In Vitro Cell Dev Biol-Plant. 2015; 51(1): 88–92.

61. Dong J, Wan G, Liang Z. Accumulation of salicylic acid-induced phenolic compounds and raised activities of secondary metabolic and antioxidative enzymes in *Salvia miltiorrhiza* cell culture J Biotechnol. 2010; 148(2-3): 99-104. doi: 10.1016/j.jbiotec.2010.05.009.

62. Hao X, Shi M, Cui L, Xu C, Zhang Y, Kai G. Effects of methyl jasmonate and salicylic acid on tanshinone production and biosynthetic gene expression in transgenic *Salvia miltiorrhiza* hairy roots. Biotechnol Appl Biochem. 2015; 62(1): 24-31. doi: 10.1002/bab.1236.

63. Sivanandhan G, Arun M, Mayavan S, Rajesh M, Jeyaraj M, Dev GK, et al. Increased production of withanolide A, withanone, and withaferin A in hairy root cultures of *Withania somnifera* (L.) Dunal elicited with methyl jasmonate and salicylic acid. Appl Biochem Biotechnol. 2012; 168(3): 681-96. doi: 10.1007/s12010-012-9809-2.

64. Jeandet P, Douillet-Breuil AC, Bessis R, Debord S, Sbaghi M, Adrian M. Phytoalexins from the Vitaceae: biosynthesis, phytoalexin gene expression in transgenic plants, antifungal activity, and metabolism. J Agric Food Chem. 2002; 50(10): 2731-41.

65. Raluca M, Sturzoiu C, Florenta H, Aurelia B, Gheorghe S. Biotic and abiotic elicitors induce biosynthesis and accumulation of resveratrol with antitumoral activity in the long - term *Vitis vinifera* L. callus cultures. Roman Biotechnol Lett. 2011; 16(6): 6683-9.

66. Krzyzanowska J, Czubacka A, Pecio L, Przybys M, Doroszewska T, Stochmal A, et al. The effects of jasmonic acid and methyl jasmonate on rosmarinic acid production in *Mentha piperita* cell suspension cultures. Plant Cell Tiss Org Cult. 2012; 108(1): 73–81.

67. Almagro L, Gutierrez J, Pedreño MA, Sottomayor M. Synergistic and additive influence of cyclodextrins and methyl jasmonate on the expression of the terpenoid indole alkaloid pathway genes and metabolites in *Catharanthus roseus* cell cultures. Plant Cell TissOrg Cult. 2014; 119(3): 543–51.

68. Silja PK, Gisha GP, Satheeshkumar K. Enhanced plumbagin accumulation in embryogenic cell suspension cultures of *Plumbago rosea* L. following elicitation. Plant Cell Tiss Org Cult. 2014; 119: 469–77.

69. Tassoni A, Fornalè S, Franceschetti M, Musiani F, Michael AJ, Perry B, et al. Jasmonates and Na-orthovanadate promote resveratrol production in *Vitis vinifera* cv. Barbera cell cultures. New Phytol. 2005; 166(3): 895-905.

70. Taurino M, Ingrosso I, D'amico L, De Domenico S, Nicoletti I, Corradini D, et al. Jasmonates elicit different sets of stilbenes in *Vitis vinifera* cv. Negramaro cell cultures. Springerplus. 2015; 4: 49. doi: 10.1186/s40064-015-0831-z.

71. Tassoni A, Durante L, Ferri M. Combined elicitation of methyl-jasmonate and red light on stilbene and anthocyanin biosynthesis. J Plant Physiol. 2012; 169(8): 775-81. doi: 10.1016/j.jplph.2012.01.017.

72. Sharma SN, Jha Z, Sinha RK, Geda AK. Jasmonate-induced biosynthesis of andrographolide in *Andrographis paniculata*. Physiol Plant. 2015; 153(2): 221-9. doi: 10.1111/ppl.12252.

73. Gangopadhyay M, Dewanjee S, Bhattacharya S. Enhanced plumbagin production in elicited *Plumbago indica* hairy root cultures. J Biosci Bioeng. 2011; 111(6): 706-10. doi: 10.1016/j.jbiosc.2011.02.003.

74. Nopo-Olazabal C, Condori J, Nopo-Olazabal L, Medina-Bolivar F. Differential induction of antioxidant stilbenoids in hairy roots of *Vitis rotundifolia* treated with methyl jasmonate and hydrogen peroxide. Plant Physiol Biochem. 2014; 74: 50-69. doi: 10.1016/j.plaphy.2013.10.035.

75. Awad V, Kuvalekar A, Harsulkar A. Microbial elicitation in root cultures of *Taverniera cuneifolia* (Roth)Arn. for elevated glycyrrhizic acid production. Ind Crops Prod. 2014; 54: 13–6.

76. Sharma P, Yadav S, Srivastava A, Shrivastava N. Methyl jasmonate mediates upregulation of bacoside A production in shoot cultures of *Bacopa monnieri*. Biotechnol Lett. 2013; 35(7): 1121-5. doi: 10.1007/s10529-013-1178-6.

12

Bacteriolyses of Bacterial Cell Walls by Cu(II) and Zn(II) Ions based on Antibacterial Results of Dilution Medium Method and Halo Antibacterial Test

Dr. Sci. Tsuneo. ISHIDA*

[1]*Life and Environment Science Research*

Corresponding author: Dr. Sci. Tsuneo. ISHIDA, Life and Environment Science Research, E-mail: ts-ishida@ac.auone-net.jp

Abstract

Bacteriolyses of bacterial cell walls by copper (II) ions and zinc (II) ions based antibacterial results of broth dilution medium method and halo antibacterial test were investigated. From dilution medium method, MIC=625mg/L, MBC=1250mg/L for Cu^{2+} solution as bactericide action were obtained against *Staphylococcus aureus*, and also from halo antibacterial test, the high antibacterial effects for Cu^{2+}, Zn^{2+} ions were obtained against *Staphylococcus epidermidis*. Bacteriolysis of *S. aureus* peptidoglycan (PGN) cell wall by Cu^{2+} ions is ascribed to the inhibition of PGN elongation due to the damages of PGN biosyntheses TG, TP and the activations of PGN autolysins. The other, bacteriolysis of *E. coli* outer membrane cell wall by Cu^{2+} ions is attributed to the destruction of outer membrane structure and to the inhibition of PGN elongation due to the damage of PGN biosynthesis TP and the activations of PGN autolysins. Furthermore, bacteriolysis of *S. aureus* PGN cell wall by Zn^{2+} ion is due to the inhibition of PGN elongation owing to the activations of PGN autolysins of amidases. The other, bacteriolysis of *E. coli* cell wall by Zn^{2+} ions is attributed to the destruction of outer membrane structure due to degradative enzymes of lipoproteins at N-, and C-terminals, whereas is dependent on the activities of PGN hydrolases and autolysins of amidase and carboxy peptidase-transpeptidase. Cu^{2+} and Zn^{2+} ions induced ROS such as O_2^-, H_2O_2, $\cdot OH$, OH^- producing in bacterial cell wall occur oxidative stress.

Keywords: MIC, MBC, CFU measurements and Halo antibacterial test, Cu^{2+} and Zn^{2+} ions, PGN cell wall, Outer membrane lipoproteins, Biosynthesis and autolysin, Reactive oxygen species(ROS).

Introduction

Silver, copper, and zinc of transition metals have highly antibacterial activities and areutilized as cheomotherapy agents. Recently, antibacterial activities of copper, zinc and these complexes call attention to potential treatments such as prevention of serious diseases [1], Exploitation during bacterial pathogenesis [2], and cancer and tumor cell [3]. Cu^{2+} ions could kill cancer cell by Cu(II)-Cu(I) redox-cycle, the other, Zn^{2+} ions may kill tumor cell by bivalent state of Zn(II), unfortunately, the killing mechanism by Cu^{2+}, Zn^{2+} ions for cancer cell remains unclear. Cancer arises from a fault in a cell. This single faulty cell then multiplies to form a cluster of cells, namely, a tumor, that these tumor cells then spread to the whole body and these metastases can eventually kill the parson [4]. Hence, the zinc homeostasis with apoptosis and necrosis for cancer cell should be eventually established. It has become apparent that zinc ions inhibit mitochondoria [5], lysosomes [6], DNA [7], and nucleus [8] of cancer cell, and that regulate cell proliferation and growth [9,10], and metastasis[11]of tumor cell. However, no confirmed common relationships of zinc ions with cancer development and progression have been identified, in which killing mechanisms of cancer cell and tumor cell by Cu^{2+}, Zn^{2+} ions are not yet definitely elucidated.

In this study, the broth dilution medium method test against *S. aureus* and *E. coli*, and the halo antibacterial susceptibility test against *Staphylococcus epidermidis* were carried out, where in it was turned out that antibacterial effects of Cu^{2+} and Zn^{2+} ions were examined. On the basis of the high antibacterial activities for these copper and zinc ions, the processes of bacteriolyses and destructions of bacterial cell walls by copper and zinc ions had been considered again st *S. aureus* peptidoglycan (PGN) and *E. coli* outer membrane cell walls. Furthermore, the bacteriolytic mechanisms by copper (II) ion and zin(II) ion solutions have been also revealed against both Gram-positive and Gram-negative bacteria.

Method

Two-fold broth dilution medium method tests for Cu^{2+} ion solutions

This method is quantitatively obtained for the antibacterial activity on the bactericidal assay. Bacteria intended for two-fold broth dilution medium method were treated as *Staphylococcus aureus* (NBRC12732) and *Escherichia coli* (ATCC25922). The other, the antibacterial copper ion of commercial copper (II) ion agent (Japan ion production Ltd. , original Cu^{2+} solution;500 mg/L) are used as bacteriostasis, and copper nitrate

($Cu(NO_3)_2 3H_2O$, Wako Pure Reagents) of special class reagent was used as bactericide action. Firstly, the sample test tube of Cu^{2+} ion concentration of 10,000 mg/L have been prepared in heart infusion agar medium(Nissui). Next, the diluted solutions of 10-stagesby two-fold dilution solution method was adjusted in tenth sample tubes for Cu^{2+} ion solution concentration of 9.8~5,000 mg/L. Afterwards, the adjustment solution within final solution of 5×10^5 cfu/mL was prepared, and then with a sterile micropipette, fungous liquid 1 mL of bacterial suspension was respectively transferred from tube No 1 to other tubes that were inoculated into the respective tubes. Finally, the tubes were incubated at 35°C for 24 hours, in which the incubated solutions were afforded to minimum inhibitory concentration (MIC), minimum bactericide concentration (MBC), colony forming unit (CFU) measurements.

Halo antibacterial susceptibility test procedure

This method is characteristics of finding of inhibitory halo-zone measurements as less qualitative antibacterial activity assay. Halo antibacterial tests have been carried out for the nitrate and sulfate aqueous solutions against *Staphylococcus epidermidis*. The other, the antibacterial reagents were prepared metallic ions 100 mM/L aqueous solutions from metallic salt reagents. The preparation method is shown in Table 1, wherein the crystalline powders of metallic salts of 0.01mol are dissolved in distilled water of 100 cc, preparing metallic ion concentration of 100 mM/L as antibacterial reagents (crystalline powders of 0.005 mol for silver sulfate and aluminum sulfate were used).

Firstly, *Staphylococcus epidermidis* that were collected from the inside of the arms, in which were incubated in physiological saline aqueous water solution of salt at a thermostat of constant temperature of 35°C through a week. And then, after the bacteria are incubated in the standard planar agar-medium, the generated colonies are incubated in gradient medium above a week. Secondly, the incubated cells were suspended in physiological saline solution in which they were painted and swabbed at 100Ml share to newly prepared planar medium. Finally, the paper discs that the metallic ion solutions are stained and placed on the center of planar medium at 35°C for a week, in which the antimicrobial liquid is swabbed and spread. Afterwards, the presence or absence of an inhibitory area (zone of inhibition, W) around the disc identifies the bacterial sensitivity to the metallic ions. The diameter of growth inhibition halo is examined and measured by ruler, and then, reports are provided as susceptible, resistant, or intermediate. Measurement of the inhibition halo must be done always with ruler. Inhibitory zone width W is represented as W=(X-8 mm)/2 (in mm) from measured inhibitory diameter X and paper-disc diameter of 8 mm, in which W is calculated from measured X.

Search and Analysis

The surface envelop cell structures of *S. aureus* as representative of Gram-positive bacterium and *E. coli* as representative of Gram-negative bacterium, molecular structures of these cell walls, molecular structure of peptidoglycan (PGN), and PGN biosyntheses and autolysins were searched in detail. Further, the reaction and the behavior of metallic ions and bacterial cell, molecular bonding manner, and zinc ion characteristics were also searched.

Table1: Preparing metallic ions 100 mM/L aqueous solutions from metallic salt reagents				
Metallic salt reagents **Nitrates and Sulfates**	**Molecular** **weight**	**Mass[g] of metallic salt reagents** **dissolved in the distilled water of 100cc.** **Metallic ion concentration ; 100 m mol/L**		
		Dissolved. **in water**	**Metallic salts**	**Metallic ion conc.**
Nitrates				
Silver nitrate ; AgNO₃	169.87	1.6987g	10.8g	10,800mg/L
Copper nitrate;Cu(NO₃)₂·3H₂O	241.60	2.4160	6.35	6,350
Zinc nitrate; Zn(NO₃)₂·6H₂O	297.47	2.9747	6.54	6,540
Aluminum nitrate; Al(NO₃)₃·9H₂O	375.13	3.7513	0.27	2,698
Lead nitrate ; Pb(NO₃)₂	331.21	3.3121	20.7	20,720
Sulfates Silver sulfate ; Ag₂SO₄	311.8	1.5590	21.6g	10,800mg/L
Copper sulfate ;CuSO₄ ·5H₂O	249.68	2.4968	6.35	6,350
Zinc sulfate ;ZnSO₄·7H₂O	287.56	2.8756	6.54	6,540
Aluminum sulfate ;Al₂(SO₄)₃· 14~18H2O	324.14	1.6207	0.54	2.698

Results

Bacteriostatic and bactericide actions of Cu^{2+} ion solution by the broth dilution medium method

Table 2 shows the bacterio stasis as disinfection agent inhibiting the bacteria growth and multiplying organism of Cu^{2+} ion, in which minimum inhibitory concentration, MIC=50mg/L

above was obtained for Cu^{2+} ion concentration range of 0.10~50 mg/L[12]. The other, table 3 indicates the results as bactericide action, in which MIC=625 mg/L and minimum bactericide concentration, MBC=1250 mg/L were obtained for Cu^{2+} ion concentration range of 9.8~5000 mg/L [13]. The killing curve of Cu^{2+} ions is shown in Figure 1(measurement's error= ±6%), in which killing effects for the copper (II) ions appear sufficiently.

Table 2: MIC measurement of commercial Cu2+ solution agents as a bacteriostatic action against *E.coli* by broth dilution medium method

Cu^{2+}solution agent Original conc 500 mg/L	Cu^{2+} solution concentration(mg/L)										MIC 50 mg/L above
	50	25	12.5	6.25	3.13	1.56	0.78	0.39	0.20	0.10	
	+	+	+	+	+	+	+	+	+	+	

(+) ; Visible bacterial growth   (−) ; No visible bacterial growth

Table 3: MIC, MBC, and CFU of Cu^{2+} in $Cu(NO_3)_2.3H_2O$ Solution as a bactericidal action againt *S.aureus* by 10- fold diluted solution medium method

Antibacterial agent Cu(NO3)23H2o	Cu^{2+} concentration (mg/L)									
	5000	2500	1250	625	313	156	78	39	20	9.8
MIC	-	-	-	-	+	+	+	+	+	+
MBC	-	-	-	+	+	+	+	+	+	+
CFU(cfu/ml)	<10	<10	<10	$1.1×10^2$	$3.1×10^8$	$4.0×10^8$	$4.5×10^8$	$5.1×10^8$	$5.5×10^8$	$5.3×10^8$

(+); Bacterial growth(Visible turbidity),   (-) No Visible Bacterial growth

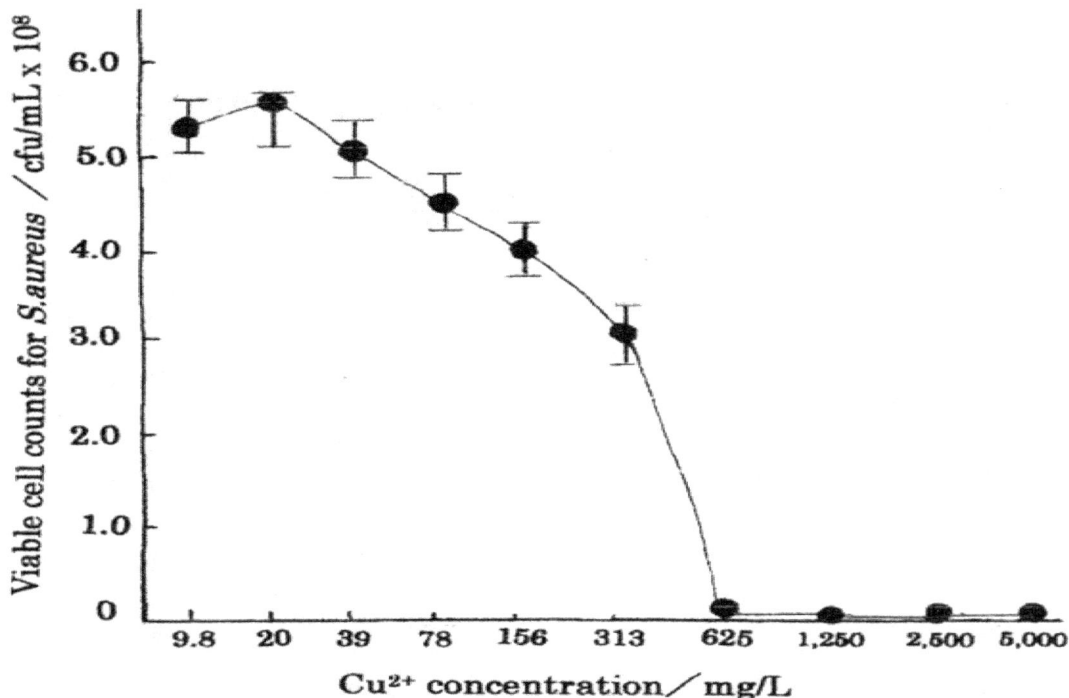

Figure 1: Relationship between increasing Cu^{2+} concentration(mg/L) and viable counts(CFU/mL) against *S.aureus*

Halo antibacterial susceptibility tests

Figure 2 indicates the bar-graphs of the relationships between various metallic ions and halo inhibitory zones width. Figure 3 shows the sample surface appearances of inhibitory zone after halo antibacterial tests for nitrate and sulfate solutions against *Staphylococcus epidermidis*. For the nitrate solutions, it is found that the antibacterial effect has nothing for alkalimetals, alkali earth metals, but, various ions of Al^{3+}, Zn^{2+}, Pb^{2+}, Cu^{2+}, Ag^+ indicate the antibacterial effects. The order of the antibacterial effect is as following; $Cu^{2+} > Zn^{2+} > Ag^+ > Pb^{2+} > Al^{3+}$. The other, in sulfate solutions, Al^{3+}, Zn^{2+}, Cu^{2+}, Ag^+ have higher antibacterial activities. From these observations, the antibacterial order is $Zn^{2+} > Cu^{2+} > Ag^+ > Al^{3+}$, in which Zn^{2+} ions indicate the highest antibacterial effect.

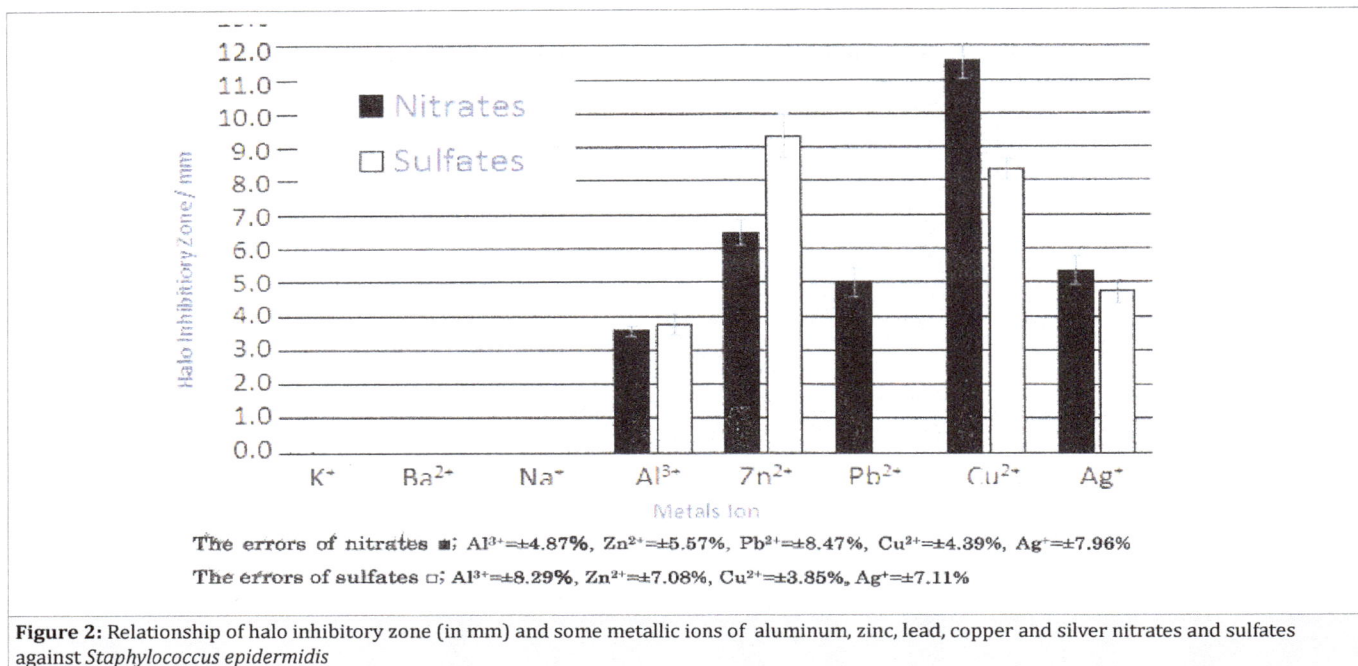

The errors of nitrates ■; $Al^{3+}=\pm4.87\%$, $Zn^{2+}=\pm5.57\%$, $Pb^{2+}=\pm8.47\%$, $Cu^{2+}=\pm4.39\%$, $Ag^+=\pm7.96\%$

The errors of sulfates □; $Al^{3+}=\pm8.29\%$, $Zn^{2+}=\pm7.08\%$, $Cu^{2+}=\pm3.85\%$, $Ag^+=\pm7.11\%$

Figure 2: Relationship of halo inhibitory zone (in mm) and some metallic ions of aluminum, zinc, lead, copper and silver nitrates and sulfates against *Staphylococcus epidermidis*

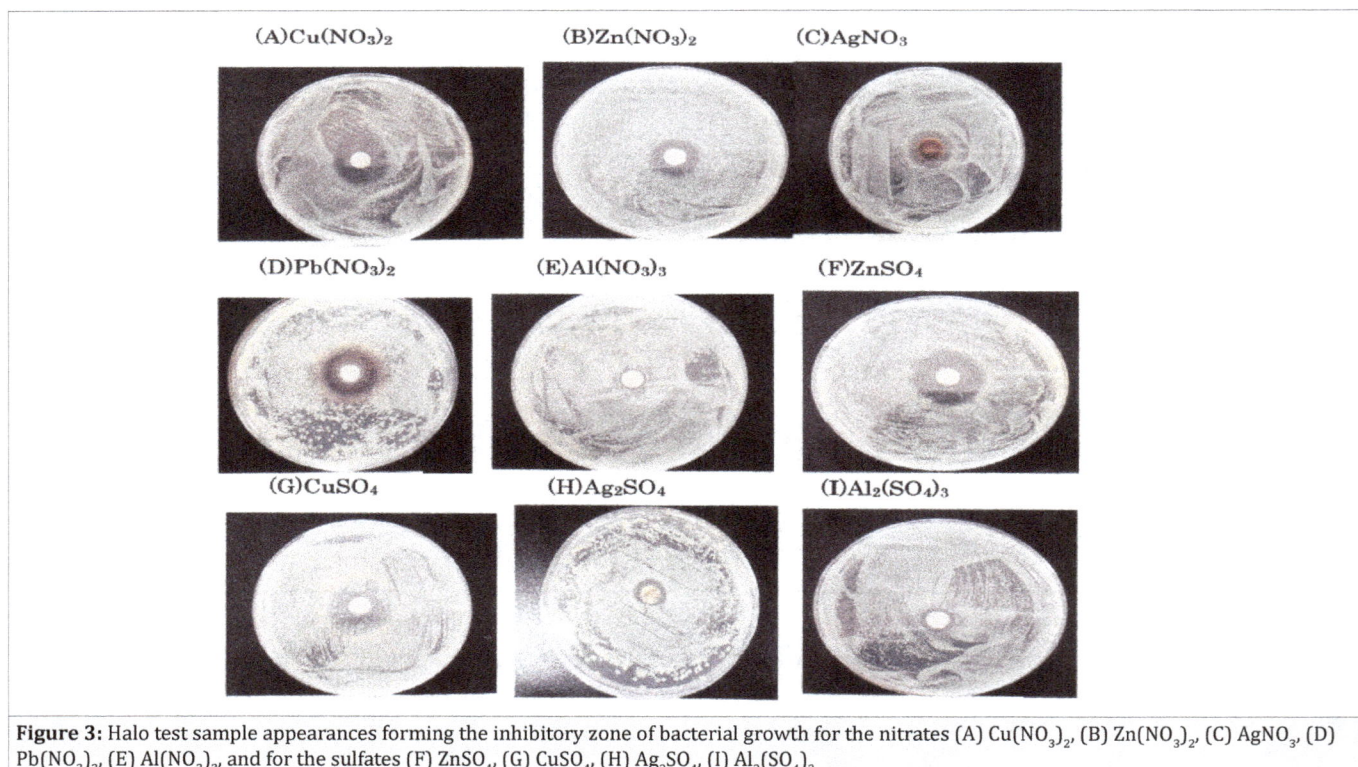

Figure 3: Halo test sample appearances forming the inhibitory zone of bacterial growth for the nitrates (A) $Cu(NO_3)_2$, (B) $Zn(NO_3)_2$, (C) $AgNO_3$, (D) $Pb(NO_3)_2$, (E) $Al(NO_3)_3$, and for the sulfates (F) $ZnSO_4$, (G) $CuSO_4$, (H) Ag_2SO_4, (I) $Al_2(SO_4)_3$

Results of Search and Analysis

(1) *S. aureus* and *E. coli* Cell walls, Action Sites of PGN biosyntheses of transglycosylase TG and transpeptidase TP and PGN autolysins

S. aureus surface cell envelop consists of teichoic acids, lipoteichoic acids, and thick peptideglycan (below PGN) cell wall [14], where as *E. coli* cell wall comprised of lipid A, lipopoly-saccharide, porin proteins, outer membrane of lipoprotein, and thinner 2-7 nm PGN layer in 30-70 nm periplasmic space[14]. Figure4 shows the molecular structure of *S. aureus* PGN cell wall, including the action sites of PGN biosynthesis enzymes of TG/TP, and PGN forth autolysins and Lysostaphin enzyme. Furthermore, figure 5 represents the molecular structure of *E. coli* cell wall and periplasmic peptidoglycan, containing the action sites of the hydrolases of lipoproteins, the peptidogly can biosynthetic enzymes TG/TP, and the autolysins. Further, interactions of PGN molecular structure, PGN syntheses and autolysins influence essentially in any event the bacteriolysis of bacterial cell walls.

(2) Characteristics of Zinc Sulfate Solution

Zinc is redox-inert and has only one valence state of Zn (II). In proteins, the coordination is limited by His, Cys, Glu, and sulfur donors from the side chains of a few amino acids. In zinc sulfate solution, $ZnSO_4$ is dissociated into aqua zinc ion $[Zn(H_2O)_6]^{2+}$ and sulfuric ion $(SO_4)^{2-}$. Aqua zinc ions are liable to be bound to ligand L having negative charge. The sulfuric ion has bactericidal inactivity [15].

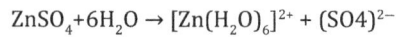

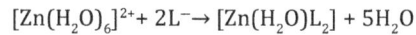

$$ZnSO_4 + 6H_2O \rightarrow [Zn(H_2O)_6]^{2+} + (SO4)^{2-}$$

$$[Zn(H_2O)_6]^{2+} + 2L^- \rightarrow [Zn(H_2O)L_2] + 5H_2O$$

$$Zn(H_2O)L_2 \rightarrow ZnL_2 + H_2O$$

By the reaction of Zn^{2+} ions with *S. aureus* surface, zinc-proteins are formed, on the ground that is due to formation of S-atom containing Zn-cysteine complex in bacteria [16].

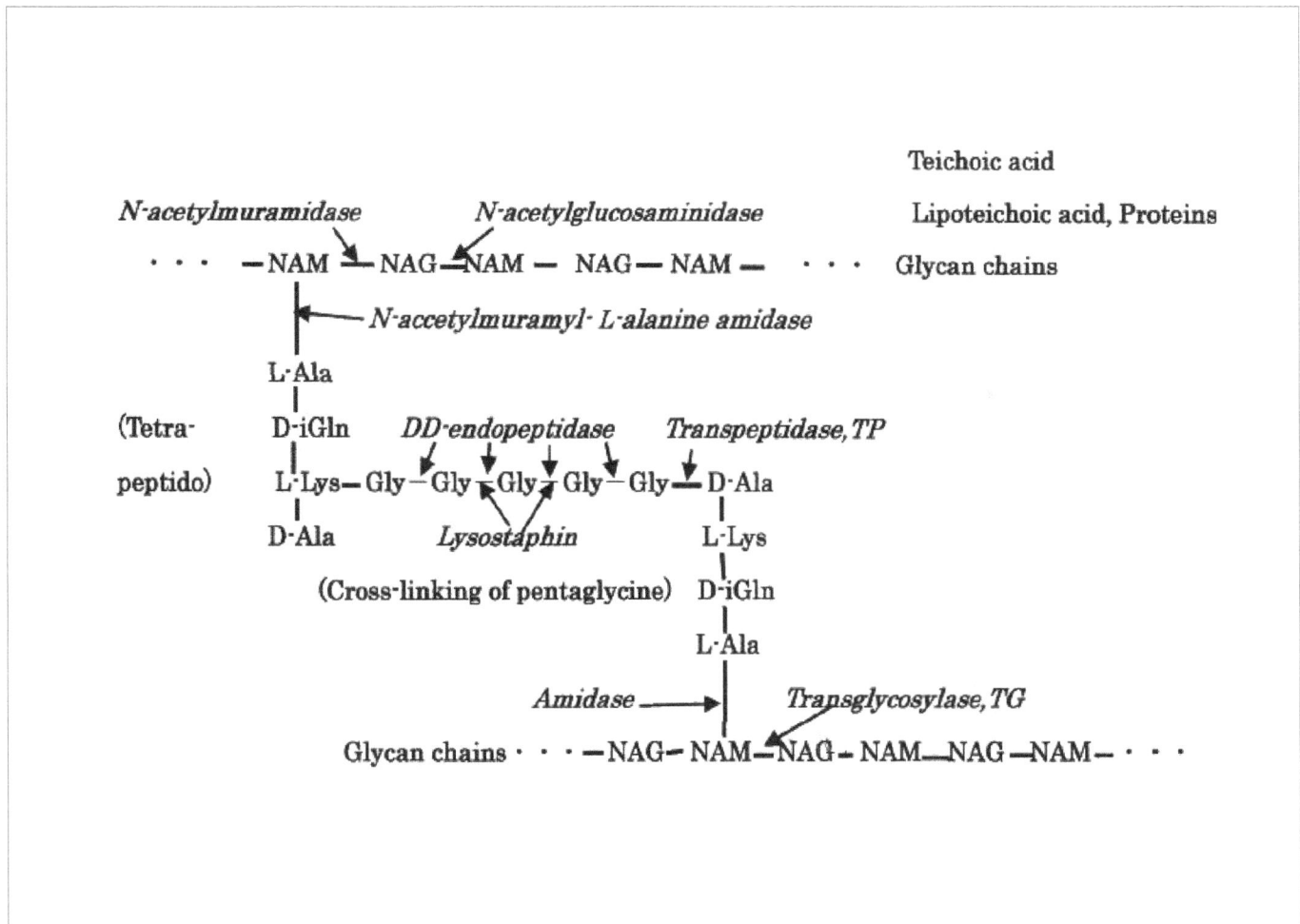

Figure.4: The molecular structure of *S.aureus* PGN cell wall, and the action sites of PGN biosynthesis enzymes of TG/TP, PGN for the autolysins, and Lysostaphin enzyme

Core oligosaccharide

Lipopolysaccaride(LPS)   Lipopolysaccaride(LPS)   O-antigen polysaccharide

Outer leaflet   Outer leaflet   Lipid A

Membrane protein CP/Porin protein PP

Inner leaflet   Inner leaflet   Fatty acid

Phospholipid   Phospholipid   Phosphoric acid base   PL

⟵—— *Endopeptidase of degradative enzyme*

Lipoprotein   Lipoprotein   LP

D-alanine   D-alanine

Meso-diaminopimelic acid   Meso-diaminopimelic acid

D-Glutamine   D-Glutamine

L-Ala   L-Ala

Muramidase, TG   ⟵—— *Amidase of degradative enzyme*

——— NAM —— NAG —— NAM ——— NAG ——   Peptidoglycan Layer(1~3 nm)

Amidase➔   *Glucosaminidase*

L-Ala

Peptidase➔

A₂pm   D-Ala   Soluble protein

Carboxypeptidase

D-Ala —(CO·NH)—— A₂pm(Diaminopimelic acid)   Periplasmic Space(PPS)

⟵—— *Carboxypeptidase*

Endopeptidase, TP   D-Glu

L-Ala

Amidase——➔

——NAG —— NAM —— NAG ——

Cell   Membrane

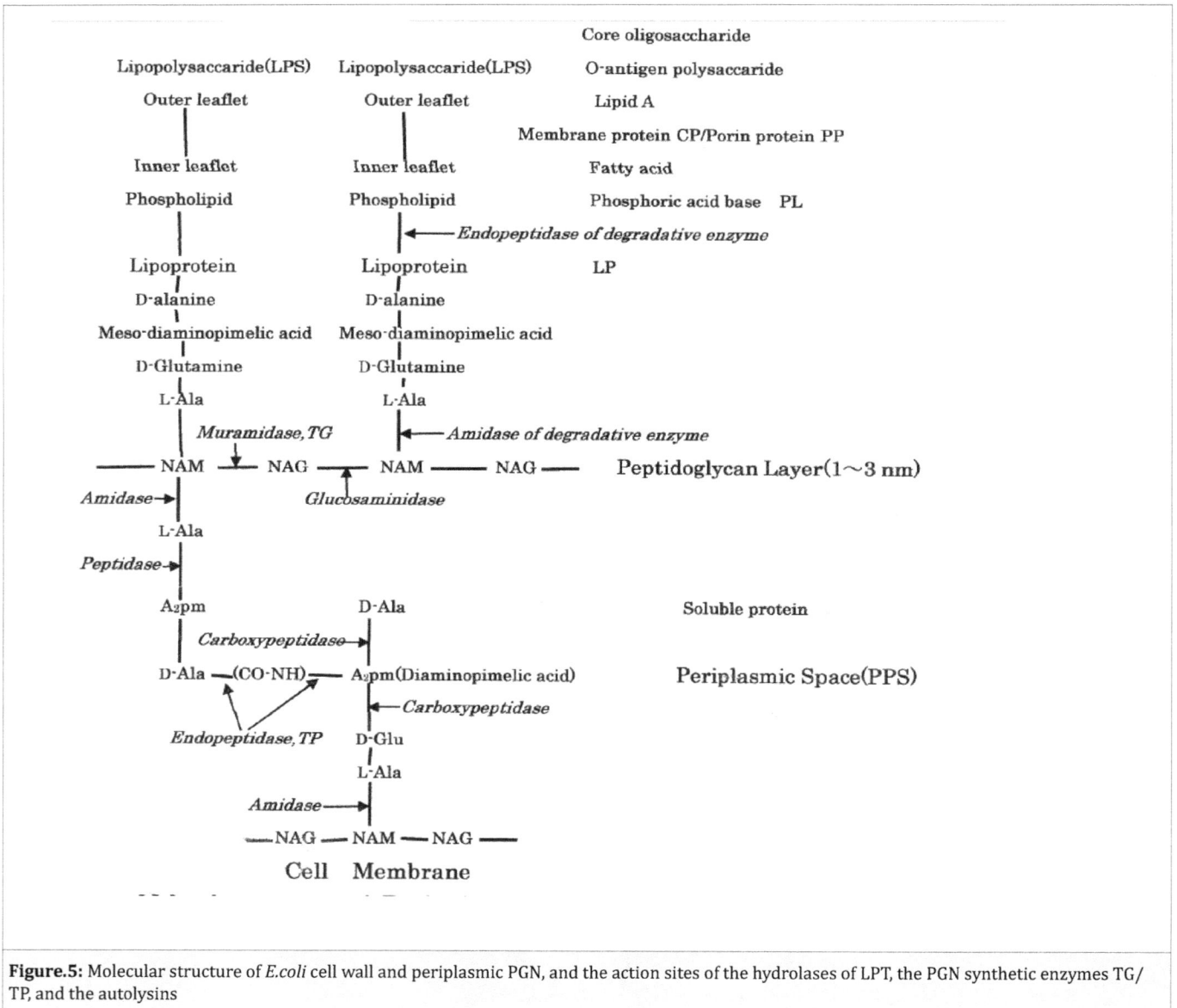

Figure.5: Molecular structure of *E.coli* cell wall and periplasmic PGN, and the action sites of the hydrolases of LPT, the PGN synthetic enzymes TG/TP, and the autolysins

Discussions

Bacteriolysis of *S. aureus* PGN cell wall by Cu^{2+} ions

(1) Bacteriolysis by balance deletion between biosynthesis enzyme and decomposition enzyme (autolysin) in PGN cell wall

For the sake of growth of *S. aureus* PGN cell wall, there is necessarily required for the adequate balance between PGN biosynthesis and PGN autolysin. When the balance is broken by Cu^{2+} penetration, Cu^{2+} ions are self-catalytically treated as coenzyme, that this is indicated that activation of autolysin is proceeded, in which bacteriolysis and killing may result. Hence, bacteriolysis of *S. aureus* PGN cell wall by Cu^{2+}ions is due to inhibition of PGN elongation owing to the damages of PGN synthetic TG/TP and the activations of PGN autolysins.

(2) Inhibition of polymerization of glycan chains bonding and cross-linking of side peptide

Cu^{2+} ions inhibit polymerization of glycan chains, forming copper complex in which is partial action sites of glycan saccharide chains. L is coordinated molecular.

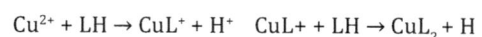

$$Cu^{2+} + LH \rightarrow CuL^+ + H^+ \quad CuL+ + LH \rightarrow CuL_2 + H$$

Copper-complexes on saccharide chains may be

• —NAG-(NAM-Cu-2O-2N-NAG)-NAM—•

The other, Cu^{2+} ions inhibit cross-linked reaction by peptide copper complex formation bonding to side-peptide chains.

$$Cu^{2+} + 2LH \rightarrow CuL_2 + H^+$$

Peptide copper complex may be 3N-Cu-O, Cu (Gly-L-Ala)H_2O.

Specially, Cu^{2+} ions react with cross-molecular penta glycine$(Gly)_5$, copper-glycine complex may be formed.

Amino acid: $Cu^{2+} + Gly^- \rightarrow Cu(Gly)^+$, $Cu(Gly)^+ + Gly^- \rightarrow Cu(Gly)_2$

Peptido: $Cu^{2+} + GlyGly \rightarrow Cu(GlyGly)$, $Cu(GlyGly) + Gly^- \rightarrow Cu(GlyGlyGly)^-$

Bacteriolysis and destruction of *E. coli* outer membrane cell wall by Cu^{2+} ions

(1) Inhibition of outer membrane cell wall

Cu^{2+} ions inactivate catalyst enzyme with forming Cu$^+$ ions.

$$Cu^{2+} + -SH \rightarrow -SCu(I) + H^+$$

By the penetration of Cu^{2+} ions, as shown in figure 5, the activations of amidase enzyme of N-terminal and endopeptidase enzyme of C-terminal are enhanced[17,18]. Accordingly, the activations of decomposition at N-, C-terminals of lipoproteins may occur with the destruction of outer membrane structure.

(2) Inhibition of biosynthesis and activation of autolysin, or regulation and deletion of autolysin.

Inhibition of *E. coli* PGN by Cu^{2+} ions is reported [19], however, the site of concrete action is not described. In *E. coli*, it is unlikely thought that Cu^{2+} ions inhibit both TG and TP [20]. The other, it is unclear that Cu^{2+} inhibit the polymerization of NAM and NAG chains. It is perhaps simpler to think that TP enzyme of cross-linked reaction is inhibited by Cu^{2+} ions and the activation of PGN autolysin occurs. By the accumulation of Cu^{2+} ions in periplasmic space, it might be possible that bacteriolysis of cell wall occur by the activation of PGN autolysin within periplasmic space. Many autolysins of *E. coli* are regulated by metals ion such as Hg^{2+}, Cu^{2+}[21]. This regulation or deletion of decomposition enzyme inhibits PGN elongation, in which the bacteriolysis of the cell wall is induced. These facts are consistent with that the destruction by bacteriolysis of cell wall had been observed against *E. coli*.

Hence, bacteriolysis of *E. coli* cell wall by Cu^{2+} ions occurs by destruction of outer membrane structure due to degradation of lipoprotein at N-, C-terminals, damage of TP enzyme and activations of PGN autolysins. Furthermore, deletion of PGN autolysin also becomes bacteriolystic factor.

(3) Antibacterial activities of cell membrane and cytoplasm

Reactive oxygen species (ROS) O_2^- and H_2O_2 generated in cell wall, permeate into cell membrane and cytoplasm, in which in cell membrane high reactive $^\cdot$OH and OH$^-$ are formed by Haber-Weiss and Fenton reactions.

Haber-Weiss reaction[22]; $H_2O_2 + O_2^- \rightarrow \cdot OH + OH^- + O_2$

Fenton reaction [23]; $Cu^+ + H_2O_2 \rightarrow \cdot OH + OH^- + Cu^{2+}$

Furthermore, new ROS productions occur by Fenton-like type. L=Ligand

$$LCu(II) + H_2O_2 \rightarrow LCu(I) + \cdot OOH + H^+$$

$$LCu(I) + H_2O_2 \rightarrow LCu(II) + \cdot OH + OH^-$$

As above-mentioned, the bactericidal processes of bacteriolysis of the *S. aureus* and *E. coli* cell walls by Cu^{2+} ions, and also the antibacterial activities of cell membrane and cytoplasm are shown in Table 4.

Bacteriolysis of *S. aureus* PGN Cell Wall by Zn^{2+} Ions

(1) PGN biosynthesis enzymes of transglycosylaseTG and transpeptidase TP

Wall teichoic acids are spatial regulators of PGN cross-linking biosynthesis TP[24], however, it is not explicit whether zinc ions could inhibit both TG and TP enzymes of the PGN, wherein is due to uncertain relation between wall teichoic acids biosynthesis and PGN biosynthesis.

(2) Inhibition of PGN elongation due to the activations of autolysins

Zn^{2+} binding Rv3717 showed no activity on polymerized PGN and but, it is induced to a potential role of N-Acetylmuramyl-L-alanine Amidase [25], PGN murein hydrolase activity and generalized autolysis; Amidase MurA [26], Lytic Amidase LytA [27], enzymatically active domain of autolysin LytM [28], Zinc-dependent metalloenzyme AmiE [29] as prevention of the pathogen growth, and Lysostaphin-like PGN hydrolase and glycylglycine endopeptidase LytM [30]. It is thought that the activations of these PGN autolysins could be enhanced the inhibitions of PGN elongation simultaneously, with bacteriolysis of *S. aureus* PGN cell wall.

(3) Production of reactive oxygen species (ROS) against S. aureus

O_2^- and H_2O_2 permeate into membrane and cytoplasm, that DNA molecular is damaged by oxidative stress [31]. For the penetration of zinc ions to PGN cell wall, the ROS production such as superoxide anion radical O_2^-, hydroxyl radical $^\cdot$OH, hydrogen peroxide H_2O_2 occurred from superoxide radical O_2^- molecular[32]. O_2^- and and H_2O_2 permeate into membrane and cytoplasm, and then, DNA molecular is damaged by oxidative stress [31].

$$O_2 + e- + H+ \rightarrow \cdot HO_2$$

$$\cdot HO_2 \rightarrow H+ + O_2$$

$$H_2O_2 + e- \rightarrow HO^- + \cdot OH$$

$$2H^+ + \cdot O_2^- + \cdot O_2^- \rightarrow H_2O_2 + O_2$$

$$H_2O \rightarrow \cdot OH + \cdot H + e- \rightarrow H_2O_2$$

Bacteriolysis and destruction of *E. coli* Cell Wall by Zn^{2+} Ions

(1) Permeability of Zinc Ions into *E. coli* Cell Wall

E. coli cell wall is constituted of lipo polysaccharide (LPS), lipoproteins (LPT), and PGN, thinner layer within periplasmic space. The first permeability barrier of zinc ions in the *E .coli* cell wall is highly anionic LPS with hydrophobic lipid A, core

Cu²⁺ ion solution	Cell wall			Cell membrane	Cytoplasm
Cu²⁺ →	*S.aureus* cell wall			·Cu-ammine complex and Cu-protein formations	·Cu-complex and Cu-3N-O complex formations
	PGN layer cell wall				
	Cu²⁺, Cu⁺				
	·Bacteriolysis of PGN cell wall by TG,TP synthesis inhibitions and activations of *S. aureus* PGN autolysins			Cu⁺, Cu²⁺	Cu²⁺
	·Cu complex of glycan saccharide chain and Cu peptide complex formations			O₂⁻	·OH
	·Reactive oxygen species O₂⁻, H₂O₂			H₂O₂	OH⁻
Cu²⁺ →	*E.coli* cell wall			·OH	(O₂⁻)
	Lipopolysaccharide (LPS)	Outer membrane(OM) Lipoprotein(LPT)	Periplasmic space(PS)	OH⁻	H₂O₂
				·OOH →	
	Corepolysaccharide Phosphorus lipid,	OmpF,A,C, Porin, Protein	PGN layer Miscible protein	· OH· formed by Haber-Weiss/Fenton reactions	·DNA/RNA synthesis inhibition
	→ Cu²⁺, Cu⁺ →	Cu²⁺, Cu⁺ →	Cu²⁺, Cu⁺		Substitution of Cu²⁺ ions into DNA hydrogen bond base pairs
	·Variation of charge properties	·Cu ion binding proteins	·Cu²⁺ ions accumulation	·Accumulation of O₂⁻ and H₂O₂	
	·Cu ion induced increase in permeability	·Damages of outer membrane structure due to the degradable enzymes of lipoprotein at N-, C-terminals and the activation of PGN hydrolases	·Inhibitions of PGN elongation by the deletions of PGN-TP enzyme and PGN autolysins		
	· O₂⁻, Cu⁺	· Cu⁺, H₂O₂	·Cu⁺,HO·, H₂O₂		

Table 4: Bactericidal action processes of Cu²⁺ solutions within the cell wall/the cell membrane/the cytoplasm against *S.aureus* and *E.coli*

polysaccharide, O-polysaccharide, in which zinc ions may be possible for the inhibition of LPS biosynthesis, owing to that promotes formation of metal-rich precipitates in a cell surface[33]. In zinc ion uptake across the outer membrane, the lipoproteins of Omp A, Omp C, Omp F porins have a role for at least some of these proteins in Zn^{2+} uptake, in which the lipoproteins have metallic cation selective and hydrophilic membrane crossing pore, to be effective for zinc transfer [34]. Zinc (II) ions react with -SH base, and then H_2 generates. Zinc bivalent is unchangeable as $^-SZn^-S^-$ bond 4-coodinated.

Zn^{2+} + 2(-SH) → -SZn(II) –S– + $2H^+$

(2) Destruction of outer membrane structure of *E. coli* cell wall by hydrolases of lipoproteins at C-, N-terminals

ZnPT (zinc pyrithione) and Tol (Tol proteins)-Pal (Protein-associated lipoprotein) complex are antimicrobial agents widely used, however, it has recently been demonstrated to be essential for bacterial survival and pathogenesis that outer membrane structure may be destroyed [35,36].

(3) Inhibition of PGN elongation due to the damage of PGN synthesis enzyme of zinc-protein amidase in periplasmic space, and the activities of PGN autolysins

The zinc-induced decrease of protein biosynthesis led to a partial disappearance of connexin-43 of protein synthesis in neurons [37], but it is unknown whether PGN biosynthesis is inhibited. Further, it is also unclear whether the both TG/TP should be inhibited by the zinc ions [38,39,40]. The other, zinc ions were accumulated in *E. coli* periplasmic space, in which the zinc ions are spent to the activation of bacteriolysis of the cell wall. Zinc depending PGN autolysin, amidase PGRPs [41], zinc

metallo enzymes AmiD[42], zinc-containing amidase; AmpD [43], zinc-present PGLYRPs[44] serve to be effective for the PGN autolysins. It is particularly worth noting that enhancement of the activities of autolysins is characterized on PGN carboxy-peptidase-transpeptidase IIW [45] requiring divalent cations. Accordingly, the inhibition of PGN elongation had occurred by zinc ion induced activities of PGN hydrolases and autolysins.

(4) ROS production and oxidative stress against *E. coli*

Zinc ions reacted with -SH, and H^+ generates. In *E. coli*, free radicals O_2^-, OH^-, $^{\cdot}OH$) and H_2O_2 are formed as follows[46]:

O_2 + e → $O_2^{\cdot-}$

$2O_2^-$ + $2H^+$ → H_2O_2 + O_2

O_2^- + H_2O_2 → OH^- + $^{\cdot}OH$ + O_2^-

In the cell wall, reacting with polyunsaturated fatty acids:

LH + OH$^{\cdot}$ → L$^{\cdot}$ + HOH

L$^{\cdot}$ + O_2 → LOO$^{\cdot}$

LH + LOO$^{\cdot}$ → L$^{\cdot}$ + LOOH

Zinc-containing Peptidoglycan Recognition Proteins (PGRPs) induce ROS production of H_2O_2, O_2^-, $HO^{\cdot}$, the ROS occur the oxidative stress, and killing by stress damage [47].

Thus, from above-mentioned results, the processes of the bacteriolysis of *S. aureus* PGN and *E. coli* outer membrane cell walls by the permeability and the antibacterial activities of Zn^{2+} ions are summarized in Table 5.

Table 5 Bacteriolytic processes of *S.aureus* PGN and *E.coli* outer membrane cell walls by the permeability and the antibacterial activities of Zn^{2+} ions

Zn^{2+} ions solution	*S.aureus* Cell Wall		
	Teichoic acid, Lipoteichoic acid, Peptidoglycan layer, Proteins		
Zn^{2+}	→ Zn^{2+} O_2^-, H_2O_2, ·OH, ·NO, $ONOO^-$ · Teichoic acids are spatial regulators of biosynthesis of PGN cross-linking TP enzyme · Zn^{2+} binding proteins · It is unknown whether zinc ions inhibit the PGN biosynthesis TG/TP enzymes · Activations of PGN autolysins of amidases · Bacteriolysis and destruction of PGN cell wall due to inhibition of PGN elongation · ROS productions and the oxidative stress		
Zn^{2+} ion solution	*E.coli* cell wall		
	Lipopolysaccharide(LPS) Lipid A, Core polysaccharide	Outer Membrane Lipoprotein, Porins OmpF,A,C	Periplasmic Space PGN layer
Zn^{2+}	→ Zn^{2+}, H^+ · Negative charge · Hydrophobic Lipid A · Inhibition of LPS biosynthesis · Zn^{2+} + 2(-SH) → -SZn—S— + H^+	→ Zn^{2+}, O_2^-, H_2O_2, · Porin proteins of hydrophilic channels · Zn binding proteins · Destruction of outer membrane structure due to degradative hydrolases of lipoprotein at C· and N-terminals · LOO·, L·(Fatty acid)	→ Zn^{2+}, O_2^-, H_2O_2, OH^-, ·OH · Zn accumulation in periplasmic space · Inhibition of PGN elongation due to activations of PGN autolysins of amidase and carboxypeptidase· transpeptidase

Conclusions

(1)From the result of antibacterial activities of Cu^{2+} ion solution by the two-fold broth dilution medium method, for bacteriostasis MIC=50 mg/L above was obtained in Cu^{2+} concentration range of 0.10~50 mg/L against *E. coli*. The other, for bactericide action MIC=625 mg/L and MBC=1250 mg/L were obtained in Cu^{2+} concentration range of 9.8~5,000 mg/L against *S. aureus*.

(2)From halo-antibacterial susceptibility tests of metallic ion concentration of 100 mM/L against *Staphylococcus epidermidis*, the order of bacterial effect for nitrate solutions is as follows: $Cu^{2+}>Zn^{2+}>Ag^+>Pb^{2+}>Al^{3+}$. The other, in the sulfate solutions, the order is $Zn^{2+}>Cu^{2+}>Ag^+>Al^{3+}$. The appearance of the highest antibacterial activity is found to be the zinc sulfate solution.

(3) Bacteriolysis of *S. aureus* PGN cell wall by Cu^{2+} ions is caused for the inhibition of PGN elongation due to damages of PGN synthetic TG/TP and activation of PGN autolysins. The other, bacteriolysis of *E. coli* outer membrane cell wall by Cu^{2+} ions is attributed tothe destruction of outer membrane structure and to the inhibition of PGN elongation due to the damage of PGN biosynthesis TP and the activation of PGN autolysins.

(4) Bacteriolysis and destruction of *S. aureus* PGN cell wall by Zn^{2+} ions are due to the inhibition of PGN elongation by the activities of PGN autolysins of amidases. The other, bacteriolysis of *E. coli* cell wall by Zn^{2+} ions are due to destruction of outer membrane structure by degrading of lipoprotein at C-, N-terminals, owing to PGN formation inhibition by activities of PGN autolysins of amidase and carboxypeptidase-transpeptidase.

(5)By the penetration of copper, orzinc ions into bacterial cell wall, productions of O_2^-, H^+, H_2O_2, $ONOO^-$ occurs. The other, in *E. coli* cell wall, the productions of O_2^-, H^+ in outer membrane, and H_2O_2, OH^-, $\cdot OH$ in periplasmic space occur. These ROS and H_2O_2 damage the cell membrane and the DNA molecules by oxidase stress.

References

1. Grabrucker AM, Rowan M, Garner CC. Brain-Delivery of Zinc-Ions as Potential Treatment for Neurological Diseases: Mini Review. Drug Deliv Lett. 2011;1(1):13-23.

2. Ma L, Terwilliger A, Maresso AW. Iron and Zinc Exploitation during Bacterial Pathogenesis, Metallomics. 2015;7(12):15421-15454. doi: 10.1039/c5mt00170f

3. D Skrajnowska, B Bobrowska, Andrzej Tokarz, Marzena Kuras, Paweł Rybicki, Marek Wachowicz. The Effect of Zinc- and Copper Sulphate Supplementation on Tumor and Hair Concentration of Trace Elements In Rats with DMBA-Induced Breast Cancer. Pol J Environ Stud. 2011;20(6):1585-1592.

4. Vaidya JS. An alternative model of cancer cell growth and metastasis. Int J Surg. 2005;5(2):73-75. doi: 10.1016/j.ijsu.2006.06.003

5. John E, Laskow TC, Buchser WJ, Pitt BR, Basse PH, Butterfield LH,et al. Zinc in innate and adaptive tumor immunity. JTranslational Medicine.

2010;8:118-134. doi: 10.1186/1479-5876-8-118.

6. Yu H, Zhou Y, Lind SE, Ding WQ. Clioquinol targets zinc to lysosomes in human cancer cells. Biochem J.2009;417(1):133-139. Doi: 10.1042/BJ20081421

7. Alam S, Kelleher SL. Cellular Mechanisms of Zinc Dysregulation: A Perspective on Zinc Homeostasis as Etiological Factor in the Development and Progression of Breast Cancer. Nutrients. 2012;4(8):875-903. Doi:10.3390/nu4080875

8. Franklin RB, Costello LC. The Important Role of the Apoptotic Effects of Zinc in the Development of Cancers. J Cell Biochem. 2009; 106(5):750-757. Doi: 10.1002/jcb.22049

9. MacDonald RS. The Role of Zinc in Growth and Cell Proliferation. J Nutr. 2000;13015:1500S-1508S.

10. van den Elsen JM, Kuntz DA, Rose DR. Structure of Golgi α-mannosidase II. a target for inhibition of growth and metastasis of cancer cells. The EMBO Journal.2001;20(12):3008-3017. Doi:10.1093/emboj/20.12.3008

11. Gumulec J, Masarik M, Krizkova S, Adam V, Hubalek J, Hrabeta J, et al. Insight to Physiology and Pathology of Zinc ions and Their Actions in Breast and Prostate Carcinoma. Curr Med Chem. 2011;18(33)5042-5051.

12. Tsuneo Ishida; Antibacterial activities of Cu^{2+} against Gram-negative bacteria, J. Copper and Copper Alloy. 2014;53:272-278. (in Japanese)

13. Tsuneo Ishida; Antibacterial susceptibility tests of Cu^{2+}solution and bacteriolysis and killing action of peptidoglycan cell wall, Chemistry and Industry. 2015;66:611-617. (in Japanese)

14. T J Silhavy, D Kahne, S Walker. The Bacterial Cell Envelope. Cold Spring Harbor Perspectives in Biology. 2010;2(5):1-14. 10.1101/cshperspect.a000414

15. Faiz U, Butt T, Satti L, Hussain W, Hanif F. Efficacy zinc as an antibacterial agent against enteric bacterial pathogens, J Ayub Med Coll Abbottabad. 2011;23(2):8-21.

16. Robert T. Crichton, Mituhiko Shioya; Japanese Translation, Biological Inorganic Chemistry, Tokyo Kagaku-Dojin Limited, 2016, 175-188.

17. Heidrich C, Ursinus A, Berger J, Schwarz H, Höltje JV. Effects of multiple deletions of murein hydrolases on viability, septum cleavage, and sensitivity to large toxic molecules in *E.coli*. J Bacteriology. 2002;184(22): 6093-6099.

18. Jean van Heijenoot. Peptidoglycan Hydrolases of *E.coli*. Microbiolgy and Molecular Biology Reviews. 2011;75(4):636-663. Doi:10.1128/MMBR.00022-11

19. Bai W, Zhao K, Asami K. Effects of copper on dielectronic properties of *E.coli* cells. Colloids and Surfaces B :Biointerfaces. 2007;58(2):105-115. Doi; 10.1016/j.colsurfb.2007.02.015

20. Vasanthi Ramachandran, B Chandrakala, Vidya P Kumar, Veeraraghavan Usha, Suresh M Solapure and Sunita M de Sousa. Screen for inhibitors of the coupled transglycosylase-transpeptidase of peptidoglycan biosynthesis in *E.coli*. Antimicrob Agents Chemother. 2006;50(4)1425-1432. 10.1128/AAC.50.4.1425-1432.2006

21. Gilad Bernadsky, Terry, J Beveridge, Anthony J Clarke. Analysis of the Sodium Dodecyl Sulfate-Stable Peptidoglycan: Autolysins of Select

Gram-Negative Pathogens by Using Renaturing Polyacrylamide Gel Electrophoresis, J.Bacteriology,1994;176(17):5225-5232. 10.1128/jb.176.17.5225-5232.1994

22. Kehrer JP. Kehrer. The Haber-Weiss reaction and mechanisms of toxicity. Toxicology. 2000;149(1):43-50.

23. Krzyszt of BARBUSINSKI. Fenton reaction-controversy concerning the chemistry, ECOLOGICAL CHEMISTRY, AND ENGINEERINGS. 2009;16(3):347-358.

24. Atilano ML, Pereira PM, Yates J, Reed P, Veiga H, Pinho MG, Filipe SR. Teichoic acid are temporal and spatial regulators of peptidoglycan cross-linking in S.aureus, PNAS. 2010;107(44):18991-18996. Doi:10.1073/pnas.1004304107

25. Prigozhin DM, Mavrici D, Huizar JP, Vansell HJ, Alber T. Structural and Biochemical Analyses of *Mycobacterium tuberculosis* N-AcetylmuramyI-L-alanineAmidaseRv3717 Point to a Role in peptidoglycan Fragment Recycling, J Biol Chem. 2013;288(44):31549-31555. Doi:10.1074/jbc.M113.510792

26. Carroll SA, Hain T, Technow U, Darji A, Pashalidis P, Joseph SW,et al. Identification and Characterization of a Peptidoglycan Hydrolase, MurA of *Listeria monocytogenes*, a Muramidase Needed for Cell Separation, J. Bacteriology. 2003;185(23):6801-6808.

27. Peter Mellrotha, Tatyana Sandalovab, Alexey Kikhneyc, Francisco Vilaplanad, Dusan Heseke, Mijoon Leee, et al. Structural and Functional Insights into Peptidoglycan Access for the Lytic Amidase LytA of *Streptococcus pneumonia*. 2014;5(1). Doi:10.1128/mBio.01120-13

28. Elzbieta Jagielsksa, Olga Chojnacka, Izabela Sabata. LytM Fusion with SH3b-Like Domain Expands Its Activity to Physiological Conditions, Microbial Drug Resistance.2016;22(6):461-469. Doi: org/10.1089/mdr.2016.0053

29. Sebastian Zoll, Bernhard Pätzold , Martin Schlag , Friedrich Götz, Hubert Kalbacher, Thilo Stehle. Structural Basis of Cell Wall Cleavage by a Staphylococcal Autolysin. PloS Pathogens. 2010;6:1-13. Doi: org/10.1371/journal.ppat.1000807

30. Ramadurai L1, Lockwood KJ, Nadakavukaren MJ, Jayaswal RK. Characterization of a chromosomally encoded glycyglycine endopeptidase of *S.aureus*. Microbiology. 1999;145(4):801-808. Doi: 10.1099/13500872-145-4-801

31. R Gaupp, N.Ledala, GA Somerville. *Staphylococal* response to oxidative stress, Frontiers in Cellular and Infection Microbiology.20132;2:1-8. Doi:10.3389/fcimb.2012.00033

32. Filis Morina, Marija Vidović, Biljana Kukavica, Sonja Veljović-Jovanović. Induction of peroxidase isoforms in the roots of two *Verbascum Thapsus L*. populations is involved in adaptive responses to excessZn2+ and Cu2+, Botanica SERBICA. 2015;39(2);151-158.

33. S Langley, TJ Beveridge. Effect of O-Side-Chain-LPS Chemistry on Metal Binding, Appl Environ Microbiol.1999;65(2):489-498.

34. Claudia A Blindauer. Advances in the molecular understanding of biological zinc transport. The Royal Society of Chemistry.2015;51:4544-4563. Doi:10.1039/C4CC10174J

35. A.J.Dinning, I.S.I.AL-Adham, P.Austin, M.Charlton and P.J.Collier,Pyrithione groups, J. Applied Microbiology. 1998;85:132-140.

36. Godlewska R, Wiśniewska K, Pietras Z, Jagusztyn-Krynicka EK. Peptidoglycan-associated lipoprotein(Pal)of Gram-negative bacteria: function, structure, role in pathogenesis and potential application in immunoprophylaxis, FEMS Microbiol Letter. 2009;298(1):1-11. Doi: 10.1111/j.1574-6968.2009.01659.x

37. Alirezaei M, Mordelet E, Rouach N, Nairn AC, Glowinski J, Prémont J. Zinc-induced inhibition of protein synthesis and reduction of connexin-43 expression and intercellular communication in mouse cortical astrocytes. Eur J Neurosci. 2002;16(6):1037-1044.

38. Alexander J F Egan, Jacob Biboy, Inge van't Veer, Eefjan Breukink, Waldemar Vollmer. Activities and regulation of peptidoglycan synthases. Philos Trans R Soc Lond B Biol Sci. 2015;370(1679). Doi:10.1098/rstb.2015.0031

39. Singh SK, SaiSree L, Amrutha RN, Reddy M. Three redundant murein endopeptidases catalyze an essential cleavage step in peptidoglycan synthesis of *E.coli* K12, Molecular Microbiology.2012; 86(5):1036-1051. Doi: 10.1111/mmi.12058

40. Ramachandran V, Chandrakala B, Kumar VP, Usha V, Solapure SM, de Sousa SM. Screen for Inhibitors of the Coupled Transglycosylase-Transpeptidase of Peptidoglycan Biosynthesis in *E.coli*. Antimicrob Agents Chemother. 2006;50(4):1425-1432. Doi: 10.1128/AAC.50.4.1425-1432.2006

41. Rivera I, Molina R, Lee M, Mobashery S, Hermoso JA. Orthologous and Paralogous AmpD Peptidoglycan Amidases from Gram-Negative Bacteria, Microb Drug Resist. 2016;22(6):470-476. Doi:10.1089/mdr.2016.0083

42. Pennartz A, Généreux C, Parquet C, Mengin-Lecreulx D, Joris B. Substrate-Induced Inactivation of the *E.coli* AmiD N-Acetylmuramoyl-L-AlanineAmidase Highlights a New Strategy To Inhibit This Class of Enzyme, Antimicrob Agents Chemother.2009;53(7):2991-2997. doi: 10.1128/AAC.01520-07

43. Carrasco-López C, Rojas-Altuve A, Zhang W, Hesek D, Lee M, Barbe S, et al. Crystal Structures of Bacterial Peptidoglycan Amidase AmpD and an Unprecedented Activation Mechanism. J Biol Chem. 2011; 286(36)9:31714-31722. Doi: 10.1074/jbc.M111.264366

44. Wang M, Liu LH, Wang S, Li X, Lu X, Gupta D, et al. Human Peptido-glycan Recognition Proteins Require Zinc to Kill Both Gram-Positive and Gram-negative Bacteria and Are Synergistic with Antibacterial Peptides J Immunol..2007;178(5):3116-3125.

45. DasGupta H Fan DP Fan. Purification and Characterization of a Carboxypeptidase-Transpeptidase of *Bacillus megaterium* Acting on the Tetra peptide Moiety of the Peptidoglycan, J. Biologycal Chemistry.1979;254(13): 5672-5682.

46. ZN Kashmiri, SA Mankar. Free radicals and oxidative stress in bacteria, International J. of Current Microbio and Appl Scienses. 2014;3:34-40.

47. Des Raj Kashyap, Annemarie Rompca, Ahmed Gaballa, John D Helmann, Jefferson Chan, Christopher J Chang, Peptidoglycan Recognition Proteins Kill Bacteria by Inducing Oxidative,Thiol,and Metal Stress. PLOS Pathogen. 2014;10(7):1-17. doi: 10.1371/journal.ppat.1004280

13

Augmentation of Antioxidant Status in the Liver of Swiss Albino Mice treated with Jamun (*Syzygium Cumini*, Skeels) Extract before Whole Body Exposure to Different Doses of γ-Radiation

Ganesh Chandra Jagetia[1]* and Prakash Chandra Shetty[2]

[1]*Department of Zoology, Mizoram University, Aizawl-796 004, India*
[2]*Department of Anatomy, Melaka Manipal Medical College, Manipal-576 004, India*

Corresponding author: *Ganesh Chandra Jagetia, Professor, Mizoram University, Tanhril, Aizawl-796 004, Mizoram, India,*
E-mail: gc.jagetia@gmail.com

Abstract

The ability of ionizing radiations to produce free radicals leads to increased oxidative stress and negative alteration of the antioxidant status in the exposed organisms. The increased oxidative stress is responsible for the induction of various deleterious effects of ionizing radiation. The introduction of phytoceuticals may reduce the radiation-induced oxidative stress. Therefore present study was designed to study the effect of oral administration of 50 mg/ kg body weight of Jamun *(Syzygium cumini)* leaf extract on the antioxidant status in the liver of mouse exposed to 0,0.5, 1, 2, 3 or 4 Gy whole body γ-radiation. Irradiation of mice to different doses of γ-radiation caused a significant and dose dependent decline in the glutathione concentration, catalase and superoxide dismutase activities at 0.5 1, 2, 4, 8, 12 and 24 h post-irradiation. A maximum decline in glutathione concentration was observed at 1 h post-irradiation, whereas superoxide dismutase activity showed a highest decrease at 2 h post-irradiation. In contrast lipid peroxidation increased in a dose dependent manner at all post-irradiation times with a greatest rise at 2 h post-irradiation in the animals exposed to 4 Gy. Administration of mice with jamun extract before exposure to 0- 4 Gy resulted in a significant elevation in the glutathione concentration and catalase and superoxide dismutase activities at all exposure doses, whereas lipid peroxidation reduced significantly when compared with the irradiated group that did not receive jamun extract at all post-irradiation times. Our study demonstrates that treatment of mice with jamun extract elevated the antioxidant status and reduced the radiation-induced oxidative stress in mouse liver.

Keywords: Mice; Radiation; Jamun; Glutathione; Catalase; Superoxide dismutase; Lipid peroxidation

Introduction

The increasing power need of the world can be met by installing nuclear power generators. However, nuclear power generation comes with a price of undesired human exposure to ionizing radiations due to increase in the back ground radiation in the vicinity of these generators. This may be compounded during nuclear accidents and or leakage of radioactivity in the environment. The Chernobyl accident has not only exposed the biota including humans in its vicinity but also irradiated the far flung areas due to massive release of radioactivity during this catastrophe leading to several physiological and genetic effects [1]. Further exposure of flora and fauna during this disaster has led to the contamination of consumable products therefore covering large human population [2]. The recent accident at Fukushima nuclear power plant in 2011 had also led to similar effects in the form of physiological, genetic, developmental and fitness effect in non-human population [3]. The exposure to low level radiation has become common due medical diagnostic and therapeutic procedures, frequent space or air travel, cosmic radiation and use of certain electronic gadgets. Other sources of radiation exposure include radon in houses, contamination from weapon testing sites, and unexpected terrorist attacks [4-7].

Ionizing radiation consists of energetic particles and electromagnetic radiation, which can penetrate living tissues or cells and transfer energy to the biological materials. This interaction of ionizing radiation with cells results in the production of ionization, free radical generation, chemical bond breakage, and oxidative stress [8,9]. These free radicals inturn react with important macromolecules including nucleic acids, membrane lipids, and proteins leading to their structural alteration [10-12]. This molecular alteration induced by ionizing radiation causes mutations, chromosomal aberrations and carcinogenesis or cell death [12,13]. The Oxidative Stress (OS) is a state of imbalance between generation of ROS and the level of antioxidant defence system. Radiation induced free radicals subsequently impair the antioxidative defence mechanism of a cell, leading to an increased membrane lipid peroxidation, which results in damage to the membrane bound enzymes [14]. It is commonly accepted that under situations of oxidative stress, reactive oxygen species including superoxide, hydroxyl, and peroxyl radicals are generated. OS and free-radical-mediated processes have been implicated in the pathogenesis of a variety of diseases including ageing, cancer, coronary heart diseases, neurodegenerative disorders, atherosclerosis, cataracts, inflammation and digestive system disorders [15-17].

Cells are well equipped to defend themselves against ROS, with a repertoire of antioxidant enzymes and molecules [7,18,19]. Antioxidant enzymes are part of the endogenous system available for the removal or detoxification of free radicals and their products formed by ionizing radiation [18,20]. The antioxidant system consists of low molecular weight antioxidants like glutathione, melatonin and various antioxidant enzymes including superoxide dismutase, catalase, glutathione peroxidase etc. [21].

Ionizing radiations induce additional oxidative stress in the irradiated organisms and the indigenous antioxidants of cells may not be able to neutralize this additional burden alone. The supplementation of exogenous substance may be required to combat the additional oxidative stress in the irradiated system [7,20-22]. The phytoceuticals may be of great use to combat the additional oxidative stress in the irradiated system or even otherwise. Jamun, *Syzygium cumini* Linn. Skeels (family Myrtaceae) is a medium sized to large tree. It has been reported to possess several medicinal properties in the folklore and Ayurvedic systems of medicine. The stem bark of jamun is astringent, anthelmintic, antibacterial, carminative, constipating, diuretic, digestive, febrifuge, refrigerant, stomachic and sweet. The fruits and seeds are used to treat diabetes, and splenopathy [23]. Jamun has been reported to be antioxidant, anti-inflammatory, antibacterial, anticancer, antidiabetic, antiviral, antifungal, antidiahorreal, antileishmanial, chemopreventive, hypoglycemic and gastro protective [24-31]. The leaf extract of jamun has been found to scavenge different free radicals *in vitro* [32]. The studies from our laboratory has indicated that the leaves of the jamun have reduced the radiation-induced DNA damage in the cultured human peripheral blood lymphocytes as well as mouse splenocytes [32,33]. Its leaf and seed extract have been reported to increase the survival of irradiated mice and it also protected against radiation-induced gastrointestinal injury [34-36]. The presence of different activities in jamun stimulated us to obtain an insight into the antioxidant status of jamun (*Syzygium cumini*, Skeels) leaf extract in the liver of mice whole body exposed to different doses of γ-radiation.

Materials and Methods

Animal care and Handling

The animal care and handling were carried out according to the guidelines issued by the World Health Organization, Geneva, Switzerland and the INSA (Indian National Science Academy, New Delhi, India). Usually ten to twelve weeks old male Swiss albino mice weighing 30 to 36 g were obtained from an inbred colony maintained under the controlled conditions of temperature (23 ± 2°C), humidity (50 ± 5%) and light (14 and 10 h of light and dark, respectively) in the institutional animal house. The animals were allowed free access to the sterile water and food. Usually, four animals were kept in a polypropylene cage containing sterile paddy husk (procured locally) as bedding throughout the

experiment. The study was approved by the Institutional Animal Ethical Committee.

Chemicals

Analytical grade cumene hydroperoxide, thiobarbituric acid (TBA), ascorbic acid, glutathione (GSH), 5,5-dithio-bis (2-nitrobenzoic acid) (DTNB), diethylenetriamine pentaacetic acid (DTPA), butylated hydroxytolune, 1-chloro-2,4-dinitrobenzene, 2,4- dinitrophenyl hydrazine, guanidine hydrochloride, ferric chloride, ferrous sulphate and tetraethoxypropane were procured from Sigma Chemicals Co, St. Louis, MO, USA. The routine chemicals were supplied by Merck India, Mumbai.

Preparation of the Extract

The fresh mature leaves of jamun (*Syzygium cumini* Linn. Skeels or Eugenia cumini Linn. Druce), family Myrtaceae were collected locally during the month of May. The tree was identified by Dr. G.K. Bhat (Department of Botany, Poorna Prajna College, Udupi, Karnataka, India) a well-known taxonomist of the region. The leaves were cleaned, dried in shade, and powdered in a mixer grinder and extracted as described earlier [37]. Briefly, the leaf powder was extracted in petroleum ether and chloroform and finally in 1:1 dichloromethane and methanol at 50-60°C using a Soxhlet apparatus. The extract was cooled and concentrated by evaporating its liquid contents in vacuo and freeze dried. The extract was stored at -70°C until further use. Henceforth the extract of *Syzygium cumini* will be called as SCE.

Preparation of drug and mode of Administration

The required amount of SCE was dissolved in 1% carboxymethyl cellulose (CMC) in sterile double distilled water immediately before use. The animals were administered orally with SCE or CMC, consecutively for 5 days.

Experimental

The effect of leaf extract of jamun was studied in the liver of irradiated mice dividing them into the following groups:-

CMC + irradiation

The animals of this group were administered with 0.01 ml/g body weight of CMC orally before irradiation to 0, 0.5, 1, 2, 3 or 4 Gy.

SCE + irradiation

The animals of this group were administered with 50 mg/ kg body weight SCE orally once daily for 5 consecutive days before exposure to 0, 0.5, 1, 2, 3 or 4 Gy of γ-radiation [32].

Irradiation

One hour after the last administration of CMC or SCE on 5th day, the prostrate and immobilized animals (achieved by inserting cotton plugs in the restrainer) were whole body exposed to 0, 0.5, 1, 2, 3 or 4 Gy of 60Co gamma radiation (Theratron, Atomic Energy Agency, Canada) in a specially designed well-ventilated acrylic box. A batch of twelve animals was irradiated each time at a dose rate of 1.66 Gy/ min at a SSD of 100 cm.

Grouping	Treatment	Exposure dose (Gy)					
CMC+irradiation	1% Carboxymethylcellulose	0	0.5	1	2	3	4
SCE+irradiation	50 mg/kg Jamun extract (SCE)	0	0.5	1	2	3	4
Post irradiation assay time (h) for each group		0.5	1	2	4	8	12 24

Experimental design

Note: 4 animals were used at each post-irradiation time (0.5, 1, 2, 4, 8, 12 and 24 h) that is 28 animals for each exposure dose in each group. In other words 28 animals were used for each exposure dose in each group 28X6 (0, 0.5,1,2,3, and 4 Gy) = 168X2 (CMC+irradiation and SCE+irradiation), which equals to 336 animals for the whole experiment.

Preparation of liver Homogenate

The animals from each group were euthanized at 0.5, 1, 2, 4, 8, 12 and 24 h post-irradiation and their livers were perfused transcardially with ice cold phosphate buffered saline. The whole liver from each animal was removed, blot dried, weighed and a 10% homogenate was prepared in ice-cold 0.2 M sodium phosphate buffer Ph 8.0 using a homogenizer (Yamato LSG LH-21, Tokyo, Japan). Four animals were used for each irradiation dose at each time interval in all concurrent groups and a total of 336 animals were used for the entire study.

Total proteins

The protein contents were determined using the modified method of Lowry.

Glutathione

Glutathione (GSH) contents were measured as described by Moron et al. [38]. Briefly, proteins were precipitated by 25% TCA, centrifuged and the supernatant was collected. The supernatant was mixed with 0.2 M sodium phosphate buffer (pH 8.0) and 0.06 mM DTNB followed by incubation for 10 min at room temperature. The absorbance of the sample/s was read against the blank at 412 nm in an ultraviolet-visible light (UV-VIS) double beam spectrophotometer (UV-260; Shimadzu Corporation, Tokyo, Japan) and the GSH concentration was calculated from the standard curve, and it has been expressed as $\mu mol/g$ tissue.

Catalase

Catalase activity was determined by catalytic reduction of hydrogen peroxide [39]. Briefly, hydrogen peroxide was added to the sample and the mixture was incubated at 37°C. Catalase activity was measured by recording the decrease in absorbance at 240 nm periodically after addition of sample in a UV-VIS spectrophotometer. The average difference in absorbance in 30 sec was calculated. A unit of catalase is defined as the amount of protein that results in a decrease in absorbance of 0.05 in 30 sec.

Superoxide dismutase

The SOD was estimated by the method of Fried [40]. Briefly, 900 µl buffer was mixed with 100 µl tissue homogenate (T), nitroblue tetrazolium (NBT), phenazine methosulphate and

NADH. The control (C) consisted of all the reagents except the homogenate, whereas, the blank (B) consisted of buffer and the homogenate without any reagents. The absorbance of T, C and B was read at 560 nm using a UV-Visible spectrophotometer and the enzyme activity has been expressed in units (1 U = 50% inhibition of NBT reduction).

Lipid peroxidation

Lipid peroxidation (LOO) was measured according to the standard protocol [41]. Briefly, the samples were incubated with a mixture of trichloroacetic acid (15%), thiobarbituric acid (0.375%), and butylated hydroxytoluene (0.01%) in 0.25 N HCl at 95°C for 25 min. The reaction mixture was cooled to room temperature and centrifuged at 8,000 g. The supernatant was collected and the absorbance was recorded against the blank using a double beam UV-VIS spectrophotometer. The LOO was measured from a standard curve and has been expressed as MDA in nM per mg protein.

Analysis of data

The statistical significance among various groups was determined using ANOVA or student-'t' test wherever necessary. Graph Pad statistical package (Graph Pad software, San Digeo, USA) was used for data analysis. All the data are expressed as mean and standard error of the mean (SEM).

Results

The results are expressed as GSH concentration ($\mu mol/g$ tissue), catalase activity ($\mu mol/mg$ protein), superoxide activity (U/mg protein) and lipid peroxidation (nM/mg protein) ± SEM (standard error of the mean) in Table [1-4] and Figures[1-8].

Glutathione

The reduced glutathione (GSH) concentration was determined in mouse liver at different post-irradiation times. The normal baseline GSH level in CMC (0 Gy) treated mice ranged between 4.18 ± 0.05 to 4.23 ± 0.06 µmol/ mg tissue. Administration of 50 mg/ kg SCE (0 Gy) orally for five consecutive days did not induce significant change in GSH concentration when compared to normal baseline concentration (4.2 ± 0.05 to 4.3 ± 0.05). Exposure of animals to different doses of γ-radiation resulted in a dose dependent decrease in the GSH concentration when compared with normal baseline GSH concentration of CMC+sham-irradiation (Figure 1). The GSH concentration declined at 0.5 h post-irradiation with a maximum decline at 1 h post-irradiation after 4 Gy irradiation, where an almost 2 fold decline in GSH concentration was observed. Thereafter the GSH concentration showed a gradual rise until 24 h post-irradiation however, normal levels could not be attained (Figure 2). The GSH concentration was significantly ($p < 0.001$) lower when compared to CMC+sham-irradiation (0 Gy). The comparison among different doses of radiation revealed that the GSH concentration was significantly reduced with increasing dose of radiation ($p < 0.05$; 0.01 or 0.001) depending on the irradiation dose and estimation time (Table 1). The pattern of decline in GSH concentration after

Table 1: The effect of jamun (SCE) pre-treatment on the radiation-induced alteration in glutathione levels in liver of mouse exposed to different doses of γ- radiation.

Dose (Gy)	Post-irradiation time (hours)													
	0.5		1		2		4		8		12		24	
	CMC+IR	SCE+IR	CMC+IR	SCE+IR	CMC+IR	SCE+IR	CMC+IR	SCE+IR	CMC+IR	SCE+IR	CMC+IR	SCE + IR	CMC+IR	SCE+IR
0	4.23±0.06	4.3±0.05	4.09±0.05	4.2±0.05	4.19±0.02	4.17±0.05	4.23±0.02	4.06±0.05	4.16±0.02	4.3±0.05	4.08±0.02	4.28±0.05	4.18±0.05	4.2±0.05
0.5	3.81±0.08	4.0 ± 0.08♣	3.67±0.04♣	4.15±0.08c	3.86±0.08	4.19±0.08b	3.96±0.01	4.06±0.08	4.01±0.01	4.15±0.08	3.98±0.01	4.16±0.08	4.01±0.01	4.18±0.05
1	3.49±0.06*	3.95±0.03$c	3.21±0.02♣*	3.65±0.03c*	3.4±0.01♣*	3.72±0.03*	3.59±0.09*	3.83±0.03	3.68±0.06$*	3.9±0.03b♣*	3.96±0.06	4.1 ± 0.03	4.04±0.06	4.06±0.03
2	2.86±0.09$*	3.65±0.05♣$*	2.5± 0.09*	3.2 ± 0.05*	2.64±0.09*	3.2 ± 0.06c*	3.09±0.09♣*	3.49±0.05$*	3.19±0.09*	3.65±0.06c♣*	3.42±0.09$*	3.87±0.06c♣*	3.78±0.09$	3.93±0.06$
3	2.69±0.15*	3.21±0.06b*	2.43±0.15*	3.16±0.06c*	2.52±0.15*	3.18±0.05*	2.99±0.15*	3.42±0.06b*	2.89±0.05♣*	3.61±0.02$*	3.01±0.15*	3.64± 0.05*	3.45±0.05$*	3.78±0.01$*
4	2.46±0.08*	2.95±0.04♣c*	2.21±0.08*	2.81±0.04$c*	2.37±0.08*	2.95±0.04c*	2.76±0.08*	2.97±0.03*	2.85±0.05$*	3.16± 0.02*	2.99±0.08♣*	3.54±0.03$*	3.22±0.05*	3.56±0.02♣*

a: *P < 0.05*, b : *P < 0.01*, c : *P < 0.001*, no symbol : non-significant. CMC+IR group compared to SCE+IR

♣: *P < 0.05*, $: *P < 0.01*, * : *P < 0.001*, no symbol : non-significant. When compared to Sham-radiation ([] Gy) group

CMC : Carboxymethylcellulose ; IR : Irradiation ; SCE : *Syzygium cumini* extract

Table 2: The effect of jamun (SCE) pre-treatment on the radiation-induced Superoxide dismutase activity in liver of mouse exposed to different doses of γ-radiation.

Dose (Gy)	Post-irradiation time (hours)													
	0.5		1		2		4		8		12		24	
	CMC + IR	SCE + IR	CMC + IR	SCE + IR	CMC + IR	SCE + IR	CMC + IR	SCE + IR	CMC + IR	SCE+IR	CMC +IR	SCE+IR	CMC+IR	SCE+IR
0	6.32± 0.02	6.21± 0.02	6.31± 0.08	6.3± 0.04	6.01± 0.02	6.45±0.02c	5.89± 0.07	6.21±0.05b	5.93± 0.07	6.01±0.08	5.99± 0.06	6.21±0.05a	5.97±0.05	6.11±0.08
0.5	5.01± 0.17*	5.86± 0.08$	4.98± 0.09*	5.62± 0.04*	4.51±0.03*	5.31±0.02c*	4.62± 0.03*	5.37±0.06c*	4.80±0.03*	5.48±0.08♣	4.91±0.03*	5.65±0.06c*	5.00±0.03*	5.72±0.08*
1	4.56± 0.04♣*	4.98±0.06c*	4.01± 0.09*	4.52± 0.02*	3.62±0.05*	4.47±0.04c*	3.89± 0.05*	4.64±0.04c*	4.21±0.06*	4.76±0.07c*	4.65±0.05*	4.80±0.04c*	4.90±0.06*	5.10±0.07c*
2	3.86± 0.06*	4.26±0.09b*	3.46± 0.09*	4.11± 0.06c*	3.21±0.06*	4.09±0.02*	3.58± 0.02*	4.22±0.06*	3.76±0.05*	4.69±0.04c*	3.97±0.02*	4.85± 0.06*	4.26±0.05*	5.30±0.04c*
3	3.21± 0.07*	3.82±0.04c*	3.02± 0.03$*	3.62± 0.02c*	2.01±0.06*	3.09±0.04c*	2.68± 0.04*	2.93±0.07*	3.02±0.03*	3.24±0.04c*	3.68±0.04*	3.97±0.07*	4.21±0.03*	5.10±0.04c*
4	3.01± 0.06*	3.52±0.02♣*	2.61±0.04♣*	2.99± 0.06c*	2.02±0.06*	2.86±0.07$c*	2.71± 0.02*	3.12±0.08♣*	2.99± 0.04*	3.41±0.03c*	3.56±0.02*	4.06±0.08♣*	3.84±0.04*	4.62±0.03c*

a: *P < 0.05*, b : *P < 0.01*, c : *P < 0.001*, no symbol : non-significant. CMC+IR group compared to SCE+IR

♣: *P < 0.05*, $: *P < 0.01*, * : *P < 0.001*, no symbol : non-significant. When compared to Sham-radiation ([] Gy) group

CMC= Carboxymethylcellulose

IR= Irradiation

SCE= *Syzygium cumini* extract

Table 3: The effect of jamun (SCE) pre-treatment on the radiation-induced catalase activity in liver of mouse exposed to different doses of γ-radiation.

Dose (Gy)	Post-irradiation time (hours)													
	0.5		1		2		4		8		12		24	
	CMC+IR	SCE+IR	CMC+IR	SCE+IR	CMC+IR	SCE+IR	CMC+IR	SCE+IR	CMC+IR	SCE+IR	CMC + IR	SCE+IR	CMC+IR	SCE+IR
0	29.21±0.25	28.21±0.45a	29.87±0.35	27.87±0.35	29.35±0.25	29.15±0.35	28.21±0.67	28.01±0.61	29.21±0.32	28.01±0.61a	29.26±0.66	28.01±0.61	29.22±0.22	28.01±0.62
0.5	24.21±0.35*	27.21±0.65c	23.27±0.35*	25.21±0.34c$	21.21±0.71*	24.21±0.35*	23.21±0.30*	24.12±0.30a$	23.11±0.75*	25.36± 0.27♣	23.21±0.30*	25.02±0.30c♣	23.11±0.75*	25.16±0.27♣
1	22.28±0.34*	23.45± 0.60$*	23.58±0.30*	25.26±0.32b$	20.21±0.39*	22.15±0.30c*	21.25±0.51*	23.21±0.29b*	21.26±0.28*	23.42±0.30c*	21.15±0.53*	23.11± 0.22*	21.16±0.25*	23.12±0.33c*
2	19.21±0.75$*	22.23±0.75a*	18.12±0.35*	21.24±0.35c*	17.28±0.62♣$*	20.25±0.75b$*	18.21±0.45*	20.01±0.75a♣$	18.26±0.85*	21.26±0.61a*	18.21±0.45*	20.01±0.75a♣*	18.26±0.85♣*	21.26±0.61a*
3	17.96±0.23*	19.24±0.35b♣*	16.21±0.78*	18.26±0.47a*	16.54±0.55*	17.86±0.35a♣*	17.01±0.82*	18.21± 0.75*	17.35±0.31$*	18.85±0.32b*	17.01±0.82*	18.21±0.75c*	17.35±0.31$*	18.85±0.32b*
4	17.01±0.42♣*	19.21±0.39b♣*	16.24±0.43*	18.24±0.41b*	15.21±0.75*	18.11±0.66b*	15.84±0.78*	17.21±0.67*	16.89±0.68*	18.93±0.75a*	15.84±0.78*	17.21±0.67c*	16.89±0.68*	18.93±0.75a*

a: *P < 0.05*, b : *P < 0.01*, c : *P < 0.001*, no symbol : non-significant. CMC+IR group compared to SCE+IR

♣: *P < 0.05*, $: *P < 0.01*, * : *P < 0.001*, no symbol : non-significant. When compared to Sham-radiation ([] Gy) group

CMC= Carboxymethylcellulose

IR= Irradiation

SCE= Syzygium cumini extract

treatment with SCE in SCE+irradiation group was identical to that of CMC+irradiation group (Figure 1) except that the SCE administration arrested the radiation-induced decline in the GSH concentration significantly at all post-irradiation times after exposure to 0.5 to 4 Gy (Table 1). However, baseline levels could not be restored even at 24 h post-irradiation except 0.5 Gy, where it was almost normal (Table 1).

Superoxide Dismutase

The MnSOD activity was measured in mouse liver at different post-irradiation times and the activity of mitochondrial SOD is expressed as units/ mg protein ± SEM (Figure 3). The MnSOD

Table 4: The effect of jamun (SCE) pre-treatment on the radiation-induced lipid peroxidation in liver of mouse exposed to different doses of γ- radiation.

Dose (Gy)	Post-irradiation time (hours)													
	0.5		1		2		4		8		12		24	
	CMC+IR	SCE+IR	CMC+IR	SCE+IR	CMC+IR	SCE+IR	CMC+IR	SCE+IR	CMC+IR	SCE+IR	CMC + IR	SCE+IR	CMC+IR	SCE+IR
0	17.23±1.25	16.54±1.21	17.54±1.04	16.89±1.32	17.61±1.32	17.21±1.51	17.3±1.74	17.12±1.62	17.61±1.04	17.01±1.61	17.43±1.74	17.12±1.62	17.59±1.04	17.01±1.62
0.5	20.21±1.25	17.65±1.21	23.87±1.35	18.21±1.25b	25.35±1.25*	21.85±1.41a	26.21±1.67$	21.89±1.68a	21.11±1.32	19.25±1.31	21.21±1.67	21.89±1.68	21.21±1.32	19.35±1.31
1	23.15±1.25	22.1 ±1.32	29.29±1.24*	24.21±1.54ᵃa	36.21±1.57*	29.58±1.65*b*	32.21±1.41*	26.09±1.58*b	26.21±1.24*	24.21±1.68	26.21±1.24	24.09±1.58*	26.21±1.24$	24.21±1.63
2	26.24±1.54$	24 ±1.26*	35.23±1.52*	27.21±1.54$b	51.89±1.31*	41.23±1.71c*	47.25±1.43*	33.01±1.32*c*	36.98±1.31*	28.85±1.87$b*	36.25±1.43$	28.01±1.32$c*	36.68±1.32*	28.71±1.87$b*
3	30.57±1.26**	24.23±1.73*b	53.21±1.87*	41.21±1.54c*	61.67±1.56$	52.21±1.32$c*	51± 1.24*	46.28±1.63*	41.21±1.48*	33.21±1.65*b*	39±1.24*	41.28±1.63*	41.32±1.48*	33.21±1.65*b*
4	39± 1.20$	33±1.63$b*	59.32±1.25*	48.21±1.35*c*	67.35±1.48*	55.21±1.54*	58.15±1.63$	49.35±1.65b*	47.21±1.42*	40.21±1.32$b*	41.15±1.63*	40.35± 1.65*	47.33±1.42*	40.21±1.32$b*

a: $P < 0.05$, b : $P < 0.01$, c : $P < 0.001$, no symbol : non-significant. CMC+IR group compared to SCE+IR

*: $P < 0.05$, $: $P < 0.01$, *: $P < 0.001$, no symbol : non-significant. When compared to Sham-radiation (0 Gy) group

CMC= Carboxymethylcellulose

IR= Irradiation

SCE= *Syzygium cumini* extract

activity ranged between 5.97 ± 0.05 to 6.32 ± 0.02 U/mg protein in CMC+sham-irradiation (0 Gy) mice. Administration of 50 mg/kg SCE orally for five consecutive days did not significantly alter the SOD activity in comparison with CMC+sham-irradiation group. Exposure of mice to different doses of γ-radiation resulted in a significant but dose dependent decrease in the SOD activity when compared with CMC + sham-irradiation (Figure 3). The earliest decline in the MnSOD activity was recorded at 0.5 h post-irradiation and a maximum decrease in the enzyme activity was observed at 2 h post-irradiation (2.02 ± 0.06) in the animals exposed to 4 Gy irradiation. The slow recovery in the SOD activity was evident from 4 h post-irradiation and SOD activity continued to rise up to 24 h post-irradiation without restoration to sham-irradiation level (Figure 4). The comparison among different doses of γ-radiation revealed that the SOD activity reduced significantly with increasing dose of radiation ($p < 0.05; 0.01$ or 0.001) depending on the radiation dose and estimation time (Table 2). The pattern of decline in SOD activity after treatment with SCE in SCE+irradiation group was almost identical to that of CMC+irradiation group (Figure 3) except that SCE administration attenuated the radiation-induced decline in the SOD activity significantly ($p < 0.05, 0.01$ or 0.001) at all post-irradiation assay times in mouse liver exposed to 0.5 to 4 Gy (Table 2). Despite this increase, the basal levels of SOD activity could not be restored to normal even by 24 h post-irradiation (Table 2).

Catalase

The catalase activity was measured in mouse liver at different post-irradiation times and it has been expressed as μmol of H_2O_2 reduced/min/mg protein ± SEM (Figure 5). The basal catalase activity in CMC+sham-irradiated (0 Gy) mice ranged between 29.21 ± 0.25 to 29.22 ± 0.22 μmol/ mg protein. Administration of 50 mg/ kg SCE orally for five consecutive days did not induce significant alteration in the catalase activity when compared to basal activity (29.21 ± 0.25 to 29.22 ± 0.22). Exposure of mice to different doses of γ-radiation resulted in a dose dependent reduction in the catalase activity when compared with CMC

+ sham-irradiation (Figure 5). The catalase activity started declining at 0.5 h post-irradiation and a maximum reduction in the catalase activity was observed at 2 h post-irradiation (15.21 ± 0.75) in the livers of mice exposed to 4 Gy (Figure 6). Thereafter the catalase activity increased gradually until 24 h post-irradiation however, normal levels could not be restored even by 24 h post-irradiation (Figure 6). The catalase activity was significantly ($p < 0.001$) lower when compared to CMC+sham-irradiation. The comparison among different doses of γ-radiation revealed that the catalase activity did reduce significantly with increasing dose of radiation ($P < 0.05; 0.01$ or 0.001) depending on the irradiation dose and estimation time (Table 3). The pattern of decline in catalase activity after treatment with SCE in SCE+irradiation group was similar to that of CMC+irradiation group (Figure 5), except that SCE administration reduced the radiation-induced decline in the catalase activity significantly ($p < 0.05, 0.01$ or 0.001) at all post-irradiation times after exposure to 0.5 to 4 Gy (Table 3). However, the catalase activity could not be restored to basal level even at 24 h post-irradiation (Table 3).

Lipid peroxidation

The CMC+sham-irradiation group showed a minimum lipid peroxidation (17.23 ± 1.25 nanomoles of MDA/ mg protein) which did not show any change with estimation time (Figure 7). Administration of mice with 50 mg/kg SCE orally once daily for five consecutive days did not significantly alter the lipid peroxide levels when compared to control (0 Gy). Exposure of animals to different doses of γ-radiation caused a significant ($p < 0.05, 0.01$ or 0.001) rise in the lipid peroxidation when compared to CMC+sham-irradiation group (Figure 7). An earliest increase in lipid peroxides was recorded at 0.5 h post-irradiation and a maximum rise (67.35 ± 1.48) in the lipid peroxidation was detected at 2 h post-irradiation. Thereafter the level of lipid peroxides gradually declined until 24 h post-irradiation without restoration to normal levels (Figure 8). The comparison among various doses of γ-radiation showed that the lipid peroxide levels were significantly higher with increasing dose of radiation and

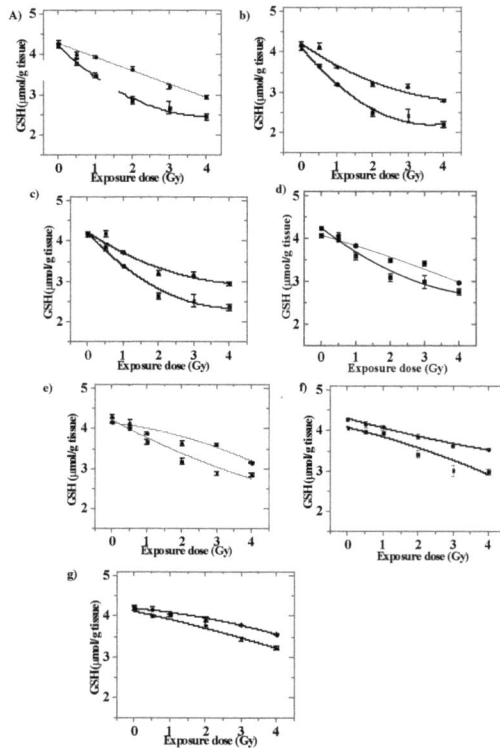

Figure 1: Effect of 50 mg/ kg body weight of SCE on the glutathione (GSH) concentration in the livers of mice exposed different doses of gamma radiation. a: 0.5h, b: 1h, c:2h, d:4h, e:8h, f: 12h and g: 24h post-irradiation. Squares: CMC+irradiation and circles: SCE+irradiation.
CMC=Carboxymethylcellulose
SCE: *Syzygium cumini* Extract

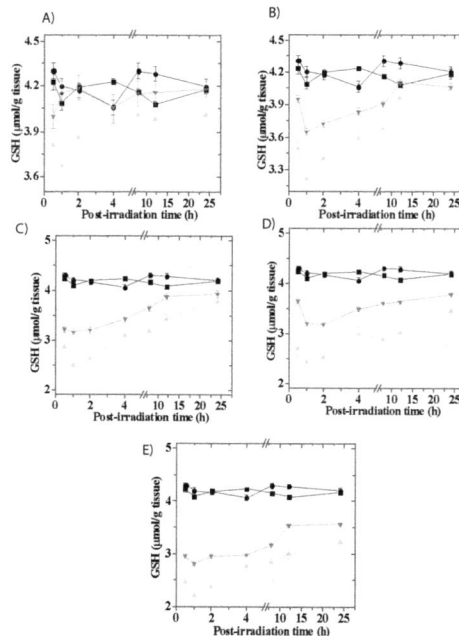

Figure 2: Effect of 50 mg/ kg body weight of SCE on the glutathione (GSH) concentration in the livers of mice exposed to different doses of gamma radiation. a: 0.5Gy, b: 1Gy, c:2Gy, d:3Gy, e:4Gy. Squares: Cmc+sham-irradiation (0 Gy); Circles: Sce+ sham-irradiation (0 Gy); Up trinagles: Cmc+irradiation and Down triangles: SCE+irradiation.
CMC: Carboxymethylcellulose
SCE: *Syzygium Cumini* Extract

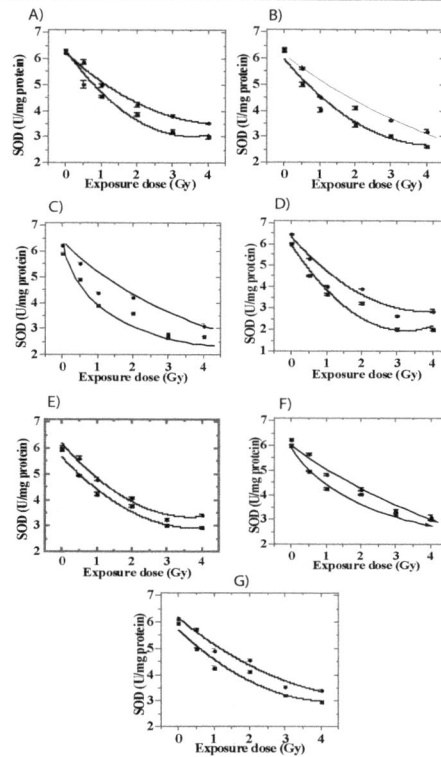

Figure 3: Effect of 50 mg/ kg body weight of SCE on the Superoxide Dismutase (SOD) activity in the livers of mice exposed different doses of gamma radiation. a: 0.5h, b: 1h, c: 2h, d: 4h, e: 8h, f: 12h and g: 24h post-irradiation. Squares: CMC+irradiation and circles: SCE+irradiation.
CMC: Carboxymethylcellulose
SCE: *Syzygium Cumini* Extract

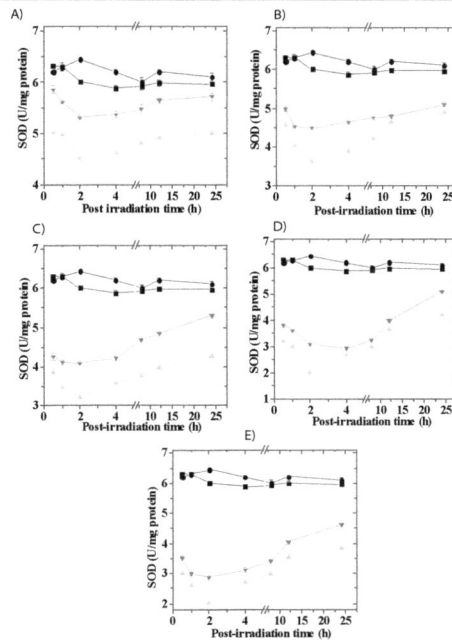

Figure 4: Effect of 50 mg/ kg body weight of SCE on the Superoxide Dismutase (SOD) activity in the livers of mice exposed different doses of gamma radiation. a: 0.5Gy, b: 1Gy, c: 2Gy, d: 3Gy, e: 4Gy. Squares: CMC+sham-irradiation (0 Gy); Circles: SCE+ sham-irradiation (0 Gy); Up trinagles: CMC+irradiation and Down triangles: SCE+irradiation.
CMC: Carboxymethylcellulose
SCE: *Syzygium Cumini* Extract

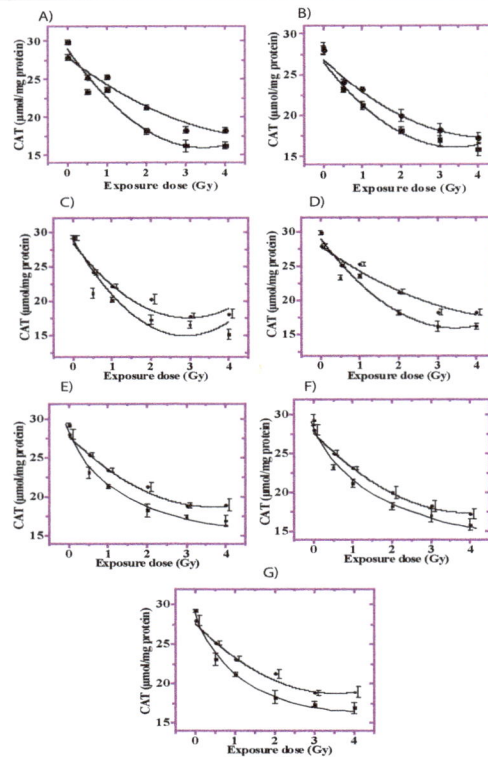

Figure 5: Effect of 50 mg/ kg body weight of SCE on the catalase (CAT) activity in the livers of mice exposed different doses of gamma radiation. a: 0.5h, b: 1h, c:2h, d:4h, e:8h, f: 12h and g: 24h post-irradiation. Squares: CMC+irradiation and circles: SCE+irradiation.
CMC: Carboxymethylcellulose
SCE: Syzygium Cumini Extract

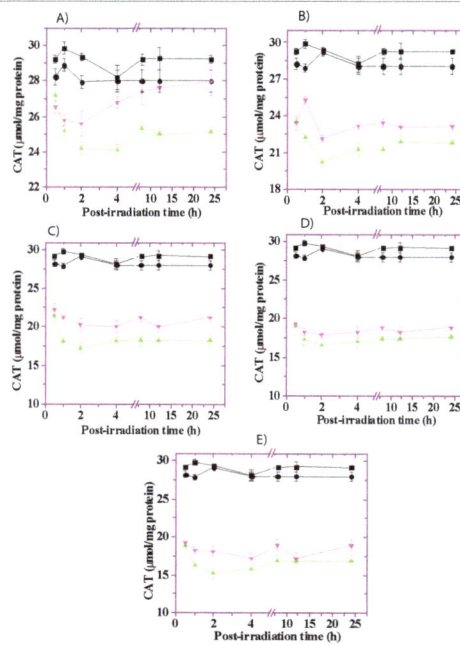

Figure 6: Effect of 50 mg/ kg body weight of SCE on the catalase (CAT) activity in the livers of mice exposed different doses of gamma radiation. a: 0.5Gy, b:1Gy, c:2Gy, d:3Gy, and e:4Gy. Squares: CMC+sham-irradiation (0 Gy); Circles: SCE+ sham-irradiation (0 Gy); Up trinagles: CMC+irradiation and Down triangles: SCE+irradiation.
CMC: Carboxymethylcellulose
SCE: *Syzygium Cumini* Extract

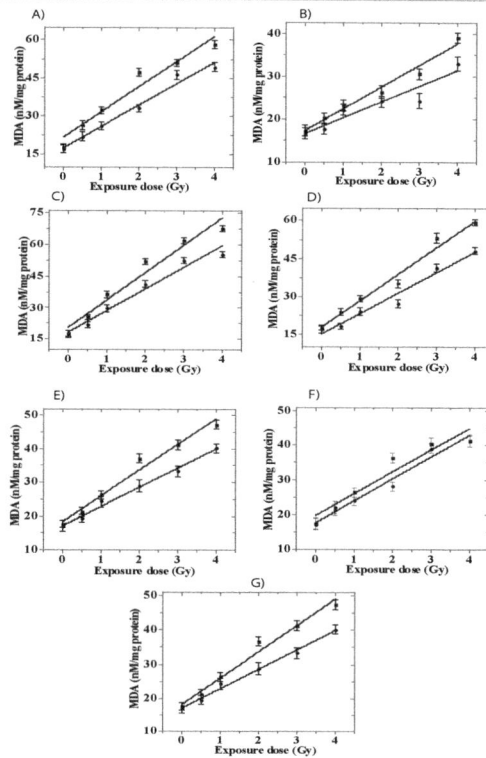

Figure 7: Effect of 50 mg/ kg body weight of SCE on the lipid peroxidation in the livers of mice exposed different doses of gamma radiation. a: 0.5h, b: 1h, c:2h, d:4h, e:8h, f: 12h and g: 24h post-irradiation. Squares: CMC+irradiation and circles: SCE+irradiation.
CMC: Carboxymethylcellulose
SCE: *Syzygium Cumini* Extract

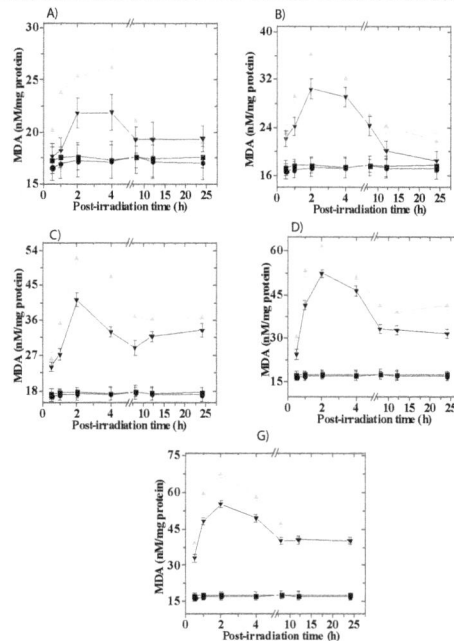

Figure 8: Effect of 50 mg/ kg body weight of SCE on the lipid peroxidation in the livers of mice exposed different doses of gamma radiation. a: 0.5Gy, b:1Gy, c:2Gy, d:3Gy, e:4Gy. Squares: CMC+sham-irradiation (0 Gy); Circles: SCE+ sham-irradiation (0 Gy); Up trinagles: CMC+irradiation and Down triangles: SCE+irradiation.
CMC: Carboxymethylcellulose
SCE: *Syzygium Cumini* Extract

also with the immediate preceding dose ($p < 0.05, 0.01$ or 0.001) depending on the post-irradiation time and irradiation dose (Table 4). Administration of SCE before irradiation resulted in a significant reduction in the lipid peroxidation when compared to CMC+irradiation group at all the post-irradiation times. The pattern of increase in lipid peroxide level after treatment with SCE in SCE+irradiation group was similar to that of CMC+irradiation group (figure 7) except that the SCE administration inhibited the radiation-induced lipid peroxidation significantly ($p < 0.05, 0.01$ or 0.001) at all post-irradiation times after exposure to 0.5 to 4 Gy (Table 4). Despite a significant decline, the level of lipid peroxide did not reach basal level even by 24 h post-irradiation.

Discussion

It a well-established fact that exposure to ionizing radiations results in a complex set of responses, whose onset, nature, and severity is a function of both radiation quality and total radiation dose received by the organism [42,43]. Since the body contains more than 75% water, ionizing radiations principally interact with water molecules abundant in the cellular milieu and generate different types of free radicals [9,44]. These radiation-induced free radicals are mainly responsible for the induction of oxidative stress in the body. The exposure to ionizing radiation negatively alters the oxidant status in different tissues of the body [45]. The increased oxidative stress by ionizing radiation has been implicated in the induction of cancer, cardiovascular, liver, lung and neurological disorders [44-49]. The use of phytoceuticals may be able to reduce the risk of radiation-induced oxidative stress. Plants have been used by humans throughout their history in different cultures for health benefits and curing diseases and have been considered nontoxic. Even today, more than 80% of the world's population is dependent on plants for handling their health related problems [50]. The biologic origin of phytoceuticals make them more attractive for the investigation as antioxidants as their biologic origin makes them more biocompatible than the synthetic drugs [6]. Therefore, present study was undertaken to elucidate the effect of SCE on the radiation-induced alteration in the antioxidant status in mouse liver exposed to different doses of γ-radiation.

The radiation-induced damage in the living cells is a consequence of generation of oxygen-derived free radicals including superoxide anion and hydrogen peroxide owing to radiolysis of water [9,44]. Once generated these reactive oxygen species react with various important biomolecules inflicting lesions in different types of cells [11]. These free radicals play an important role in the production of indirect biological damage induced by low LET ionizing radiations [51]. Organisms possess comprehensive and integrated endogenous enzymatic repair systems to cope up with ROS induced-damage. Glutathione (GSH), vitamin E and C, -carotene and uric acid are some of the important non-enzymatic antioxidants that are taken up with food or synthesized endogenously, whereas SOD (Cu Zn, or Mn), catalase, and glutathione peroxidase (GSHpx) represent endogenous enzymatic antioxidants that will neutralize or mitigate effect of oxidative stress induced by physical or chemical agents [52]. Usually ROS responses after irradiation are diminished by a number of cellular defenses including GSH, GSHpx, catalase, SOD etc. [5,7,8,19,53]. The enzymes including superoxide dismutase (SODs) and glutathione peroxidase (GPx) help to neutralize the radiation-induced superoxide anion or hydrogen peroxide (H_2O_2) and defend cells from radiation-induced oxidative stress.

Glutathione is one of the most prevalent intracellular thiols that exert its antioxidant action by scavenging hydroxyl radicals and singlet oxygen, and protect the cell against oxidative stress induced by them [54]. GSH offers protection against oxygen-derived free radicals and cellular lethality following exposure to ionizing radiation [55]. The alleviation of GSH leads to formation of reactive oxygen species and oxidative stress that affect functional as well as structural integrity of cell and organelle membranes [56]. In the normal conditions, the cells are intact and healthy and GSH is restored by synthesis [54] however, availability of GSH after irradiation may decrease due to reduced GSH synthesis enhanced efflux or inefficient reduction of GSSG [57]. Moreover, GSH may be also utilized to neutralize the radiation-induced free radicals and in the formation of thiyl radicals that associate to produce GSSG [58]. This explains a dose dependent reduction in the GSH concentration after exposure to 0 to 4 Gy and its maximum depletion at 1 h post-irradiation in the present study. These results are consistent with earlier reports where irradiation has been found to reduce GSH concentration [5,7,20,59]. A number of natural or synthetic radioprotectors can alter the balance of endogenous protective systems, such as glutathione and antioxidant enzyme systems after irradiation, and reduce the effect of radiation [5,7,20,22]. Administration of jamun extract reduced the radiation-induced decline in the GSH concentration, which could be partially due to abatement of radiation-induced free radicals or upregulation of GSH synthesis. Likewise, ascorbic acid, naringin, *Agele marmelos*, *Zingiber officinale*, *Phyllanthus amarus*, *Nigella sativa*, curcumin, and hesperidin have been reported to increase the GSH concentration after irradiation [5,7,19,20,22,60-62].

The enzyme superoxide dismutase eliminates superoxide anion, which forms part of the H_2O_2 in cells. Irradiation of mice resulted in a dose dependent reduction in the SOD activity that was below spontaneous level even after 24 h post-irradiation. Irradiation has been reported to decrease superoxide dismutase activity in vivo and in vitro earlier [5,7,19,20,22,59,63]. Treatment of mice with SCE caused a significant elevation in the liver SOD activity after irradiation that would have helped in the efficient neutralization of radiation-induced superoxide anion and thus reduced the radiation-induced oxidative stress. An identical effect has been observed earlier in mice treated with α- tocopherol, ascorbic acid, naringin, curcumin and hesperidin before γ-irradiation [5,7,19,20,22,63]. Our observations are also supported by the reports that indicate that over expression of Mn-SOD protected cells from the damage induced by reactive oxygen species [64].

Catalase is one of the important enzymes, which is involved in the conversion of highly reactive and toxic H_2O_2 into nontoxic byproducts such as water and molecular oxygen. Irradiation of mice with different doses of gamma radiation resulted in a dose dependent decline in catalase activity in the mice liver. This decline may to due to the participation of catalase into the detoxification of H_2O_2 generated after irradiation. Irradiation has been reported to reduce the activity of enzyme catalase in various tissues [5,19,22]. Administration of SCE caused a significant rise in the activity of catalase in the mouse liver and a maximum elevation was observed at 2 h post-irradiation. This increase in catalase by SCE may have helped to abate the radiation-induced generation of H_2O_2 and thus protecting the mouse against the deleterious effect of radiation. Likewise, catalase has been reported to be elevated after treatment with various radioprotectors in the irradiated animals earlier [5, 19, 22,59,63].

Lipid peroxidation is another important consequence of irradiation and is one of the measures to determine the cellular toxicity [65]. Earlier studies have indicated that the peroxidation of membrane lipids might be the main cause of membrane damage induced by radiation [66], therefore studies on lipid peroxidation can provide valuable information about the damage caused by the ionizing radiations. The exposure of animals to different doses of γ-radiation increased lipid peroxidation in a dose dependent manner in mouse liver and this rise in lipid peroxidation may be due to the attack of radiation-induced free radicals on the fatty acid component of membrane lipids, that may lead to cell death [5,7,19,22,67,68]. A similar increase in radiation-induced lipid peroxidation has been reported earlier [5,7,19,22]. Administration of mice with SCE reduced the formation of lipid peroxides. Likewise, ascorbic acid, melatonin, and other botanicals including ginger, *Agele marmelos, Nigella sativa*, naringin, mangiferin, curcumin and hesperidin have also been reported to protect the mice against gamma radiation by elevating glutathione levels and reducing lipid peroxidation both in vivo and in vitro [5,7,19,20,22]. The flavonoids quercetin, myricetin, and kaempferol are known to activate glutathione-synthesizing enzyme [69] that would have helped to increase the GSH concentration in the present study and also reduce lipid peroxidation. The increase in GSH, catalase, and SOD by SCE may have neutralized the reactive oxygen species and thus protected against the radiation-induced lipid peroxidation after exposure to different doses of γ-radiation.

The exact mechanism by which the jamun extract has up regulated the antioxidant status is not known. However, it is speculated that pretreatment of mice with jamun extract may have up regulated the transcriptional activation of Nrf2 gene, that may have stimulated the activation of various genes related to GSH, SOD, catalase and increasing their synthesis and thus reduced the negative effect of radiation on the antioxidant status. Ionizing radiation has been reported to down regulate the Nrf2 gene earlier [70].

The present study demonstrates that SCE pretreatment elevates the GSH, SOD and catalase in mouse liver, and this increase in antioxidants and reduced lipid peroxidation may be due to the transcriptional activation of Nrf2 gene in the irradiated mouse liver leading to its radio protective activity.

Acknowledgements

The authors are thankful to Dr. M.S. Vidyasagar, Professor and Head, and Dr. J.G.R. Solomon, Department of Radiotherapy and Oncology, Kasturba Medical College Hospital, Manipal, India for providing the necessary irradiation facilities and dosimetric calculations, respectively.

Conflict of interest statement

Authros do not have any conflict of statement to declare.

References

1. Møller AP, Mousseau TA. Biological consequences of Chernobyl: 20 years on. Trends Ecol Evol. 2006;21(4):2007.

2. I.I. Kryshev. Radioactive contamination of aquatic ecosystem following the Chernobyl accident. J. Environm. Radioact. 1995;27(3):207-219. Doi:10.1016/0265-931X(94)00042-U.

3. Steen TY, Mousseau T. Outcomes of Fukushima: Biological Effects of Radiation on Nonhuman Species. J Hered. 2014;105(5):702-3. Doi:10.1093/jhered/esu049.

4. Turner ND, Braby LA, Ford J, Lupton JR. Opportunities for nutritional amelioration of radiation-induced cellular damage. Nutrition.2002;18(10):904-12.

5. Jagetia GC, Reddy TK. Modulation of radiation-induced alteration in the antioxidant status of mice by naringin. LifeSci.2005;77(7):780-94.

6. C Jagetia G. Radioprotective potential of plants and herbs against the effects of radiation. J Clin Biochem Nutr. 2007;40(2):74-81. Doi:10.3164/jcbn.40.74.

7. G.C. Jagetia, K.V.N.M. Rao Hesperidin, A Citrus Bioflavonoid Reduces the Oxidative Stress in the Skin of Mouse Exposed to Partial Body γ-Radiation. Transcriptomics 3(2) (2015). Doi:10.4172/2329-8936.1000111.

8. Riley PA. Free Radicals in Biology: Oxidative stress and the effect of ionizing radiation. Int J Radiat Biol. 1994;65(1):27-33.

9. S.Le Caër, Water radiolysis: influence of oxide surfaces on H2 production under ionizing radiation. Water (20734441);2011;3(1):235.

10. Dainiak N, Tan BJ. Utility of biological membranes as indicators for radiation exposure: alterations in membrane structureandfunctionovertime.StemCells.1995;13Suppl1:142-52.

11. Lomax ME, Folkes LK, O'Neill P. Biological consequences of radiation-induced DNA damage: relevance to radiotherapy.ClinOncol(RCollRadiol).2013;25(10):578-85.Doi:10.1016/j.clon.2013.06.007.

12. O. Desouky N. Ding G. Zhou. Targeted and non-targeted effects of ionizing radiation. J. Radiat. Res. Appl. Sci. 8 2015;247-254.

13. Shah DJ, Sachs RK, Wilson DJ. Radiation-induced cancer: a modern view. Br J Radiol. 2012;85(1020):e1166-73. Doi: 10.1259/bjr/25026140.

14. B. Halliwell, J.M.C. Gutteridge. Free radicals in biology and medicine, second ed. Clarendon Press, Oxford. 1989.

15. Burns J, Gardner PT, Matthews D, Duthie GG, Lean ME, Crozier A. Extraction of phenolics and changes in antioxidant activity of red wines during vinification. J Agric Food Chem. 2001;49(12):5797-808.

16. Giustarini D, Dalle-Donne I, Tsikas D, Rossi R. Oxidative stress and human diseases: Origin, link, measurement, mechanisms, and biomarkers. Crit Rev Clin Lab Sci. 2009;46(5-6):241-81. Doi: 10.3109/10408360903142326.

17. Kryston TB, Georgiev AB, Pissis P, Georgakilas AG. Role of oxidative stress and DNA damage in human carcinogenesis. MutatRes.2011;711(1-2):193-201.Doi:10.1016/j.mrfmmm.2010.12.016.

18. Marklund SL, Westman NG, Roos G, Carlsson J. Radiation resistance and the CuZn superoxide dismutase, Mn superoxide dismutase, catalase, and glutathione peroxidase activities of seven human cell lines. Radiat Res. 1984;100(1):115-23.

19. Chandra Jagetia G, Rajanikant GK, Rao SK, Shrinath Baliga M. Alteration in the glutathione, glutathione peroxidase, superoxide dismutase and lipid peroxidation by ascorbic acid in the skin of mice exposed to fractionated γ -radiation. Clin Chim Acta. 2003;332(1-2):111-21.

20. Jagetia GC, Rajanikant GK. Curcumin Stimulates the Antioxidant Mechanisms in Mouse Skin Exposed to Fractionatedγ-Irradiation.Anti oxidants(Basel).2015;4(1):25-41.Doi:10.3390/antiox4010025.

21. Halliwell B. Biochemistry of oxidative stress. Biochem Soc Trans. 2007;35(Pt 5):1147-50.

22. Jagetia GC, Ravikiran PB. Acceleration of Wound Repair and Regeneration by Nigella Sativa in the Deep Dermal Excision Wound of Mice Whole Body Exposed to Different Doses of γ-radiation. Am. Res. J. Med. Surg. 2015;1(3):1-17.

23. Lal BN, Choudhuri KD. Observations on Momordica charantia Linn, and Eugenia jambolana Lam. as oral antidiabetic remedies. Indian J. Med. Res. 2(1968):161.

24. Chandrasekaran M, Venkatesalu V. Antibacterial and antifungal activity of Syzygium jambolanum seeds. J Ethnopharmacol.2004;91(1):105-8.

25. Muruganandan S, Pant S, Srinivasan K, Chandra S, Tandan SK, Lal J, et al. Inhibitory role of Syzygium cumini on autacoid-induced inflammationinrats. Indian J Physiol Pharmacol. 2002; 46(4):482-6.

26. Ramirez RO, Roa CC Jr. The gastroprotective effect of tannins extracted from duhat (Syzygium cumini Skeels) bark on HCl/ ethanol induced gastric mucosal injury in Sprague-Dawley rats. Clin Hemorheol Microcirc. 2003;29(3-4):253-61.

27. Ravi K, Rajasekaran S, Subramanian S. Antihyperlipidemic effect of Eugenia jambolana seed kernel on streptozotocin induced diabetesinrats. Food Chem Toxicol. 2005; 43(9): 1433-9.

28. A. Banerjee, N. Dasgupta, B De. In vitro study of antioxidant activity of Syzygium cumini fruit. Food Chem. 90(2005) 727-733.

29. Braga FG, Bouzada ML, Fabri RL, de O Matos M, Moreira FO, Scio E, et al. Antileishmanial and antifungal activity of plantsused intraditional medicine in Brazil. J Ethnopharmacol. 2007;111(2):396-402.

30. S.B. Swami, N.S.J. Thakor, M.M. Patil, P.M. Haldankar Jamun (Syzygium cumini (L.)): A Review of its food and medicinal uses. Food Nutr. Sci. 2012;1100-1117.

31. Chagas VT, França LM, Malik S, Paes AM. Paes Syzygium cumini (L.) skeels: a prominent source of bioactive molecules against cardiometabolic diseases. Front Pharmacol. 2015;6:259. Doi:10.3389/fphar.2015.00259.

32. Jagetia GC, Shetty PC, Vidyasagar MS. Inhibition of radiation-induced DNA damage by jamun, Syzygium cumini, in the cultured splenocytes of mice exposed to different doses of γ-radiation. Integr Cancer Ther. 2012;11(2):141-53. Doi:10.1177/1534735411413261.

33. Jagetia GC, Baliga MS. Syzygium cumini (Jamun) reduces the radiation-induced DNA damage in the cultured humanperipheralbloodlymphoc ytes:apreliminarystudy.Toxicol Lett. 2002;132(1):19-25.

34. Jagetia GC, Baliga MS. Evaluation of the radioprotective effect of the leaf extract of Syzygium cumini (Jamun) in miceexposedtoalethaldoseofgamma-irradiation. Nahrung. 2003;47(3):181-5.

35. Jagetia GC, Baliga MS, Venkatesh P. Influence of seed extract of Syzygium cumini (jamun) on mice exposed to differentdosesofγ-radiation.JRadiatRes.2005;46(1):59-65.

36. G.C. Jagetia, P.C. Shetty, M.S.Vidyasagar.Treatment of mice with leaf extract of jamun (Syzygium cumini Linn. Skeels) protects against the radiation-induced damage in the intestinal mucosa of mice exposed to different doses of γ-radiation. Pharmacol. Onl. 1;2008;169-195.

37. Suffness M, Douros J. Drugs of plant origin. Meth. Cancer Res. 1979;26:73-126.

38. Moron MS, Depierre JW, Mannervik B. Levels of glutathione, glutathione reductase and glutathione S-transferase activities in rat lung and liver. Biochim Biophys Acta. 1979;582(1):67-78.

39. Aebi H. atalaseinvitro. In: Packer, L. (Ed.), MethodsEnzymol.1984;105:121-6.

40. Fried R. Enzymatic and non-enzymatic assay of superoxide dismutase. Biochimie. 1975;57(5):657-60.

41. Buege JA, Aust SD. Microsomallipidperoxidation. Methods Enzymol.1978;52:302-10.

42. Feinendegen LE, Pollycove M, Sondhaus CA. Responses to Low Doses of Ionizing Radiation in Biological Systems. Non linearity Biol Toxicol Med. 2004;2(3):143-71. Doi:10.1080/15401420490507431.

43. Blakely EA. Taylor lecture on radiation protection and measurments: what makes particle radiation so effective? Health Phys.2012;103(5): 508-28. Doi: 10.1097/HP.0b013e31826a5b85.

44. Kreipl MS, Friedland W, Paretzke HG. Time- and space resolved Monte Carlo study of water radiolysis for photon, electron and ion irradiation. Radiat Environ Biophys. 2009;48(1):11-20. Doi: 10.1007/s00411-008-0194-8.

45. Z. Hui, Z. Naikun, Z. Rong, L. Xiumin,C. Huifang. Effect of ionizing radiation bio-oxidase activities in cytoplasm of mouse blood and liver cells. Chin J Radiol Med Prot.1996;16 (3):179-82.

46. L. Zhang H. Yang Y. Tian Radiation-induced cognitive impairment. Therap. Targets Neurol. Dis. 2 (2015) e837.

47. Mattonen SA, Tetar S, Palma DA, Louie AV, Senan S, Ward AD. Imaging texture analysis for automated prediction of lung cancer recurrence after stereotactic radiotherapy.J Med Imaging (Bellingham). 2015;2(4):041010. Doi: 10.1117/1.JMI.2.4.041010.

48. Benson R, Madan R, Kilambi R, Chander S. Radiation induced liver disease: A clinical update. J Egypt Natl Canc Inst.2016;28(1):7-11. Doi:10.1016/j.jnci.2015.08.001.

49. Gujral DM, Lloyd G, Bhattacharyya S. Radiation-induced valvular heart disease. Heart. 2015. pii: heartjnl-2015-308765. Doi:10.1136/heartjnl-2015-308765.

50. L.Shantabi, G.C.Jagetia, Vabeiryureilai M. and Lalrinzuali K. Phytochemical Screening of Certain Medicinal Plants of Mizoram, India and their Folklore Use. Journal Biodiversity, Bioprospecting and Development.2014;1(4).

51. J.E. Biaglow, M.E. Varnes, E.R Epp, E.P Clark. Anticarcinogenesis and radiation protection, Plennum Press, New York.1987.

52. Carocho M, Ferreira IC. A review on antioxidants, prooxidants and related controversy: Natural and synthetic compounds, screening and

analysis methodologies and future Perspectives. Food Chem Toxicol. 2013;51:15-25. Doi: 10.1016/j.fct.2012.09.021.

53. Peltola V, Parvinen M, Huhtaniemi I, Kulmala J, Ahotupa M. Comparison of effects of 0.5 and 3.0 Gy X-irradiation on lipid peroxidation and antioxidant enzyme function in rat testis and liver. J Androl. 1993;14(4):267-74.

54. MeisterA, Anderson ME.Glutathione. AnnuRevBiochem. 1983;52:711-60.

55. Hall EJ, Giaccia AJ. Radiobiology for the Radiologist. Lippincott Williams & Wilkins: Philadelphia USA; 2012.

56. Leong PK, Ko KM. Induction of the Glutathione Antioxidant Response/ Glutathione Redox Cycling by Nutraceuticals: Mechanism of Protection against Oxidant-induced Cell Death. Curr Trends Nutraceut. 2016;1:1-2.

57. Jones DP. The Role of Oxygen Concentration in oxidative stress: hypoxic and hyperoxic models. In: Sies, H. (Ed), Oxidative Stress. Academic Press. New York: 1985;152-196.

58. Bogdanovic V, Bogdanovic G, Grubor-Lajsic G, Rudic A, Baltic V. Effect of irradiation on enzymes of antioxidant defense system in L 929 cell culture in the presence of α-tocopherol acetate. Arch Oncol. 2000;8(4):157-59.

59. Navarro J, Obrador E, Pellicer JA, Aseni M, Viña J, Estrela JM. Blood glutathione as an index of radiation-induced oxidative stress in mice and humans. Free Radic Biol Med.1997;22(7):1203-9.

60. Han Y, Son SJ, Akhalaia M, Platonov A, Son HJ, Lee KH, et al. Modulation of radiation-induced disturbances of antioxidant defense systems by ginsan. Evid Based Complement Alternat Med. 2005;2(4):529-36.

61. Jagetia GC, Baliga MS, Venkatesh P, Ulloor JN. Influence of ginger rhizome (Zingiber officinale Rosc) on survival, glutathione and lipid peroxidation in mice after whole-body exposure to gamma radiation.

Radiat Res. 2003 Nov;160(5):584-92.

62. Jagetia GC, Venkatesh P, Baliga MS. Evaluation of the radio protective effect of bael leaf (Aegle marmelos) extract in mice. Int J Radiat Biol. 2004;80(4):281-90.

63. Kumar KB, Kuttan R. Protective effect of an extract of Phyllanthus amarus against radiation-induced damage in mice. JRadiatRes. 2004;45(1):133-9.

64. Hirose K, Longo DL, Oppenheim JJ, Matsushima K. Over expression of mitochondrial manganese superoxide dismutase promotes the survival of tumor cells exposed to interleukin-1, tumor necrosis factor, selected anticancer drugs, and ionizing radiation. The FASEB J.1993;7(2):361-368.

65. Girotti AW. Lipid hydroperoxide generation, turnover, and effector action in biological systems. J Lipid Res. 1998;39(8):1529-42.

66. Wills ED, Wilkinson AE. Effects of Irradiation on Sub-cellular Components. Int J Radiat Biol Relat Stud Phys Chem Med.1970;17(3):229-36.

67. Raleigh JA, Kremers W, Gaboury B. Dose-rate and oxygen effects in models of lipid membranes: linoleic acid. Int J Radiat Biol Relat Stud Phys Chem Med. 1977;31(3):203-13.

68. Jagetia GC, Venkatesha VA. Effect of mangiferin on radiation-induced micronucleus formation in cultured human peripheralbloodlymphocytes.EnvironMolMutagen.2005;46(1):12-21.

69. Moskaug JØ, Carlsen H, Myhrstad MC, Blomhoff R. Polyphenols and glutathione synthesis regulation. Am J Clin Nutr.2005;81(1):277S-283S.

70. Khan A, Manna K, Das DK, Kesh SB, Sinha M, Das U, et al. Gossypetin ameliorates ionizing radiation-induced oxidative stress in mice liver-a molecular app. Free Radic Res. 2015;49(10):1173-86. Doi: 10.3109/10715762.2015.1053878.

14

Microbial Decolorization of Various Dyes by a *Bacillus subtilis* Strain Isolated from an Industrial Effluent Treatment Plant

Gorla V. Reddy1, Maulin P Shah[2*]

[1]*K Scientific Solutions Private Limited, Gachibowli, Hyderabad 500 032, India*
[2]*Industrial Waste Water Research Lab, Division of Applied & Environmental Microbiology, Environ Technology Limited, India*

Corresponding author: Maulin P. Shah, Industrial Waste Water Research Lab, Division of Applied & Environmental Microbiology, Environ Technology Limited, India, Email: shahmp@uniphos.com

Abstract

A *Bacillus subtilis* strain exhibiting laccase activity was isolated from an industrial effluent treatment plant. M9 medium containing Cu^{2+} was used for enrichment and isolation of bacterial strains capable of oxidizing syringaldazine, a known laccase substrate. An isolated strain was identified as *Bacillus subtilis* based on the results of physiological and biochemical tests and sequence analysis of the 16S rRNA gene. The strain could grow at temperatures ranging from 20 to 55°C and showed optimal growth temperature and pH at 25°C and 7.0, respectively. The rate of strain sporulation clearly correlated well with laccase activity. The half-life of the spore laccase was 2.5 h at 80°C and the pH half-life is 15 days at pH 9.0. The spore laccase could discolor 50-90% Remazol brilliant blue R, Alizarin, Congo red, methyl orange and methyl violet, suggesting the possible application of the spore laccase in the treatment of dyestuff.

Keywords: Bacterial laccase; *Bacillus subtilis*; Spore; Decolorization

Introduction

Laccases are multi-copper proteins that can oxidize a wide range of inorganic and aromatic compounds, especially phenols, while reducing molecular oxygen to water [1]. Laccases catalyze the removal of one hydrogen atom from phenolic substrates and aromatic amines by an abstraction of electrons. Free radicals formed during the reaction are also capable of undergoing depolymerization, further repolymerization, demethylation, or quinone formation [2-4]. The low substrate specificity of laccases and their ability to oxidize various pollutants suggest their industrial-technological and biotechnological applications [5,6]. Laccases are widely distributed in fungi and plants [7]. However, it has been found that laccases are also widespread in bacteria [1]. To date, laccases have mostly been isolated and characterized from plants and fungi, but only fungal laccases are currently used in biotechnology applications. In contrast, only a few bacterial laccases have been characterized. Bacterial laccases have the ability to oxidize syringaldazine and 2, 6-dimethoxyphenol, which are typical substrates for laccases, and also possess the canonical four areas for the binding of copper. Nevertheless, overall sequences of bacterial laccases show little resemblance to fungal laccases. Therefore, they are often called "multicopper oxidases" or "(poly) phenol oxidases" and their activity is generally defined as "laccase-like" [8]. The first report of bacterial laccase was from the strain *Azospirillum lipoferum*, which was isolated from the rhizosphere of rice [9]. This enzyme has been identified as a laccase using a combination of substrates and inhibitors [9] [10]. Laccase activities have also been found in *Bacillus sphaericus* [11], *Escherichia coli* [12], *Bacillus halodurans* [13], and *Streptomyces psammoticus* [14] to name a few. CotA, the *Bacillus subtilis* endospore layer component, is the most studied bacterial laccase [15]. Since spores allow microorganisms to survive in harsh conditions, spore coat enzymes can also withstand high temperatures or extreme pH values. As most fungal laccases are unstable at pH values greater than 7.0, their detoxification efficiency for pollutants often decrease under alkaline conditions. This limits the potential industrial application of fungal laccases as many processes are performed under alkaline conditions. Alternatively, spore laccases that are active in the alkaline pH range could be used for bioremediation or application in membrane reactors [4]. Compared to fungal laccases, bacterial laccases have the advantage of being less sensitive to halides and alkaline conditions and the producing strain typically exhibits a rapid growth rate [16]. Despite the importance of bacterial laccase in the degradation of pollutants, only a few new bacterial strains with "laccase-like" activity have been discovered. The lack of a robust and inexpensive commercially available laccase is a major obstacle to the widespread application of laccase in various industrial sectors [17]. Since bacterial genetic tools and biotechnological processes are well established, the development of bacterial laccases would be of great significance [18]. The present study was therefore conducted to isolate and characterize the strain, *Bacillus* sp. ETL 1979, which was isolated from an industrial textile effluent treatment plant. The spore laccase of this strain was characterized and used to decolorize various synthetic dyes.

Materials and Methods

Sample collection

Soil samples used in this study were collected from textile

effluents from the textile industry, Ankleshwar,   Gujarat, India. Collected soil samples were stored at 4°C aerobically.

Isolation of microorganisms

For the isolation and enrichment of bacterial strains with the ability to produce laccase, 250 ml flasks containing 100 ml M9 culture medium supplemented with 0.2 mmol/l Cu^{2+} were inoculated with 10 g of soil and incubated at 37°C on a rotary shaker (130 rpm) for 2 days. Then 5 ml of the cultures were transferred to 100 ml of Luria-Bertani medium (LB) culture medium containing 0.2 mmol/l Cu^{2+} and incubated at 37°C at 130 rpm for 7 days. Stable enrichment cultures were obtained after sub culturing. To isolate pure cultures, cultivated fortified products were appropriately diluted with a sterile saline solution (0.9% NaCl) before spreading onto LB/Cu^{2+} plates. The plates were incubated at 37°C for 3 days. Bacterial colonies from individual plates were flooded with a 0.1% (w/v) syringaldazine solution to determine whether any of the isolates exhibited laccase activity. Colonies with pink halos were streaked onto new LB/Cu^{2+} plates for purification. Re-inoculation was performed after identification of syringaldazine-positive colonies as described above. The isolation process was repeated several times until the isolates were shown to be pure.

Characterization of isolates

Gram staining was performed according to standard protocol. The characteristics of Gram and cell morphology of the isolated strain were determined by microscopy. For the use of carbon sources, the pure cultures were seeded respectively in peptone culture medium - water containing 1% substrate, and incubated at 37°C for 24 h. The results were determined by varying both turbidity and color of the culture medium. Selected biochemical metabolic capacity properties were determined by inoculating isolated bacteria on media.

Molecular Characterization

The bacterial cells were collected by centrifugation at 10 000 rpm for 2 minutes and incubated with 100 μg/ml lysozyme at 37°C for 1 h, followed by treatment with the lysis solution (1% SDS, 1 mmol/L EDTA, 20 mmol/L CH_3COONa, and 40 mmol/L Tris-HCl (pH 8.0). After addition of 5 mmol/L NaCl to the lysis solution, the mixture was extracted with phenol/chloroform/isoamyl alcohol (25:24:1). The supernatant was harvested and subsequently precipitated with absolute ethanol. The genomic DNA obtained was dissolved in sterile deionized water and stored at -20°C for later use. For the polymerase chain reaction (PCR), specific primers for eubacterial 16S rRNA gene sequence amplification 27F: 5'-GAGTTTGATCMTGGCTCAG-3 '(H = A or C) 1492R: 5'-TACGGYTACCTTGTTACGACTT-3' (Y = C or T) were used [19]. PCR was performed in a Gene Amp PCR System 9700 (Applied Biosystems, Singapore). The amplification reaction consisted of an initial denaturation at 93°C for 5 min, followed by 30 cycles of 94°C for 18 s, 56°C for 15 sec and 72°C for 78 s, and an extension step final at 72°C for 7 min. PCR products were analyzed by electrophoresis in 1.0% (w/v) agarose gel and photographed using a Bio Imaging System (Gene Genius, USA). The amplicons

were cloned using a commercially available cloning vector pMD18-T kit and transformed into competent E. coli JM109 cells. Positive clones were identified by PCR amplification with the 16S rRNA gene primers specified above.

Nucleotide sequencing, alignment, and phylogeny

16S rRNA gene sequencing of the isolated strain was performed by Bangalore Genei Company, India. Related sequences were obtained from the GenBank database after using the BLAST online tool [20]. Multiple sequence alignment was performed using Clustal X 1.81 [20]. PHYLIP package [20] was used to calculate the similarity values and to build a phylogenetic tree.

Optimization of growth conditions

The optimal growth conditions with regards to pH and temperature were determined. The strain was inoculated in LB media which have been adjusted to various pH values and incubated at 15-55°C. The optical density of growing cultures was observed at 600 nm using a UV spectrophotometer - 1800 (Shimzadzu, Japan) to determine the optimum growth conditions. All assays were performed in triplicate.

Effect of metals and saline solution on bacterial growth

To study the effect of metals on the growth of the laccase-producing strain, 200 μg/ml Zn^{2+}, Fe^{3+}, Ca^{2+}, Mn^{2+}, Mg^{2+} or Cu^{2+} was added to the LB culture medium, respectively. Cultures were grown in 25 ml medium in 100 ml conical flasks at 37°C for 24 h. Cultures grown in the absence of metals was used as a control. Growth was determined by measuring the absorbance at 600 nm against the blank. In addition, the strain was inoculated in LB medium supplemented with 1, 2, 4, 6, 8, 10, or 12 % (w/v) NaCl. The turbidity of the cultures in the growth medium was observed at 600 nm using a UV spectrophotometer - 1800 (Shimzadzu, Japan) to determine the growth state. All assays were performed in triplicate.

Sporulation rate and laccase activity relationship

B. subtilis ETL 1979 was inoculated onto LB plates containing 0.2 mmol/L Cu^{2+} and incubated at 30°C. The amount of spores was calculated daily and the activity of the laccase was determined at the same time. The sporulation rate was determined by the percentage of the quantity of spores opposed to all cells. The spores were removed from the agar with 1 mol/L KCl, washed with 0.5 mol/L NaCl and resuspended in 0.1 mol/L citrate phosphate buffers (pH 6.8). The spore suspension was prepared for the determination of the activity of laccase. All assays were performed in triplicate for each sample.

Spore laccase activity assay

Laccase activity of the spores was determined at 40°C using syringaldazine (dissolved in absolute ethanol, Sigma) as substrate. The oxidation of syringaldazine was detected by measuring the increase in absorbance at 525 nm ($\varepsilon_{525} = 65$ mmol^{-1} cm^{-1} L) after 3 min using a spectrophotometer (UV spectrophotometer - 1800 Shimzadzu, Japan). The reaction mixture (3 mL) contained

100 µl of spore suspension (10 mg wet spores), 2.4 ml of citrate phosphate buffer (0.1 mol/L, pH 6.8) and 0.5 ml of 0.5 mmol/L syringaldazine. One unit of enzyme activity is defined as the amount of enzyme required to oxidize one µmol of substrate per minute. All assays were performed in triplicate for each sample. The standard deviation does not exceed 5% of the average values.

Effect of pH and temperature

Determination of the effect of pH on laccase activity was performed in 0.1 mol/L citrate buffer - phosphate in the range of pH 4.0 to 8.0 using syringaldazine as substrate. The effect of temperature on the spore laccase activity was determined in the range of 0 to 100°C at the optimum pH value. Syringaldazine was used as the substrate as described before. All assays were performed in triplicate. The thermal stability of the spore laccase was determined by pre-incubation of the spores in 0.1 mol/L citrate buffer phosphate (pH optimum) at 60 and 80°C and the remaining activity was measured by the test described above. pH stability was examined similarly by incubating the spores in different buffers ranging from pH 4.0 to 9.0, at 30°C. All assays were performed in triplicate.

Determination of dye decolorization efficiency

Remazol Brilliant Blue R (RBBR), Alizarin red, Congo red, methyl orange, and methyl violet, were individually prepared at a concentration of 25 mg/L in sterilized distilled water. The prepared dye solution was mixed with 100 g/L spores and incubated at 37°C under mild conditions, shaking for 5 days. Dye samples without spores, which received the same treatment, were designated as the controls. The spectrum of each dye between 200 and 800 nm absorption was measured with a UV spectrophotometer - 1800 (Shimadzu, Japan). Dye decolorization was evaluated by the decrease in absorbance at the maximum wavelength of the dye. All assays were performed in triplicate.

Results

Isolation of the bacterial strain with the highest laccase activity

One hundred and forty colonies were selected from M9 agar plates supplemented with 0.2 mmol/L Cu^{2+}. After a secondary screening, 46 bacterial strains were selected based on the color development reaction to syringaldazine. One potential strain with high levels of laccase activity was named ETL 1979 and selected for further studies. Strain ETL 1979, which formed pink colonies on LB agar, is a gram-positive bacterium, spore-forming, rod-shaped, 1 to 2 µm long, motile, and formed white colonies on LB agar supplemented Cu^{2+} 0.2 mmol/L. The optimum pH for growth was determined to be 7.0 and the optimal temperature 25°C. The 16S rRNA gene amplicon was approximately 1.5 kb (Figure 1). The biochemical, physiological, morphological characteristics (Table 1), and the comparative analysis of the 16S rRNA gene sequence with the available database (GenBank) showed that the isolated strain is a *B. subtilis*. The similarity of the sequence (100%) and phylogeny based on Clustal X indicate that the ETL 1979 strain is a *B. subtilis* (Figure 2).

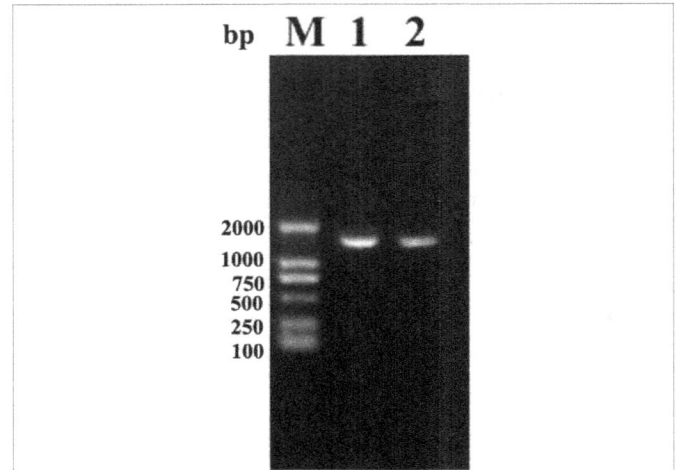

Figure 1: PCR product of the 16S rRNA gene of *B. subtilis* ETL 1979. Lane M: Molecular weight marker (DL2000), lanes 1 & 2: *B. subtilis* ETL 1979.

Characteristics	B. subtillis ETL-1979
Colony diameter	1 – 3 mm
Colony color	White
Cell morphology	Rod
Motility	+
Gelatin hydrolysis	+
Urase	+
Lipase	–
Oxidase	+
Catalase	+
Casein protease	+
Amylase	+
NO_3^- reduction to NO_2^-	+
M-R reaction	–
V-P reaction	+
Utilization of:	
Mannite	+
Phaseomannite	+
Sorbierite	+
L-rhamnose	–
Melibiose	+
Lactose	–
Glucose	+
Maltose	+
Xylose	–
Sucrose	+
Gum sugar	–
Fructose	+

Table 1: The morphological and biochemical characteristics of *B. subtilis* ETL 1979.

Effect of metals and saline solution on bacterial growth

Metal cations Zn^{2+}, Fe^{3+}, Ca^{2+}, Mn^{2+}, Cu^{2+}, and Mg^{2+} (200 µg/ml) all showed some degree of inhibition of the growth of the strain. Among them all, Zn^{2+} showed the highest degree of inhibition.

The relationship between sporulation rate and laccase activity

The positive correlation of the activity of laccase and the percentage sporulation was observed in the 10 day old culture as seen in (Figure 3). The result shows that the activity of laccase

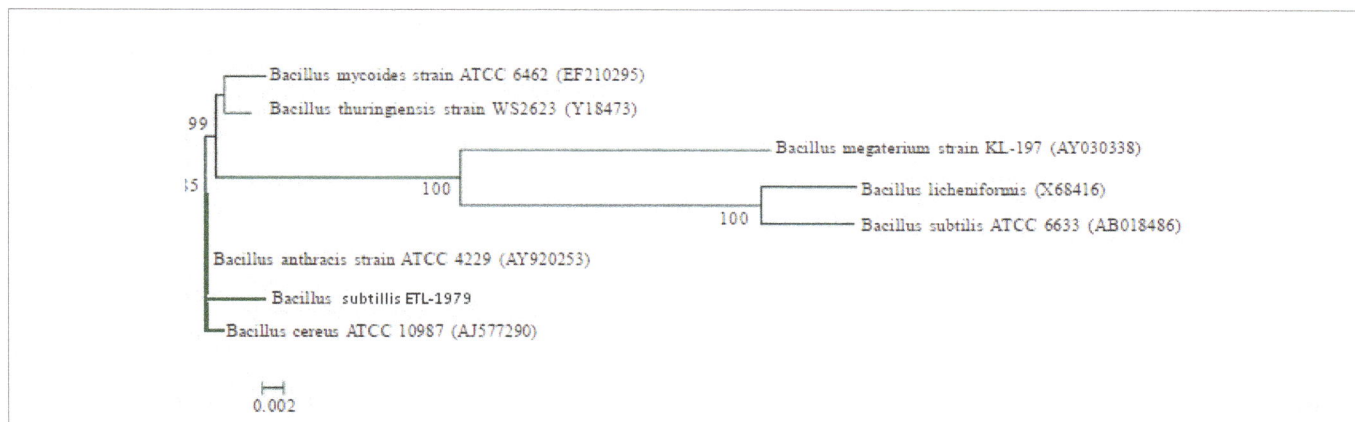

Figure 2: Phylogenetic analysis of the 16S rRNA gene sequences of *B. subtilis* ETL 1979 and related taxa.

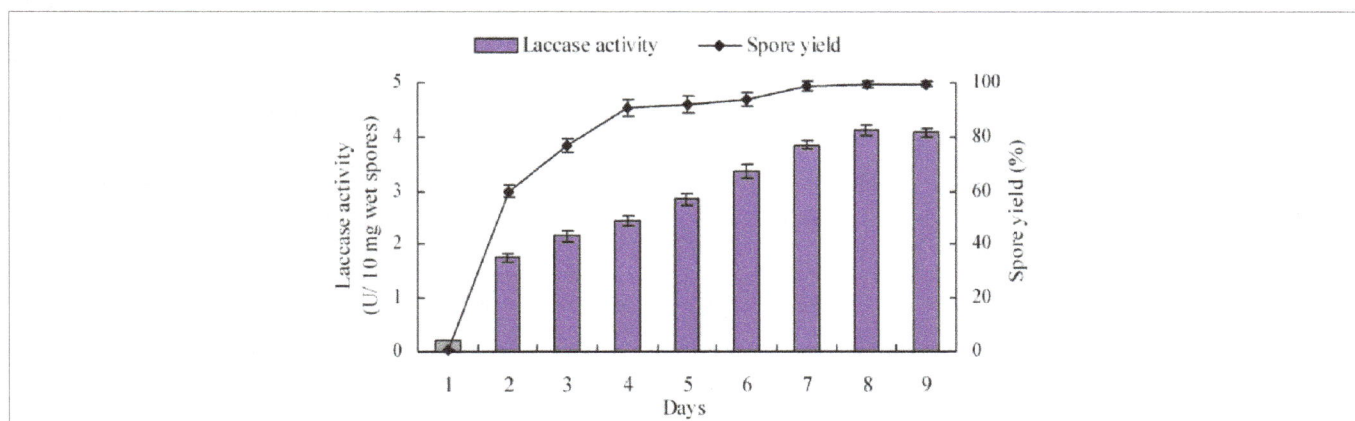

Figure 3: The relationship between sporulation rate and laccase activity of *B. subtilis* ETL 1979.

was derived from spores.

Effect of pH and temperature on the activity and stability of spore laccase

Highest laccase activity was detected at a pH of 6.8 and the optimum temperature was observed at 60°C. Laccase spores showed greater stability under conditions of high temperature and under alkaline conditions that most fungal laccases. The half-life of laccase was 2.5 h at 80°C, while the half-life of the laccase was 15 days at a pH of 9.0.

Efficiency of dye decolorization

To demonstrate the potential application of this bacterium for the treatment of wastewater containing a dye, the spores were used for bleaching RBBR, alizarin, Congo red, methyl orange, and methyl violet. The bleaching rate was 90% in the treatment of RBBR and alizarin red, and 50 to 70% in the treatment of the other dyes (Figure 4). These results indicate that the spore laccase has the ability to decolorize the selected dyes without the need for redox mediators.

Discussion

In this study, a new strain of *B. subtilis*, strain ETL 1979, was

isolated from soil collected at an industrial effluent treatment plant. This strain was unable to use xylose and sugar gum, while the type strain of *B. subtilis* according to Bergey's manual Unlike other *B. subtilis* strains in our laboratory which showed little laccase activity, strain ETL 1979 exhibited high laccase activity. Laccases as biocatalysts have received much attention because of their great capacity to oxidize phenolic and other aromatic compounds. This advantage makes laccases highly suited for certain biotechnological applications, such as the biodegradation of xenobiotics, including aniline, methoxyphenols and benzenethiols [21, 22]. In contrast to fungal laccases, bacterial laccases are very active and much more stable at high temperatures and high pH levels. As indicated above, most of the effluents from textile industries are characterized by a neutral to alkaline pH (about 7-11) [23, 24]. For many industrial applications it is necessary that catalysts such as laccases are kept active throughout the process or via immobilization onto intermediate membrane reactors [18]. The spore laccase of *B. subtilis* ETL 1979 has a high thermal stability and high stability under alkaline conditions. These characteristics could be of great importance for biotechnological applications.

It is well known that copper ions are toxic to a number of bacteria, even when present at low concentrations. However,

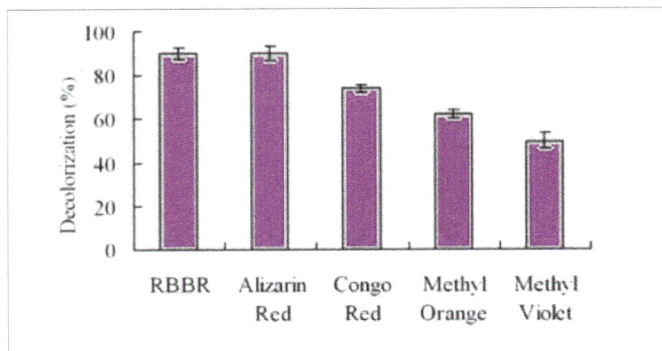

Figure 4: Decolorization of dyes with spore-bound laccase from *B. subtilis* ETL 1979.

certain bacterial laccases, such as CopA and CueO play a role in copper tolerance [20]. The regulation of copper homeostasis in *E. coli* was analyzed, and although the mechanism is still uncertain [25], it has been postulated that CueO is involved in the main mechanism of copper resistance and involves the oxidation of Cu^+ to Cu^{2+} [26]. This method is effective for copper resistance because Cu^+ is more harmful than Cu^{2+} [27]. The present study showed that *B. subtilis* ETL 1979 can survive in a medium containing copper. However, the ETL 1979 strain is unable to form melanin in the medium containing copper ions. CotA in *B. subtilis* has been associated with the formation of a brown pigment [28, 29]. Strain ETL 1979, also showed strong resistance to high concentrations of NaCl; it can survive in 10% NaCl. This advantage makes it potentially useful for dealing with saline wastewater and reduces the pretreatment time. To date, bacterial laccases have only been found in *A. lipoferum*, *Alteromonas* sp. MMB-1, *Pseudomonas* sp. KU03, *E. coli*, and some species of *Streptomyces* and *Bacillus*. There is little information on the use of enrichment culture methods and the sampling of soil from industrial effluent treatment plants for the prospecting of bacterial laccases. In this study, soil samples were taken at the factory of an industrial effluent treatment plant. Other reports on the isolation of bacterial species with laccase activity was focused on the rhizosphere of rice [9], seawater [21], river mud or top-soil containing organic waste [13], contaminated soil with dye and textile industry, and lignocellulosic waste effluents [30]. In our study, the spore laccase was used for the bleaching of an anthraquinone dye and azo dyes without the addition of nutrients or redox mediators. Our results indicate that the spore laccase can decolourize the dyes effectively in 5 days. This result is similar to that of the spore laccase from *Bacillus* sp. SF [4]. However, few spore laccases can be reused because it is difficult to separate the laccase from the decolorized solution. Immobilized enzymes are highly effective in bleaching stains, because immobilization can improve the utilization rate of the enzyme despite the reduction in enzyme activity. Future studies would therefore focus on the immobilization of the ETL 1979 spore laccase.

The results presented here demonstrate that the spore laccase has a potential application in the treatment of aqueous solutions that contains dye. In summary, the *B. subtilis* ETL 1979 strain that exhibited laccase activity was isolated from soil collected from an industrial effluent treatment plant and was characterized during the course of this study. The strain showed the ability to catalyze a substrate that is considered to be a typical laccase substrate (syringaldazine), it exhibited good growth at 55°C, and its spore laccase can decolorize selected dyes without the addition of redox mediators.

References

1. Alexandre G, Zhulin IB. Laccases are widespread in bacteria. Trends Biotechnol. 2000;18(2):41-2.

2. D'Annibale A, Stazi SR, Vinciguerra V, Giovannozzi Sermanni G. Oxirane-immobilized *Lentinula edodes* laccase: stability and phenolics removal efficiency in olive mill wastewater. J Biotechnol. 2000;77(2-3):265-73.

3. Ullah MA, Bedford CT, Evans CS. Reactions of pentachlorophenol with laccase from *Coriolus versicolor*. Appl Microbiol Biotechnol. 2000;53(2):230-4.

4. Held C, Kandelbauer A, Schroeder M, Cavaco-Paulo A, Guebitz GM. Biotransformation of phenolics with laccase containing bacterial spores. Environ. Chem. 2005;Lett. 3(2):74-77. DOI: 10.1007/s10311-005-0006-1.

5. Witayakran S, Ragauskas AJ. Synthetic Applications of Laccase in Green Chemistry. Adv. Synth. Catal.2009; 351(9):1187-1209. DOI: 10.1002/adsc.200800775.

6. Aaron Steevensz, Mohammad Mousa Al-Ansari, Keith E Taylor, Jatinder K Bewtra and Nihar Biswas. Comparison of soybean peroxidase with laccase in the removal of phenol from synthetic and refinery wastewater samples. J. Chem. Technol. Biotechnol. 2009;84:761-769. DOI: 10.1002/jctb.2109.

7. Ben Younes S, Mechichi T, Sayadi S. Purification and characterization of the laccase secreted by the white rot fungus *Perenniporia tephropora* and its role in the decolourization of synthetic dyes. J Appl Microbiol. 2007; 102(4):1033-42.

8. Solano F, Lucas-Elío P, López-Serrano D, Fernández E, Sanchez-Amat A. Dimethoxyphenol oxidase activity of different microbial blue multicopper proteins. FEMS Microbiol Lett. 2001 Oct 16;204(1):175-81.

9. Givaudan A, Effosse A, Faure D, Potier P, Bouillant ML, Bally R. Polyphenol oxidase in *Azospirillum lipoferum* isolated from rice rhizosphere: evidence for laccase activity in non-motile strains of *Azospirillum lipoferum*. FEMS Microbiol.1993; Lett. 108(2):205-210.

10. Diamantidis G, Effosse A, Potier P, Bally R. Purification and characterization of the first bacterial laccase in the rhizospheric bacterium *Azospirillum lipoferum*. Soil Biol. Biochem. 2000;32(7):919-927. Doi:10.1016/S0038-0717(99)00221-7.

11. Dubé E, Shareck F, Hurtubise Y, Daneault C, Beauregard M. Homologous cloning, expression, and characterisation of a laccase from Streptomyces coelicolor and enzymatic decolourisation of an indigo dye. Appl Microbiol Biotechnol. 2008;79(4):597-603. Doi: 10.1007/s00253-008-1475-5.

12. Grass G, Rensing C. CueO is a multi-copper oxidase that confers copper tolerance in *Escherichia coli*. Biochem Biophys Res Commun. 2001;286(5):902-8.

13. Ruijssenaars HJ, Hartmans S. A cloned *Bacillus halodurans* multicopper oxidase exhibiting alkaline laccase activity. Appl Microbiol Biotechnology. 2004;65(2):177-82.

14. Niladevi KN, Prema P. Immobilization of laccase from *Streptomyces psammoticus* and its application in phenol removal using packed bed reactor. World J. Microbiol. Biotechnol.2008; 24(7):1215-1222. DOI: 10.1007/s11274-007-9598-x.

15. Oh P, Goel R, Capalash N. Bacterial laccases. World J. Microbiol. Biotechnol. 2001;23(6):823-832. DOI: 10.1007/s11274-006-9305-3.

16. Jimenez-Juarez N, Roman-Miranda R, Baeza A, Sánchez-Amat A, Vazquez-Duhalt R, Valderrama B, et al. Alkali and halide-resistant catalysis by the multipotent oxidase from *Marinomonas mediterranea*. J Biotechnol. 2005;117(1):73-82.

17. Dubé E, Shareck F, Hurtubise Y, Daneault C, Beauregard M. Homologous cloning, expression, and characterisation of a laccase from *Streptomyces coelicolor* and enzymatic decolourisation of an indigo dye. Appl Microbiol Biotechnol. 2008;79(4):597-603. Doi: 10.1007/s00253-008-1475-5.

18. Sharma P, Goel R, Capalash N. Bacterial laccases. World J. Microbiol. Biotechnol.2007; 23(6): 823-832. DOI: 10.1007/s11274-006-9305-3.

19. Wang YL, Yang RH, Mao AJ, Wang JQ, Dong ZY. Phylogenetic diversity analyse of Rumen Bacteria using culture independent method. Wei Sheng Wu Xue Bao. 2005;45(6):915-9.

20. Rensing C, Grass G. *Escherichia coli* mechanisms of copper homeostasis in a changing environment. FEMS Microbiol Rev. 2003;27(2-3):197-213.

21. Solano F, Garcia E, Perez D, Sanchez-Amat A. Isolation and characterization of strain MMB-1 (CECT 4803), a novel melanogenic marine bacterium. Appl Environ Microbiol. 1997;63(9):3499-506.

22. Xu F. Oxidation of phenols, anilines, and benzenethiols by fungal laccases: correlation between activity and redox potentials as well as halide inhibition. Biochemistry. 1996;35(23):7608-14.

23. Manu B, Chaudhari S. Anaerobic decolorisation of simulated textile wastewater containing azo dyes. Bioresour Technol. 2002;82(3):225-31.

24. Jahmeerbacus MI, Kistamah N, Ramgulam RB. Fuzzy control of dye bath pH in exhaust dyeing. Coloration Technol. 2004;120: 51-55. DOI: 10.1111/j.1478-4408.2004.tb00206.x.

25. Nakamura K, GO N. Function and molecular evolution of multicopper blue proteins. Cell Mol Life Sci. 2005;62(18):2050-66.

26. Singh SK, Grass G, Rensing C, Montfort WR. Cuprous oxidase activity of CueO from Escherichia J Bacteriol. 2004;186(22):7815-7.

27. Outten FW, Huffman DL, Hale JA, O'Halloran TV. The independent cue and cus systems confer copper tolerance during aerobic and anaerobic growth in *Escherichia coli*. J Biol Chem. 2001 ;276(33):30670-7.

28. Endo K, Hosono K, Beppu T, Ueda K. A novel extracytoplasmic phenol oxidase of *Streptomyces*: its possible involvement in the onset of morphogenesis. Microbiology. 2002;148(Pt 6):1767-76.

29. Sanchez-Amat A, Solano F. A pluripotent polyphenol oxidase from the melanogenic marine *Alteromonas* sp. share catalytic capabilities of tyrosinases and laccases. Biochem Biophys Res Commun. 1997;240(3):787-92.

30. Senan RC, Abraham TE .Bioremediation of textile azo dyes by aerobic bacterial consortium: aerobic degradation of selected azo dyes by bacterial consortium. Biodegradation, 2004;15(4):275-280. DOI: 10.1023/B:BIOD.0000043000.18427.0a.

15

Optimization of growth conditions for zinc Solubilizing Plant Growth associated Bacteria and Fungi

Shabnam S Shaikh and Meenu S Saraf*

Department of Microbiology and Biotechnology, School of sciences, Gujarat University, Ahmedabad

Corresponding author: Meenu S Saraf, Department of Microbiology and Biotechnology, School of sciences, Gujarat University, Ahmedabad, E-mail: sarafmeenu@gmail.com

Abstract

Zinc (Zn) is an essential element necessary for plant, humans and microorganisms required in little quantities to compose a complete array of physiological functions. Rhizospheric microbes are known to influence plant growth by various direct and indirect mechanisms and have some additional properties such as multiple metal solubilization. In the current investigation, we have isolated zinc solubilizing microbes and optimized their growth condition for further application in agriculture industry. Seven isolates amongst which four fungi and three fungi were studied for their Plant growth promoting ability, Zinc solubilization and optimization of growth. Isolate MSSZB4 and MSS-ZF3 were showing significant Plant promoting abilities and shows best optimization with 0.1% ZnO concentration, dextrose as carbon source, Ammonium Sulphate as nitrogen source and the optimum pH and Temperature was found between 6 to 6.5 and 28 to 30°C respectively. The present study demonstrates the optimum growth conditions for zinc solubilizing microbes, which can further be used for their potential applications, such as biofortification and bioremediations.

Keywords: Zinc; Solubilization; Plant growth promoting properties; Optimization

Introduction

Plant growth promoting rhizobacteria can affect plant growth by different direct and indirect mechanisms [1]. PGPR influence direct growth promotion of plants by fixing atmospheric nitrogen, solubilizing insoluble phosphates, secreting hormones such as IAA, GAs, and Kinetics besides ACC deaminase production, which helps in regulation of ethylene. Induced systemic resistance (ISR), antibiosis, competition for nutrients, parasitism, production of metabolites (hydrogen cyanide, siderophores) suppressive to deleterious rhizobacteria are some of the mechanism that indirectly benefit plant growth. Zinc (Zn) is an essential element necessary for plant, humans and microorganisms [2,3]. Human and other living things require Zn throughout requires in little quantities to compose a complete array of physiological functions. Zinc is a vital mineral of "exceptional biological and public health importance" [4]. Furthermore 100 specific enzymes are found in which zinc serves as structural ions in transcription factors and is stored and transferred in metallothioneinsand typically the 2nd most abundant transition metal in organisms, after iron and it is the only metal which appears in all enzyme classes [3].

Zinc is important micronutrient for plant which plays numerous functions in life cycle of plants [5]. Crop growth, vigor, maturity and yield are very much reliant upon essential micronutrient (Zn). To address the problem of Zn deficiency, micronutrient biofortification of grain crop is increased interest in developing countries [6]. Several approaches have been projected and practiced for fortification of cereals [7]. Enhancing Zn concentration of cereal grain has been recognized as an approach of tackling human Zn deficiency [8]. Plant scientists are formulating different methodologies to tackle the Zn deficiencies in crop through fertilizes applications and/ or by means of plant breeding strategies to augment the adsorption and or bioavailability of Zn in grain crops [6].

Plant growth promoting rhizobacteria (PGPR) is multifunction microbes functioning in sustainable agriculture. PGPR are a diverse group of bacteria that can be found in the rhizosphere on root surfaces as well as in association with roots [9]. These bacteria move around from the bulk soil to the living plant rhizosphere and antagonistically colonize the rhizosphere and roots of plant [10]. Soil bacteria which are important for plant growth are termed as plant growth promoting rhizobacteria (PGPR) [10]. In addition to phosphate mobilization they are responsible to play key role in carrying out the bioavailability of soil phosphorus, potassium, iron, zinc and silicate to plant roots [11].

Viable application of PGPR are been tested and are repeatedly promising; however, good understanding of microbial interactions will significantly raise the success rate of field application [12].

Material and Methods

Physical and chemical characterization of soil sample

Three soil samples were collected from rhizosphere region of Agriculturall and were collected from the different region of Gujarat. Physical characteristics, various chemical tests likes alinity, pH, total carbon, phosphates, total dissolved solids, edoxpotential (mV), conductivity, chlorides, Sulphate, potassium, nitrates as well as micro metals present in the soil like Few were also characterize for soil samples.

Qualitative and quantitative phosphate solubilization

Phosphate Solubilization was studied using tricalcium phosphateas in soluble phosphate. The strains were spotinoculated on Pikovskaya's agar medium. The plates were incubated with 30 °C for 48 to 72 h for bacteria and 3 to 6 days for fungi. The clear halo around the colony indicates the zone of phosphate Solubilization due to the production of organic acids as possible mechanism of the phosphate solubilization. Quantitative phosphate Solubilization was carried out in liquid Pikovskaya's medium in 250 ml flasks for14d.

The concentration of the soluble phosphate in the supernatant was estimated every 7dayby Stannous Chloride (SnCl2.2HO) method [13]. A simultaneous change in the pH was also recorded in the supernatant on Systronics digital pH meter (pH system 361).

Qualitative and quantitative production of Siderophore

Siderophore production was checked by using Chrome azurols (CAS) agar medium by the method described by Schwyn and Neilands, [14]. Actively growing cultures were spot inoculated on the CAS blue agar plate. These plates were then incubated at 37 °C for 48 to 72 h for Bacteria and at 28 °C for 3-6 days for fungi. Formation of yellow-orange halo around the colony indicated production and release of the siderophores on the agar plate.

Indole Acetic Acid production

Auxin production was studied in trypton yeast medium. Bacteria were grown in 50 ml yeast extract broth supplemented with 50 mgL^{-1} of L-Tryptophan and incubated in dark on orbital shaker at 200 rpm for 72 h. Hormone production was checked in supernatant using Salkowsky's reagent method [15]. The amount of IAA produced was calculated from the standard graph of pure indole acetic acid. Study was carried out every 24 h for up to 120 h and the pattern of IAA production was recorded.

Ammonia and HCN production

Each strain was tested for the production of ammonia in peptone water. Cultures (100 μl inoculum with approximately 3 x 10^8 c.f.u. ml -1) were inoculated in 10 ml peptone water and these plates were then incubated at

37 °C for 48 to 72 h for Bacteria and at 28 °C for 3-6 days for fungi. After Incubation Nessler's reagent (1 ml) was added to each tube. Development of brown to yellow colour was recorded as a positive test for ammonia production [16]. Production of hydrocyanic acid (HCN) was checked on nutrient agar slants streaked with the test isolates. Filter paper strips dipped in picric acid and 2 % sodium carbonate were inserted in the tubes. HCN production was checked based on changes in colour from yellow to light brown, moderate brown or strong brown of the yellow filter paper strips [17].

Exopolysaccharide (EPS) production

Normally EPS production is studied in basal medium of all different organisms. Ascarbohydrate source 5% of sucrose is to be added as polysaccharide in to the medium [18]. 10 ml of culture suspension was collected after 5-6 days and centrifuge at 30,000 rpm for 45 minutes add thrice the volume of chilled acetone. EPS will be separated from the mixture in the form of a slimy precipitates.

Zinc solubilization

The BTG medium was mixed with thorough stirring to obtain a homogeneous suspension. Experiments in liquid culture were performed in a defined Mineral Salt Medium (MSM), with glucose (10 g) as the sole carbon source and, when required, 0.1% insoluble zincoxide [19]. The dilution medium for viable counts was sterile NaCl solution, 8 g /liter. All glass ware used was soaked for 1 h in 1 M HCl and rinsed three times in distilled deionized water prior to use. Inoculation was carried out by using pure colony of a bacteria and fungi. It was inoculated to medium and allowed to grow. (For Bacteria at 37 °C and for fungi at 28 °C) for 14 days respectively [20]. The Zone of Soublization was Observed and measured in millimeter (mm).

Zinc solubilization in PVK (Pikovskaya's medium)

Zinc solubilization was checked using zinc oxide as insoluble zinc source. Spot inoculation of the isolates was done in the centre of the Pikovskaya's agar medium. These plates were then incubated at 37 ° C for 48 to 72 h for Bacteria and at 28oC for 3-6 days for fungi. Phosphate solubilization was checked in the form of a clear halo formed around the colony representing the production of organic acids as a possible mechanism of the zinc solubilization. Quantitative zinc solubilization was carried out in liquid Pikovskaya's medium in 250 ml flasks for 14 d [21].

Optimization of Media and Growth Condition for Zinc Solubilization: Zinc solubilizing ability of bacterial strains was tested in four different types of agar media. Composition of different media is given in table. Among them PVK (Pikovskaya'smedium) media with 0.1% Zinc Oxide was selected based on proper zone formation, opacity of medium and growth of isolates [21].

Effect of various Zinc source on efficiency of Zinc Solubilization: Effect of various Zinc sources like Zinc Carbonate, ZincSulphate and Zinc Oxide, were studied in PVK Broth. The isolates were checked for solubilization activity in PVK broth amended with different Zinc source. Inoculation was carried out by using pure colony of a bacteria and fungi. It was inoculated to medium and allowed to grow. (for Bacteria at 37oC and for fungi at 28 °C) for 14 days respectively [20]. The Zone of Solubilization was observed and measured in millimetre (mm). Zinc oxide was selected as the optimum zinc source for the further optimization, based on proper zone formation and opacity of the medium.

Effect of different concentration of Zinc Oxide on efficiency of Zinc Solubilization: Effect of different concentration of Zinc Oxide was added in the PVK agar medium which was 0.1%, 0.2%,

0.3%, 0.4% and 0.5%. Nitrogen sources like (NH4)2SO4, Urea, Casein, and NaNO3 were studied in PVK Broth. Inoculation was carried out by using pure colony of a bacteria and fungi. It was inoculated to medium and allowed to grow. (for Bacteria at 37 °C and for fungi at 28 °C) for 14 days respectively [20]. The zone of solubilization was Observed and measured in millimetre (mm).

Effect of various Carbon sources on efficiency of Zinc solubilization: Effect of various carbon sources like glucose, fructose, sucrose, lactose, glycerol and xylose, were studied in PVK agar plate. The isolates were checked for solubilization activity in PVK agar medium amended with 0.1% Zinc Oxide. Inoculation was carried out by using pure colony of a bacteria and fungi. It was inoculated to medium and allowed to grow. (For Bacteria at 37 ° C and for fungi at 28 °C) for 14 days respectively [20]. The Zone of Solubilization was Observed and measured in millimetre (mm).

Effect of various Nitrogen sources on efficiency of Zinc solubilization: Effect of various Nitrogen sources like (NH4)2SO4, Urea, Casein, and NaNO3 were studied in PVK Broth. The isolates were checked for solubilization activity in PVK broth amended with 0.1% Zinc Oxide. Inoculation was carried out by using pure colony of a bacteria and fungi. It was inoculated to medium and allowed to grow. (For Bacteria at 37 °C and for fungi at 28 °C) for 14 days respectively [20]. The zone of solubilization was Observed and measured in millimetre (mm).

Effect of temperature on efficiency of Zinc solubilization: Media composition to which the bacteria responded best was used as substrate. Inoculation was carried out by using pure colony of a bacterial grown on Basal medium of isolates and allowed to grow and maintained at 8 ° C, 15 ° C, 28 ° C, Room Temperature, 37 °C, and 55 °C for 14 days respectively [20]. The zone of solubilization was observed and measured in millimetre (mm).

Effect of pH on efficiency of Zinc Solubilization: Optimal media and temperature was used, but the pH of the media was set at pH 4, pH 6, pH 6.5, pH 7, pH 9 using NaOH or HCl and grown for 14 days respectively [20]. The Zone of solubilization was Observed and measured in millimetre (mm).

Effect of different Salinity on efficiency of Zinc Solubilization: Optimal media and Conditions were used, but the saline concentration was added as NaCl (0.2%, 0.4%, 0.6%, 0.8% and 1%) and KCl (0.02%, 0.04%, 0.06%, 0.08% and 0.1%) in the media was set and grown for 14 days respectively [20]. The zone of solubilization was Observed and measured in millimetre (mm).

Results and Discussion

Physical characterization of soil samples

Physical characteristics of all the different soil samples shows that soil from rhizosphere was brown in colour and its texture was granular to loamy. The chemical characterization of soil samples is shown in (Table 1). The observed variation in the pH could be due to heterogenous composition of soil at all the

three sites. Similar results were also reported by Amanul where variations were observed in soil samples of different agricultural soils [22]. Higher Organic and nitrogen content.

Isolation and Microbiological characterization of the soil samples

Total 20 isolates from three soil samples were screened for phosphate, zinc solubilization and giving maximum growth (fast grower). Among them Seven isolates were selected (four bacteria and three fungi) for further studies. They were purified on their respective medium.

Qualitative and quantitative phosphate solubilization

Phosphate solubilization results show that all the seven isolates were significant Phosphate solubilizers and showing zone of phosphate solubilization on solid Pikovskyaya's medium after 3 days of incubation at 30 ± 2 °C. Maximum zone was observed in isolate MSS-ZF3 (49 mm). Significant zones were also seen in MSS-ZF2 (45mm), MSS-ZB4(43 mm), MSS-ZF1 (40 mm), MSS-ZB2(35 mm) MSS-ZB1 (33 mm) and MSS-ZB3(30 mm) after eight days of incubation (Figure 1).

Maximum TCP (Tricalcium phosphate) solubilization in liquid medium was observed in MSS-ZF3 (29 µg/ml) followed by MSS-ZF2 ((24 µg/ ml), MSS-ZB4 (37 µg/ ml), MSS-ZF1 (18 µg/

Table 1: Chemical characteristics of soil samples.

Parameters	Sample 1	Sample 2	Sample 3
pH	8.25	8.36	7.89
E.C	0.21	0.17	0.15
Organic carbon (%)	1.26	1.56	1.77
Available Nitrogen (%)	0.10	0.13	0.15
P_2O_5 (ppm)	1.211	1.197	1.156
K_2O (ppm)	19	14	16
Available Fe (ppm)	4.3	4.06	5.09
Available K (kg/hec)	538	202	409
Available Zn (ppm)	3.62	3.70	4.20
Available cu (ppm)	1.00	1.14	1.19
Available Mn (ppm)	3.62	3.70	4.02

HCN Production by the Isolates

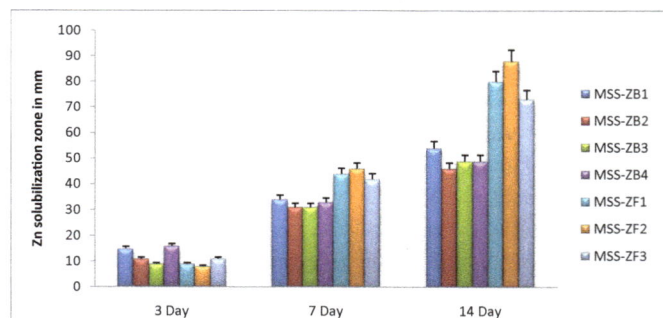

Figure 1: Zinc solubilization by the selected isolates.

ml) and MSS-ZB2 (16 µg/ ml), MSS-ZB1 (12 µg/ ml) and MSS-ZB3 (10 µg/ ml) in 3.2). The result observed was that the isolates showed maximum zone of solubilization on solid medium, are also showing similar phosphate solubilization in liquid medium. The pH of the broth having fungal isolates has been decreased from 7.0 to 4.0. The observed result shows that in bacterial culture there was no decrease in pH, but in all the fungal isolates it shows the decrease in broth pH. The results showing no correlation between Phosphate solubilization and pH reduction are also published by many researchers (Tank and Saraf 2003). This drop-in pH may also be an attribute of glucose utilization by the isolates (Arora et al. 2008). Plant growth is frequently limited by an insufficiency of phosphates, an important nutrient in plants next to nitrogen. Although all isolates showed similar decline in pH, 3.3 - 4.5, amount of phosphate solubilization was different in different PGPR's isolated. This indicates that there is no relation between degree of phosphate solubilized and change in pH [13].

Qualitative and quantitative siderophore production

The universal assay described by Schwyn and Neilands was used for the detection of siderophore by different microorganisms (fungi and bacteria) in solid medium. Siderophore on solid CAS blue agar plate shows a clear zone of decolourization representing iron chelation by the isolate in the medium. Highest zone of dye decolourization was observed in MSS-ZF3 (20 mm), MSS-F2 (18 mm), MSS-ZF1 (16 mm), MSS-ZB1(14 mm) and MSS-ZB3(10 mm) after 120 h of Incubation. And no zone was observed in MSS-ZB2.Siderophore was detected by the formation of orange halos surrounding bacterial colonies on CAS agar plates after 48 hour at 28 ° C [23].

Indole Acetic Acid Production

All the seven selected isolates showed significant production of IAA. Highest IAA production was reported in MSS-ZF1 (44 µg/ ml), MSS-ZF2 (40 µg/ml), MSS-ZB1 (24 µg/ml), MSS-ZB2 (19 µg/ ml), MSS-ZF3 (18 µg/ml) and MSS-ZB4 (14 µg/ml). All the isolates showed a continuous increase in the IAA production within the incubation period of 6 days.

Different isolates showed different optimum incubation time for highest IAA production. It is estimated that about 80 % of soil bacteria possess IAA producing potential [24].Though reports reveal that IAA production reaches maximum after 120 h (5 d) of incubation many of our isolates did not follow this pattern and showed maximum IAA production even after 240 h (10 d) [25]. However, reports of other researchers showed that IAA production was not detected after 5 d. Though it is reported that there is continuous decrease in IAA production after reaching the peak production, this pattern was also followed by our isolates. IAA production curves of the isolates showed continuous increase and decrease up to 12 d. These types of curves are in agreement with the IAA production curves reported by Torres-Rubioet al [26,27]. The reason for such fluctuations could be the utilization of IAA by the cells as nutrient during late stationary phase or production of IAA degrading enzymes by the cells which are inducible enzymes in presence of IAA [26].

Ammonia and HCN production

Ammonia production was studied up to 42- 72 h of incubation as per method given. Maximum concentration of ammonia production was observed in isolates MSS-ZB4 (32 µg/ml) followed by MSS-ZB3 (29µg/ml), MSS-ZB1 (27 µg/ml), MSS-ZF3 (24µg/ml), MSS-ZF2 (22µg/ml), MSS-ZSF1(21 µg/ml) and MSS-ZB2 (19 µg/ml).

Ammonia released by diazotrophs is one of the most important traits of PGPR's which benefits the crop [25]. This accumulation of ammonia in soil may increase in pH creating alkaline condition of soil at pH 9-9.5. It suppresses the growth of certain fungi and nitrobacteria due to it potent inhibition effect. Christiansen et al. have reported that level of oxygen in aerobic conditions was same as the level of ammonia excretion under oxygen limiting conditions. However, Joseph et al. reported ammonia production in 95% of isolates of *Bacillus* followed by *Pseudomonas* (94.2%), *Rhizobium* (74.2%) and *Azotobacter* (45%) [3,4,7,29-32].

HCN production was checked in all isolates the results are showed in table 6. Presence or absence and intensity of HCN production can play a significant role in antagonistic potential of bacteria against phytopathogens. Similar results were also reported by Cattelan et al. who reported that production of cyanide was an important trait in a PGPT in controlling fungal diseases in wheat seedlings under in-vitro conditions. Chandra et al. reported production of HCN by the PGPR which was inhibitory to the growth of *S. sclerotium*. Kumar et al. also reported in vitro antagonism by HCN producing PGPR against sclerotia germination of *M. phaseolina*. Production of HCN along with siderophore production has been reported as the major cause of biocontrol activity for protection of Black pepper and ginger [30,33-35].

Table 2: HCN production by the Isolates.(+ positive) and (– negative).

Isolates	Result
MSS-ZB1	+
MSS-ZB2	-
MSS-ZB3	+++
MSS-ZB4	++
MSS-ZF1	-
MSS-ZF2	-
MSS-ZF3	-

Table 3: pH change in the liquid medium.

Isolate	2nd day pH	4th day pH	6th day pH
MSS-ZB1	7.0	6.0	5.4
MSS-ZB2	6.9	6.1	5.1
MSS-ZB3	6.8	6.2	5.2
MSS-ZB4	7.0	6.3	5.3
MSS-ZF1	6.5	6.0	5.0
MSS-ZF2	6.5	5.9	4.9
MSS-ZF3	6.5	5.8	4.8

Exopolysaccharide (EPS) production by selected isolates

From all the Seven culture, the three bacterial isolates shows EPS Production, maximum amount of EPS production was observed in isolate MSS-ZB1(44.5 mg/ ml) followed by MSS-ZB3 (30.5 mg/ ml) and MSS-ZB4(20.0 mg/ ml) after five days of incubation.

Maximum of EPS production occurs during early stationary phase than in the late stationary of culture [18]. The highest EPS production was recorded in *P. aeruginosa* (226 µg/ ml) grown in nitrogen free medium followed by *S. mutans* and *B. subtilis* (220 and 206 µg/ ml respectively) in nitrogen free medium after 7 days of incubation at 37 ° C reported that production of EPS by *Burkholderia gladioli* IN-26 a strain of PGPR reduced bacterial speck on tomato. Similarly, Alami et al. reported that EPS produced by root associated saprophytic bacterium (rhizobacterium) *Pantoeaagglomerans*YAS34 was associated with plant growth promotion of sunflower reported that *Paenibacillus polymyxa* produces a large amount of polysaccharide possessing high activity against crown rot disease caused by *Aspergillus niger* in plants[29,33,36].

Solubilization of insoluble zinc by the isolates

Three media were selected to study the solubilization zinc oxide (BTG, Minimal salt medium and Pikovskaya medium) from these media Pikovs kaya medium was selected for further studies. Zinc phosphate-supplemented medium, where bacterial cells belonging to this strain are small Gram-negative rods, are able to grow in a simple mineral-glucose medium, with colonies being UV fluorescent. However, Appanna and Whitmore found that the production of protein-rich, zinc-binding moieties by *P. fluorescens* ATCC 15325 accounted for a mechanism of zinc tolerance in this strain. Although a similar mechanism may also occur in our strain during the phase of increase in free Zn, alternatively, the protein overproduction may be a factor involved in the solubilization process and/or observed Zn toxicit [1,19].The absence of detectable chelated zinc suggested that the solubilization process is an indirect consequence of an increase in hydrogen ion activity in the solution [19]. The observed acidification of the medium, both in the zinc supplemented and in the control cultures, initially occurred without correlation with the release of organic acids. A cause of such an increase in the proton concentration may be the depletion of ammonia, required for protein synthesis. Only when zinc phosphate was present was there a secondary production of gluconic acid (and/or keto-derivatives) which caused a further decrease in pH, accounting for the observed high levels of Zn [19].

Zinc solubilization in PVK (Pikovskaya's medium)

Zinc solubilization was studied in Pikovskaya's agar and liquid medium, a zone of inhibitions was obtained. Maximum zinc solubilization zone was observed in isolate MSS-ZF3 and MSS-ZF1 (90 mm), followed by MSS-ZF2 (80 mm), MSS-ZB1 (56 mm) MSS-ZB2 (47mm) MSS-ZB4 (45 mm) and MSS-ZB3 (44 mm) after incubation 14 days at 37 ° C for Bacteria and at 28 °C fungi (Figure1, 2).

The ability to dissolve appreciable amounts of zinc phosphate is not a common feature amongst the culturable bacteria of the surface soil samples. In contrast, many fungal isolates were able to produce visible clear haloes on the zinc phosphate-amended solid medium, but in only one case was the solubilization a result of bacterial activity. However, it is difficult, and not within the scope to extrapolate what the significance of this process is in the soil as it is widely recognized that only a small number of the members of bacterial soil communities are culturable with traditional isolation methods.

Optimization of Media and Growth Condition for Zinc Solubilization: Zinc solubilizing ability of bacterial strains was tested in four different types of agar media. The media selected were PVK, AYG, NBRIY and NBRiP, the maximum zone of solubilization was observed in PVK Medium, Followed by NBRIY, NBRiP, and AYG (Figure 3). From these observations as PVK medium was giving the optimum results among all four media, so PVK medium was selected for the further studies [21].

Effect of various Zinc source on efficiency of Zinc Solubilization: Various Zinc sources like Zinc Carbonate, Zinc Sulphate and Zinc Oxide, were studied in PVK Broth. The maximum zinc solubilization zone was observed in isolate MSS-ZF1 (90 mm), followed by MSS-ZF3 (89 mm), MSS-ZF2 (79 mm), MSS-ZB1 (57 mm) MSS-ZB2 (45 mm) MSS-ZB3 (45 mm) and MSS-ZB4 (44 mm) after incubation of 14 days at 37 °C for Bacteria and at 28 °C fungi.

Zinc solubilizing potential varied with each isolate, the ZSB-O-1 (*Bacillus* sp.) obtained from the zinc ore exhibited the highest potential in Sphalerite (ZnS) containing medium, producing a clearing zone of 2.80 cm. Its performance in zinc

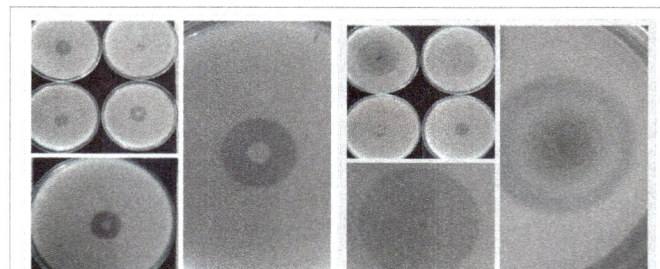

Figure 2: Zinc solubilization by the selected microbes.

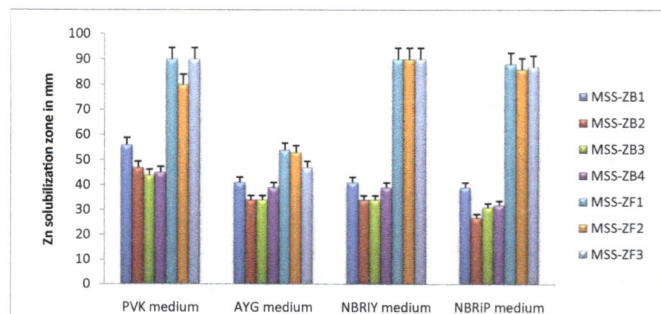

Figure 3: Effect of different medium on Zinc solubilization

carbonate and zinc oxide was 1.50 cm of clearing zone with zinc oxide, respectively. The ZSB-S-2 (*Pseudomonas* sp.) produced maximum clearing zone of 3.30 cm with zinc oxide and performed poorly in zinc carbonate, with a clearing zone of 2.00 cm. The ZSB-S-4 (*Pseudomonas* sp.) showed the highest potential in zinc carbonate, with a clearing zone of 4.00 cm [36].

Effect of different concentration of Zinc Oxide on efficiency of zinc Solubilization: Different concentration of Zinc Oxide was added in the PVK agar medium, 0.1%, 0.2%, 0.3%, 0.4% and 0.5%. The maximum zinc solubilization zone was observed at concentration 0.1% of ZnO, MSS-ZF1 (90 mm), followed by MSS-ZF3 (89 mm), MSS-ZF2 (79 mm), MSS-ZB1 (57mm) MSS-ZB2 (45 mm) MSS-ZB3 (45 mm) and MSS-ZB4 (44 mm) after incubation of 14 days at 37 ° C for Bacteria and at 28 °C fungi.

0.2% of concentration shows, the maximum zinc solubilization zone was observed at concentration 0.1% of ZnO, MSS-ZF3 (55 mm), MSS-ZF2 (52 mm), MSS-ZF1 (39 mm), MSS-ZB4 (39 mm) MSS-ZB1 (37 mm) MSS-ZB2 (34 mm) and MSS-ZB3 (34 mm) after incubation of 14 days at 37 ° C for Bacteria and at 28 °C fungi. No zone of solubilization was observed in concentration 0.3%, 0.4% and 0.5%. from the result, it is observed that the concentration above 0.2% ZnO seems to be inhibitory for the isolates so no zone of inhibition was observed (Figure 4).

The results have been obtained by Saravanam et al. that Even at 25 mg/kg concentration, there was reduction in population within 24 hours and afterwards population remained stable up to 8 days. At zinc concentration above 100mg/ kg, a further reduction in population was observed, which was more pronounced at 200 mg/ kg. The results showed the inherent capacity of the isolates to tolerate various levels of zinc. At 500 mg/kg level, ZSB-S-2 was completely inhibited at the 8th day, while ZSB-O-1 recorded 2 x 10^4 cells ml-1 at the 8th day after inoculation compared to 180 x 10^6 cells ml-1 observed just after inoculation [36].

Effect of various Carbon sources on efficiency of Zinc Solubilization: Various carbon sources like dextrose, glucose, sucrose, lactose, glycerol and xylose, were studied in PVK agar plate. The maximum zinc solubilization zone was observed in dextrose, followed by glucose, fructose, sucrose, lactose and glycerol.

In 1% Dextrose, Maximum zone of solubilization was observed in isolate MSS-ZF2 (57mm), followed by MSS-ZF1 (56mm), MSS-ZF3 (54 mm), MSS-ZB1(54 mm), MSS-ZB4 (49 mm), MSS-ZB3 (49 mm) and MSS-ZB2 (46 mm), after incubation of 14 days at 37 ° C for Bacteria and at 28 °C fungi (Figure 5).

In 1% Glucose, Maximum zone of zinc solubilization was observed in isolates MSS-ZF3 (54mm), followed by MSS-ZF1 (49 mm), MSS-ZF2 (47 mm), MSS-ZB4 (42 mm) MSS-ZB3 (36 mm) MSS-ZB1 (34 mm) and MSS-ZB2 (34 mm) after incubation of 14 days at 37 °C for Bacteria and at 28 °C fungi (Figure 5).

In 1% Sucrose zone of zinc solubilization was observed only in two isolates MSS-ZB2 (21 mm) and MSS-ZB2 (20 mm) after incubation of 14 days at 37 ° C for Bacteria and at 28°C fungi (Figure 5).

In 1% Lactose, Maximum zone of zinc solubilization was observed in isolate MSS-ZF1 (65 mm), MSS-ZF2 (45 mm), MSS-ZB4 (21 mm), MSS-ZB1 (19 mm) MSS-ZF3 (18 mm) MSS-ZB2 (18 mm) and MSS-ZB4 (12 mm) after incubation of 14 days at 37 °C for Bacteria and at 28 °C fungi (Figure 5).

In 1% Glycerol, Maximum zone of zinc solubilization was observed in isolate MSS-ZF1 (38 mm), MSS-ZF3 (29 mm), MSS-ZB4 (21 mm), MSS-ZB1 (19 mm), MSS-ZF2 (17 mm) and no zone of zinc solubilization was observed in MSS-ZB2 and MSS-ZB3, after incubation of 14 days at 37 ° C for Bacteria and at 28 ° C fungi (Figure 5).

In 1% Xylose, Maximum zone of solubilization was observed in isolate MSS-ZF3 (55 mm), MSS-ZF2(20 mm), MSS-ZB1 (10 mm), MSS-ZB3 (9 mm) after incubation of 14 days at 37 °C for Bacteria and at 28°C fungi, and no zone of zinc solubilization was observed in other isolates (Figure 5) [37].

The effect of different carbon sources on zinc phosphate dissolution by *Pseudomonas fluorescens* showed that the Glucose was found to be the only suitable carbon source for the occurrence of a clear halo around colonies on solid zinc phosphate-containing medium some solubilization was also observed with mannose [19].

Effect of various Nitrogen sources on efficiency of Zinc solubilization: Various Nitrogen sources like $(NH_4)_2SO_4$, Urea, Casein, and NaNO3 were studied in PVK Broth. Amongst all the nitrogen source zone of zinc solubilization was observed in $(NH_4)_2SO_4$ and no zone of solubilization was observed in other nitrogen source [37].

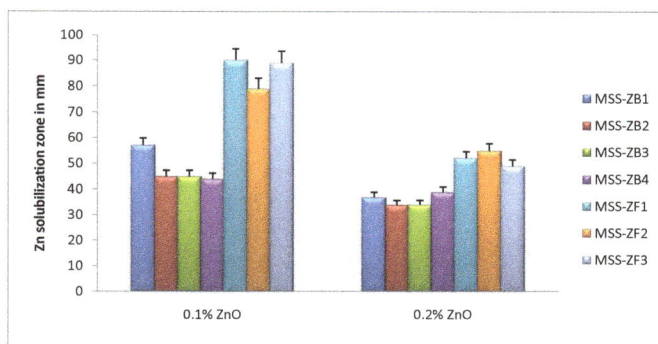

Figure 4: Effect of different ZnO Concentration on Zinc solubilization.

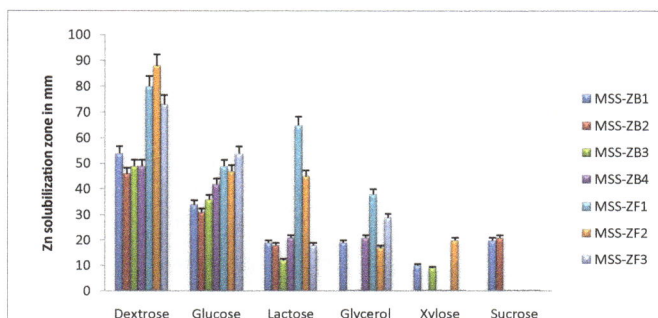

Figure 5: Effect of different carbon source on Zinc solubilization.

Effect of temperature on efficiency of Zinc solubilization: Media composition to which the bacteria responded best was used as substrate. Inoculation was carried out by using pure colony of a bacterial grown on Basal medium of isolates and allowed to grow and maintained at 8 °C, 15 °C, 28 °C, Room Temperature, 37 °C, and 55 °C for 14 days respectively [20]. The Zone of Solublization was Observed and measured in millimeter (Figure 6).

Effect of pH on efficiency of Zinc Solubilization: Optimal media and temperature was used, but the pH of the media was set at pH 4, pH 6, pH 6.5, pH 7, pH 9 using NaOH or HCl and grown for 14 days respectively [19]. The zone of solubilization was Observed and measured in millimeter (Figure 7).

Effect of different Salinity on efficiency of Zinc Solubilization: Optimal media and Conditions were used, but the saline concentration was added as NaCl(0.2%, 0.4%, 0.6%, 0.8% and 1%) and KCl (0.02%, 0.04%, 0.06%, 0.08% and 0.1%) in the media was set and grown for 14 days respectively (Figure 6,7) [20]. The Zone of solubilization was Observed and measured in millimeter (Figure 8,9).

Based on the optimization of media and growth condition results it was found that both bacterial and fungal cultures were able to grow and solubilize zinc optimum on carbon source 1% dextrose, nitrogen source Ammonium Sulphate and with ZnO (0.1% ZnO). The temperature of incubation was different; 37 °C was found optimum for bacteria and 30 ± 2 for fungi. From the pH study, it was observed that 7 pH was optimum for bacteria and 6.5 was optimum for fungi. Further Zinc estimation was performed

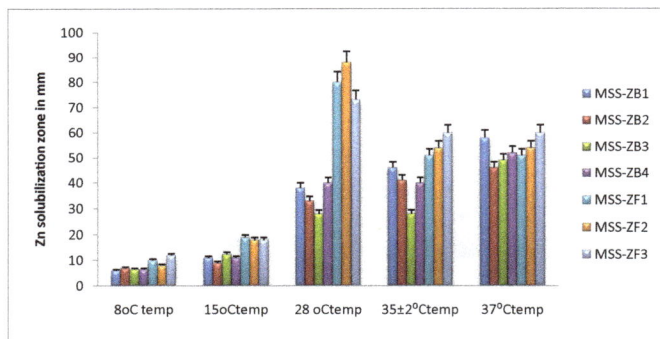

Figure 6: Effect of different temperature on Zinc solubilization.

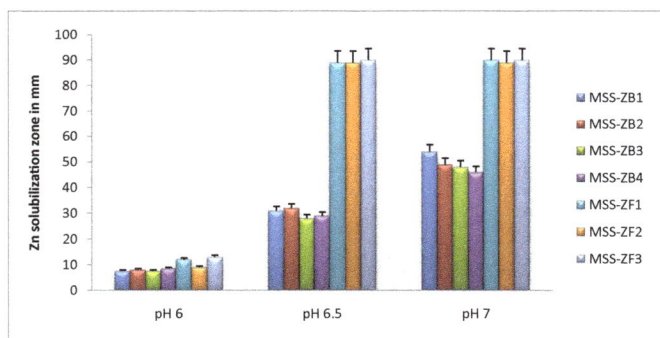

Figure 7: Effect of different pH on Zinc solubilization.

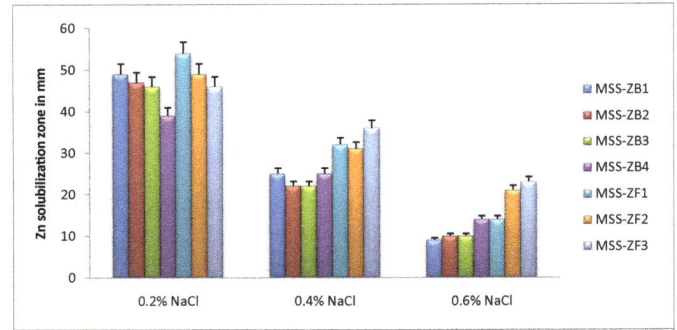

Figure 8: Effect of different NaCl concentration on Zinc solubilization.

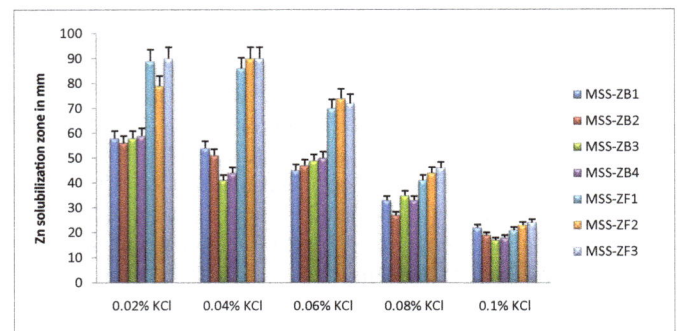

Figure 9: Effect of different KCl concentration on Zinc solubilization.

in the optimized media and growth condition in Pikovskaya's liquid broth amended with 0.1% ZnO.

The observations recorded during the growth of liquid cultures of *P. fluorescens* 3a by Di Simine et al are reported. Although the solid and liquid media contained different nitrogen sources, the microorganism was able to dissolve zinc phosphate in liquid medium, consistent with the observations on solid medium. Analysis of the supernatants, performed by AAS and voltammetry, showed an increase in the concentration of soluble Zn up to values of about 7 mM [19]. Such an increase occurred without a meaningful difference between the free Zn and the total zinc concentrations, suggesting the absence of complexation phenomena involving the Zn in solution [19].

Di Simine et al also reported that during the time course of the experiment, bacterial proliferation occurred concurrently with a drop in the pH of both the control and the zinc phosphate-supplemented cultures, growth of the cultures was complete within 24 h, the pH at this time reaching a value of about 4.5 [19]. The pH subsequently remained constant in the control culture, whereas a further slow decrease to values closer to pH 4 was observed in the zinc phosphate-supplemented culture [18]. Same decrease in the pH was also observed in cultures which showed a shift in pH after growth in the broth. After 15 days, the pH of the broth was acidic in all cultures. The pH shifted from 7-7.3 to 4.8-6.5. The ZSB-S-4 culture showed the lowest pH value (4.8) on 15th day after inoculation, indicating a higher acidity due to growth [37].

Conclusion

In the present investigation, the application of PGPR had been studied for the Zinc solubilizing ability. The ability to dissolve appreciable amounts of zinc oxide was not a common feature amongst the cultivable bacteria of the surface soil samples examined in the present investigation. In contrast, three fungal isolates and four bacterial isolates were able to produce visible clear haloes on the zinc oxide amended solid medium. Amongst all the seven isolates (MSS-ZB1, MSS-ZB2, MSS-ZB3, MSS-ZB4, MSS-ZF1, MSS-ZF2 and MSS-ZF3) MSS-ZB4 and MSS-ZF3 were showing best suitable PGPR characteristics for the plant growth promotion. Considering the plant growth promoting abilities and zinc solubilizing abilities of strains, biofertilizer preparation is possible. Thus, our strains MSS-ZB4 and MSS-ZF3 can be further use as Plant growth promoting rhizobacteria for improvement of micronutrient deficiency will be promising due to its ecological, economic and ecofriendly nature.

Acknowledgement

We are thankful to UGC-Maulana Azad National Fellowship for their financial assistance.

Reference

1. Glick BR. The enhancement of plant growth by free-living bacteria. Can J Microbiol, 1995;41(2):109-117.

2. Mayak S, Tirosh T, Glick B R. Effect of wild-type and mutant plant growth-promoting rhizobacteria on the rooting of mung bean cuttings. J plant growth regul. 1999;18(2): 49-53.

3. Broadley MR, White PJ, Hammond JP, Zelko I, Lux A. Zinc in plants. New Phytol. 2007;173(4):677-702.

4. Hambidge KM, Krebs NF. Zinc deficiency: a special challenge. J Nutr. 2007;137(4): 1101-1105.

5. Hirschi K. Nutritional improvements in plants: time to bite on biofortified foods. Trends Plant Sci. 2008;13(9):459-463. doi: 10.1016/j.tplants.2008.05.009

6. Cakmak I. Enrichment of cereal grains with zinc: agronomic or genetic biofortification? Plant and soil.2008;302(1):1-17. doi: 10.1007/s11104-007-9466-3

7. Pfeiffer WH, McClafferty B. HarvestPlus: breeding crops for better nutrition. Crop Science.47:88 doi:10.2135/cropsci2007.09.0020IPBS

8. Pahlavan Rad M R., &Pessarakli M. Response of wheat plants to zinc, iron, and manganese applications and uptake and concentration of zinc, iron, and manganese in wheat grains. Commun Soil Sci Plant Anal. 2009; 40(7-8):1322-1332. doi: 10.1080/00103620902761262

9. Wani P A, Khan M S, Zaidi A. Effect of metal-tolerant plant growth-promoting Rhizobium on the performance of pea grown in metal-amended soil. Arch Environ Contam Toxicol. 2008;55(1):33-42.

10. Hafeez F Y, Naeem F I, Naeem R, Zaidi A H, & Malik K A. Symbiotic effectiveness and bacteriocin production by Rhizobium leguminosarum bv. viciae isolated from agriculture soils in Faisalabad. Environmental and experimental botany. 2005;54(2):142-147.

11. Abaid Ullah M, Hassan MN, Jamil M, Brader G, Shah MKN, Sessitsch A,et.al. Plant growth promoting rhizobacteria: an alternate way to improve yield and quality of wheat(Triticum aestivum). Int. J. Agric. Biol.2015; 17:51-60.

12. Saharan B S, Nehra V. Plant growth promoting rhizobacteria: a critical review. Life Sci Med Res. 2011;21(1):30.

13. Gaur, A. C. (1990). Phosphate solubilizing micro-organisms as biofertilizer. Omega scientific publishers.

14. Schwyn B, Neilands JB. Universal chemical assay for the detection and determination of siderophores. Anal Biochem.19871;60(1): 47-56.

15. Sarwar M, Kremer R J. Determination of bacterially derived auxins using a microplate method. Letters in applied microbiology.1995;20(5):282-285. doi: 10.1111/j.1472-765X.1995.tb00446.x

16. Cappucino JC, Sherman N. (1992). Nitrogen cycle. Microbiology: a laboratory manual, 4th edn. Benjamin/Cumming, New York, 311-312.

17. Askeland R A, Morrison S M. cyanide production by Pseudomonas fluorescens and Pseudomonas aeruginosa. Appl Environ Microbiol.1983;45(6):1802-1807.

18. Mody B, Bindra M, Modi, V. Extracellular polysaccharides of cowpea rhizobia: compositional and functional studies. Arch Microbiol.1989;153(1):38-42. doi: 10.1007/BF00277538

19. Di Simine CD, Sayer JA, Gadd GM. Solubilization of zinc phosphate by a strain of Pseudomonas fluorescens isolated from a forest soil. Biol Fertil Soils. 1998;28(1):87-94.

20. Fasim F, Ahmed N, Parsons R, Gadd GM. Solubilization of zinc salts by a bacterium isolated from the air environment of a tannery. FEMS microbiology letters. 2002;213(1):1-6.

21. Bapiri A, Asgharzadeh A, Mujallali H, Khavazi K, Pazira E. Evaluation of Zinc solubilization potential by different strains of Fluorescent Pseudomonads. Journal of Applied Sciences and Environmental Management. 2012: 16(3):295-298.

22. Amanul HN. Substrate utilization profiles and functional diversity of microbial fractions in organic farming soil of Tharad. 2015.

23. Gyaneshwar P, Kumar G N, Parekh, LJ, Poole PS. Role of soil microorganisms in improving P nutrition of plants. In Food Security in Nutrient-Stressed Environments: Exploiting Plants' Genetic Capabilities. 2002 133-143.

24. Zimmer W, Roeben K, Bothe H. An alternative explanation for plant growth promotion by bacteria of the genus Azospirillum. Planta.1988;176(3):333-342. doi: 10.1007/BF00395413

25. Bhattacharyya R N, Pati BR. Growth behaviour and indole acetic acid (IAA) production by a Rhizobium isolated from root nodules of Alysicarpus vaginalis DC. Acta Microbiol Immunol Hung. 2000;47(1):41-51.

26. Torres-Rubio MG, Valencia-Plata SA, Bernal-Castillo J, Martínez-Nieto P. Isolation of Enterobacteria, Azotobacter sp. and Pseudomonas sp., producers of indole-3-acetic acid and siderophores, from Colombian rice rhizosphere. Rev Latinoam Microbiol. 2000;42(4):171-176.

27. Christiansen-Weniger C, Van Veen JA. NH4+-excreting Azospirillum brasilense mutants enhance the nitrogen supply of a wheat Host. Appl Environ Microbiol. 1991;57(10):3006-3012.

28. Alami Y, Achouak W, Marol C, Heulin T. Rhizosphere soil aggregation and plant growth promotion of sunflowers by an exopolysaccharide-producing Rhizobiumsp. Strain isolated from sunflower roots. Appl Environ Microbiol. 2000;66(8):3393-3398. doi: 10.1128/AEM.66.8.3393-3398.2000

29. Cattelan AJ, Hartel PG, Fuhrmann JJ. Screening for plant growth–promoting rhizobacteria to promote early soybean growth. Soil

Science Society of America Journal. 1999;63(6):1670-1680. doi:10.2136/sssaj1999.6361670x

30. Chandra S, Choure K, Dubey RC, Maheshwari DK. Rhizosphere competent Mesorhizobiumloti MP6 induces root hair curling, inhibits Sclerotinia sclerotiorum and enhances growth of Indian mustard (Brassica campestris). Braz J Microbiol. 2007;38(1):124-130. doi:10.1590/S1517-83822007000100026

31. Diby P. Chapter 2.In: Phisiological, biochemical and moleculer studies on the root rot (caused by Phytophthora capsici) suppression in black pepper (Piper nigrum L.) by rhizosphere bacteria. Disertasi. University Calicut. India. 2004.

32. Haggag WM, Mohamed HAA. Biotechnological aspects of microorganisms used in plant biological control. American-Eurasian Journal of Sustainable Agriculture. 2007;1(1):7-12.

33. Kumar A, Sharma S. An evaluation of multipurpose oil seed crop for industrial uses (Jatropha curcas L.): a review. Ind Crops Prod. 2008;28(1):1-10. doi:10.1016/j.indcrop.2008.01.001

34. Joseph B, Ranjan Patra R, Lawrence R. Characterization of plant growth promoting rhizobacteria associated with chickpea (Cicer arietinum L.). International Journal of Plant Production.2007;1(2):141-152. doi: 10.22069/ijpp.2012.532

35. Saravanan V S, Subramoniam S R, Raj SA. Assessing in vitro solubilization potential of different zinc solubilizing bacterial (zsb) isolates. Braz J Microbiol.2004;35(1-2):121-125. doi: 10.1590/S1517-83822004000100020

36. Sagervanshi A, Kumari P, Nagee A, Kumar A. Media optimization for inorganic phosphate solubilizing bacteria isolated from Anand agriculture soil. IJLPR.2012;2(3):245-255.

37. Kloepper JW, Leong J, Teintze M, Schroth MN. Enhanced plant growth by siderophores produced by plant growth-promoting rhizobacteria. Nature.1980;286: 885-886.

38. Mody B, Bindra M, Modi, V. Extracellular polysaccharides of cowpea rhizobia: compositional and functional studies. Arch Microbiol.1989;153(1):38-42. doi: 10.1007/BF00277538

16

Small RNA Extraction using Fractionation Approach and Library Preparation for NGS Platform

Spandan Chaudhary[1]*, Pooja S. Chaudhary[2] and Toral A. Vaishnani[3]

[1]*Department of Medical Genetics, Xcelris Labs Limited, Old Premchandnagar Road, Opp. Satyagrah Chhavani, Bodakdev, Ahmedabad-380015, Gujarat, India*
[2]*NGS department, Xcelris Labs Limited, Old Premchandnagar Road, Opp. Satyagrah Chhavani, Bodakdev, Ahmedabad-380015, Gujarat, India*
[3]*Bioinformatics department, Xcelris Labs Limited, Old Premchandnagar Road, Opp. Satyagrah Chhavani, Bodakdev, Ahmedabad-380015, Gujarat, India*

*Corresponding author: Dr. Spandan Chaudhary, Department of Medical Genetics, Xcelris Labs Limited, Old Premchandnagar Road, Opp. Satyagrah Chhavani, Bodakdev, Ahmedabad-380015, Gujarat, India, E-mail: spandan.chaudhary@gmail.com

Abstract

Small RNA isolation is a herculean task; it requires lots of standardization at each and every step. In present study, we have combined two protocols of total RNA isolation and standardized the protocol for small RNA isolation using flash page fractionation method. Due to difficulty in isolating high quality small RNA, most technologies use total RNA as a starting material for preparing small RNA library. Quality of both total and small RNA was determined using scientifically proven technology like Agilent bioanalyzer. In present study we have demonstrated a protocol for total and small RNA isolation from rice, followed by small RNA library preparation using Solid total RNA seq kit protocol for NGS platform (For SOLiD platform). Small RNA library was of good quality as per the parameters given in the protocol. We have validated the procedure four times which resulted good data, it concludes these procedures can be use to isolate high quality small RNA to be used for deep sequencing.

Keywords: miRNA; Small RNA; Flash page fractionators; Small RNA library; NGS; SOLiD

Introduction

Small RNA pairs with their target messenger RNA molecules and to suppress the gene expression in order to control the gene expression. miRNAs play a major role in cell proliferation, cell cycle, cell differentiation, metabolism, apoptosis, developmental timing, neuronal cell fate, neuronal gene expression, brain morphogenesis, muscle differentiation and stem cell division. The miRNAs are small, highly conserved RNA molecules that act as key regulators of development, cell proliferation, differentiation, and the cell cycle. Emerging evidence also implicates miRNAs in the pathogenesis of human diseases such as cancers, metabolic diseases, neurological disorders, infectious diseases and other illnesses [1-2]. Small RNAs have been classified into at least six groups, which are microRNAs (miRNAs), heterochromatic small interfering RNAs (hc-siRNAs), trans -acting small interfering RNAs (tasiRNAs), natural antisense small interfering RNAs (nat-siRNAs), repeat-associated small interfering RNAs (ra-siRNAs), and in metazoans, the piwi-interacting RNAs (piRNAs) [3-7]. The active mature miRNAs are typically 21-24 nucleotides,

single stranded RNA molecules expressed in eukaryotic cells. Small RNAs are 21-24 nucleotides in length and are known to play a major role in the activation of mRNAs and genomic DNAs [8]. 21-nucleotide microRNAs (miRNAs) and 24-nucleotide Pol IV-dependent small interfering RNAs (p4-siRNAs) are the most abundant types of small RNAs in angiosperms [9]. Some miRNAs are well conserved among different plant lineages; whereas others are less conserved, and it is not clear whether less-conserved miRNAs have the same functionality as the well conserved ones [1]. Recent whole genome sequencing data indicates that the 5'- and 3'- ends of miRNAs are variable, in which 5'end is less variable than 3' end [10]. These alternative length miRNAs are called isomers, and their biological function is unknown, miRNAs are known to affect the translation and/ or stability of the target messenger RNAs [11]. Each miRNA is believed to regulate multiple genes, and it is currently thought that greater than one third of all human genes may be regulated by miRNA molecules.

Small RNA was first discovered by David's Baulcombe's group at the Samsbury laboratory in Norwich, England, as a part of post transcriptional gene silencing in plants. RNA interference invariably leads to gene silencing via remodeling chromatin to thereby suppress transcription, degrading complementary miRNA or blocking protein translation [12]. Small RNAs are naturally produced as part of the RNA interference (RNAi) pathway by the enzyme dicer. SiRNAs are short double stranded RNA with 2 nucleotide overhangs on either end including a 5'-phosphate group and a 3'- hydroxyl(-OH) group [13].

Rice is a staple food in India and a very important part of appetite of the entire world. It is one of the most important cereal and model monocot plant [14,15]. About three billion of the population depends on rice for their daily calorie needs [16]. Plenty of research work is going on and rice genome is also available, so it's comparatively easy to identify and study the siRNAs. There are two species: *Oryza sativa and Oryza glaberrima*, first being the most commonly grown throughout Asia, Australia, the Americas and Africa; and second species is grown on a small scale in western Africa [17]. In present study, we have standardized the protocol for small RNA extraction from 4 rice samples, suitable for downstream application like small RNA sequencing using next generation sequencing technologies like Miseq, NextSeq, HiSeq, Solid, Ion PGM etc. But, we have demonstrated the standardized protocol of preparing libraries for Solid analyzer. In any of the next generation platform for getting the best data from small RNA sequencing, quality of small RNA is very crucial criteria. Because of the complexity of the small RNA isolation protocols, Illumina is referring poly acryl amide gel based method to extract small RNA from adapter ligated total RNA [18].

Small RNA can be isolated from any tissue or plants directly using commercially available kits like mirVana™ miRNA Isolation Kit (Thermo scientific), miRNeasy Mini Kit (Qiagen) etc uses dual column based method in which small RNA is bound to second column. There is also a kit available which directly captures small RNA from tissue like mirPremier microRNA Isolation Kit (Sigma Aldrich). Common chemistry shared by all this kit is pH based binding of nucleic acid to the silica column and these columns are made to bind certain specific size of nucleic acids. On the other hand, flash PAGE Fractionator works on the principle of gel electrophoresis which isolates the molecules on the bases of the size which helps in efficient small RNA extraction. Column based methods isolate small RNA upto 200 nucleotides which captures data in deep sequencing technologies where as small RNAs like piRNA or siRNA are of 21-24 nucleotides long which are the desired target for any study, can be isolated precisely using flash page system.

Materials and Methods

Sample collection

Prior to the harvesting, juvenile leaves of rice were washed with DEPC (Diethylpyrocarbonate- D5758, Sigma) treated water to remove surface contamination at sample collection site Jetalpur, Ahmedabad (Gujarat) India and were transported in TMS RNA stabilizer solution (XGgtms-100) with dry ice followed by storage at -80°C.

Total RNA isolation

Rice leaves were thawed at room temperature 100 mg and and taken for RNA isolation using pure Link miRNA isolation kit (Invitrogen- cat. no. K1570-01) and pure Link RNA Micro kit (Invitrogen- cat. no. 12183-016) as per manufacturer's protocol. Further, combination of both these kit were used to get high quality and yield of total RNA. Total RNA isolation was done in triplicate using three different methods. Combination of both these kit's methods are as follow:

100 mg of plant leaves were grounded to fine powder in motor and pestle with liquid nitrogen. 300 µl of binding buffer (L3) (supplied with miRNA isolation kit) was added to powder in mortar pestle and ground it gently to mix properly till it become complete homogenous solution. Lysate was centrifuged at 12000 g for 2 minutes at room temperature to remove any particular material. Supernatant was transferred to another sterile micro centrifuge tube and 300 µl of 70% of ethanol was added to lysate

and vortexed to mix well. Whole lysate volume 600 µl was loaded on a new spin cartridge and centrifuged it at 12000 g for 1 minute at room temperature. 350 µl of Wash Buffer I was added to spin cartridge and centrifuged at 12000 g for 15 seconds at room temperature and flow through was discarded. In a separate 0.2 ml tube 70 µl of DNase buffer and 20 µl of DNase-I enzyme (Thermo scientific cat. no. ENO525) was taken (Total 80 µl) and mixed by pipetting and kept on ice. DNase I mixture (80 µl) was added to the center of the cartridge and kept on ice for 15 minutes. After 15 minutes again 350 µl of Wash Buffer I was added to the center of the spin cartridge and centrifuged at 12000 g for 15 seconds at room temperature and flow through was discarded. Spin cartridge was transferred to a clean RNA wash tube, provided with the kit and 500 µl of Wash Buffer II with ethanol was added. Spin cartridge was centrifuged at 12000 g for 15 seconds at room temperature and flow through was discarded. Above two steps were repeated and spin cartridge was centrifuged at 12000 g for 1 minute at room temperature to dry the membrane. Collection tube was discarded and cartridge was inserted into RNA recovery tube supplied with the kit. 30 µl of RNase- free water was added to the center of spin cartridge and incubated at room temperature for 1 minute then centrifuged for 2 minute at ≥ 12000 g at room temperature to elute the total RNA. Above step was repeated with 30 µl of RNase- free water with same collection tube.

Quantitative and qualitative determination of RNA

Concentration and purity of total RNA samples were measured using the Nano Drop ND3.0 spectrophotometer (NanoDrop Technologies Inc, Wilmington, DE). For preparing 1% denaturing agarose gel, 0.5 g of agar powder was added to 50 ml of DEPC treated water and boiled till it melted, followed by adding 8.75 ml of formaldehyde and 5 ml of 10X mops. Gel was run at 90Volts for about 45 minutes (Figure 1). RNA integrity was assessed using nano chip on Agilent 2100 Bio analyzer (Agilent Technologies, Palo Alto, CA) (Figure 2). The gel image and bio analyzer profile with RIN value were analyzed to proceed with small RNA enrichment step.

Small RNA enrichment

Flash page fractionator (Ambion cat no.13100) (Figure 3) is a miniaturized version of poly acryl amide gel electrophoresis for isolating small RNA from total RNA samples. For the enrichment of small RNA using flash page fractionator system, four other materials are required.

1. Flash page Pre-Cast gel (Type A) (Cat no. 10010- ambion)

2. Flash page buffer kit (Type A) (Cat no. 9015- ambion)

3. ElectroZap™ Electrode Decontamination Solution (Cat #9785)

4. Flash page Reaction Clean-up Kit (cat no.Am12200).

60µg of high quality (RIN value above 6) total RNA was taken with loading buffer and dye, heat denatured and loaded onto the upper gel surface in the flash page fractionator, electrophoresed at 80V constant voltage until the blue dye begins to exit the gel. As per manufacture's protocol this time should be around 12

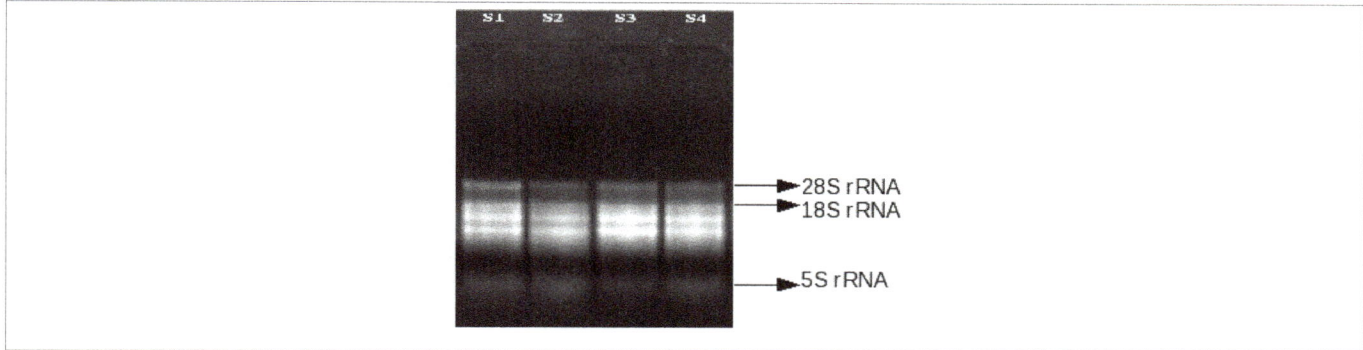

Figure 1: Agarose denaturing gel profile of total RNA isolated by combined pureLink miRNA isolation kit (invitrogen) and pureLink RNA Micro kit. Sample-1, 2, 3, 4 were loaded in lane 1, 2, 3, 4 respectively.

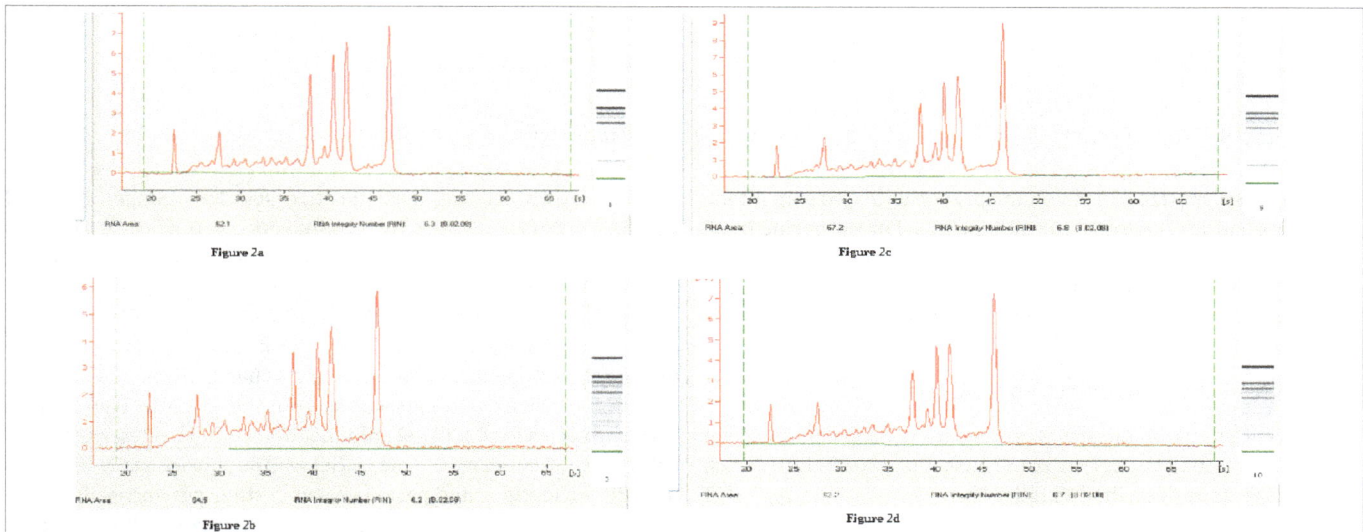

Figure 2: Total RNA profiling of samples 1, 2, 3 and 4 as figures 2a, 2b, c and 2d respectively, run on Agilent bioanalyzer nano chip.

minutes to obtain small RNAs with length of 15-140 nucleotides in the lower reservoir of the flash page system. As the selection of the small RNA is very important aspect of small RNA library preparation this step has also similar importance. The time described in the manufacturer protocol is mainly depends upon the amount and quality of total RNA used, we had standardized the time for small RNA isolation is 13-14 minutes with the supplied pre-cast gels. The time required to obtain small RNA in the lower reservoir depend upon age of the pre-cast gel, as it is near to the expiry it will take more time which can be ranging from 12-15 minutes. Further, small RNAs were purified using flash page reaction clean-up kit (Ambion) as per manufacturer's protocol and small RNA profile was checked on Agilent 2100 bio analyzer using small RNA chips (cat no.5067-1548) (Figure 4). Handling of small RNA reagents for running on bio analyzer chip is difficult because it uses highly concentrated gel to separate small RNA fragments which are 21-24 nucleotides in length. Usually bioanalyzer protocols recommend use of filtered gel matrix within one month of time but it is not suitable with small RNA gel matrix. If stored gel matrix is used it will not detect the chip on bioanalyzer which was very usual case with small RNA chips the reason behind this is electrode cannot sense the gel matrix of the chip. To overcome this we used silica column (Nanosep centrifugal Devices with Omega Membrane, cat. no.29300-610) to prepare fresh gel matrix as agilent kit provides

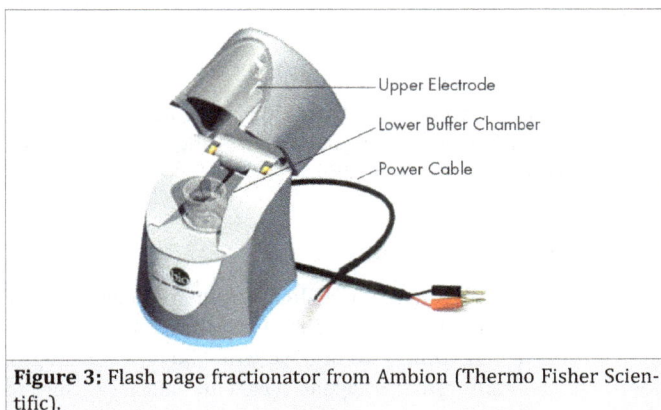

Figure 3: Flash page fractionator from Ambion (Thermo Fisher Scientific).

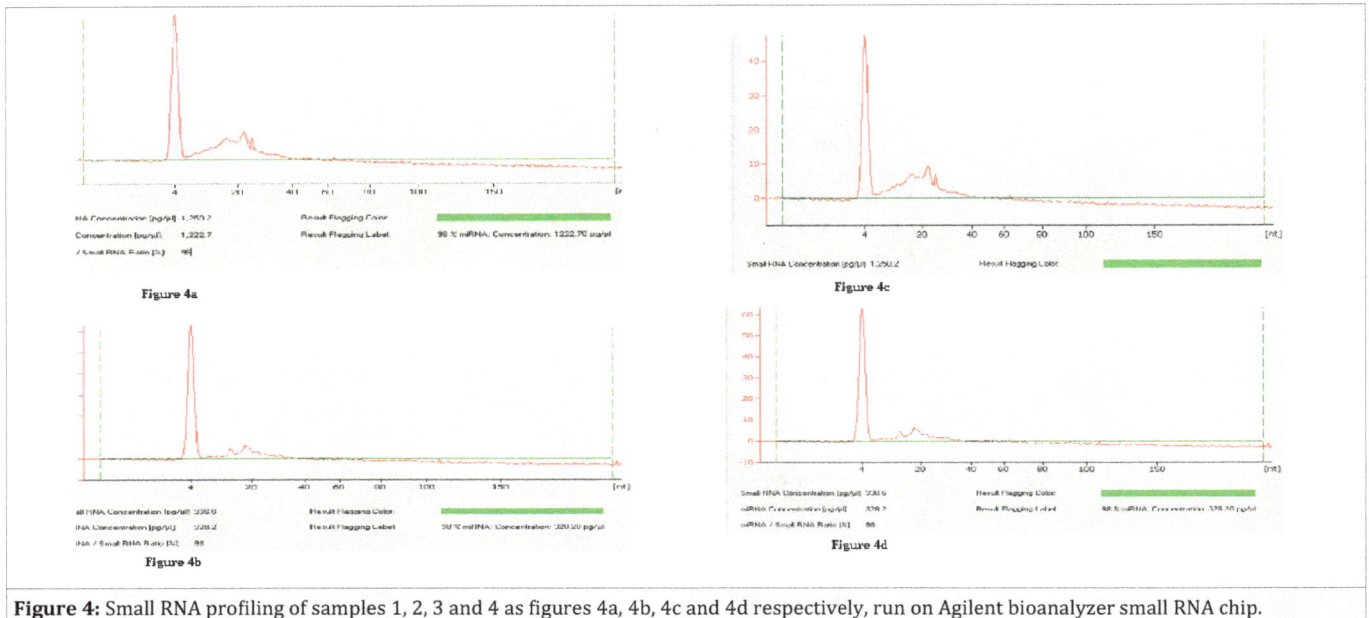

Figure 4: Small RNA profiling of samples 1, 2, 3 and 4 as figures 4a, 4b, 4c and 4d respectively, run on Agilent bioanalyzer small RNA chip.

two vials of gel matrix 650 µl volume in each with two spin filters, however quantity is enough to run 25 chips. For preparing fresh gel matrix 45 µl of gel was filtered each time through the column. Two micro liter dye (provided in the kit) was added to 40 µl of filtered gel and rest protocol was followed as per manufacturer's instructions [19].

Small RNA cDNA library preparation

Small RNA cDNA library was prepared using Solid™ Total RNA-Seq Kit (Part Number 4452437 Rev. A 01/2010) [20] which involve five steps described as follows:

Hybridization and adapter ligation

The 5'- adapter ligation reaction was carried out in 20 µl reaction containing 80-100 ng (as per nanodrop readings) of enriched small RNA in 3 µl volume, 3 µl hybridization solution, 2 µl solid adapter mix, 8 µl of 2X ligation buffer and 2 µl of ligation enzyme. Reaction mixture was incubated at 16 degree for 16 hrs in a thermal cycler with heated lid open.

RT-PCR of adapter ligated small RNAs

After 16 hours of incubation, reverse transcription was carried out using 20 µl reaction, containing 4 µl 10X RT buffer, 2 µl dNTP mixes, 2 µl solid RT primer and 11 µl nuclease free water. The reaction mixture was heat denatured at 70 degree for 5 minutes followed by snap chilling then 1 µl array script reverse transcriptase enzyme was added in the reaction mixture and was incubated at 42 degree for 30 minutes resulting into the formation of single stranded cDNA followed by purification by using minelute PCR purification kit (Qiagen). The single stranded cDNA was quantified by using nanodrop 8000 spectrophotometer.

Size selection for library preparation

10% TBE urea gel 1.0 mM, 10 wells (invitrogen cat

no.EC68752BOX) was used for size selection. Single stranded cDNA samples; 100ng of 10bp ladder (Cat no.10821015 invitrogen) at the concentration of 40 ng/ µl was loaded in the gel. The gel was run in x cell sure lock mini cell gel assembly using 1X TBE running buffer at 180 volts for 47 minutes instead of 45 minutes recommended in the protocol. Gel was stained for 10 minutes with gentle agitation using syber gold nucleic acid gel stain. A continuous smear was seen in the SS cDNA well (Figure 5). The cycles for the in gel amplification would be determined on the basis of intensity of smear. Gel was cut horizontally between 60-80 nucleotides and gel slices was further cut vertically into 4 pieces of around 1mm X 6mm using clean blade. All four pieces of the gel slices were transferred to fresh 0.2ml RNAs free tubes.

In Gel PCR amplification and purification

Out Of four gel slices, 2nd and 3rd gel slices were used first for in gel PCR amplification. This 100 µl reaction contains 10 µl of PCR buffer, 8 µl 2.5 mm dNTP, 2 µl solid 5' primer, 2 µl solid 3' primer, 1.2 µl amplitaq DNA polymerase and 76.8 µl nuclease free water and the gel slice as template DNA. 18 cycles of amplification was carried out and checked on E-gel 2% size select gel (cat. no.G6610-02 Invitrogen).

Quantification and qualification of cDNA library

This cDNA library was purified using pure link PCR micro kit (Invitrogen) and further, quantified and qualified using 2100 bio analyzer with agilent high sensitivity (HS) DNA kit (cat no. 5067-4626) (Figure 6).

Result and Discussion

Ethanol precipitation method along with nucleic acid carrier such as glycogen is recommended for extracting total RNA with sizes of RNA having from ~ 20 nt to several kilo bases in length. For RNA isolation, either reagent based or silica membrane

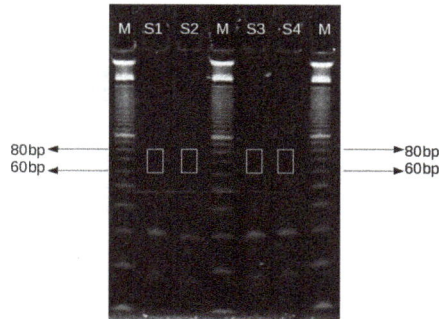

Figure 5: Gel image of size selection using10% TBE urea gel 1.0 mM, 10 wells invitrogen cat no. EC68752 along with 10bp Invitrogen ladder in well labeled as M, samples are labeled as S1, S2, S3 and S4.

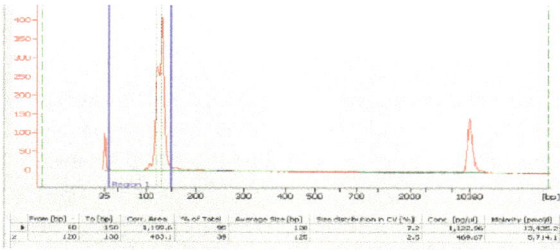

Figure 6a

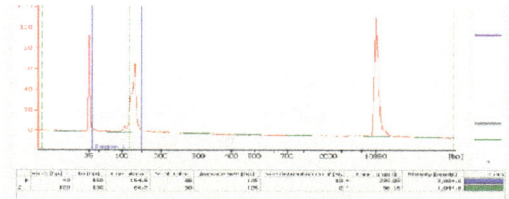

Figure 6c

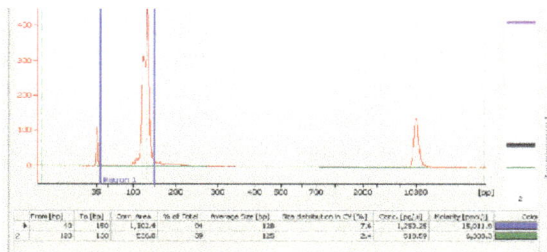

Figure 6b

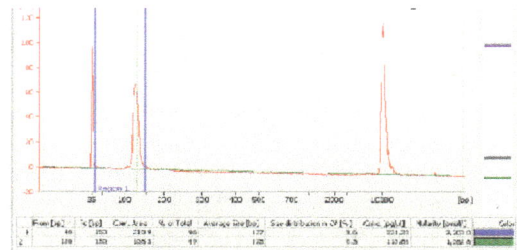

Figure 6d

Figure 6: Small RNA library profile of samples 1, 2, 3 and 4 as figures 6a, 6b, 6c and 6d respectively, run using high sensitivity DNA chip on Agilent bioanalyzer.

based methods are generally used but it was found that small RNAs cannot be isolated using silica column based method instead trizol reagent based method is more suitable for the same [8], but we extracted total RNA using a modified silica column based method and isolated high quality total RNA with RIN value above 6, from this total RNA samples small RNA was isolated and checked on Agilent 2100 bioanalyzer. (Agilent Technologies, Santa Clara, Palo Alto, CA, USA). For getting good data by deep sequencing the most important parameter is the good quality library preparation which depends on the quality of total RNA and small RNA. RNA degradation during isolation procedure or loss of small RNA is the main reason for failure to obtain a good library. Absorbance 260/280 ratio was found to be 1.9, with the total yield of 90 micrograms, total RNA was loaded on 1% denaturing agarose gel, 2 distinct bands were observed, first band was of 28s rRNA and second band was of 18s rRNA (Figure 1). In Bio-analyzer profile, distinct peaks of 28s and 18s rRNA were visible along with a small peak near the lower marker

indicating 5s rRNA region.

The quality of total RNA and selection of small RNA enrichment procedure play a momentous role in the small RNA cDNA library profiling. Total RNA having RIN value below 6 normally could not be used for small RNA enrichment, because such RNAs are found to be degraded. Several kits are available in the market that isolate small RNA directly from the tissues which don't require total RNA isolation to be done separately like Invitrogen miRNA isolation kit. But a drawback of this small RNA isolation method is that along with small RNA mRNAs and some other RNAs get also isolated which hinders in library preparation process and grab the data in the deep sequencing process. Flash page fractionators works as a vertical gel unit, it has negative electrode on the top and positive electrode at the bottom, RNA will migrate from upper side to the lower side. Same like normal gel electrophoresis principle, smaller fragments will migrate faster than the longer ones. When total RNA degradation occurs, 28s ribosomal RNA 18s ribosomal RNA band degrades

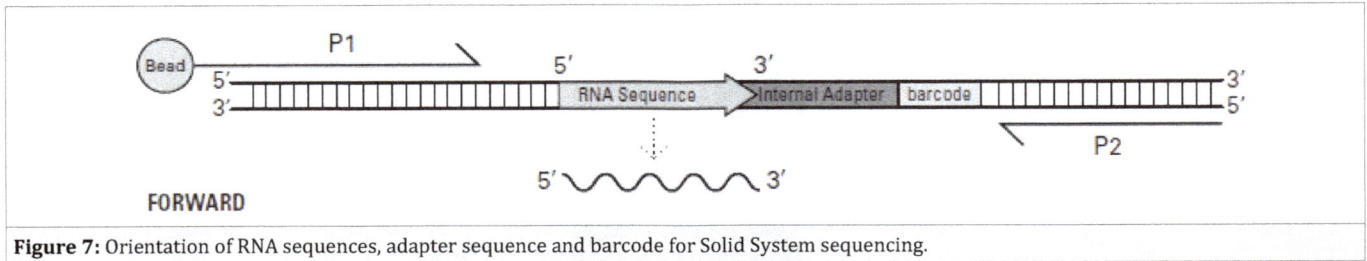

Figure 7: Orientation of RNA sequences, adapter sequence and barcode for Solid System sequencing.

Table 1: Input amount of starting material and amount of resulted product.

No.	Steps	Starting Material	End Product (Average)
1	Total RNA isolation	100 mg leaves	60 μg total RNA
2	Small RNA enrichment	60 μg total RNA	350 ng small RNA
3	Hybridization and adapter ligation	100 ng of enriched small RNA	155 ng single stranded cDNA
4	cDNA library	Gel slices	81 ng

and produces small size ribosomal RNA fragments [21] which comes out with small RNA when pass through flash gel which ultimately leads to failure of the library or contamination of small RNA library with ribosomal reads. In order to avoid this, intact total RNA without any degradation is advised to use for preparing small RNA libraries even with flash page fractionators system.

NGS library preparation process requires optimization at each and every step from nucleic acid isolation to library quality check. Here, we have tried to provide the accurate data for how much starting material is required at each step and how much end product will be generated which is the principle missing point in all the available protocols, these will help the readers to repeat the same experiment. For small RNA enrichment by flash page fractionators, up to 100 μg of high quality total RNA with RIN value of above 6 is advice. In present study, we started with 60 μg of high quality total RNA having RIN value 6.2 to 6.8, and after purification we obtained 350 ng of enriched small RNA. Small RNA profiles were analyzed using Agilent small RNA chip (shown in figure 4.) and the amount of small RNA was calculated and 100 ng small RNA was taken for adapter ligation.

The RNA samples are hybridized and ligated with the adapters containing set of oligonucleotides with a single-stranded degenerate sequence at one end and defined sequence required sequencing at the other end. The adapters ligate in a manner that it provides template for sequencing from the 5' end of the sense strand. The downstream emulsion PCR primer alignment and the resulting products of template bead preparation for Solid System sequencing are illustrated in (figure 7) Accurate size selection is very important for any library preparation for deep sequencing. Size selection was done using 10 bp ladders as a reference; the gel was cut between 60-80 nucleotides. The insert length of the adapters is 18-38 bp, so the length of the amplified PCR product will be 110-130 bp but ideally small RNA library bands obtained within the range of 120-130 bp.

According to the protocol Solid TM Total RNA-Seq kit, criteria for small RNA cDNA library to be used for sequencing run

depends upon the values obtained from the smear analysis done on DNA 1000 chip run. In present study we have used DNA High sensitivity chip so we modified the smear analysis criteria also. As per protocol, the library will be used for sequencing only if the ratio of 120-130 bp area / 25-150 bp area is more than 50%, if this ratio is less than 50%, second round of size selection will be required. But as we have used high sensitivity chip, in which lower marker peak is at 35 bp instead of 15 bp in DNA 1000 chip, we have calculated the ratio 40-150 bp area. High sensitivity chip was used instead of DNA 1000 chip. In case of plants, it is very difficult to obtain small RNA libraries with >50% ratio but in the present study we have achieved 52% ratio in one of our four samples.

Conclusion

Total RNA isolation protocol prepared by combining two different kits worked efficiently and provided high quality total RNA. Small RNA isolated from total RNA using flash page fractionator was of the good quality to be used for preparing small RNA libraries for all the NGS platforms. The libraries prepared by modified method were used for sequencing using Solid analyzer and generated good quality data (data not provided) which proves the modification applied at each stage of library preparation are acceptable.

Acknowledgment

Authors' are thankful to the management of Xcelris Labs for providing financial support.

References

1. Nobuta K, McCormick K, Nakano M, Meyers BC. Bioinformatics analysis of small RNAs in plants using next generation sequencing technologies." Methods Mol Biol. 2010;592:89-106. Doi: 10.1007/978-1-60327-005-2_7.

2. Ha TY. MicroRNAs in human diseases: from cancer to cardiovascular disease. Immune Netw. 2011;11(3):135-54. Doi: 10.4110/in.2011.11.3.135.

3. Hofmann NR. MicroRNA evolution in the genus Arabidopsis. Plant

Cell. 2010;22(4):994. Doi: 10.1105/tpc.110.220411.

4. O'Donnell KA, Boeke JD. Mighty Piwis defend the germline against genome intruders. Cell. 2007;129(1):37-44.

5. Vazquez F, Vaucheret H, Rajagopalan R, Lepers C, Gasciolli V, Mallory AC, et al. Endogenous trans-acting siRNAs regulate the accumulation of Arabidopsis mRNAs. Mol Cell. 2004;16(1):69-79.

6. Voinnet O. Origin, biogenesis, and activity of plant microRNAs. Cell. 2009;136(4):669-87. Doi: 10.1016/j.cell.2009.01.046.

7. Siomi H, Siomi MC. On the road to reading the RNA-interference code. Nature. 2009;457(7228):396-404. Doi: 10.1038/nature07754.

8. Lu C, Meyers BC, Green PJ. Construction of small RNA cDNA libraries for deep sequencing." Methods. 2007;43(2):110-7.

9. Ma Z, Coruh C, Axtell MJ. Arabidopsis lyrata small RNAs: transient MIRNA and small interfering RNA loci within the Arabidopsis genus. Plant Cell. 2010;22(4):1090-103. Doi: 10.1105/tpc.110.073882.

10. Nygaard S, Jacobsen A, Lindow M, Eriksen J, Balslev E, Flyger H, Identification and analysis of miRNAs in human breast cancer and teratoma samples using deep sequencing. BMC Med Genomics. 2009;2:35. Doi: 10.1186/1755-8794-2-35.

11. Pillai RS. MicroRNA function: multiple mechanisms for a tiny RNA?. RNA. 2005;11(12):1753-61.

12. Ahmad, Parvaiz, and Saiema R, eds. Emerging Technologies and Management of Crop Stress Tolerance: Volume 1-Biological

Techniques. Vol. 1. Academic Press, 2014.

13. Chiu YL , Rana TM. siRNA function in RNAi: a chemical modification analysis." RNA. 2003;9(9):1034-48.

14. Goff SA, Ricke D, Lan TH, Presting G, Wang R, Dunn M, et al. A draft sequence of the rice genome (Oryza sativa L. ssp. japonica). Science. 2002;296(5565):92-100.

15. Yu J, Hu S, Wang J, Wong GK, Li S, Liu B, et al. A draft sequence of the rice genome (Oryza sativa L. ssp. indica). Science. 2002;296(5565):79-92.

16. Slayton T and Peter CT. Japan, China and Thailand can solve the rice crisis-But US Leadership is needed. Center for Global Development Notes. Accessed August 22: 2008. http://www.cgdev.org/files/16028_file_Solve_the_Rice_Crisis_UPDATED.pdf

17. Israt Nadia, AKM Mohiuddin, Shahanaz Sultanaand, Jannatul Ferdous. Diversity analysis of indica rice accessions (Oryza sativa L.) using morphological and SSR markers. Annals of Biological Research, 2014;5 (11):20-31.

18. Truseq small RNA sample preparation guide, Catalog # RS-930-1012 Part # 15004197 Rev. A November 2010.

19. Agilent Small RNA Kit . Reorder number 5067-1548.

20. Solid" Total RNA-Seq Kit. Part Number 4452437. Rev. A 01/2010.

21. Schroeder A, Mueller O, Stocker S, Salowsky R, Leiber M, Gassmann M, et al. The RIN: an RNA integrity number for assigning integrity values to RNA measurements." BMC Mol Biol. 2006;7:3.

17

Microbiological Examination of Animal Fertilizer and effects of associated pathogens on the health of Farmers and Farm Animals

D. O. Akeredolu[1,2]* and AO. Ekundayo[2]

[1]Department of Microbiology, University of Benin, Benin-City, Edo State

[2]Ambrose Alli University, Ekpoma, Edo State

*Corresponding author: D. O. Akeredolu, Department of Microbiology, University of Benin, Benin-City, Edo State; E-mail: dennisakeredolu@gmail.com

Abstract

The aim of this study was to determine the microbial load and examine the individual types of microorganisms found in animal fertilizer. A microbiological analysis of animal fertilizer was carried out. The samples were collected from commercially prepared animal and poultry fertilizer. The presence of bacteria and fungi were detected. The bacteria spp isolated were *Salmonella* spp, *Staphylococcus aureus*, *Listeria monocytogenes*, *Bacillus* sp., *Proteus* sp., *Escherichia coli*, *Klebsiella* spp. and *Enterobacter* spp. while the fungal isolates were *Aspergillus niger*, *Aspergillus flavus*, *Penicillium* spp. and *Mucor* spp. The microbial burden of animal fertilizer analysed was high with an average of 7.2×10^6 cfu/g. some of these organisms such as *Salmonella* sp., *Listeria monocytogenes*, *Escherichia coli* and *Aspergillus* and *Penicillium* spp. have been associated with pathogenic conditions and are likely to cause salmonellosis, listeriosis, gastroinstinal trac infection, and diseases of the lungs if not properly handled. Survival of these microorganisms in animal fertilizer is affected by the source, pH, dry matter content, age, and chemical composition of the manure as well as microbial characteristics. The results of this study show that animal fertilizer contains very dangerous and harmful microorganisms which can cause ill-health to humans and other living organisms.

Keywords: Bacteria; Fungi; Animal fertilizer; Microbial load

Introduction

Animal fertilizer (manure) is a biologically active material, alive with bacteria and other microorganisms that depend on the energy contained in animal fertilizer. Animal manure is an energy-rich feedstuff produced as animal waste that is filled with potentially active microorganisms (1). The use of animal fertilizer energy by microorganisms is a natural process of decomposition. Except in extreme cases of cold, ph, or lack of water, biological decomposition is inevitable. By storing, handling, or treating manure in various ways, farmers can control the by-products produced by this biological activity. This is important to a farmer desiring to manage nutrients, control or create a marketable product (1). Maintaining profitability in farming while protecting quality and the health of humans and animals is one of the challenges that farmers face. Run-off of sediments,

pesticides and nutrients such as phosphorus and nitrogen has been considered the greatest environmental threats to water quality posed by animal agriculture (2). When considering the problem of pathogen transmission through animal fertilizer it is important to recognize that animal fertilizer, which consist of animal excreta (feces and urine), beddings, and dilution water, blood, vagina, mammary gland, skin, and placenta. Microbes in these secretions, as well as those in the excreta, can potentially accumulate on the barn floor. Unless the manure is handled appropriately, the pathogens may infect other animals or humans (Strauch and Ballarini, 1994). These microorganisms will use the energy in manure, causing the manure to change. The challenge of biological manipulation is to manage the environment in which these changes take place in order to produce (or not to produce) specific by-products (1). Before the widespread use of chemical fertilizers, animal fertilizers (manures) were used as a primary source of nutrients in crop production. In addition to supplying plant nutrients to the soil, manure also improves soil health by increasing soil organic matter and promoting beneficial organisms. Incorporating manure by improving soil structure (3). Animal manure contains all the essential micro and macro elements required for plant growth. Land application of animal manure increases soil organic matter and improves a number of soil properties including soil tilth, water holding capacity, oxygen content, and soil fertility. It also reduces nutrient leaching, and increases water infiltration rates, reduces nutrient leaching, and increases crop yields. In general, results of research indicate that manure is a valuable bio-resource that should be utilized (Pratt, 1982, 4,5).

Livestock manure contains a broader range of nutrients than most commercial fertilizers. This is because a large portion of the plant nutrients initially ingested by the animals, generally 80% of the phosphorus, 90% of the potassium and 75% of the nitrogen are still present in the manure (5). Nutrient availability, however, is determined by the manure handling system, as well as by climate and soil characteristics. Nutrient values also vary with different types of livestock and the animal feed ratios, which vary with the season. Generally, poultry manure tends to be high in Nitrogen (N) and phosphorus (P), while dairy manure tends to be high in Potassium (K) (6). The actual nutrient value of manure from a particular operation will differ considerably with the method of collection, storage facilities, and species of animal (Zhang, 2005). Nutrients in waste may be lost or converted to other forms during treatment or storage and handling, affecting their availability for use by growing plants. The type of animal housing system and/or waste handling method is known to affect the final nutrient composition of the waste (Zhang, 2005). Bedding and waste have a diluting effect on the final nutrient concentration of waste and result in less nutrient value per ton. In addition, the type of housing and waste handling system can decrease the final nutrient composition of waste materials. For instance, there can be considerable loss of nitrogen to the air, and there is potential for runoff and leaching when animal waste is exposed to weather conditions in an open lot system.

Manure application rates should be specific to the crops and soil, and applications should be scheduled to fit the farming operation and the season. The application rate should be carefully estimated because if it is excessive, it can cause pollution of surface and ground water, toxicity to livestock consuming the crops, and contribute to problems with plant growth. Application should also be timed to avoid spreading in winter and early spring; frozen ground and rainfall which may lead to run off and leaching of nutrients into water resources where they become pollutants (Peterson, 1995). Methods of application include a conventional beater-spreader, liquid tank wagon, large bore irrigation nozzle (for liquid), and/or a shovel and pitch fork. The "big-gun" sprinkler applications should be avoided because they offer the least accuracy and control when applying liquid manure to a specific area. Tank-type spreaders on the other hand, provide the most accuracy and control (Peterson, 1995). Manure (animal fertilizer) should be applied to flat land whenever possible and be incorporated into soil shortly after spreading to reduce loss of nitrogen. The rate of manure application should decrease as slope increase. Vegetative buffer strips at the base of the slope can prevent run off of nutrients on sloped land. Whenever manure is applied to land, it is important to keep good records of the data, amount applied, nutrient content, soil test results, weather as well as any other notes that would be helpful to have in the future. These records can be used to determine future application rates and also to provide documentation of application if questions or issues arise in the future (Peterson, 1995). Biological manipulation can be used to manage odor, nutrients, consistency, and stability of the treated manure product. For example, manure, combined with a carbon rich material such as sawdust and sufficient air, can be transformed into stable compost. On the other hand, by eliminating all air and adding heat, raw manure that contains little bedding can be transform into biogas and a low-odor, nutrient rich liquefied, stable effluent. Biological manipulation involves providing the proper "diet" and environment for the specific microorganisms that will use the manure energy (1).

The commission of the European Communities identified reportable bacteria that are of particular concern for animal and human health (7) included in their list are *Salmonella* spp, *Escherichia coli, Bacillus anthracis, Mycobacterium* spp, *Brucella* spp,, (especially *Brucella abortus*), *Leptospira* spp,, *Chlamydia* spp, and *Rickettsia* spp,. In addition to these organisms, other potential bacterial pathogens in manure include; *Listeria monocytogenes, Yersinia enterocolitica, Clostridium perfringes,* and *Klebsiella* spp, (8). To assess the threat posed by different microorganisms in manure, bacteria survival in manure as it is usually handled on farms must be evaluated. Survival is affected by the source, pH, dry matter content, age, and chemical composition of the manure as well as by microbial characteristics. Manure that is well mixed with bedding is more likely to undergo aerobic fermentation with accompanying temperature increases than in slurry with minimal amounts of beddings. Problems are also posed by viruses, which are obligate intracellular parasites (Snowdon et al., 1989; Strauch, 1991) that often have limited host range. *Giardia* spp, and *Cryptosporidium parvum* are protozoans that cause severe diarrhea in both animals and humans. In 1993 to 1994, one third of the outbreaks associated with drinking water for which the causative agent was identified were due to these two pathogens (9). *Cryptosporidium parvum,* first identified in 1975 (Rose, 1990). In healthy mature animal (including humans), the infections caused by both *Giardia* and *C. parvum* are usually self limiting and although they cause significant discomfort, do not pose serious long-term health risks. Infected animals may shed as many as 1.0×10^9 oocysts daily for 1 to 12 days (Ridley and Olsen, 1991). Farmers, animal handlers, veterinarians, others who work with animals are more likely to be infected than the general population (2). Whenever manure is applied to food crops, safety precautions should be taken to avoid contamination that might result in human illness. The pathogens of most concern that can be found in livestock manure are *E. coli* and *Samonella* sp. to avoid the risk of contamination, fresh manure should not to be applied within 60 days of harvesting food crops (10). Xylanases are used in the pretreatment of forage crops to improve the digestibility of ruminant feeds and to facilitate composting along with glucanases, pectinases, cellulases, proteases, amylases, galactosidases, and lipases. Phytase is an enzyme that makes the phosphorus from phytin available for animal digestion. Up to now, phytase is increasingly used in animal feeds, science and technology related to this enzyme are rapidly evolved. The benefits of phytase are its double effects on reducing the use of expensive inorganic phosphorus in animal diets and the environment pollution from excessive manure phosphorus runoff. The aim of this study was to determine the microbial load and examine the individual types of microorganisms found in animal fertilizer and their possible pathogenic effects to farm animals and farmers.

Materials and Methods

Collection of Samples and Media Preparation

The samples were collected from commercially prepared animal/bird fertilizer. All glasswares were sterilized by heating at

160⁰C for 1hour after washing them. This was done by the hot air oven. All plastics were washed and disinfected with formaldehyde. The following media were used during the course of study; Nutrient agar, MacConkey, Peptone water, Potato dextrose agar.

The Nutrient agar was prepared by dissolving 14g in 500ml of distilled water and sterilized with pressure cooker at 121⁰C for 15minutes. MacCokey agar was prepared by dissolving 24g in 500ml of distilled water and sterilized at 121⁰C for 15minutes with the pressure cooker. Peptone water was prepared by dissolving 7.5g of powder in 500ml of distilled water and was sterilized with pressure cooker for 15minutes at 121⁰C.

Sample Preparation and Culturing

One gram of the animal/ bird dung fertilizer was weighed using a beam balance and introduced into one of the ten test tubes containing 9ml each of normal saline to form the stock solution of the sample. Using a sterile pipette, 1ml was taken from the stock to test tube II, from this, 1ml was taken from the II after shaking into test tube III. This procedure was repeated until a 10^{-10} dilution was obtained. The pour plate method was used to determine the appropriate dilution to be adopted. This was done by pouring molten nutrient agar on 1ml of sample. The plates were allowed to solidify and incubated at 37⁰C for 24hours. After 24hours, colonies from plate 10^5 .10^7, and 10^9 dilution were countable while those of plates 10^1 and 10^3 were too many to be counted. The different colonies were purified by sub culturing to obtain discrete colonies by streaking on new plates containing Nutrient and MacConkey agar.

Identification of Bacteria Isolates

Identification was based on morphology, cultural and biochemical characteristics of the colonies.

Colonies having the same morphology were used to form slants, after Gram staining and other biochemical tests were carried out such as gram staining reaction, motility, catalyst test, indole test, sugar fermentation test, citrate test, oxidase test and coagulase test:

Identification of Fungi

With a sterile inoculation needle the fungus was removed from its pure culture on a grease free slide. It was teased out with the sterile needle and then stained with methylene blue. A cover slip was placed on the stained portion. It was then observed under the light microscope for cultural characteristics such as hyphae type, spores etc.

Analysis of disease outbreak

A health accessment was carried out for all farmers who were using the animal fertilizers. A questioner was developed to find out if there were any diseases out break as a result of the use of the animal fertilizers.

Results

The total viable bacterial count at 37⁰C for 24 hours was enumerated and had a mean count of 7.2x10⁶cfu/g (Table 1). Table 3 showed the biochemical characteristics and Gram reaction of the bacteria isolates. The following bacteria were *isolated: Escherichia coli, Staphylococcus aureus, Bacillus* spp, *Proteus* spp, *Klebsiella* spp, *Listeria monocytogenes, Enterobacter* spp and *Salmonella* spp. the following fungal isolates were identified and Characterized: *Aspergillus niger, Aspergillus flavus, Penicillium species* and *Mucor* species (Table 2).

Table 1: Enumeration of total viable bacterial count at 37⁰C for 24 hours

Samples	Dilution used	Colony count	Mean bacteria count (cfu/g)
Commercially prepared animal/bird dung fertilizer	5-Oct	72	7.2x106

Table 2: Identification and Characterization of fungal isolates

Isolates	Cultural and Morphological characteristics	Fungi
A	Light green fluffy with black spot	*Aspergillus niger*
B	Thick green not widely spreading	*Aspergillus flavus*
C	Center green surrounded with edges	*Penicillium spp*
D	Black sporty growth with no rhizoids	*Mucor spp*

Discussion

This study reveals the presence of eight bacterial species and four fungal species. The bacterial species include; *Escherichia coli, Staphylococcus aureus, Bacillus* spp, *Proteus* spp, *Klebsiella* spp, *Listeria monocytogenes, Enterobacter* spp and *Salmonella* spp While the fungal species include; *Aspergillus niger, Aspergillus flavus, Penicillium* species and *Mucor* species. Past research has shown some of these bacteria isolates as potential bacterial pathogens in animal fertilizer. Work by Mawdsley et al, (5) reveals the presence of *Listeria monocytogenes, Salmonella* spp, *Escherichia coli* and *Klebsiella* spp in livestock waste. Dairy farms have been identified as reservoirs of *E.coli* (10) and *Listeria monocytogenes* lives naturally in plant and soil environments and poorly fermented silage often contains high numbers of *Listeria monocytogenes* (11). Animal fertilizer has been shown to contain *Salmonella* spp, even from the feces of apparently healthy animals (12). The gastro intestinal tracts of animals also harbor most of these organisms such as *Enterobacter* spp, *Proteus* spp, and *Staphylococcus aureus*. While *Bacillus* spp, are environment friendly because of their spores (11). The presence of these bacteria in animal fertilizer is a cause for public health concern, since these bacteria especially *E. coli, Staphylococcus aureus, Bacillus* spp, *Salmonella* spp and Listeria monocytogenes are known to be pathogenic.

Table 3: The biochemical characteristics and Gram reaction of the bacteria isolates

Isolates	Gram reaction	Morphology	Motility	Catalase	Indole	Oxidase	Coagulase	Citrate	Lactose	Glucose	Sucrose	Dulcitol	Xylose	Manitol	Maltose	Suspected organisms
A	-	rods	+	+	-	-	-	+	A	A	-	±	-	+	+	*Salmonella spp*
B	+	cocci	-	+	-	-	+	-	+	+	+	-	+	+	+	*Staphylococcus aureus*
C	+	rods	+	+	-	-	-	-	+	+	±	-	+	-	+	*Listeria monocytogenes*
D	-	rods	+	+	-	-	-	-	-	+	+	-	-	+	+	*Bacillus spp*
E	-	rods	+	+	+	-	-	+	-	+	±	+	-	-	+	*Proteus spp*
F	-	rods	+	-	+	-	-	-	+G	+G	+	-	-	+	+	*Escherichia coli*
G	-	rods	-	+	-	-	-	+	+	+	+	ND	ND	+	+	*Klebsiella pneumoniae*
H	-	rods	+	-	-	-	-	+	+	+	+	+	-	±	+	*Enterobacter spp*

From the observation made from this research work, moulds such as *Aspergillus flavus, Aspergillus niger, Penicillium* spp and *Mucor* spp were responsible for the decomposition of this animal fertilizer, since they are common fungi found in the environment. This shows that animal fertilizer contains an incredible amount of pathogenic microorganisms. The survival of these microorganisms in animal fertilizer is affected by the source, pH, dry matter content, age, and chemical composition of the manure as well as the individual microbial characteristics. From of the health accessmet from the farmers, there was no significant disease out break as a result of the use of animal fertilizers. It could be that the farmers may not have reported acurrately on their questioners or that they have applied very good health safty measures while using the animal fertilizers. It is therefore very important to use hand gloves and other safty garjets while appling animal fertilzer to avoid microbial infection that are associated with animal fertilizer.

Conclusion

The results of this study show that animal fertilizer contains very dangerous and harmful microorganisms which can cause ill-health to humans and other living organisms if they get to source of easy intake like water. There should be proper management procedures taken to ensure that these animal fertilizers do not get to any source of drinking water or food. Before applying manure, it is important to have both the manure and the soil tested for nutrient content. Food safety considerations should be a priority when utilizing manure to fertilize food crops, especially fruits and vegetables and appropriate environmental safeguards, such as filter strips are to be used.

References

1. Leggett JA.1996. Biological Manipulation of Manure: getting what you want from animal manure. Department of Agriculture and Biological Engineering. College of Agriculture Sciences. Pennesylvania University Cooperative extension.

2. Mac Kenzie WR, Hoxie NJ, Proctor ME, Gradus MS, Blair KA, Peterson DE, et al. A massive outbreak in Milwaukee of Cryptosporidium infection transmitted through the public water supply. N Engl J Med. 1994;331(3):161-167. Doi: 10.1056/NEJM199407213310304

3. Hermason, R. E. 1996. Manure Sampling For Nutrient Analysis with Worksheets for Calculating Fertilizer Values. WSU Extention Bulletin No.1819.

4. Araji A A, Stodick LD. "The Economic potential of feedlot wastes utilization in Agricultural production". Biological Manure. 1990;32(2):111-124. Doi:10.1016/0269-7483(90)90076-5

5. Cassman KG, Steiner R, Johnson AE. "Long Term Experiments and Productivity Indexes to Evaluate the Sustainability of Cropping System". In: Agricultural Sustainabilit: Economic, Environmental and Statistical Considerations. 1st edition, John Wiley and Son. 1995;348.

6. Brandy N, Ray W. The Nature and properties of soils. Prentice Hall, New Jersery. 1996:1-3.

7. Kelly WR. Animal and animal health hazards associated with the utilization of animal effluents. EUR. 6009. Office. Publ. Eur. Commun. Luxembourg. 1978.

8. Mawdsley JL, Bardgett RD, Merry RJ, Pain BF, and Theodorou MK. Pathogens in Livestock Waste, their Potential for movement through soil and environmental pollution. Appl Soil Ecol. 1995;2(1):1-15.Doi: 10.1016/0929-1393(94)00039-A

9. Kramer MH, Herwaldt B L, Craun GF, Calderon RL, Juranek DD. Waterborne Disease: 1993,1994,1996. J AM Water Works Assoc. 88:66

10. Hancock DD, Besser TE, Kinsel ML, Terr PI, Rice Dh, Peros M G. The Prevalence of Escherichia coli: 0157.H7 in dairy and beef cattle in Washington State. Epidermiol infect. 1994;113(2):199-207.

11. Husu JR. Epidemiology Studies on the Occurrence of Listeria monocytogenes in the feces of dairy cattle. JVet Med Ser. 1990;37:276-282. Doi:10.1111/j.1439-0450.1990.tb01059.x

12. Jones P W. Health Hazards Associated with the handling of Animal Wastes. Vet Rec. 1980;106(1):4-7

18

Screening and Characterization of *Achromobacter xylosoxidans* isolated from rhizosphere of *Jatropha curcas* L. (Energy Crop) for plant-growth-promoting traits

Preeti Vyas[1], Damendra Kumar[2], Anamika Dubey[1] and Ashwani Kumar[1*]

[1]*Metagenomics & Secretomics Research laboratory, Department of Botany, Dr.Harisingh Gour University (A Central University), Sagar (M.P.), India*

[2]*Department of Biotechnology, Dr. Harisingh Gour University (A Central University), Sagar (M.P.), India*

__Corresponding author:__ Ashwani Kumar, Assistant Professor, Metagenomics & Secretomics Research laboratory, Department of Botany, Dr.Harisingh Gour University (A Central University), Sagar (M.P.) India, E-mail id: ashwaniiitd@hotmail.com

Abstract

Plant growth promoting rhizobacteria (PGPR) colonizes almost all the ecological niches in and around the plant roots and enhances plant growth and show profound impact upon plants productivity. In the present study we have isolated large number of bacterial isolates from the rhizosphere of non-edible oil seed plant *Jatropha curcas* (Common name: Physic nut; Family: Euphorbiaceae). Out of large number of isolates we have selected only four bacterial isolates (AKDJ1, AKDJ2, AKDJ3, and AKDJ4) on the basis of their multifarious PGP traits (bioflim production, ammonia production, indole acetic acid (IAA), phosphate solubilization, catalase enzyme and cellulase enzyme production). Out of four, the isolate AKDJ2 was characterized by various biochemical utilization tests (Citrate, lysine, ornithine, urease, phenylalanine, H_2S production, nitrate reduction, glucose, lactose, adonitol, sorbitol, arabinose, and 35 different carbohydrate sources) and identified as *Achromobacter xylosoxidans* (Gene bank Accession no KX698100) which showed 99% similarity with *Achromobacter xylosoxidans* strain NBRC15126 by using 16S rDNA sequencing. We conclude that, bacterial isolates screened from the rhizosphere of plant could serve as a source of potential biofertilizer for improving the production of same and other crops under variety of stress conditions.

Keywords: Plant growth promoting rhizobacteria; Biofertilizer; 16S rDNA; Energy crop

Introduction

Jatropha curcas L. belongs to family Euphorbiaceae is a perennial, drought resistant, and multipurpose oil seed plant. It is often recognized as an potential source for future biodiesel production [1–4]. *Jatropha curcas* is a tropical plant and grow in wastelands, areas with low precipitation. *Jatropha* can be grown in boundaries to protect agriculture fields from grazing and soil erosion or they can be planted in the farms as a commercial crop. To improve its growth and production for biofuel generation, a number of agricultural management practice have been used by several researchers in the past. Kumar et al.[2,5–9] used different bioinoculants to improve its growth and yield under saline and alkaline soil conditions. The bioinoculants used in previous studies are commonly screened and isolated from rhizosphere and commonly known as plant growth promoting rhizobacteria (PGPR). These PGPR promote plant growth and suppresses disease incidence which is solely resulted due to the synergistic effect of nutrients and phytohormones produced by these bacteria [10–12]. Bioinoculation of selected PGPR with seeds of *Jatropha* can improve growth of plant by providing resistance to plants towards different abiotic and biotic stress conditions. PGPR are often used for improving fertility and facilitates establishment of plant [13–15]. Great efforts have been made to investigate the beneficial role of PGPR on crop production under variety of stress conditions [16–20]. PGPR can stimulate plant growth by using direct and indirect mechanism of action. Direct mechanism of PGPR action includes fixation of atmospheric nitrogen, phosphate solubilization, Siderophore production, and production of plant hormones (like Auxins and Cytokines). Indirect mechanism of plant growth stimulation includes synthesis of some plant growth substances or facilitating the uptake of certain nutrients [21].

Soil is most dynamic and complex system that supports overall growth of the plant. The abiotic and biotic stresses are the major constrain for sustainable agricultural production. Most of these microbes are dependent upon different root exudates secreted by plants for their survival [21]. Evidences supports the fact that plants utilizes greater amount of nutrients that are present in the soil in modern intensive cultivation and often needs replenishment of the nutrients. Under these conditions microbes offers a good alternative strategy to replenish various nutrients. Kumar et al. (2015b) in his study isolated one hundred and six PGPR bacteria from the rhizosphere and endosphere of *Hippophae rhamnoides* L. (Sea-buckthorn). Theses bacterial isolates were then screened for different PGP traits. Results of their study showed 76.41 % of bacterial isolates, depict IAA or auxin production activity, 43.39 % of bacterial isolates depicts siderophore activity and 19.4 % of bacterial isolates shows HCN production activity.

The objectives of the present study was to isolate bacterial strains from the rhizosphere of the *Jatropha curcas*, and characterized them for morphological and physiological attributes as well as identify them by using 16S rDNA sequencing. Graphical representation of work done is presented in Figure 1.

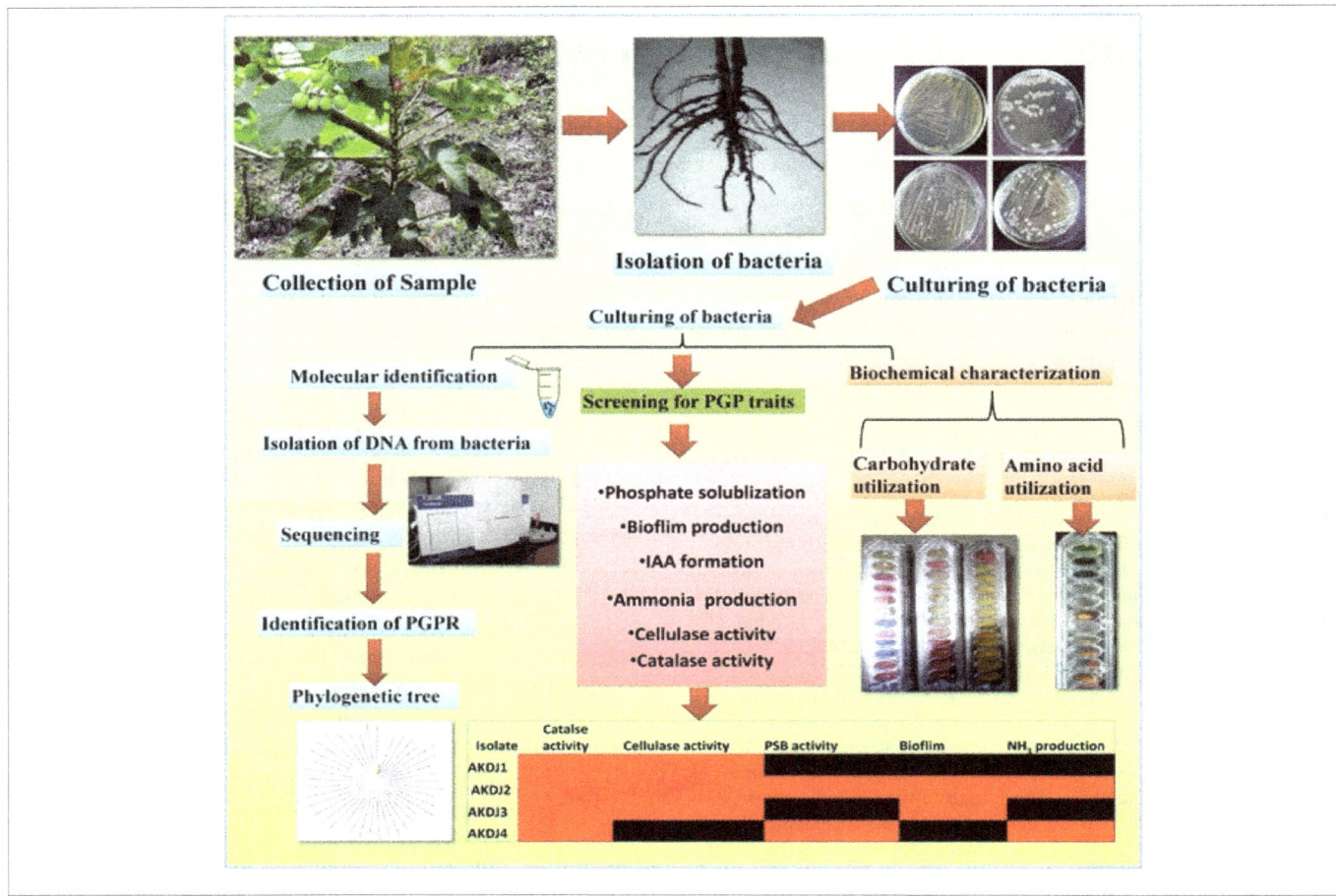

Figure 1: Flow chart of experimental work

Materials and methods

Collection of samples

The rhizosphere soil sample was collected from the Botanical garden located at Dr. Harisingh Gour University Sagar (M.P), India. The location of the site is at 23°49'34 N latitude and 78°46'35 E longitude as shown in Figure 2. The rhizosphere soil was often collected after digging in depth up to 15 cm. These samples were placed in sterile polythene bags and brought to the lab. And stored at 4ºC in refrigerator until use

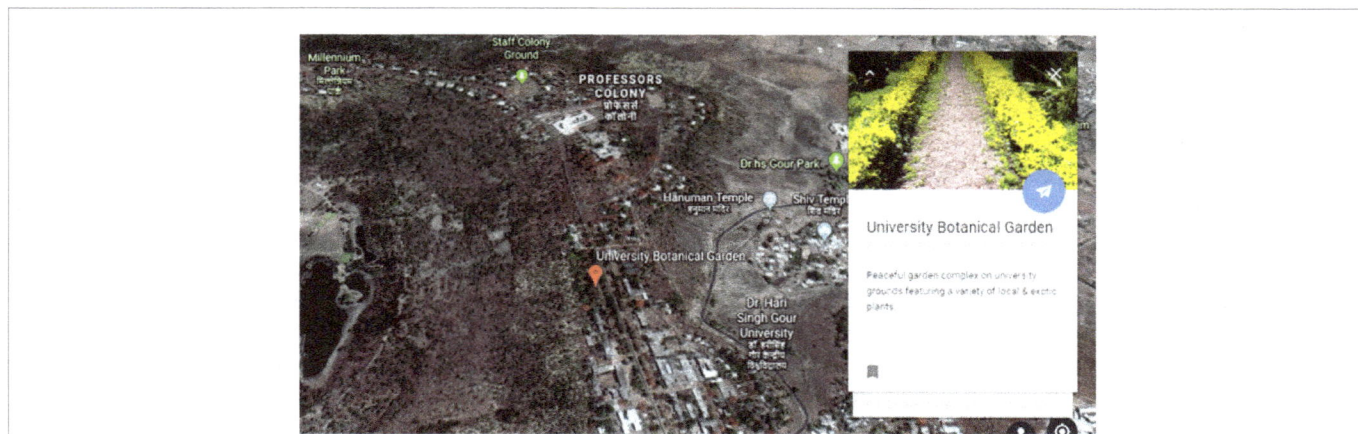

Figure 2: Sample collection site (Google earth image)

Isolation of Rhizobacteria from the rhizosphere

For isolation of bacteria, 1g of rhizospheric soil was used and was plated on nutrient agar media (Peptone 10g/L, NaCl 5.0g/L, Yeast extract 10g/l, 1 M NaOH, 10ml/l, Agar15g/l at pH 7). The plates were then incubated for about 24 hrs at 28 °C for further experiment. These bacterial isolates were further maintained at liquid nutrient broth and preserved in glycerol at -20°C

Morphological and biochemical characterization

Morphological characterization of each bacterial isolates was examined on nutrient agar plates. Three days old culture of bacterial isolates were used for determining the size, color, shape, surface, elevation, and margin of colonies. The Grams staining of the isolated strains was also carried out to find out the gram positive and gram negative strain as described by Vincent and Humphrey [23]. Light microscopy was used to observe the size and motility of the bacterial cell. Biochemical and carbohydrate test was conducted by using kits (KB002 HiAssortedTM Biochemical Test Kit and HiCarbohydrateTM Test kit KB009), respectively.

Molecular identification of potential bacteria and bioinformatic analysis

Total genomic DNA was successfully extracted from bacteria isolates by using Insta Gene TM Matrix Genomic DNA isolation kit. Final concentration of DNA was determined by using nano drop and visualized by running DNA gel electrophoresis. Isolated DNA was PCR amplified by using universal 27F forward primers with sequence (AGAGTTTGATCMTGGCTCAG) and 1492R reverse primer with sequence (TACGGTACCTTGTTACGACTT). The PCR reaction was performed by using the method presented in Vyas et al. (2018). PCR products were then purified and sequenced using an ABI Big Dye Terminator v3.1 cycle sequencing kit (Applied Biosystems, Grand Island, NY, USA). Obtained sequences were then compared with other sequences through NCBI BLAST at http://www.ncbi.n1m.nih.gov/blast/Blast.cgi. Sequences were submitted to NCBI GenBank data base and obtained accession numbers (Accession number: KX698100). The phylogenetic analysis of sequences with the closely related sequence of NCBI blast results was performed by following multiple sequence alignment. A phylogenetic tree for these bacterial sequences were constructed by using iTOL (Interactive tree of life) after establishing relationship among the similar sequences analysis generated from Mega 5.05 software [27, 28].

Biochemical Assays

Solubilization of Insoluble Phosphate

200 µL of bacterial suspension was spot inoculated at the centre phosphate solubilizing agar plates or Pikovaskay's plate [24]. The plates were then incubated at 28°C for about 5 days and halos' zones produced were measured indicating varying levels of phosphate solubilisation.

Ammonia production

This test is based on the production of urease which break urea into ammonia and which in turn increase the pH of the medium. Freshly grown cultures of bacterial isolates were inoculated into urea broth containing peptone and incubated at 37°C for 24 hrs. Bacterial culture was then centrifuged. 1 ml of Nessler's reagent was added to the supernatant and change in color yellow to brown was a positive test for ammonia production.

Production of Indole-3-Acetic Acid

Bacterial cultures were grown on Dev tryptophan broth [25] on rotatry shaker at 37°C for 5 days. Bacterial suspension was then centrifuged at 1000 rpm for 20 min. 1ml of Salkowsky's reagent was added to 1 ml of supernatant and incubated in dark incubator for 1h. Then, development of pink color considered positive for IAA production and further measured at 536 nm by using microplate reader.

Bioflim formation

In this, ability of PGPR to form biofilm on root surface will be assayed using CV (crystal violet) by following standard protocol. For this assay bacterial isolates will be grown in nutrient broth and incubated at 37°C for 3 days. After incubation, samples were further stained by using 1% crystal violet solution and extracted with ethanol. The crystal violet stain will be then spectro-photometrically quantified by measuring the absorbance at 690 nm.

Catalase activity

1ml of bacterial culture taken into tubes and add few drop of H2O2 separately. The evolution of oxygen in the form of bubble indicates positive reaction for catalase production.

Cellulase activity

Cellulase production was determined by using the standard protocols. Agar medium (NaNO$_3$, K$_2$HPO$_4$, MgSO$_4$, KCl, Sodium CMC, peptone and Agar) with yeast extract plates were inoculated with individual bacterial isolates and incubated for 3-5 days at 28°C. Bacterial growth surrounded by clear halos was considered as positive indication of cellulose production. The incubated CMC agar plates were then flooded with grams iodine solution allowed to stand for 1 min at room temperature. 1M NaCl was thoroughly used for counter staining the plates. Clear halos zones were observed around growing bacterial colonies indicating hydrolysis of cellulose [26].

Results and Discussion

In this present study total of 4 bacterial isolates were successfully isolated from the rhizosphere of non-edible oil seed plant Jatropha curcas. These four bacterial isolates (AKDJ1, AKDJ2, AKDJ3, and AKDJ4) were identified morphologically on the basis of size, color, shape, surface, elevation, and margin, of colonies. The Grams staining was performed most of the bacterial species were grams negative. These bacterial isolates were further screened for different plant growth promoting traits (bioflim production, ammonia production, indole acetic acid (IAA), phosphate solubilization, and catalase and cellulase production). Out of the four isolates (AKDJ1, AKDJ2, AKDJ3, AKDJ4), two isolates formed biofilm (AKDJ2, AKDJ3). Biofilm formation by rhizobacteria is an important trait, with respect to their beneficial activity. Two bacterial isolates (AKDJ3, AKDJ4) showed positive result for ammonia production as shown in Figure 3. Two bacterial isolates(AKDJ2, AKDJ4) shows formation of clear halos zone when inoculated at PSB agar medium as shown in Figure 4. Two isolates showed positive result for IAA activity (AKDJ3, AKDJ4). All the four bacterial isolates (AKDJ1, AKDJ2, AKDJ3, AKDJ4), showed catalase activity as shown in Figure 5. Three bacterial isolates (AKDJ1, AKDJ2, and AKDJ3) showed cellulase producing activity as shown in Figure 6.

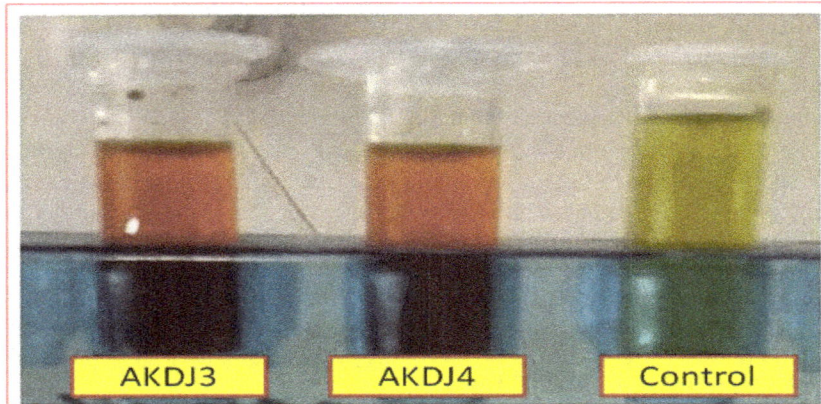

Figure 3: Bacterial isolates showing ammonia production activity

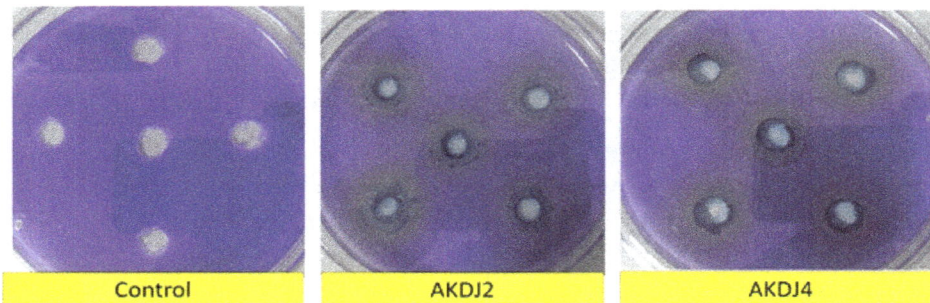

Figure 4: Bacterial isolates showing phosphate solubilizing activity

Figure 5: Bacterial isolates showing catalase activity

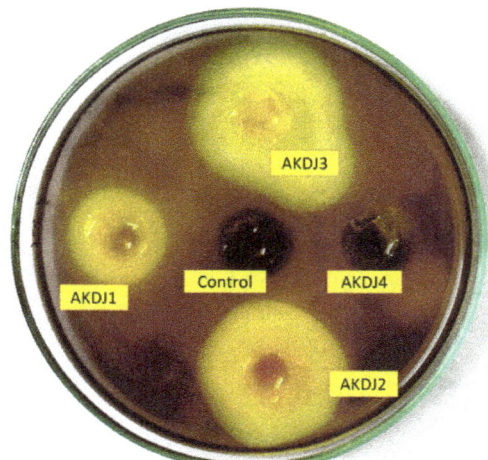

Figure 6: Bacterial isolates showing cellulase producing activity

Biochemical characterization test were performed for isolates AKDJ2. This isolate show positive result for orinithine, urease, nitrate reductase, Adonitol, lactose and sorbitol utilization and show negative result for citrate utilization, lysine utilization, phenylalanine deamination, H_2S, glucose and arabinose utilization. This bacterial isolates (AKDJ2) characterized biochemically for 35 carbohydrates sources and showed positive test for 24 carbohydrate sources and negative for 11 carbon sources and utilized 7 biochemicals out of 12, as shown in Table 1 and Table 2.

Table 1: Utilization of carbohydrates by AKDJ2

S No.	Carbohydrate	AKDJ2
1.	Lactose	Negative
2.	Xylose	Positive
3.	Maltose	Positive
4.	Fructose	Positive
5.	Dextrose	Positive
6.	Galactose	Positive
7.	Raffinose	Positive
8.	Trehalose	Positive
9.	Melibiose	Positive
10.	Sucrose	Positive
11.	L-Arabinose	Positive
12.	Mannose	Negative
13.	Inulin	Negative
14.	Sodium gluconate	Positive
15.	Glycerol	Positive
16.	Salicin	Positive
17.	Dulcitol	Positive
18.	Inositol	Positive
19.	Sorbitol	Positive
20.	Mannitol	Negative
21.	Adinitol	Negative
22.	Arabitol	Negative

23.	Erythritol	Negative
24.	Methyl-D-glucoside	Positive
25.	Rhamnose	Positive
26.	Cellobiose	Negative
27.	Melezitose	Negative
28.	Methyl-D-mannoside	Positive
29.	Xylitol	Negative
30.	ONPG	Positive
31.	Esculin hydrolysis	Negative
32.	D-Arabinose	Positive
33.	Citrate Utilization	Positive
34.	Malonate utilization	Positive
35.	Sorbose	Positive
36.	Control	Negative

Table 2: Biochemical Utilization by AKDJ2

S No.	Test	AKDJ2
1	Citrate Utilization	Negative
2	Lysine Utilization	Negative
3	Ornithine Utilization	Positive
4	Urease	Positive
5	Phenylalanine Deamination	Negative
6	Nitrate reduction	Positive
7	H_2S Production	Negative
8	Glucose	Positive
9	Adonitol	Positive
10	Lactose	Positive
11	Arabinose	Negative
12	Sorbitol	Positive

AKDJ2 was identified as Achromobacter xylosoxidans (Gene bank Accession no KX698100) which showed 99% similarity with Achromobacter xylosoxidans strain NBRC15126 (Accession number: KX698100) by using 16S rDNA sequencing. Phylogenetic tree was prepared by using iTOL tool for establishing relationship of this isolate with other closely related genera (Figure 7).

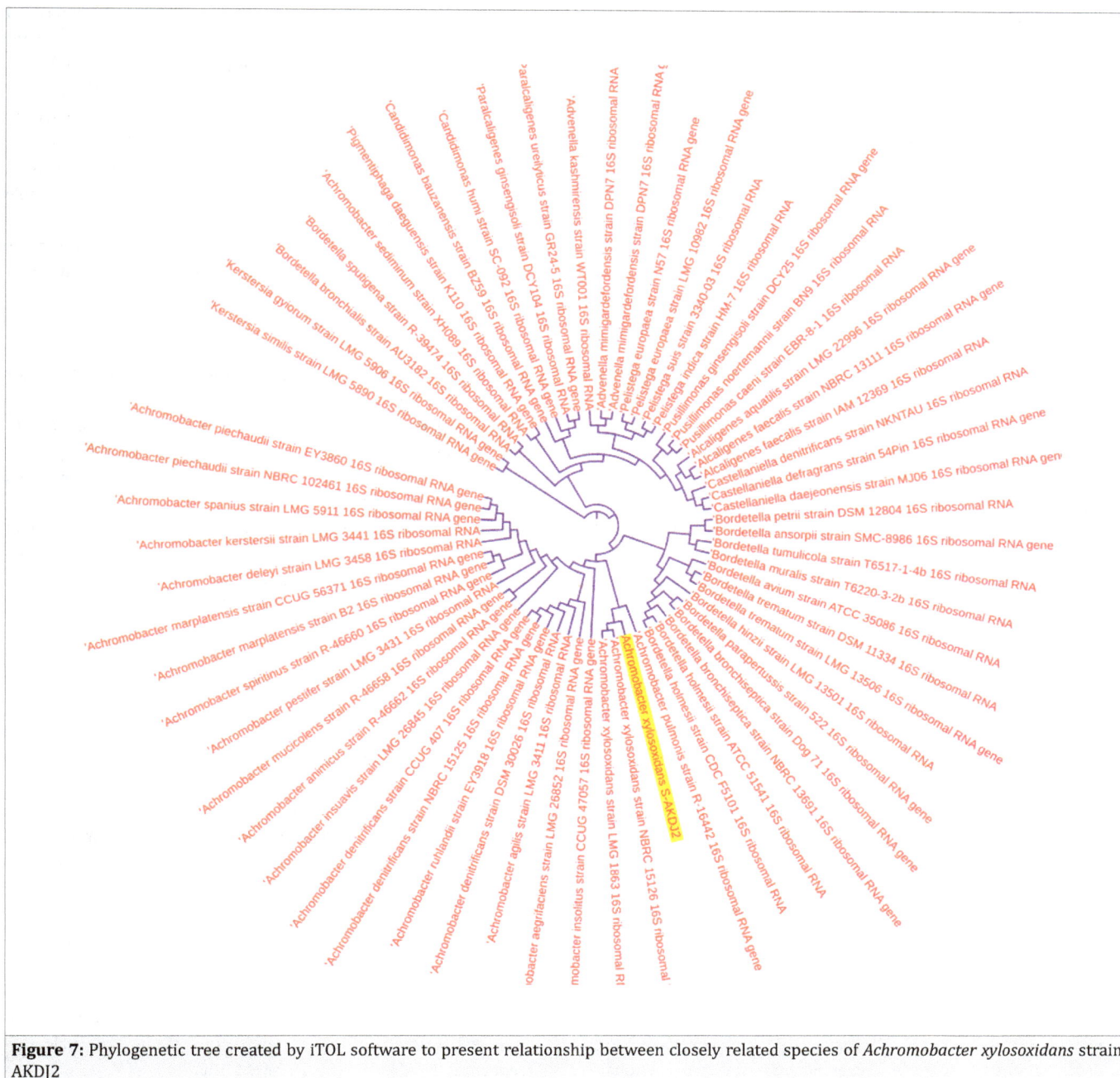

Figure 7: Phylogenetic tree created by iTOL software to present relationship between closely related species of *Achromobacter xylosoxidans* strain AKDJ2

Conclusions

PGPR are a group of bacteria that play an important role in plant growth promotion. Screening PGPR from rhizosphere of plants may be viable option to enhance the biomass production on limited soil conditions/marginal land. Additionally the potential isolates may be further utilized as tailor made biofertilizer for promoting growth of the other plants. The application of PGPR instead of chemical fertilizers offers a sustainable, safe, and eco-friendly approach to increase crop production and soil health.

Acknowledgement

PV acknowledges University fellowship and AK would like to acknowledge the UGC Startup grant (Awarded to AK) for the financial support.

References

1. Kumar A, Sharma S. An evaluation of multipurpose oil seed crop for industrial uses (Jatropha curcas L.): A review. Industrial Crops and Products. 2008;28(1):1-10.

2. Kumar A, Kumar K, Kaushik N, Sharma S, Mishra S. Renewable energy in India: Current status and future potentials. Renew Sustain Energy Rev. 2010;14(8):2434–2442. doi: 10.1016/j.rser.2010.04.003

3. Maghuly F, Laimer M. Jatropha curcas, a biofuel crop: Functional genomics for understanding metabolic pathways and genetic improvement. Biotechnology Journal. 2013;8(10):1172–1182.

4. Kumar A, Sharma S. Potential non-edible oil resources as biodiesel feedstock: An Indian perspective. Renew Sustain Energy Rev. 2011;15(4):1791–1800. doi: 10.1016/j.rser.2010.11.020

5. Kumar A, Sharma S, Mishra S. Influence of arbuscular mycorrhizal (AM) fungi and salinity on seedling growth, solute accumulation, and mycorrhizal dependency of Jatropha curcas L. J Plant Growth Regul. 2010;29(3):297–306. doi: 10.1007/s00344-009-9136-1

6. El-Naby SKMA. Effect of banana compost as organic manure on growth, nutrients status, yield and fruit quality of Maghrabi banana. Assiut J Agric Sci. 2000;31(3):101–114.

7. Kumar A, Dames JF, Gupta A, Sharma S, Gilbert JA, Ahmad P. Current developments in arbuscular mycorrhizal fungi research and its role in salinity stress alleviation : a biotechnological perspective. Crit Rev Biotechnol. 2014;8551:1–14.

8. Kumar A, Sharma S, Mishra S, Dames JF. Arbuscular mycorrhizal inoculation improves growth and antioxidative response of Jatropha curcas (L.) under Na2SO4 salt stress. Plant Biosyst. 2015;149(2):260-269.

9. Kumar A, Sharma S, Mishra S. Evaluating effect of arbuscular mycorrhizal fungal consortia and Azotobacter chroococcum in improving biomass yield of Jatropha curcas. Plant Biosyst. 2016;150(5):1056–1064.

10. Hashem A, Fathi E, Allah A, Alqarawi AA, Kumar A. Plant defense approach of Bacillus subtilis (BERA 71) against Macrophomina phaseolina (Tassi) Goid in mung bean. J Plant Interact. 2017;12(1): 390-401.

11. Doornbos RF, Van Loon LC, Bakker PAHM. Impact of root exudates and plant defense signaling on bacterial communities in the rhizosphere. A review. Agron Sustain Dev. 2012;32(1):227–243.

12. Barnawal D, Bharti N, Pandey SS, Pandey A, Chanotiya CS, Kalra A. Plant growth-promoting rhizobacteria enhance wheat salt and drought stress tolerance by altering endogenous phytohormone levels and TaCTR1/TaDREB2 expression. Physiol Plant. 2017;161(4):502–514. doi: 10.1111/ppl.12614

13. Igiehon NO, Babalola OO. Rhizosphere Microbiome Modulators : Contributions of Nitrogen Fixing Bacteria towards Sustainable Agriculture Int J Environ Res Public Health. 2018;15(4). doi: 10.3390/ijerph15040574

14. Kumar A, Sharma S, Mishra S, Dames JF. Arbuscular mycorrhizal inoculation improves growth and antioxidative response of Jatropha curcas (L.) under Na2So4 salt stress. Plant Biosyst - An Int J Deal with all Asp Plant Biol [Internet]. 2013;149(2):260–269.

15. Dubey A, Kumar A, Abd_Allah EF, Hashem A, Khan ML. Growing more with less: Breeding and developing drought resilient soybean to improve food security. Ecol Indic [Internet]. 2018.

16. Porcel R, Aroca R, Ruiz-Lozano JM. Salinity stress alleviation using arbuscular mycorrhizal fungi. A review. Agronomy for Sustainable Development. 2012;32(1):181–200.

17. Babalola OO. Beneficial bacteria of agricultural importance. Biotechnology Letters. 2010;32(11):1559–1570. doi: 10.1007/s10529-010-0347-0

18. Lugtenberg B, Kamilova F. Plant-Growth-Promoting Rhizobacteria. Annu Rev Microbiol [Internet]. 2009;63(1):541–556.

19. Miransari M. Plant Growth Promoting Rhizobacteria. J Plant Nutr [Internet]. 2014;37(14):2227–2235.

20. Rengasamy P. Soil processes affecting crop production in salt-affected soils. Functional Plant Biology. 2010;37:613–620

21. Glick BR. Plant Growth-Promoting Bacteria: Mechanisms and Applications. Scientifica. 2012:1–15.

22. Kumar A, Guleria S, Mehta P, Walia A, Chauhan A, C.K. S. Plant growth-promoting traits of phosphate solubilizing bacteria isolated from Hippophae rhamnoides L. (Sea-buckthorn) growing in cold desert Trans-Himalayan Lahul and Spiti regions of India. Acta Physiologiae Plantarum. 2015b;37:1-12.

23. Vincent JM, Humphrey B. Taxonomically significant group antigens in Rhizobium. J Gen Microbiol. 1970;63(3):379–382. doi: 10.1099/00221287-63-3-379

24. Sylvester-Bradley R, Asakawa N, La Torraca S, Magalhães FM, Oliveira LA, Pereira RM. Levantamento quantitativo de microrganismos solubilizadores de fosfatos na rizosfera de gramíneas e leguminosas forrageiras na Amazônia. Acta Amaz. 1982;12(1):15–22.

25. Frankenberger WT, Arshad M. Yield Response Of Watermelon And Muskmelon To L-Tryptophan Applied To Soil. Hortscience. 1991;26(1):35–37.

26. Chand R, Richa K, Dhar H, Dutt S, Gulati A. A Rapid and Easy Method for the Detection of Microbial Cellulases on Agar Plates Using Gram' s Iodine. Curr Microbiol. 2008;57(5):503-507. doi: 10.1007/s00284-008-9276-8

27. Kimura M. A simple method for estimating evolutionary rates of base substitutions through comparative studies of nucleotide sequences. J Mol Evol. 1980;16(2):111–120.

28. Letunic I, Bork P. Interactive tree of life (iTOL) v3 : an online tool for the display and annotation of phylogenetic and other trees. 2016;44:242–245. doi: 10.1093/nar/gkw290

19

Probiotics as Therapeutics

Shiwangi Morya, Gauri Aeron*

Department of Biotechnology, SD College of Engineering &Technology, Muzaffarnagar-25100, India

*****Corresponding author:** *Gauri Aeron, Department of Biotechnology, S.D. College of Engineering &Technology, Muzaffarnagar-25100, India,*
E-mail: gauriaeron@gmail.com

Abstract

There is an increasing scientific and commercial interest in the use of valuable microorganisms, or "probiotics," for the averting and treatment of disease. The microorganisms most frequently used as probiotic agents are lactic-acid bacteria such as *Lactobacilli species,* and many other species are also used such as *Bifidobacteria* species, *saccharomyces cerevisie var. bouladi* and many more which has been extensively studied as probiotic species. Multiple mechanisms of action have been hypothesized, including lactose digestion, production of antimicrobial agents (such as bacteriocin protein compound), competition for space or nutrients, and immunomodulation. We have reviewed recent studies of probiotics for the treatment and manage of infectious diseases. Studies of pediatrics diarrhea show considerable evidence of clinical benefits from probiotic therapy in patients with viral gastroenteritis, and data treatment for *Clostridium difficile diarrhea* appear promising. However, data to support use of probiotics for prevention of traveler's diarrhea are more limited. New research suggests eventual applications in vaccine expansion and prevention of sexually transmitted diseases. Further studies are needed to take full advantage of this conventional medical approach and to apply it to the infectious diseases of the new millennium.

Introduction

Probiotics are living microorganisms that help us to stay healthy which when ingested have valuable effects on the equilibrium and the physiological functions of the human intestinal microflora by various ways. Probiotics have been recently defined as "live microbes which transit the gastro-intestinal tract and in doing so profit the health of the consumer differing from the earlier definitions which focused on probiotic interactions with aboriginal intestinal microbes. These definitions of probiotic bacteria generally agree that probiotic bacteria should be living organisms to furnish health benefits. Probiotics have been reported to play a therapeutic role by enhancing immunity, lowering cholesterol, improving lactose tolerance and preventing some types of cancers such as colon cancer and many more. In the recent past, there has been an surge of probiotic-based health products mostly in the form of fermented dairy products as well as nutritional supplements. The markets for probiotic products and supplements are growing worldwide because of various benefits. Today there are more than 80 "Bifidus"- and "Acidophilus"-containing products worldwide, including a number of fermented dairy products. Capability of probiotic bacteria (such as *lactobacilli, bifidibateria* and many other species of different bacteria) in a product at the point of operation is an most important consideration for their effectiveness, as they have to survive during the processing and shelf life of food and supplements, transit through high acid conditions of the stomach and enzymes and bile salts in the small intestine very efficiently. The consumption of probiotics at a level of 10^8-10^9 cfu/g per day is a generally quoted figure for ample probiotic consumption, equating to 100 g of a food product with 10^6-10^7 cfu/g. Analysis of probiotic products in many different countries has established that probiotic strains exhibit poor survival in conventional fermented dairy products. The probiotic preparations such as tablets, powders etc. may contain lesser viable counts. Of the 15 feed supplements examined, viable probiotic counts varied to a great extent with 3 products containing no lactobacillus species at all, although the supplements were supposed to contain L. acidophilus. Probiotic survival in products is affected by a range of factors including pH, acidification (during storage) in fermented products, hydrogen peroxide production, oxygen toxicity (oxygen permeation through packaging), storage temperatures, stability in dried or frozen form, poor growth in milk, lack of proteases to break down milk protein to simpler nitrogenous substances and compatibility with traditional starter culture during fermentation (Dave and Shah, 1997a, b, c; Kailasapathy and Rybka, 1997; Shah, 2000). Oxygen plays a vital role in the poor survival of probiotic bacteria. Encapsulation of probiotic bacteria is an alternative that provides protection for living cells bare to an adverse environment. It also helps food materials to oppose processing and packaging conditions, improving flavou, aroma, stability, nutritional value and product appearance.

How antibiotics imbalance gut microbiota

The gastrointestinal tract is one of the chief interfaces between the human internal atmosphere and the outside world. Its function is to digest and take up the vital nutrients offer by food. At the equivalent time, it also provides a barrier that prevent health intimidating molecules to pass through the intestinal mucosa and access the systemic circulation. Intestinal dysfunctions are now supposed to be contributing factors to many chronic diseases such as allergies, autoimmune disorders, inflammatory disorders and degenerative diseases.

Intestinal dysbiosis

Intestinal microflora signify an ecosystem of the highest intricacy. The microflora not only has a decisive role in the digestion and absorption of nutrients, in the amalgamation of vitamins and fatty acids, in the detoxification of ingested chemicals but also in the command of the immune system. Alterations in the composition of the microflora may therefore have severe penalty for the hosts' health. A frequent disorder of the intestinal function is dysbiosis. This is an overgrowth/ overproduction of pathogenic bacteria in the intestine. The adult human intestinal tract is estimated to host up 55 different genera of bacteria, accounting for more than 501 different species.

• Antibiotic use is a common cause of major modification. Dosage, length of administration, spectrum of activity will conclude the impact on the microbial flora.

• Psychological stress can also affect the composition of the flora, including a noteworthy diminish in beneficial bacteria (*Lactobacilli* and *Bifidobacteria*) and an increase in pathogenic *E. coli*. Stress may affect bacterial growth by significantly tumbling the mucosal production of *mucopolysaccharides* and *mucins,* which are important for hinder the adherence of pathogenic organisms, and by decreasing the production of immunoglobulin A (IgA), which play a crucial role in their elimination. Neurochemicals twisted upon psychological stress can also nonstop enhance the growth of pathogenic organisms: norepinephrine stimulates the growth of *Y. Enterocolitica, P. Aeruginosa,* and gram-negative bacteria such as *E. coli.*

• Another factor that may have an impact on the human intestinal flora is diet. Some diets encourage the enlargement of beneficial microorganisms, while others promote harmful microfloral activities. For instance, diets rich in sulfur compounds (dairy products, eggs, certain vegetables, dried fruits...) promote the growth of sulfate-reducing bacteria. Globally it emerge that populations consuming the typical Western diet have more anaerobic bacteria, less *Enterococci,* and less types of yeasts than populations uncontrollable a vegetarian or high complex-carbohydrate diet.

Intestinal permeability: leaky gut syndrome

The healthy intestinal mucosa normally take up small molecules that result from complete digestion. Intestinal cells articulate specialized carrier protein that bring nutrients through the intestinal wall and into the bloodstream. Bigger molecules will not be transported by these systems and are normally kept within the gut for the reason that the intestine mucosal cells are tightly packed together. Leaky gut syndrome (LGS) is a condition in which the ability of the intestinal wall to keep out huge and unwanted molecules, is reduced. When the places between the cells of the intestinal wall become enlarged, macromolecules, antigens and toxins will make their way into the bloodstream.

What causes Leaky Gut Syndrome

A large numeral of factors can lead to leaky gut

• Dietary components: Fermentation of certain dietary components (proteins, refined carbohydrates) leads to potentially injurious end-products: ammonia, amines, phenols, sulfides... these compounds diminish the life-span of mucosal cells. Food additives, alcohol, caffeine also annoy the gut wall.

• Gut dysbiosis: Production of toxic compounds through fermentation also depends on the type of bacteria present in the bowel. In case of dysbiosis, overgrown pathogenic bacteria create toxins and compounds that are very detrimental to intestinal cells. For instance, sulfate-reducing bacteria produce toxic hydrogen sulfide.

• Food allergies and intolerances: Intolerance to certain food (lactose...) can lead to destructive gut inflammation.

• Chronic stress: In addition to favoring dysbiosis, stress diminish blood flow to the gut leaving it unable to repair itself. Stress also causes the cells of the intestine to contract which results in larger gaps stuck between cells.

Consequences of Leaky Gut

Altered permeability of the intestinal wall can have highly harmful effects, including

• **Nutritional deficiencies:** The carrier systems that usually transfer the nutrients through the intestinal wall are less active in damaged or inflamed mucosal cells.

• **Increased absorption of environmental toxins:** The gut mucosa is usually an efficient barrier against environmental chemicals that are present in food (food additives, pesticides, PCBs...). When allowed to pass into the circulation, these toxins can cause harm to all organs, notably the liver and the brain. Multiple Chemical Sensitivities may develop as the nervous system becomes sensitized.

• **Development of allergies and auto-immune reactions:** Undigested, large molecules pass into the bloodstream. The immune system identify these molecules as foreign and raises immunoglobulins against them. As a result, affected patients will develop allergies to many types of foods, which actually initiates a vicious cycle, since allergies will cause gut inflammation that conduct to more intestinal permeability...In addition, some of the molecules that pass into the blood may share homologies with proteins that are normally present in the body. Antibodies against these molecules will therefore attack the body's own cells, leading to auto-immune diseases.

• **Chronic activation (inflammation) of the immune system:** One bacterial compound that can easily make its way to the blood is lipopolysaccharide (LPS). Present in the bloodstream LPS will encourage a strong pro-inflammatory response in monocytes and macrophages, involving recognition by a receptor (Toll-like receptor-4) and the subsequent secretion of cytokines such as IL-1, IL-6, TNF-alpha. Such chronic inflammatory condition is observed Chronic Fatigue Syndrome (CFS/ME). LPS also induces the NK-kB-mediated production of nitric oxide. Because NO is amplified, NK function is inhibited and opportunistic infections such as mycoplasma infections are often observed. Herpesviruses, which tend to reactivate in a context of immune activation, will also be frequently detected.

How do probiotics work ?

The screening process of probiotics which has been explained by the following mechanism:

• **Exertion for nutrients** – Within the gut valuable and pathogenic micro-organisms will be utilising the equivalent types of nutrients. This results in a general competition between various types of bacteria for these nutrients. When a probiotic is taken there is an overall decline in nutrients available/ competition of nutrients for pathogenic bacteria and as a result this minimises or decreases the levels of pathogenic/ infectious micro-organisms effectively.

• **Competition for adhesion sites** – Beneficial bacteria can affix to the gut wall and form colonies at various places throughout the gut. This ruin pathogenic/ infectious bacterium from gaining a foothold, ensuing in their eviction from the body.

• **Augmentation in digestion** – Probiotics have been exposed to increase the efficacy of digestion and therefore provide an step up in digestion.

• **Lactic acid fabrication** – Probiotics produce lactic acid which take action to reduce the gut pH, inhibiting the growth of pathogenic bacteria, which prefer a additional alkaline environment.

• **Effect on immunity** – Probiotics have been shown to boost the levels of cell-signalling chemicals and the efficacy of infection-fighting cells (white blood cells)

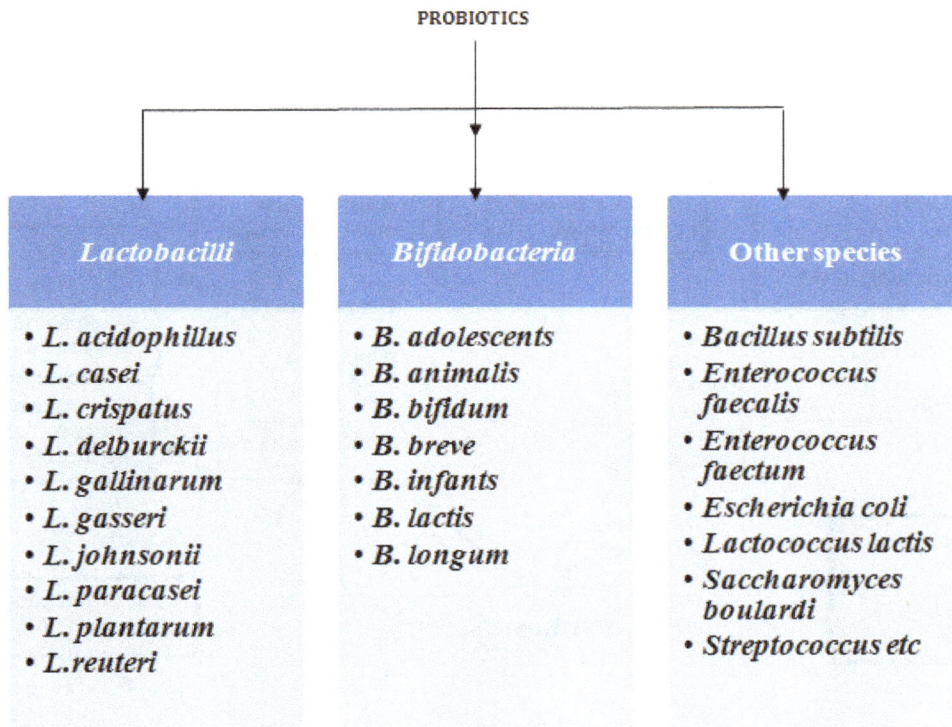

PROBIOTICS

Lactobacilli	Bifidobacteria	Other species
• L. acidophillus	• B. adolescents	• Bacillus subtilis
• L. casei	• B. animalis	• Enterococcus faecalis
• L. crispatus	• B. bifidum	• Enterococcus faectum
• L. delburckii	• B. breve	• Escherichia coli
• L. gallinarum	• B. infants	• Lactococcus lactis
• L. gasseri	• B. lactis	• Saccharomyces boulardi
• L. johnsonii	• B. longum	• Streptococcus etc
• L. paracasei		
• L. plantarum		
• L.reuteri		

Health Benefits / Efficacy of Probiotics

Probiotics have a number of health benefits for humans and animals, such as reducing lactose intolerance indication, enhancing the bioavailability of nutrients, endorsement lactose digestion, maintaining gut motility and many more. Probiotics help control intestinal microflora and improve immune function. Probiotics also decrease the pervasiveness of allergies in individuals, inhibit the inflammatory responses in the gut, and have antagonistic effects adjacent to intestinal and food-borne pathogens such as *E.coli., Staphylococcus aureus* and many more (Holzapfel WH, Haberer P et al. "Overview of gut flora and probiotics". Int J of Food Microbiol 1998;41:85-101)

1. Increased nutritional assessment (better digestibility, increased absorption of vitamins and minerals);

2. Endorsement of intestinal lactose digestion;

3. Encouraging influence on intestinal and urogenital flora (antibiotics and radiation induced colitis, yeast infections and vaginitis in women);

4. Regulation of gut motility (constipation, irritable bowel syndrome);

5. Decreased frequency and duration of diarrhea (antibiotic associated, Clostridium difficile, travelers, and rotaviral);

6. Maintenance of mucosal veracity;

7. Enhancement of immune system;

8. Anti-carcinogenic, anti-mutagenic and anti-allergic activities;

9. Sensation of well-being;

10. Anti-Candida properties;

11. Assist in the maintaining of inflammatory digestive conditions such as Inflammatory Bowel Disease (IBD), Crohn's disease and Ulcerative Colitis and Interstitial Cystitis

12. Relieving urinary region infections;

13. Optimistic influence on autistic children;

14. Provides antagonistic environment for pathogens for e.g. E.coli;

15. Blocking bond sites from pathogens; and

16. Inactivating enterotoxins.

Bacteria typically colonize the intestinal tract first and then reinforce the host defense systems by inducing generalized mucosal immune responses. Reports indicate that lactic acid bacteria (LAB) as Lactobacillus and Bifidobacterium and their fermented products are effective at enhancing innate and adaptive immunity, prevent gastric mucosal lesion development, alleviate allergies, and put up defense against intestinal pathogen infection.

Efficacy of Probiotics depends upon several factors

1 Dose administration

2 Mode of administration

3 Strains

4 Duration of treatment

5 Health status of individual

The diagram below is a good representation of various functions and health benefits of probiotics. (Holzapfel WH, Haberer P et al. "Overview of gut flora and probiotics". Int J of Food Microbiol)

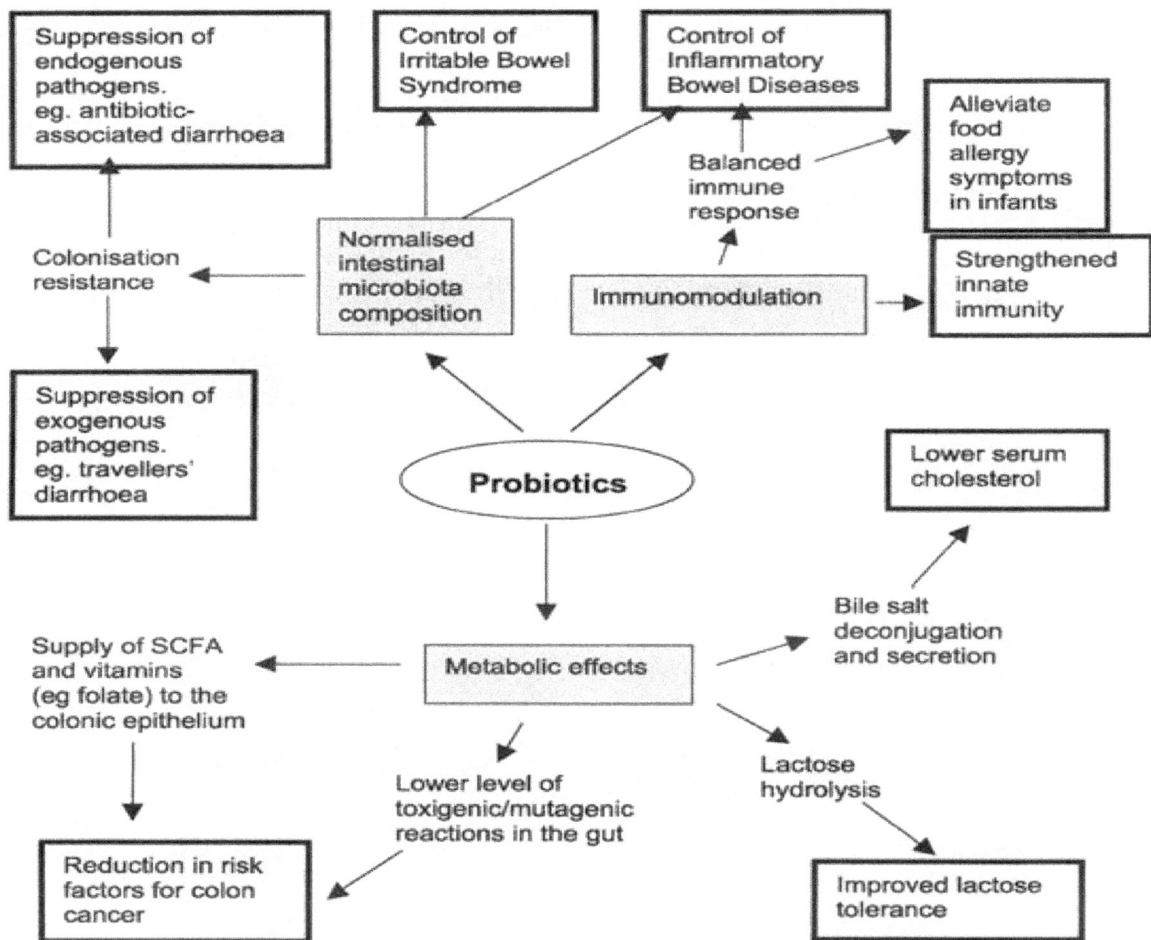

Proposed health benefits stemming from probiotic consumption.

Probiotics play an vital role in human nutrition and health, and in balancing the intestinal microflora naturally. Health benefits attributed to the consumption of probiotics include: protection of the normal gut flora, improvement of lactose (milk-sugar) intolerance, improvement of digestive processes and assimilation of nutrients, and stimulation of body's immune system. Probiotics are widely used in a lot of countries by consumers and in clinical practice and their health benefits are being examine widely, especially in the last few decades. The need to find alternative therapeutic approaches to overcome side effects associated with the current pharmacological treatments, and the require for new antimicrobials due to the overuse of antibiotics has drive forward research on probiotics against a mass of disorders of varying severity given their favorable safety profiles (Mackie R., Gaskins H.R., "Gastrointestinal microbial ecology". Science & Medicine Nov./Dec. 1999)

It May Help Reduce Cholesterol

High cholesterol levels may increase the menace of heart disease. This is especially true for "bad" LDL cholesterol. Fortunately, studies suggest that certain probiotics can assist reduce cholesterol levels and that L. acidophilus may be more valuable than other types of probiotics. Some of these studies have examine probiotics on their own, whereas others have used milk drinks fermented by probiotics. One study found that taking L. acidophilus and another probiotic for six weeks considerably lowered total and LDL cholesterol, but also "good" HDL cholesterol. A similar six-week study establish that L. acidophilus on its own had no effect. However, there is evidence that stick together Lactobacillus. acidophilus with prebiotics, or heavy carbohydrates and other sources that help healthy bacteria plough can help boost HDL cholesterol and lower blood sugar. This has been confirmed in studies using probiotics and prebiotics, both as supplements and in fermented milk drinks.

It May Prevent and Reduce Diarrhea

Diarrhea affect people for a number of reasons, including bacterial infections. It can be dangerous if it last a long time, as it results in watery loss and, in some cases, dehydration. A number of observations have shown that probiotics may assist prevent and reduce diarrhea that's connected with various types of diseases. Evidence on the ability of L. acidophilus to treat acute diarrhea in children. Some studies have shown a beneficial effect. One meta-analysis involving more than 200 children found that probiotics help diminish diarrhea, but not only in hospitalized brood. What's further when harried in combination with another probiotic, L. acidophilus may help reduce diarrhea caused by radiotherapy in mature cancer patients. Similarly, it may help decrease diarrhea associated with antibiotics and a common infection called Clostridium difficile, and Staphylococcus aureus and many more. Diarrhea is also common in people who take a trip to different countries and are exposed to new cuisine and environments. A review of 12 studies found that probiotics are successful at preventing traveler's diarrhea and that Lactobacillus acidophilus, in grouping with another probiotic, was most effectual on doing so.

It Can Improve Symptoms of Irritable Bowel Syndrome

Irritable bowel syndrome (IBS) affects up to one in five people in many countries. Its sign include abdominal pain, bloating and curious bowel movements. Some research suggests that IBS it might be caused by certain types of bacteria in the intestines. Therefore, a numeral of studies have examined whether probiotics can help recover its symptoms. In a study in 70 people with functional bowel disorders including IBS, taking a combination of probiotic for one to two months improved bloating. A similar study found that Lactobacillus acidophilus alone also reduced abdominal pain in IBS patients. On the other hand, a study that examined a mixture of Lactobacillus acidophilus and other probiotics found that it had no effect IBS symptoms. This might be explained by another study signifying that taking a low dose of single-strain probiotics for a short duration may recover IBS symptoms the most. Specifically, the study indicate that the best way to take probiotics for IBS is to use single-strain probiotics, rather than a mix, for less than eight weeks, as well as a dose of less than 11 billion colony-forming units (CFUs) per day. However, it's important to choose a probiotic supplement that has been scientifically proven to profit IBS.

It Can Help Treat and Prevent Vaginal Infections

Vaginosis and vulvovaginal candidiasis are the most common types of vaginal infections. There is virtuous evidence that Lactobacillus acidophilus can help treat and prevent such infections. Lactobacilli are classically the most common bacteria in the vagina. They generate lactic acid, which avoid the growth of other harmful bacteria. However, in cases of certain vaginal disorders, other species of bacteria begin to outnumber lactobacillus species. A number of studies have found taking probiotic supplement can prevent and treat vaginal infections by increasing lactobacilli in the vagina. Nevertheless, other observations have found no effect. Eating yogurt that enclose various probiotic bacteria may also prevent vaginal infections. Until now both of the studies that scrutinize this were quite small and would need to be simulated on a larger scale before any assumption could be made.

It May Help Prevent and Reduce Symptoms of Eczema

Eczema is a situation in which the skin becomes red-looking, resulting in itchiness and pain. The most ordinary form is called atopic dermatitis. Verification suggests that probiotics can lessen the symptoms of this inflammatory condition in both adults and children. Several study set up that giving a mix of certain probiotics bacteria (such as Lactobacillus acidophilus and many others suchas bifidobacteri species) to pregnant women and their infants during the first three months of life reduced the occurrence of eczema by 25% by the time the infants reached approximately one year of age . A comparable study found that L. acidophilus, in combination with traditional medical therapy, significantly improved atopic dermatitis symptoms in children. However, not all studies have shown optimistic effects. A large study in 235 newborn children given Lactobacillus acidophilus for the first six months of life found no valuable effect in cases of atopic dermatosis. In fact, it increased sensitivity to allergen

It's Good for Your Gut Health

Your gut is lined with trillions of bacteria that play an significant role in your health. Generally, lactobacillus species are very good for gut health. They produce lactic acid, which may prevent injurious bacteria from inhabit the intestines. They also ensure the lining of the intestines stays intact. *L. acidophilus* can amplify the amounts of other healthy bacteria in the gut, including other *lactobacilli* and *Bifidobacteria.* It can also enlarge levels of short-chain fatty acids, such as butyrate and many more, which promote gut health. Another study carefully examined the effects of *Lactobacillus acidophilus* on the gut. It found that taking it as a probiotic increased the appearance of genes in the intestines that are involved in immune response. These results suggest that probiotics may support a healthy immune system. A separate study examined how the amalgamation of *L. acidophilus* and a prebiotic affected human gut health. It found that the combined supplement increased the amounts of lactobacilli and *Bifidobacteria* in the intestines, as well as branched-chain fatty acids, which are an significant part of a healthy gut.

Conclusion

The microbiome is an assortment of all microbial species that coexist within an individual. These organisms manipulate several aspects of individual body functions. Probiotic organisms are generally beneficial components of microflora and confer normal health statusin the well being. Usually, probiotics should be provided from the exterior in the diet for maintaining suitable health status in today,s world. Probiotics are extensively used worldwide due to advancement in the affiliation between nutrition and health besides their promising benefits and approximately, negligible sideeffects. Indian probiotic industry is achieving its rapidity at steady state with oppurtinities for quick growth in near future. The future of probiotics / probiotic foods is even shows a potential, as current consumers/people are worried to sustain their personnel health and expect that they eat to be healthy and capable of preventing illness. This paper offers a brief overview of the health benefits of probiotics as therapeutic agents.

Reference

1. Green BK and Schleicher L: US patent, 2800457, CA 1957,51;15842d 1957;13-627.
2. Ansel HC, Pharmaceutical dosage form and drug delivery system. Lippincott Williams and Wilkins. 2000;233-234.
3. Yazici E, Oner, Kas HS, Hincal AA. Phenytoin sodium microcapsules: bench scale formulation, process characterization and release kinetics. Pharmaceutics Dev Technol. 1996;1:175-183. Doi:10.3109/10837459609029892
4. Blair HS, Guthrie J, Law T, Turkington P Chitosan and modified chitosan membranes I, preparation and characterisation. J App Poly Sci. 1987;33:641-656. Doi: 10.1002/app.1987.070330226
5. Nack H, Microencapsulation techniques, application and problems. J.Soc.Cosmetic Chemists. 1970;21:85-98.
6. Swapan Kumar Ghosh. Functional Coatings and Microencapsulation: A General Perspective. WILEY-VCH Verlag GmbH & Co. KGaA, Weinheim. 2006. Doi: 10.1002/3527608478.ch1
7. Finch CA, Polymers for microcapsule walls. Chem. Ind. 1985;22:752-756.
8. Li SP, Kowarski CR, Feld KM and Grim WM Recent advances in microencapsulation technology and equipment. Drug Dev Ind Pharm. 1988;14(2-3):353-376. Doi: 10.3109/03639048809151975
9. Lehman Leon, Lieberman A. Herbert and Kanig L.Josep. The Theory and Practice of Industrial Pharmacy. 3rd edition, Vargehese Publishing House. 1976;412.
10. Schwendeman SP. Recent advances in the stabilization of proteins encapsulated in injectable PLGA delivery systems. Crit Rev Ther Drug Carrier Syst. 2002;19(1):73-98.

20

MicroRNAs' Gene Expression Technology for the Evaluation of Disease Modulation by Mild Nutrients

Farid E Ahmed[1], Mostafa M Gouda[2,3], Laila Hussein[2], Paul W Vos[4], Mohamed Mahmoud[5], and

Nancy C Ahmed[1]

[1]GEM Tox Labs, Institute for Research in Biotechnology, 2905 South Memorial Drive, Greenville, NC 27834, USA.

[2]Department of Nutrition & Food Science, National Research Center, El-Bohooth Street, Dokki, Cairo, Egypt

[3]National Research and Development Center for Egg Processing, College of Food Science and Technology, Huazhong Agricultural University, Wuhan, Hubei, PR, China

[4]Department of Biostatistics, College of Allied Health Sciences, East Carolina University, 600 Moye Boulevard, Greenville, NC 27858, USA

[5]USDA/ARS Children's Nutrition Research, 1100 Bates Street, Houston, TX 77030, USA

Corresponding author: *Farid E Ahmed, GEM Tox Labs, Institute for Research in Biotechnology, 2905 South Memorial Dr, Greenville, NC 27834, USA, Tel. 252-375-9656, E-mail: gemtoxconsultants@yahoo.com*

Abstract

This article demonstrates the use of melt curve analysis (MCA) for the interpretation of mild nutrogenomic micro (mi) RNA gene expression data, by measuring the magnitude of the expression of key miRNA molecules in stool of healthy human adults, used as molecular markers in excrements, following the intake of polyphenol-rich Pomegranate juice (PGJ), functional fermented sobya (FS), which is rich in potential Probiotic lactobacilli, or their combination. Total small RNA was isolated from stool of 25 volunteers before and following a three week dietary intervention trial. Expression of 88 miRNA genes was evaluated using Qiagen's 96 well plate RT2 miRNA qPCR arrays. Employing parallel coordinates plots, no significant separation for the gene expression (C q) values was observed using Roche 480® PCR Light Cycler instrument used in this study, and none of the miRNAs showed significant statistical expression after controlling for the false discovery rate. On the other hand, melting temperature profiles produced during PCR amplification run, found seven significant genes (miR-184, miR-203, miR-373, miR-124, miR-96, miR-379 and miR-301a), which separated candidate miRNAs that could function as novel molecular markers of relevance to oxidative stress and immunoglobulin function, for the intake of polyphenol (PP)-rich, functional fermented foods rich in lactobacilli (FS), or their combination. We elaborate on these results, and present a detailed review on use of melt curves for analyzing nutigenomic miRNA expression data, which initially appear to show no significant expressions, but are actually more subtle than this simplistic view, necessitating the appreciation of the important role of MCA for a comprehensive understanding of what the collective expression and MCA data collectively imply. We have correlated miRNA genes with messenger (m) RNA genes using bioinformatics' methods.

Key Words: biomarkers, DNA, melt curve, fermented sobya, miRNA, pomegranate, PCR, RNA

Introduction

Gene expression and its control by miRNAs

Cell's gene expression profile determines its function, phenotype and cells' response to external stimuli, and thus helps elucidate various cellular functions, biochemical pathways and regulatory mechanisms [1]. Several gene expression profiling methods at the mRNA level have emerged during past years, and have been successfully applied to cancer research. Profiling by microarrays [2, 3] allows for the parallel quantification of thousands of genes from multiple samples simultaneously, using a single RNA preparation, and has become valuable because microarrays are convenient to use, do not require large-scale DNA sequencing, gives a clear idea of cells' physiological state, and is considered a comprehensive approach to characterize cancer molecularly, as seen in studies on colon cancer [1-3].

Control of gene expression has been studied by miRNA molecules, a small non-coding RNA molecules (18–24 nt long), involved in transcriptional and post-transcriptional regulation of gene expression by inhibiting gene translation, and the discovery of this molecule resulted in a 2006 Noble Prize in Physiology & Medicine to Andrew Z Fire and Caraig C. Menlo for their work on RNA interference (RNAi) [https://science.howstuffworks.com/environmental/conservation/issues/nobel-prize-rnai.htm]. MiRNAs silence gene expression through inhibiting mRNA translation to protein, or by enhancing the degradation of mRNA. Since first reported in 1993 [4], the number of identified miRNAs in June 2014, version 14.0, the latest miRBase release (v20) [5] contains 24,521 miRNA loci from 206 species, processed to produce 30,424 mature miRNA products. MiRNAs are processed by RNA polymerase II to form a precursor step which is a long primary transcript. Pri-miR is converted to miRNA by sequential cutting with two enzymes belonging to a class of RNA III endonuclease, Drosha and Dicer. Drosha converts the long primary transcripts to ~70 nt long primary miRNAs (pri-miR), which migrate to the cytoplasm by Exportin 5, and converted to mature miRNA (~22 net) by Dicer [6]. Each miRNA may control multiple genes, and one or more miRNAs regulate a large proportion of human protein-coding genes, whereas each single gene may be regulated by multiple miRNAs [7]. MiRNAs inhibit gene expression through interaction with 3-untranslated regions (3 UTRs) of target mRNAs carrying complementary sequences [7]. Thus, the tumors had figured out a shrewed way to turn on the miRNAs, creating a growth process that is impossible to stop.

Effect of antioxidant polyphenols --abundant in Mediterranean diets-- on gene expression unraveled by the availability of molecular biology techniques, reveals our adaptation to environmental changes [8]. Efforts to study the human transcriptome have collectively been applied to tissue, blood, and urine (i.e., normally sterile materials), as well as stool (a non-sterile medium). Extraction protocols that employ commercial reagents to obtain high-yield, reverse-transcribable (RT) RNA from human stool in studies performed on colon cancer have been reported [1, 2, 9]

Micro (mi) RNAs as biomarkers, and their roles in disease processes

A biomarker is believed to be a characteristic indicator of normal biological processes, pathogenic processes, or pharmacological responses to therapeutic interventions [10, 11]. In contrast, clinical endpoints are considered as variables representing a study subject's health from his/her perspective [12-22]. A variety of biomarkers exist today as surrogates to access clinical outcomes in diseases, predict the health of individuals, or improve drug development. An ideal biomarker should be safe and easily measured, is cost effective to follow up, is modifiable with treatment, and is consistent across genders and various ethnic groups. Because we never have a complete understanding of all processes affecting individual's health, biomarkers need to be constantly reevaluated for their relationship between surrogate endpoints and true clinical endpoints [12-14]. MiRNAs have been used herein as biomarkers for assessing the effect of intake of PP-rich or fermented foods on the expression of 88 miRNA genes known to influence cancer.

Disease modulation by nutrients

Cardiovascular diseases due to hypercholesterolemia are considered a risk factor for Chronic Heart Disease (CHD), and chronic degenerative diseases --caused wholly or partially by dietary patterns-- represent the most serious threat to public health [23]. Moreover, nearly one-third of all cancer deaths are due to poor nutrition, lack of physical activity, and obesity; and these risk factors account for nearly 80% of large intestine, breast, and prostate cancers [24]. Chronic inflammation is considered a common factor that contributes to the development and progression of these illnesses, which are caused by and/or modified by diet [25].

Pomegranate juice (PGJ) and derived products are considered the richest sources of polyphenolic compounds [26], with positive implication on total serum cholesterol (TC), low-density lipoprotein cholesterol (LDL-C) and triglyceride (TG) plasma lipid profile [27]. Moreover, anthocyanin and ellagitannins pigments, mainly punicalagins, inhibit the activities of enzymes 3-hydroxy-3-methylglutaryl-CoA reductase and sterol O-acyltransferase, important in cholesterol metabolism [28]. Probiotic bacteria also contribute to lowering plasma hyper cholestrolemia due to the above mechanism, caused by the Probiotic bile salt hydrolases (BSH) activity. This Probiotic enzyme hydrolyses conjugates both glycodeoxycholic and taurodeoxycholic acids to hydrolysis products, inhibiting cholesterol absorption and decreasing reabsorption of bile acid [29].

Colonic micro biota is a central site for the metabolism of dietary PP and colonization of Probiotic bacteria. A dietary intervention study with Probiotic strains from three Lactobacillus species (L. acidophilus, L. casei and L. rhamnosus) given to healthy adults, showed that bacterial consumption caused the differential expression of from hundreds to thousands of genes in vivo in the human mucosa. The interaction of PP with the gut micro biota influences the expression of some human genes (i.e., nutritional transcriptomics), which mediates mechanisms underlying their beneficial effects [30]. Similar in vivo mucosal transcriptome findings have been reported when adults were given the Probiotic L. plantarum, illustrating how probiotics modulate human cellular pathways, and show remarkable similarity to responses obtained for certain bioactive molecules and drugs [31].

Materials and Methods

Participants

Study subjects were 25 healthy adults, 20 to 34 years old; exclusion was: absence of metabolic diseases, no use of medication for the last 6 weeks, and no signs of allergy or hypersensitivity to food or ingested material.

Exclusion criteria at the time of the screening were as follows: History of diabetes, hypertension, heart disease, or endocrine disorders; abnormal blood chemistry profile, fasting LDL-cholesterol concentration > 3.37 m mol/L (> 130 mg /d L), or fasting triacylglycerol concentration > 3.39 mmol/L (> 300 mg/dL), taking antioxidant or fish oil supplements. Female subjects were neither pregnant nor lactating. To minimize the potential confounding effects of consuming fluctuating amounts of foods and beverages that are high in dietary flavonoids, all subjects avoided the intakes of purple grapes, cocoa and chocolate during the entire three week dietary intervention trial. The volunteers were instructed to continue to eat their normal diet and not to alter their usual dietary or fluid intake with the exception of the previously mentioned food restrictions.

Compliance with the supplementation in all subjects was satisfactory, as assessed daily, and all subjects continued their habitual diets throughout the study. The research protocol was approved by the institution review board at Egypt's NRC, and all subjects have given written consent prior to their participation in the study.

Design of the study

Figure 1 shows the design of the nutrigenomic randomized

study. Estimated dietary intake was assessed by 3 repeated food records, one week before they were enrolled in the trial. The average portion sizes consumed, as well as composition data values from nutrient composition of the food were combined to assess average daily energy and nutrient intakes by the "nutrisurvey" software program. The characteristics of the voluntary subjects who were enrolled in the study, the mean daily energy intake, as well as selected macro nutrients are presented in Table 1.

Isolate RNA→Reverse transcribe total RNA into cDNA→ Carry out NGS or microarray studies →Analyze NGS or microaray data→ Chose a smaller number of samples to run RT-qPCR reactions→Analyze quantitative PCR data

Figure 1: miRNA Quantitation Experimental Workflow

Table 1: Composition of the supplements

Parameter	Unit	Dietary supplements			
		Control (portion served)	FS (portion served)	PGJ (portion served)	FS+ PG (portionserved)
Portion Size	g	-	170	250	150+10
Total Solids	g	-	40.01	17.75	48
Carbohydrate	g	-	51.10	32.75	59
Dietary Fiber	g	-	54	0.25	48.6
Energy	kcal	-	263	135	290
Lactobacillus	cfu/	-			
diverse	serving size	-	5.1 x 109	-	4.5 x109
Yeast	cfu/serving size	-	2.77 x1010	-	2.44 x1010
Total PP	mg*/portion g	-	-	519.1±8.75	207.65±3.5
Antioxidant		-			
activity	(AEAC)**	-	7.74±1.33	11.35±2.2	11.37±2.2

Supplements

Pomegranate was obtained in bulk from the Obour Public Market, Cairo, Egypt. Pomegranate fruits were peeled and the juice was extracted using a laboratory pilot press (Braun, Germany). The juice was distributed in aliquots of 100 or 250 grams in air tight, light-proof polyethylene bottles, and frozen at −20°C, where pomegranate polyphenols remained stable. Soursobya, a fermented rice porridge containing per gram 3×10^7 cfu diverse lactic acid bacteria (LAB) and 1×10^7 cfu *Sacharomyces cerivisiae*. With added ingredients such as milk, sugar and grated coconut, was purchased twice a week from the retail market, and saved in the refrigerator. Sobya is fermented rice. Table 2 illustrates the proximate initial and final mean urinary polyphenols, plasma and urinary ant oxidative activity, urinary thiobarbituric acid reactive species (TBARS), and erythrocytic glutathione-S-transferase (GST).

Urine, blood and stool collection & storage

One day before starting the trial, urine and stool samples were collected from all volunteers, processed according to our standard operating methods, and saved frozen at -70°C.

Urine samples: Collection begins in the early morning after the subjects had fasted 10–12 h both on −1, and +21 d nutritional intervention and aliquots (2 mL) were immediately frozen at −20°C for biochemical analysis.

Blood samples: Blood was drawn by vein-puncture and collected in sodium citrated tubes. The plasma was separated from blood cells by centrifugation at 3000 rpm for 15 min at 4°C and the separated plasma was stored at -70⁰ C for later biochemical analysis. The red blood cells (RBCs) were washed using cold physiological saline solution and stored at -20⁰C.

Table 2: Initial and final mean urinary polyphenols, plasma and urinary ant oxidative activity, urinary TBARS and erythrocytic GST

Parameter	Unit	Control				Sobya				Pomegranate				Sobya+Pomegranate			
		Baseline		Final		Baseline		Final		Baseline		Final		Baseline		Final	
		X±SE		X±SE p		X±SE		X±SE p		X±SE		X±SE p		X±SE		X±SE p	
Urinary	GAE/mg	10.36		8.11		11.84		9.86		5.70		55.23 <0.05		10.40		21.62	
polyphenol	creat	±1.8		±2.2		±6.2		±1.8		±1.4		±21.7		±3.2		±7.3	
Urinary	AEAC/mg	9.74		8.13		3.89		10.30		7.18		46.57 <0.05		10.90		20.25	
antioxidant activity	creat*	±2.0		±2.7		±09		±2.3		±0.9		±18.0		±2.4		±3.9	
Urinary	ug/mg	83.04		75.17		82.77		29.97		173.93		51.48 <0.05		157.70		40.62	
TBARS	creat	±12.1		±15.3		±27.8		±4.4		±44.8		±8.2		±47.8		±8.3	
Plasma	AEAC/	6.36		5.99		3.70		4.55		3.64		5.92 <0.05		2.78		4.49	
antioxidant	1oo ml	±2.81		±2.66		±0.33		±0.27		±0.30		±0.68		±0.11		±0.58	
E-GST	IU/g Hb	5.94		5.45		4.26 <0.05		7.21		4.73		8.34		4.56		6.90	
activity		±3.3		±4.1		±0.5		±0.8		±1.0		±1.0		±1.0		±1.0	

X±SE: Mean ± Standard Error, *:mmol ascorbic acid equivalent antioxidant capacity/mg creatinine; GAE: gallic acid equivalent, Mean values are significantly different if the P-values are less than 0,05 (P<0.05) [13].

Stool samples: Feces (excrement) were obtained from the 25 healthy adults, twice at day 0 and three weeks after the dietary intervention. All stools were collected with sterile, disposable wood spatulas in clean containers, after stools were freshly passed, and then placed for storage into Nalgene screw top vials (Thermo Fisher Scientific, Inc., Palo Alto, CA, USA), each containing 2 ml of the preservative RNA later (Applied Biosystems/Ambion, Austin, TX, USA), which prevents the fragmentation of the fragile mRNA molecule [1], and vials were stored at – 70 °C until samples were ready for further analysis. Total small RNA, containing miRNAs, was extracted from all frozen samples at once, when ready, and there was no need to separate mRNA containing small miRNAs from total RNA, as small total RNA was suitable daytracting small total RNA from stool was carried out using a guanidinium-based buffer, whand saved frozen atich comes with the RNeasy isolation Kit®, Qiagen, Valencia, CA, USA, as we have previously detailed. DNase digestion was not carried out, as our earlier work demonstrated no difference in RNA yield or effect on RT-PCR after DNase digestion [1]. The time to purify aqueous RNA from all of the 25 frozen stool samples was ~ three hours. Small RNA concentrations were measured spectrophotometrically at λ 260 nm, 280 nm and 230nm, using a Nano-Drop spectrophotometer (Themo-Fischer Scientific). The integrity of total RNA was determined by an Agilent 2100 Bio analyzer (Agilent Technologies, Inc., Palo Alto, CA, USA) utilizing the RNA 6000 Nano Lab Chip®. RNA integrity number (RIN) was computed for each sample using instrument's software [1].

Preparation of ss-cDNA for molecular analysis

The RT² miRNA First Strand Kit® from SABiosciences Corporation (Frederick, MD, USA) was employed for making a copy of ss-DNA in a 10.0 µl reverse transcription (RT) reaction, for each RNA samples in a sterile PCR tube, containing 100 ng total RNA, 1.0 µl miRNA RT primer & ERC mix, 2.0 /µl 5X miRNA RT buffer, 1.0 µl miRNA RT enzyme mix, 1.0 µl nucleotide mix and Rnase-free H_2O to a final volume of 10.0 µl. The same amount of total RNA was used for each sample. Contents were gently mixed with a pipettor, followed by brief centrifugation. All tubes were then incubated for 2 hours at 37°C, followed by heating at 95°C for 5 minutes to degrade the RNA and inactivate the RT. All tubes were chilled on ice for 5 minutes, and 90 µl of Rnase-free H_2O was added to each tube. Finished miRNA First Strand C DNA synthesis reactions were then stored overnight at -20°C [1].

Use of cancer RT² miRNA PCR array 96-well plate to study miRNAs' expressions

We used a SABiosciences RT² miRNA qPCR Array Plate System for Human (Qiagen) to analyze miRNA expression using real-time, reverse transcription PCR (RT-qPCR) as a sensitive and reliable quantitative method for miRNA expression analysis.

The arrays employ a SYBR Green real-time PCR detection system, which has been optimized to analyze the expression of many mature miRNAs simultaneously. Each 96-well array plate contains a panel of primer sets for 88 relevant miRNA focused pathways (one universal primer and one gene-specific primer for each miRNA sequence), plus four housekeeping genes (Human SNORD 48, 47 and 44, and U6), and two RNA and two PCR quality controls. Duplicate RT Controls (RTC) to test the efficiency of the miRNA RT reaction, with a primer set that detects the template synthesized from the built in miRNA External RNA Control (ERC).

There are duplicate RT controls (RTC) to test the efficiency of the miRNA RT process, with a primer set to detect the template synthesized from the kit's built-in miRNA External RNA Control (ERC). There is also duplicate positive PCR controls (PPC) to test the efficiency of the PCR process, using a per-dispensed artificial DNA sequence and the primer set that detects it. The two sets of duplicate control wells (RTC and PPC) also test for inter-well, intra-plate consistency. The human RT² miRNA PCR Arrays reflect miRNA sequences annotated by the Sanger miRBase Release 14. Figure 2 shows the layout of the MAH-102F array.

Layout for Human cancer RT² miRNA qPCR Array (MAH-102A)

	1	2	3	4	5	6	7	8	9	10	11	12
A	let-7a	miR-133b	miR-122	miR-20b	mir-335	miR-196a	miR-125a-5p	miR-142-5p	miR-96	miR-222	miR-148b	miR-92a
B	miR-184	miR-214	miR-15a	miR-18b	miR-378	let-7b	miR-205	miR-181a	miR-130a	miR-199a-3p	miR-140-5p	miR-20a
C	miR-146b-5p	miR-132	miR-193b	miR-183	miR-34c-5p	miR-30c	miR-148a	miR-134	let-7g	miR-138	miR-373	let-7c
D	let-7e	miR-218	miR-29b	miR-146a	miR-212	miR-135b	miR-206	miR-124	miR-21	miR-181d	miR-301a	mir-200c
E	miR-100	miR-10b	miR-155	miR-1	miR-363	miR-150	let-7i	miR-27b	miR-7	miR-127-5p	miR-29a	miR-191
F	let-7d	miR-9	let-7f	miR-10a	miR-181b	miR-15b	miR-16	miR-210	miR-106a/17	miR-98	miR-34a	miR-25
G	miR-144	miR-128a	miR-143	miR-215	miR-19a	miR-193a-5p	miR-18a	miR-125b	miR-126	miR-27a	miR-372	miR-149
H	miR-23b	miR-203	miR-32	miR-181c	SNORD48	SNORD47	SNORD44	RNU6-2	miRTC	miRTC	PPC	PPC

Source: SABiosciemces, a Qiagen Company, Fredrick, Maryland, USA

Figure 2: Array lay out for Qiagen's human cancer RT2 miRNA qPCR Plate Array (MAH-102A).

Performing real-time quantitative polymerase chain reaction (qPCR)

We used RT² SYBR Green qPCR Master Mix (SBA Biosciences) to obtain accurate results from our qPCR arrays. The following components were mixed in a 15-ml tube for 96-well plate format: 1275 µl of 2X RT² SYBR Green PCR Master Mix, 100 µl of diluted first strand reaction, 1175 µl of ddH$_2$O (total volume 2550 µl, of which 2400 was needed for 96 reactions, each well having 25 µl, with150 µl cocktail remaining.

We employed a Roche Light Cycler 480® 96-well block PCR Machine (Roche, Mannheim, Germany) to carry out quantitative real-time miRNA expressions. When ready, we removed the needed miRNA qPCR Arrays, each wrapped in aluminum foil, from their sealed bags, added 25 µl of the same cocktail to each well, adjusted the ramp rate to 1°C/sec. We used 45 cycles in the program, and employed the Second Derivative Maximum method, available with the Light Cycler 480® software for data analysis [32]. We first heated the 96 well plate for 10 min at 95°C to activate the Hot Start DNA polymerase, then used a three-step cycling program (a 15 seconds heating at 95°C to separate the ds DNA, a 30 seconds annealing step at 60oC to detect and record SYBR Green fluorescence at each well during each cycle, and a final heating step for 30 seconds at 72°C). Each plate was visually inspected after the run for signs of evaporation from the wells. Data were analyzed using the 2-ΔΔCt method [33]. Resulting threshold cycle values for all wells were exported to a blank

Excel sheet for analysis. We also ran a Dissociation (Melt) Curve Program after the cycling program [34], and generated a first derivative dissociation curve for each well in the plate, using the LC (Light cycler's®) software.

Use of transcriptomic markers for colon cancer screening in stool, blood and tissue

Using molecular methods to carry out quantitative gene expression studies on stool, blood and tissue samples, allows for sensitive and specific routine monitoring and diagnostic screening of colon cancer at the early malignant tags for ten genes: HPRT, IGF2, FLNA, TGFßigh, CKS2, CSE1L, KLK10, CXCL3, GUCA2B, IL-12] in a two-step quantitative PCR end-point, on an Applied Bio system 9600 thermo cycler (Foster City, CA) on single-stranded (s s)-cDNA at >90% sensitivity and >95% specificity, using a master mix containing final concentrations of 1X Hugh fidelity PCR buffer, 0,2 mM dNTP, 2 mM MgSO$_4$, 0.4 µm forward and reverse primers, 0.1 ng ss-cDNA template and 1 U of "hot start" Platinum High Taq DNA polymerase (Invitrogen) in a final volume of 25 ul, in a 100 ul PCR tube. Running conditions were: one cycle at 94°c for 3 min to activate the hot start Taq, 35 cycles of 94°C denaturation for 45 sec, 55°C annealing for 1 min and 72°C elongation for 1 mimn, followed by one elongation/extension cycle at 72°C for 7 min. Reactions were placed in a well of a 1% agarose gel immersed with 1X Tris-acetate EDTA (TAE) gel running buffer in an electrophoresis apparatus (5 V per cm), gel stained with ethedium bromide (0.25 u g/ml final concentration), and visualized using an Alpha Innotech charge-coupled device (CCD) based imaging system (San Leandro, CA). A semi-quantitative real-time PCR, which employed the comparative cross point (CP) method, also named E-method [36] was employed on a Roche's Light Cycler (LCTM) model 2.0 PCR using the LC Relative Quantification Software TM; which employs the LC Relative Quantification Software using standard curves in which standard concentrations are plotted versus the threshold cycle to calculate the unknown sample, without user input [36-38], Figure 2 [1]. However, we later on employed the more stable miRNAs' genes, because miRNA molecules are more stable, and easier to handle and work with than the more fragile mRNA molecules. We have bioinformatically correlated the mRNA with miRNA genes using Target Scan algorithm, and employing the DAVID program [39].

Role of biomarker miRNAs in various diseases

MiRNA functions were shown to regulate development [35] and apoptosis [36], and dysregulation of miRNAs has been associated with many diseases such as various cancers [37], heart diseases [38], and kidney diseases [39]. Nervous system diseases [40], alcoholism [41], obesity [42], auditory diseases [43], eye diseases [44], skeletal growth defects [45], as well as key role in host–virus pathogenesis of viral diseases [46].A negative correlation was found between tissue specificity of interactions and miRNA in a number of diseases, and an association between miRNA conservation and disease, and predefined miRNA groups allow for identification of novel disease biomarkers at the miRNA level [47]. Specific miRNAs are crucial in oncogenesis [48], effective in classifying solid [49] and liquid tumors [50], and function as oncogenes or tumor suppressor genes [51]. MiRNA genes are often located at fragile sites, as well as minimal regions of loss of heterozygosity, or amplification of common breakpoints regions, suggesting their involvement in carcinogenesis [52]. MiRNAs have shown to serve as biomarkers for cancer diagnosis, prognosis and/or response to therapy [53, 54]. Profiles of miRNA expression differ between normal tissues and tumor types, and evidence suggests that miRNA expression profiles can cluster similar tumor types together more accurately than expression profiles of protein-coding messenger (m) RNA genes [55]. Besides, small miRNAs (~18-22 nt long) are stable molecules than the fragile mRNA [56].

Melt Curve analysis

MCA has been an effective and economical way for identification of virus stains [57], genes [58], bacterial strains [59], insect species [60], temperature validation of PCR cyclers [61], detection of translocations in lymphomas [62] and RNA interference/gene silencing [63]. Thus, the presence of double peaks during MCA is not always indicative of non-specific amplification, and other methods such as agarose gel electrophoresis and use of melt curve prediction software [64, 65] are also needed in odder to determine the purity of an amplicon. For example, Figure 5A shows a single peak for exon 17b of CFTE

(Cystic Fibrosis Transmembrane Conductance Regulator) gene, whereas the melt curve for an amplicon from exon 7 of CFTR shows two peaks, which could be interpreted as indicative of two separate amplicons (Figure 5B). However, analysis by agarose gel electrophoresis showed only one peak. To solve this conflict, an understanding of how melt curves are produced is needed. It should be emphasized that intercalating dyes used in qPCR, such as SYBR Green, will fluoresce only when the dye is bound to ds DNA, but not in the presence of a ssDNA, or when the DNA is free in solution. After the amplification cycle in qPCR, the instrument starts at a preset temperature above the primer Tm, and as the temperature increases dsDNA denatures becoming ssDNA and the

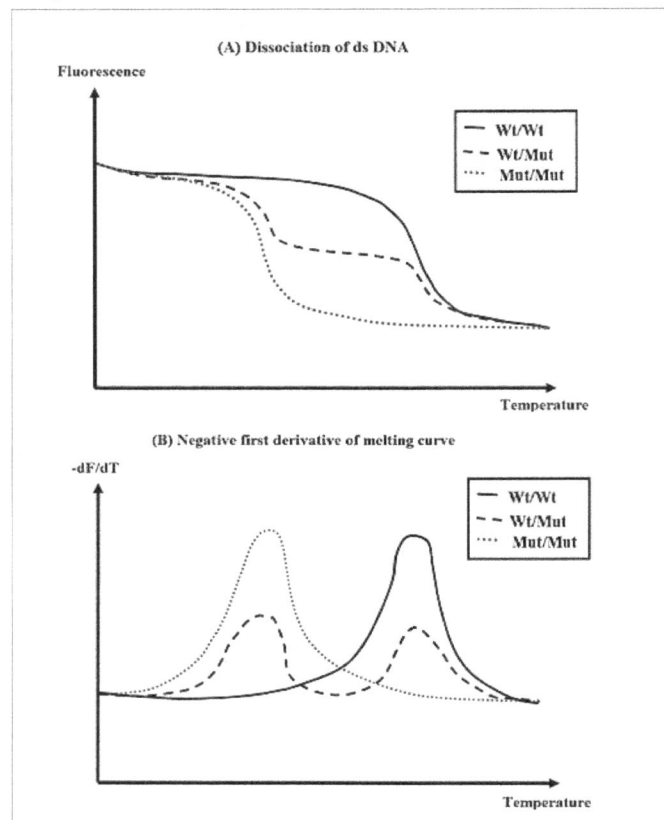

Figure 3A: Gene expression for stool samples taken from 60 patients. The condition of the patient is indicated by the bottom row of the panel and by the type of line. There were 20 normal patients (gray lines), 10 with IBD (dashed lines), and 30 with cancer (black lines). Instances of high expression appear on the right and those with low expression on the left. Expression was measured by CP and scales were chosen so that minimum values line up on the Min mark labeled at the top of the panel. The same is true for the maximum values which line up under the mark labeled Max.

Figure 3 B: This panel displays gene expression for stool samples taken from 30 cancer patients. Stage of cancer is indicated by the bottom row of the panel and by the type of line. There were 20 patients with stage 0 or 1 (gray lines), 5 with stage 2 (dashed lines), and 5 with stage 3 (black lines) cancer. The 30 noncancerous patients (stage NA) are not shown. C) Gene expression for tissue samples taken from 60 patients. Conditions of the patient are the same as in panel A. D) This panel displays gene expression for tissue samples taken from 30 cancer patients. Stages of cancer are indicated as in panel B.

A. CFTR Exon 17b

B. CFTR Exon 7

Figure 4A: Graph illustrating the relation between fluorescence and and temperature for labeled probe designed for a wild type (Wt) sequence, homozygous Wt, heterozygous(Wt/Mut), and homozygous mutant (Mut/Mut) combination.

Figure 4 B: Graph of the first derivative of melting curve (-df/dT) that pinpoints the temperature of dissociation, defined as 50% dissociation, by formed peaks. Figure courtesy of Integrated DNA Technologies **(www.idtdna.com)**; Downey N. (2014) Interpreting melt curves: An indicator, not a diagnosis. [Online] Coralville, Integrated DNA Technologies. [Accessed May, 26, 2017]

(A)

(B)

Figure 5A: Melt curves from qPCR of the CFTR (Cystic Fibrosis Transmembrane Conductance Regulator)gene. A. An amplicon from CFTR exon 17b reveals a single peak following melt curve analysis.

Figure 5B: An amplicon from CFTR exon 7 reveals two peaks. Figure courtesy of Integrated DNA Technologies (**www.idtdna.com**);Downey N. (2014) Interpreting melt curves: An indicator, not a diagnosis. [Online] Coralville, Integrated DNA Technologies. [Accessed May, 26, 2017]

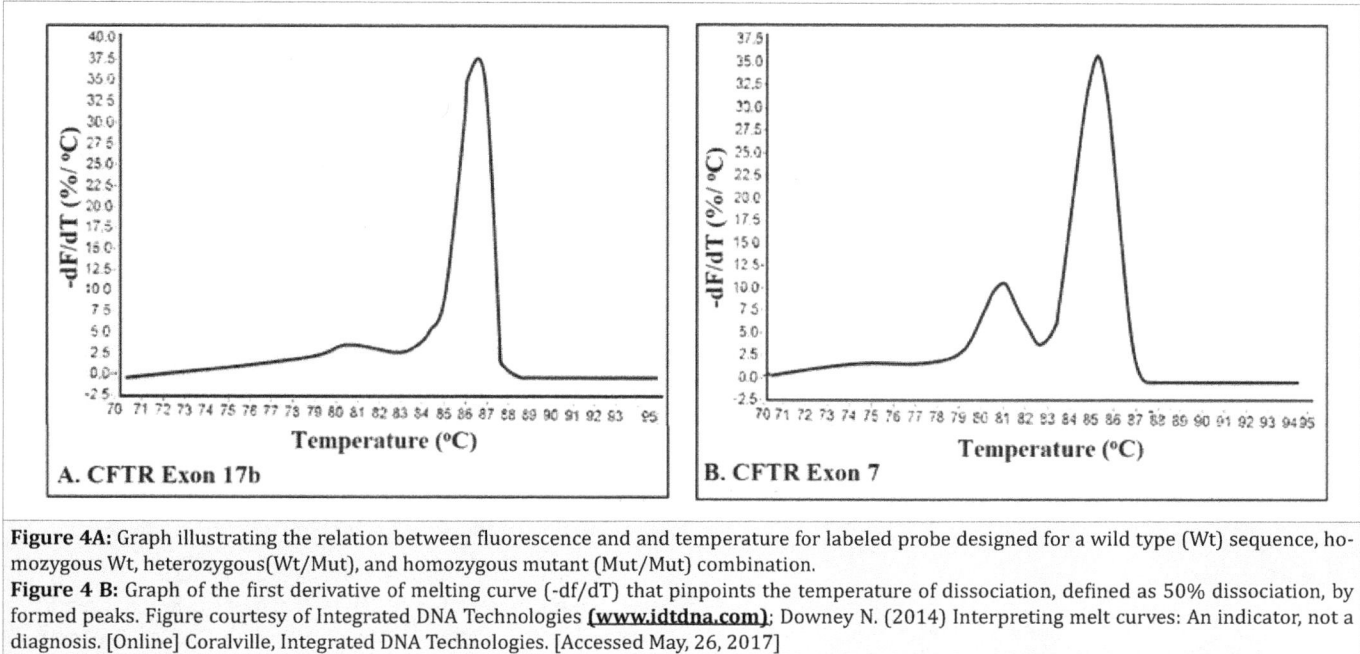

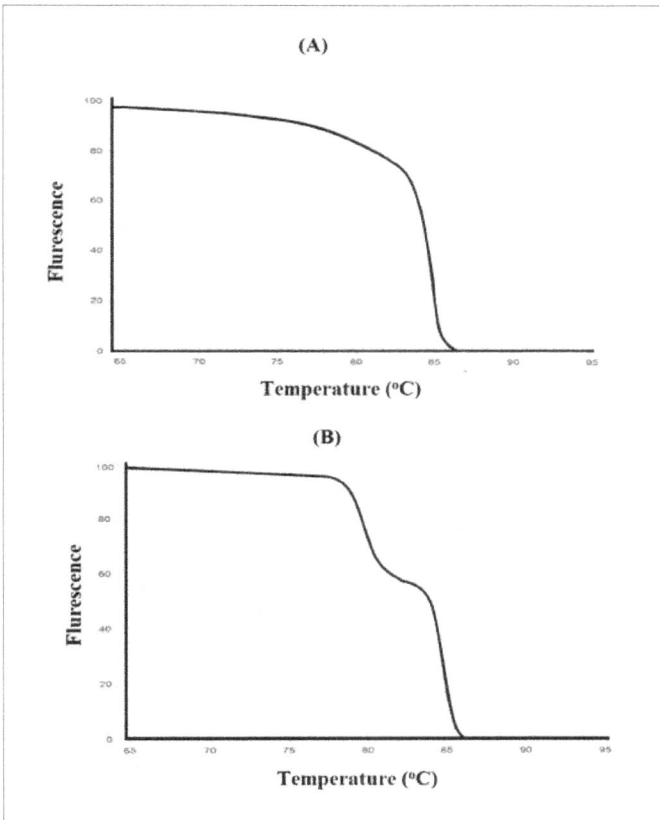

dye therefore dissociates from the ssDNA (Figure 3A). The change in slope of this curve when blotted as a function of temperature to obtain a melt curve for CFTR exom 17b (Figure 4A) However, if we allow for the possibility that DNA my assume an intermediate state that is neither dsDNA or ssDNA, raw date from CFTR exon 7 melt will look Figure 4B. This could happen when there are regions of the amplicon that are more stable (e.g., G/C rich), which do not melt immediately, but maintain their ds configuration until the temperature becomes sufficiently high to melt it, which results in two phases (Figure 4B). Additional sequence factors, such as amplicon misalignment in A/T rich regions, and designs that have secondary structure in the amplicon region, can also produce products that melt in multiple phases.

An advancement of MCA, referred to as High Resolution Melt (HRM), discovered and developed by Idaho Technology and the University of Utah [66, http://www.dna.utah.edu/Hi-Res/TOP_Hi-Res%20Melting.html], which has been useful for mutation detection and SNPs, enabling differentiation of homozygous wildtype, heterozygous and homozygous mutant alleles from the dissociation patterns. HRM has been used to identify variation in nucleic acid sequences, enabled by use of a more advanced software, and is therefore less expensive than probe-based genotyping methods, and allows for identification of variants quickly and accurately [67]. This method has been widely used in molecular diagnosis and for detection of mutations [68-74].

MCA is an assessment of dissociation characteristics of dsDNA during heating, leading to rise in absorbance, intensity and hyperchromicity. The temperature at which 50% of DNA is denatured is referred to as melting point, Tm.

Gathered information can be used to infer the presence of single nucleotide polymorphism (SNP), as well as clues to molecule's mode of interaction with DNA, such as intercalator slots in between base pairs through pi stacking and increasing salt concentration, leading to rise in melt temperature, whereas pH can affect DNA's stability, leading to lowering of its melting temperature [75,76]. Originally, strand dissociation was measured using UV absorbency, but now techniques based on fluorescence measurements using DNA intercalating fluorophores such as SYBR Green I, Eva Green, or Fluorophore-labelled DNA probes (FRET probes) when they are bound to ds DNA [75] are now common. Specialized thermal cyclers that run the qPCR, such as Roche Light Cycler (LC) 480®, used in this study, is programmed to produce the melt curve after the amplification cycles are completed. As the temperature increases, dsDNA denatures becoming ss and the dye dissociates, resulting in decrease in fluorescence. The graph of the negative first derivative of the melting-curve (-dF/dT) represents the rate of change of fluorescence in the amplification reaction, and allows pin-pointing the temperature of dissociation (50% dissociation) using formed peaks to obviate or complement sequencing efforts [77].

The melting temperature (Tm) of each product is defined as the temperature at which the corresponding peak maximum occurs. The MCA confirms the specificity of the chosen primers, as well as reveals the presence of primer-dimers, which usually melt at lower temperatures than the desired product, because of their small size, and their presence severely reduce the amplification efficiency of the target gene as they compete for reaction components during amplification, and ultimately the accuracy of the data. The greatest effect is observed at the lowest concentrations of DNA, which ultimately compromises the dynamic range. Moreover, nonspecific amplifications may result in PCR products that melt at temperatures above or below that of the desired product. Optimizing reaction components (Mg2+, detergents, SYBR Green I concentration) and annealing temperatures aid in decreasing nonspecific product formation [78-80]. Adequate product design, however, is considered to be the best method to avoid nonspecific products' formation. Including a negative control will determine if there is a co amplified genomic DNA [81,82]. The formula for Tm calculation is shown by the equation:

$$Tm = \underline{\quad} \frac{\Sigma\Delta Ho_{n-n}}{\quad} --- 273.15 \quad , \Sigma\Delta So_{n-n} + RLnC_T$$

where thermodynamic parameter ΔHo is Enthalpy changes, ΔSo parameter is Entropy changes, and C_T is total strand concentration; these free-energy parameters predict T_m of most oligonucleotide duplexes to within 5°C; and permit prediction of DNA, as well as RNA duplex stabilities. It should be noted that Tm depends on the conditions of the experiment, such as oligonucleotide concentration, salts' concentration, mismatches and single nucleotide polymorphisms (SNPs) [83]. OligoAnalyzer® Tool [www.idtdna.com/analyzer/Applications/Oligoanalyzer] allows for calculating the T_m of employed nucleotides.

Statistical and bioinformatics analysis

Gene expressions were standardized by dividing the SNORD48 value while raw melting temperatures were used. Analysis were done using the software R (version 3.1.3), with the package MASS [84]. One individual had so many missing values that this case was not used in the analysis so that the number of individuals is 24. For each standardized gene and each melting temperature, a one way ANOVA was used to obtain a p-value. There were four levels of the explanatory variable: Control, Sobya, Pom, and Both. Parallel coordinate plots (parcoord command in R) [85] were used to visualize the data for each gene and each melting temperature. Coordinates were ordered using the magnitude of the p-value. The two sample t-test was used on gene expression to compare control to sobya and control to both (t.test command in R with var.equal=FALSE) [86]. P-values were adjusted to control for false discovery rate. The method is outlined in [87] Benjamini and Yekutieli (p.adjust command in R with method='BY').

We have bioinformatically correlated the 2-7 or 2-8 complement nucleotide bases in the mature miRNAs with the untranslated 3' region of target mRNA (3' UTR) of a message using a basic algorithm such as Broad's Institute's Target Scan [88] http://www.targetscan.org/archives.html, which provides a precompiled list for their prediction.

Results

At base line, all participants in the trial excreted urinary total polyphenols; however, the inter individual variation was considerably high (4.89-12.59 mg GAE/100 ml urine).

Composition of the three supplements (FS, PGJ and FS + PGJ served to the volunteers is presented in Table 1. The initial and final mean urinary polyphenols, plasma and urinary ant oxidative activity, urinary TBARS and erythrocytic GST, as well as the daily portion of PGJ provided 21 mg PP /day, and the combination of PGJ – FS was 9 mg PP /day, as presented in Table 2.

Figure 2 is a layout of RT² miRNA PCR Array Human Cancer microRNA (MAH-102A). Figure 6 is a graphical representation of the parallel plot coordinates of the studied miRNA genes for melting temperature curve analysis. The genes were ordered using the p-values of a one way ANOVA based on groups. Genes with the smallest p-values are presented first. Figures 3 through 5 represent characteristics of melt curve analysis protocols.

Figure 6a show eight employed control genes (Snord48, Snord47, Snord44, RNUU6-2. MiRTC1, miRTC2, and PPC1, PPC2). In Figure 6b, five miRNA genes (miR-184, miR-203, miR-124, miR-96 and miR-378) show clear separation. Gene miR-184 has the highest separation from the control gene. MiR-203 genes are hardly amplified in Sobya, while it is highly expressed in Pomegranate. For miR-373 gene, the control group is different from the other three treatment groups. For genes miR-124, miR-96 and miR-378, Pomegranate is well separated from other three groups. In Figure 6c, for gene miR-301a, the control is separated from the other three groups. Additional miRNA genes are not

Discussion

Suitability of stool as a medium for developing a sensitive molecular biomarker screen

Stool represents a challenging environment, as it contains many substances that may not be consistently removed in PCR, in addition to the existence of certain inhibitors [91-94], which all must be removed for a successful PCR reaction. Our results [56, 75, 78, 95-97] and others [9] have shown that the presence of non-transformed RNA and other substances in stool do not interfere with measuring miRNA expressions, because of the use of suitable PCR primers, and the robustness of the real-time qPCR method. Besides, stool colonocytes contain much more miRNA and mRNA than that available in free circulation, as in plasma [97, 98], all factors that facilitate accurate and quantitative measurements.

PCR amplification and the effect of Inhibitory Substances

PCR has been used for miRNAs quantification because of its extreme sensitivity. This method, however, could lead to errors because of the presence of inhibiting substances, representing diverse compounds with different properties and mechanisms of action, which induce their effects by direct interaction with DNA that will be amplified, or through interference with the employed thermos table DNA polymerase. Agents that reduce Mg2+ availability, or interfere with the binding of Mg2+ to DNA polymerase could inhibit the PCR reaction [92]. Calcium ion is another inorganic inhibiting substance, although most PCR inhibitors are organic compounds (e.g., bile salts, sodium dodecyl sulphate, urea, phenol, ethanol, polysaccharides), as well as proteins (e.g., collagen, hemoglobin, immunoglobin G and proteinase) [99]. The existence of polysaccharides in stool could decrease the capacity to resuspend precipitated RNA, or disrupt the enzymatic reaction by mimicking the structure of nucleic acid. The DNA template of the PCR, as well as primers binding to DNA template can be inhibited by nucleases and other inhibitors, [93]. Remedial strategies for removal of inhibitors in stool, such as additional extraction steps, sephadex G-200 chromatography, heat treatment before the PCR, chloroform extraction, treatment with activated carbon, adding BSA, or dilution of sample, have been suggested [94]. We found the dilution method, in which the extracted ribonucleic acid (RNA) is diluted in the reaction mixture with distilled water or an isotonic buffer, to be the most practical method for preventing PCR inhibition using a commercially available diluent [100-102].

Melting curve analysis (MCA)

Microscale thermophoresis is a method that determines the stability, length, conformation and modifications of DNA and RNA. It relies on the directed movement of molecules in a temperature gradient that depends on surface characteristics of the molecule, such as size, charge and hydrophobicity. By measuring thermophoresis of nucleic acids over a temperature gradient,

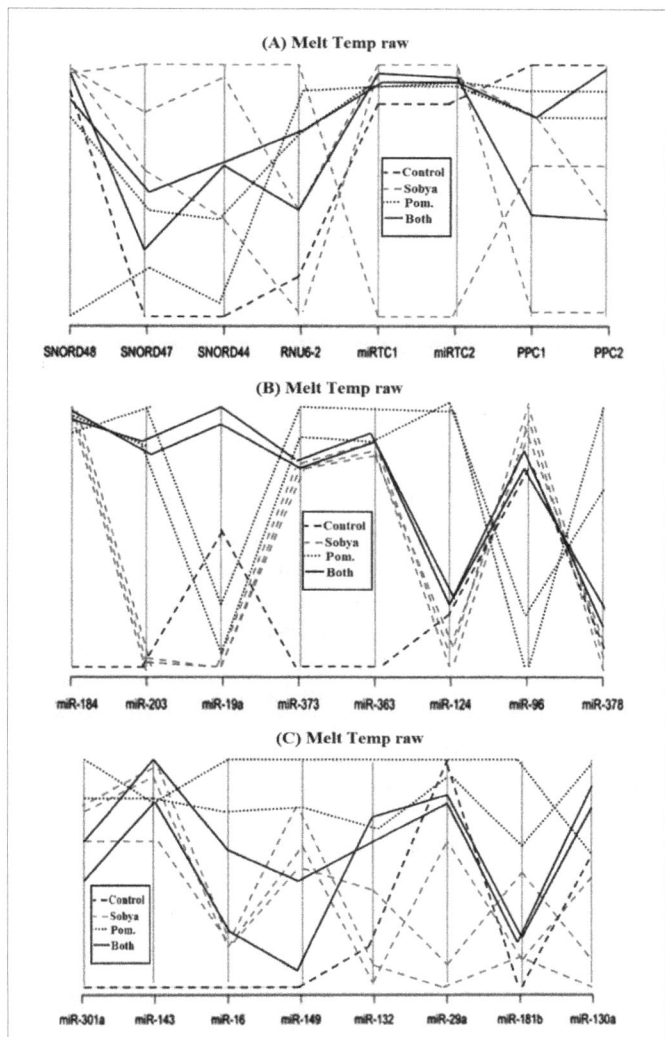

Figure 6 A: Change in fluorescence with increasing temperature is measured; as the temperature is increased from 65oC is measured, the two strands of the amplicon separate to form a ssDNA, causing the fluorescent intercalating dye to dissociate from DNA and stop fluorescing.
Figure 6 B: The shoulder in the curve between 80oC and 85oC suggests the presence of an intermediate state where the DNA is in both ds and ss configurations. Figure courtesy of Integrated DNA Technologies (www.idtdna.com);Downey N. (2014) Interpreting melt curves: An indicator, not a diagnosis. [Online] Coralville, Integrated DNA Technologies [Accessed May, 26, 2017]

shown, as their p-values are greater (less significant), and the graphs did not show any meaningful separations.

Bioinformatics analysis using the Target Scan algorithm [88] for up-regulated and down-regulated mRNAs genes is shown in Table 3. The program yielded 21 mRNA genes encoding different cell regulatory functions. The first 12 of these mRNAs were found with the DAVID program [89] to be active in the nucleus and related to transcriptional control of gene regulation. For down regulated miRNAs, the DAVID algorithm found the first four of these mRNAs to be clustered in cell cycle regulation categories. [90].

Table 3: Up-regulated and down-regulated target mRNA genes detected by a DAVID Bioinformatics algorithm.

Up-regulated target mRNA genes
BCL11B, CUGBP2, EGR3, DLHAP2, NUFIP2, KLF3, MECP2, ZNF532, APPLI1, NFIB, SMAD7, SNF1LK, ANKRD52, C17orf39, FAM13A1, GLT8D3, KIAA0240, PCT, SOCS6, TNRC6B and UHRF1BP1.
Down-regulated target mRNA genes
TGFB1, CKS2, IGF2, KLK10, FLNA, CSE1L, CXCL3, DPEP1 AND GUCA2B

one finds clear melting transitions, and can resolve intermediate conformational states (Figure 3). These intermediate states are indicated by an additional peak in the thermophoretic signal preceding most melting transitions (Figure 3B) [73, 77-79]. Agarose gel visualization is the gold standard for analyzing PCR products. Alternatively to reduce the number of gels needed to conform the presence of a single amplicon, "uMelt" melting curve prediction software (http://www.dna.utah/umelt/umelt.html) can be used to confirm that a single amplicon is generated by PCR [80]. This program predicts melt curves and their derivatives for qPCR-length amplicons, and is suited to test for multiple peaks in a single amplicon product.

...Because SYBR Green I dye has several limitations, including inhibition of PCR, preferential binding to CG-rich sequences and effects on MCA, two intercalating dyes SYTO-13 and SYTO-82 were tried and did not show these negative effects, and SYTO-82 demonstrated a 50-fold lower detection limit [81], as well as best combinations of time-to threshold (Tt) and signal-to-noise ratio (SNR) [82]. To optimize performance of the buffer, a PCR mix supplemented with two additives, 1M 1, 2-propanediol and 0.2 M trehalose, were shown to decrease Tm, efficiently neutralize

PCR inhibitors, and increase the robustness and performance of qPCR with short amplicons [83]. "UAnalyzeSM"is another web-based tool, similar to uMELT, for analyzing high-resolution melting PCR products' data, in which recursive nearest neighbor thermodynamic calculations are used to predict a melt curve. Using 14 amplicons of CYBB [cytochrome b-245 heavy chain, also known as cytochromae b(558) subunit], the main +/- standard deviation, the difference between experimental and predicted fluorescence at 50% helicity was -0.04 +/- 0.48oC [64].

In our study, we found the melt curve analysis to be a useful and an informative method because after the statistical analysis carried on our miRNA expression samples showed no preferential expression of any of the 88 miRNA genes, a melt curve analysis on the same samples found that we could distinguish 7 miRNA (miR-184, miR-203, miR-373, miR-124, miR-96, miR-373 and miR-301a), due to different separation melting profiles (Figure 6). Thus, we believe that it is imperatives for investigators to run this kind of analysis on samples that particularly may not show expression differences in their mRNA or miRNA studied genes, such as nutritional samples.

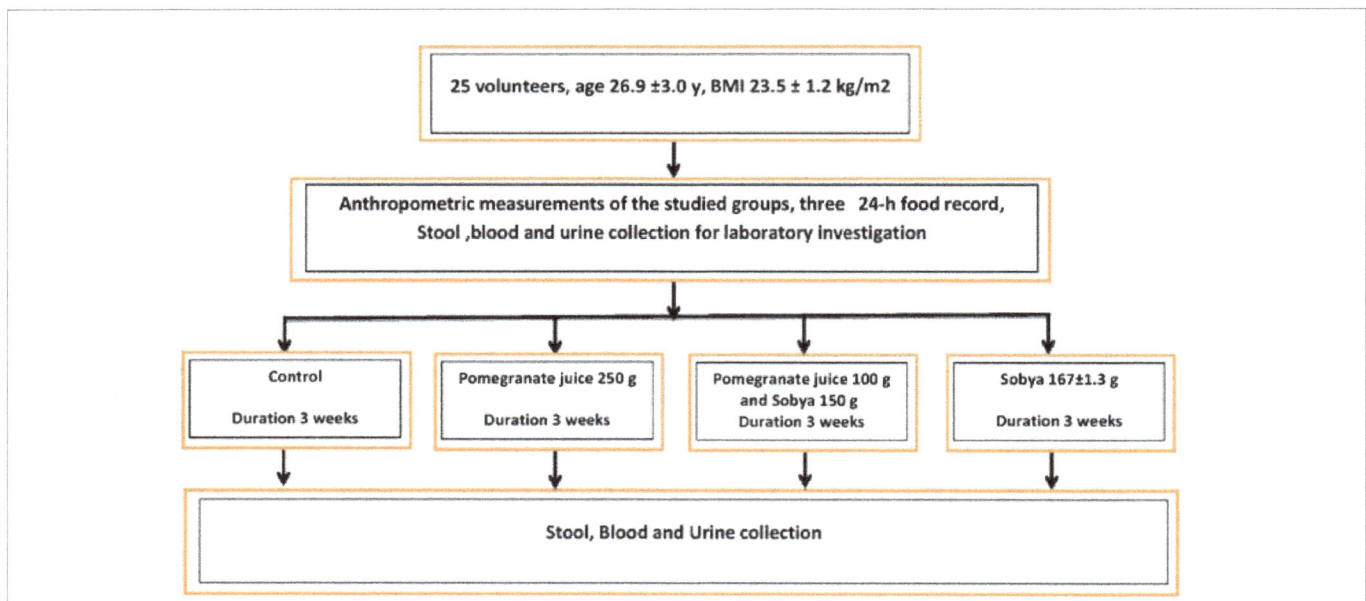

Chart 1. *A chart illustrating the design of the randomized controlled trial.*

Figure 7: A graphical representation of the parallel plot coordinates of the studies miRNA genes. The genes were ordered using the p-values of a one way ANOVA based on groups. Genes with the smallest p-values are presented first.
Figure 7a: show control genes. In
Figures7b, c: five miRNA genes show separation.

Bioinformatics methods to correlate seed miRNA Data with messenger (m) RNA Data

To provide information about complex regulatory elements, we correlated miRNA results with our available mRNA data [1], as well as those data available in the open literature using computer model Target Scan [88, 89]. The authenticity of functional miRNA/mRNA target pair, once identified was validated by fulfilling four basic criteria: a) miRNA/mRNA target interaction can be verified, b) the predicted miRNA and mRNA target genes are co-expressed, c) a given miRNA must have a predictable effect on target protein expression, and d) miRNA-mediated regulation of target gene expression should equate to altered biological function. Bioinformatics showed 21 up regulated mRNA genes encoding different cell regulatory functions, and 12 of these mRNAs were found to be active in the nucleus and related to transcriptional control of gene regulation. For down-regulated miRNAs, four of the mRNAs appeared to be clustered in cell cycle regulation categories (Table 3) [56].

Clinical Significance

The clinical significance of the study presented above is that using melting temperatures for analyzing nutrient-gene data is a promising new approach for identifying key regulatory miRNA genes related to metabolites rich in polyphenols, Probiotic lactobacilli, or combinations of the two metabolites. Melt curve analysis is a powerful novel approach because after the statistical analysis carried on our miRNA samples produced negative gene expression (Cq) results, running melt curve analysis on the same samples identified 7 of the 88 miRNA genes imprinted on the highly sensitive focused PCR arrays (~ 8% of the genes), and using parallel coordinates plots showed noticeable separation of melt curve profiles. Thus, we believe that it is imperatives for investigators to run this kind of MCA on nutrition samples that are mild in nature, and many not normally show significant differences in the expression of studied miRNA genes, particularly at the early stages (0-1) of the cancer, before metastasis, so the cancer can be cured. The same analysis can also be envisioned for messenger mRNA amplifications, using mRNA arrays, and then using bioinformatics resources to correlate mRNA with miRNA data.

We are also planning to validate these initial results by carrying out additional miRNA nutrigenomic expression studies, with much more observations using PP, FS and their combinations, and collectively the obtained results would fully demonstrate the sensitivity/specificity of this powerful systemic molecular approach for analyzing nutrient-gene data.

Research efforts for the management of cancer are directed to identify new strategies for its early detection. Stable miRNAs are a new promising class of circulating biomarkers for cancer detection. However, the lack of consensus on data normalization, using relative PCR quantification methods, has affected the diagnostic potential of circulating miRNAs. There is thus a growing interest in techniques that allow for an absolute quantification of miRNAs, which would be more precise, and therefore more useful for early diagnosis of this curable cancer, if the cancer can be detected at the early premalignant disease stage (stage 0-1).

Recently, digital PCR, mainly based on droplets generation, emerged as an affordable technology for the precise and absolute quantification of nucleic acids (103, 1`04). Given its reproducibility and reliability, as chip-digital absolute PCR quantification technique has becomes more established, it would be a robust tool for the quantitative assessment of miRNA copy number necessary for the diagnosis of the cancer, as well as for the identification and quantification of miRNAs in other biological samples such as circulating exosomes or protein complexes.

Acknowledgments

We express our deep thanks to those volunteers who participated in this research by completing a dietary questionnaire, and providing urine, blood and stool samples for the study. We thank Dr. Clark D. Jeffries at Renaissance Computing Institute, University of North Carolina at Chapel Hill for his bioinformatics analysis, Funds for this research were provided by NIH Grant 1R43CA144823, and by GEM Tox Labs operating funds.

References

1. Ahmed FE, Vos P, iJames S, Lysle DT, Allison RR, Flake G, Naziri W, et al. Transcriptomic molecular markers for screening human colon cancer in stool and tissue. Cancer Genom Proteom. 2007;4(1).1-20.

2. Ahmed FE. Microarray RNA transcriptional profiling: Part I. Platforms, experimental design and standardization. Exp Rev Mol Diag. 2006;6(4):535-550.

3. Ahmed FE. Microarray RNA transcriptional profiling: Part II. Analytical considerations and annotations. Exp Rev Mol Diag. 2006;6(5):703-715.

4. Lee RC, Feinbaum RL and Ambros V. The C. elegans heterochronic gene lin-4 encodes a small RNAs with antisense complimentarity to lin-14. Cell. 1993;75(5).843-854.

5. Kozomara A and Griffithis JS. miRBase. annotating high confidence microRNAs using deep sequencing data. Nucleic Acids Res. 2014;42(Database issue).D68-73. doi. 10.1093/nar/gkt1181

6. Lund E and Dahlberg JE. Substrate selectivity of exportin 5 and Dicer in the biogenesis of microRNAs. Cold Spring Harb Symp Quant Biol. 2006;71.59-66.

7. Lim LP, Lau NC, Garrett-Engele P, Grimson A, Schelter JM, Castle J, Bartel DP A, et al. Microarray analysis shows that some microRNAs downregulate large numbers of target mRNAs. Nature. 2005;433(7027).769-773.

8. De Caterina R and Madonna R. Nutrients and gene expression. World Rev Nutr Diet. 2004;93.99-133.

9. Link A, Balaguer F, Shen Y, Nagasaka T, Lozano JJ, Boland CR, Goel A., et al. Fecal MicroRNAs as novel biomarkers for colon cancer screening. Cancer Epidemiol Biomarkers Prev. 2010;19(7).1766-1774. doi. 10.1158/1055-9965

10. Biomarkers Definition Working Group. Biomarkers and surrogate endpoints: preferred definitions and conceptual framework. Clin Pharmacol Therapeutics. 2001;69(3):89–95.

11. Strimbu K and Tavel JA. What are biomarkers?. Curr Opin HIV AIDS. 2010;5(6).463-466.

12. Roever L. Endpoints in Clinical Trials: Advantages and Milestones. Evidence Based Medicine and Practice. 2016;1:2. doi: 10.4172/ebmp.1000e111

13. Endpoints: How the Results of Clinical Trials are Measured. 2004.

14. Clinical endpoints: Principles of Translational Science in Medicine. 2nd Ed. 2015.

15. Bautista RR, Gomez AO, Miranda AH, Dehesa AZ, Villarreal-Gmza C, Avila-Moreno F, Arrieta, et al. Long-coding RNAsi: Implications in targeted diagnosis, prognosis, and improved therapeutic strategies in human non- and triple-negative breast cancer. Clin Epigenet. 2018;10(88). doi: 10.1186/s13148-018-0514-z

16. Klasić M, Markulin D, Vojta A, Samaržija I, Biruš I, Dobrinić P, Ventham NT,et al. Promoter methylation of the MGAT3 and BACH2 genes correlates with the composition of the immunoglobulin G glycome in inflammatory bowel disease. Clin Epigenetics. 2018;10(75):1-14.

17. Research (CBER). Guidance for industry: Clinical trial endpoints for the approval and of cancer drugs and biologics. 2007.

18. Werner RM and Pearson TA. LDL-Cholestrol. A risk factor for Coronary Artery Disease from Epidemiology to Clin Trials. Canad J Cardiology. 1998;14 Suppl B.3B-10B.

19. Bikdeli B, Pumanithinot N, Akram Y, Lee L, Desai NR, Ross J, Krumholz HM, et al. Two decades of cardiovascular trials with primary surrogate endpoints 1990-2011. J Am Heart Assoc. 2017;6(3): e005285. doi: 10.1161/JAHA.116.005285

20. Waladkhani AR. Conducting clinical trials: A theoretical and practical guide. 2008.

21. Wenting and Sargent D J. Statistics and Clinical Trials: Clinical Radiation Oncology. 3rd Ed. 2012.

22. Hussein L, Abdl-Rehim EA, Afifi AMR, El-Arab AE and Labib E. Effectiveness of apricots (Prunus ameniea) juice and lactic acid fermented sobya on plasma levels of lipid profile parameters and total homocysteine among Egyptian adults. Food Nutr Sci. 2014; 5(22):11.

23. World Health Organization (WHO). Cardiovascular Diseases (CVDs). WHO Fact Sheet No. 317, Geneva. 2011.

24. Bernedetti S, Catalani S, Palma F and Trari FC. The antioxidant protection of CELLFOOD® against oxidatice change in vitro. Food Chem Toxicol. 2011;49(9).2292-2298. doi. 10.1016/j.fct.2011.06.029

25. Bengmark S. Advanced glycation and lipoxidation end products-- amplifiers of inflammation. the role of food. J Parenter Enteral Nutr. 2007;31(5).430-440.

26. Gouda M, Moustafa A, Hussein L, and Hamza M. Three week dietary intervention using apricots, pomegranate juice or/and fermented sour sobya and impact on biomarkers of antioxidative activity, oxidative stress and erythrocytic glutathione transferase activity among adults. Nutritional J. 2016;15.52. doi. 10.1186/s12937-016-0173-x

27. Esmaillzadeh A, Tahbaz F, Gaieni I, Alavi-Majd H. Cholestrol-lowering effect of concentrated pomeg- ranate juice consumption in Type II diabetic patients with hyperlipidemia. Int J Vit Nutr Res. 2006;76(3).147-151.

28. Zhuang G, Liu XM, Zhang QX, Tian F W. Research advances with regards to clinical outcome and potential mechanisms of the cholestrol-lowering effects of probiotics. Clinical Lipidology. 2012;7(5).501-50.

29. van Baarlen P, Troost F, van der Meer C, Hooiveld G. Human mucosal in vivo transcriptome responses to three lactobacilli indicate how probiotics may modulate human cellular pathways. Proc Natl Acad Sci USA.2011;108(Suppl 1).4562-4569.

30. Richards AL, Burns MB, Alazizi A, Barreiro LB, Roger Pique-Regi, Ran Blekhman. Genetic and transcriptional analysis of human host response to healthy gut microbiota. MSystems. 2016;1(4).e00067-16. doi.10.1128/mSystems

31. Noori-Daloii MR and Nejatizdeh A. Nutritional transcriptomics: An Overview. Eds Debasis Bagehi, Anand Swarooop and Manashi Bagehi. 2015:545-556.

32. Luu-The V, Paquet N, Calvo E, Cumps J. Improved real-time RT-PCR measurements using second derivative calculation and double correction. BioTechniques. 2005;38(2).287-293.

33. Livak K J, Schmittgen T D. Analysis of relative gene expression data using real-time quantitative PCR and thwe 2- ΔΔCt method. Methods. 2001;25(4).402-408.

34. Reinhart BJ, Slack FJ, Basson M, Pasquinell AE, Bettinger JC, Rougvie AE, et al. RNA regulates developmental timing in Caenorhabditis elegans. Nature. 2000;403(6772).901-906.

35. Tellman G. The E-method: a highly accurate technique for gene-expression analysis. NatureMethods. 2006.

36. http://www.lightcycler-online.com/lc_sys/soft_nd.htm#quant.

37. LightCycler. Roche Moecular Biochemicals. Manneheim, Germany. 2001:64-79.

38. Huang da W, Sherman BT, Lempicki RA. Systematic and integrative analysis of large gene lists using DAVID bioinformatics resources. Nat Protocol. 2009;4(1).44-57. doi. 10.1038/nprot.2008.211

39. Ririe KM, Rasmussen RP and Wittwer CT. Product determination by analysis of DNA melting curves during the polymerase chain reaction. Anal Biochem. 1997;245(2).154-160.

40. Xu P, Guo M and Hay BA. MicroRNAs and the regulation of cell death. Trend Genet. 2004;20(12).617-624.

41. Malumbres M. miRNAs and cancer. an epigenetics view. Mol Aspects Med. 2013;34(4).863-874. doi. 10.1016/j.mam.2012.06.005

42. Zhao Y, Ransom JF, Li A, Vedantham V, von Drehle M, Muth AN, et al. Dysregulation of cardiogenesis, cardiac conduction, and cell cycle in mice lacking miRNA-1-2. Cell. 2007;129(2).303-317.

43. Phua YL, Chu JY, Marrone AK, Bodnar AJ, Sims-Lucas S, Ho J. Renal stromal miRNAs is required for normal nephrogenesis and glomerular mesangial survival. Physiological Reports. 2015;3(10).pii.e12537. doi.10.14814/phy2.12537

44. Maes OC, Chertkow HM, Wang E and Schipper HM. MicroRNA. Implications for Alzheimer Disease and other Human CNS Disorders. Current Genomics. 2009;10(3).154-168. doi. 10.2174/138920209788185252

45. Li J, Li J, Liu X, Qin S, Guan Y, Liu Y, et al. MicroRNA expression profile and functional analysis reveal that miR-382 is a critical novel gene of alcohol addiction. EMBO Mol Med. 2013;5(9):1402-1414. doi: 10.1002/emmm.201201900

46. Romao JM, Jin W, Dodson MV, Hausman GJ, Moore SS, Guan LL. MicroRNA regulation in mammalian adipogenesis. Exp Biol Med. 2011;236(9):997-1004. doi: 10.1258/ebm.2011.01110

47. Mencía A, Modamio-Høybjør S, Redshaw N, Morín M, Mayo-Merino F, Olavarrieta L, et al. Mutations in the seed region of human miR-96 are responsible for nonsyndromic progressive hearing loss. Nat Genet. 2009;41(5):609-613. doi: 10.1038/ng.355

48. Hughes AE, Bradley DT, Campbell M, Lechner , Dash DP, Simpson DA, et al. Mutation Altering the miR-184 Seed Region Causes Familial Keratoconus with Cataract. Am J Hum Genet. 2011;89(5):628-633. doi: 10.1016/j.ajhg.2011.09.014

49. de Pontual L, Yao E, Callier P, Faivre L, Drouin V, Cariou S, et al. Germline deletion of the miR-7-92 cluster causes kel skeletal and growth defects in humans. Nat. Genet. 2011;43(10):1026-1030. doi: 10.1038/ng.915

50. Tuddenham L, Jung JS, Chane-Woon-Ming B, Dölken L, Pfeffer S. Small RNA deep sequencing identifies microRNAs and other small noncoding RNAs from human herpesvirus 6B. J Virol. 2012;86(3):1638-1649. doi: 10.1128/JVI.05911-11

51. Lu M, Zhang Q, Deng M, Guo Y, Wei Gao, Qinghua Cui, et al. An analysis of human microRNA and disease association. Plos One. 2008;3(10):e3420.

52. Gregory RI and Shiekhattar R. MicroRNA biogenesis and cancer. Cancer Res. 2005;65(9):3509-3512.

53. Calin GA, Ferracin M, Cimmino A, Dileva G, Shimizu M, Wojcik SE, Iorio MV, et al. A microRNA signature associated with prognosis and progression in chronic lymphocytic leukemia. N Eng J Med. 2005;353(17):1793-1801.

54. Chang-Zheng C. MicroRNAs as oncogenes and tumor supressors. N Eng J Med. 2005;353(17):1768-1771.

55. Calin GA, Sevignai C, Dumitru CD, Hyslop T. Human microRNA genes are frequently located at fragile sites and genomic regions involved in cancers. Proc Natl Acad Sci USA. 2004;101(9):2999-3004.

56. Schepler T, Reinert JT, Oslenfeld MS, Christensen LL , Silahtaroglu AN, Dyrskjøt L, Wiuf C, et al. Diagnostic and prognostic microRNAs in Stage II colon cancer. Cancer Res. 2008;68(15):6416-6424. doi:10.1158/0008-5472.CAN-07-6110

57. Schetter AJ, Leung SY, Sohn JJ, Zanetti KA, Elise D. Bowman, Nozomu Yanaihara, Siu Tsan Yuen, et al. MicroRNA expression profile associated with progression and therapeutic outcome in colon adenocarcinoma. J Am Med Assoc. 2008;299(4):425–436. doi: 10.1001/jama.299.4.425

58. Calin GA and Croce CM. MicroRNA signatures in human cancers. Nat Rev Cancer. 2006;6(11):857-866.

59. Ahmed FE, Ahmed NC, Vos PW, Bonnerup C, Atkins JN, Casey M, Nuovo GJ, et al. Diagnostic microRNA markers to screen for sporadic human colon cancer in stool. I. Proof of principle. Cancer Genom Proteom. 2013;10(3).93-113.

60. Varga A and Delano J. Effect of amplicon size, melt rate, and dye translocation. J Virol Meth. 2006;132(1-2):146-153.

61. Mendes RE, Kiyota KA, Monteiro J, Castanheira M, Andrade SS, Gales AC, et al. Rapid detection and identification of metallo-β-Lactamase-encoding genes by multiplex real-time PCR asay and melt curve analysis. J Clin Microbiol. 2007;45(2):544-547.

62. Guion CE, Ochoa TJ, Walker CM, Barletta F, Cleary TG. Detection of diarrheagenic Escerichia coli by use of melting-curve analysis and real-time multiplex PCR. J Clin Microbiol. 2008;46(5):1752-1757. doi: 10.1128/JCM.02341-07

63. Winder L, Phillips C, Richards N, Ochoa-Corona F, Hardwick S, Vink CJ, et al. Evaluation of DNA melting analysis as a tool for species identification. Meth Ecology Evol. 2011;2(3):312-320.

64. Von Keyserling H, Bergmann T, Wiesel M, Kaufmann AM. The use of melting curves as a novel approach for Validation of real-time PCR instruments. BioTechniques. 2011;51(3):179-184. doi: 10.2144/000113735

65. Bohling SD, Wittwer CT, King TC, Elenitoba-Johnson KS. Fluorescence melting curve analysis for the detection of the bcl-1/JH translocation in mantle cell lymphoma. Lab Invest. 1999;79(3):337-345.

66. de Fdippes FF, Wang J-W and Weigel D: MIGS: miRNA-induced gene silencing. The Plant J 70(3): 541-547, 2011.

67. Dwight Z, Palais R and Wittwer CT. μMELT:prediction of high-resolution melting curves and dynamic melting profiles of PCR products in a rich web application. Bioinformatics. 2011;27(7):1019-1020. doi: 10.1093/bioinformatics/btr065

68. Downey N. Interpreting melt curves. An indicator, not a diagnosis.

69. Farrar JS, Reed GH, Wittwer CT. High-resolution melting curve analysis for molecular diagnosis. In Molecular Diagnostics. 2014;229-245. doi. 10.1016/B978-0-12-374537-8.00015-8

70. Krypuy M, Ahmed AA, Etemadmoghadam D, DeFazio A, Fox SB, Brenton JD, et al. High resolution melting for mutation scanning of TP53 exon 5-8. BMC Cancer. 2007;7:168.

71. Pasay C, Arlian L, Morgan M, et al. High -resolution melt analysis for the detection of a mutation associated with permethrin resistance in a population of scabies mites. Med Vet Entomol. 2008; 22(1):82-88.

72. Montgomery JL, Sanford LN, Twitter CT. High-resolution DNA melting analysis in clinical research and diagnosis. Expert Rev Mol Diagn. 2010;10(2):219-240. doi: 10.1586/erm.09.84

73. Ansevin AT, Vizard DL, Brown BW and McConathy J. High-resolution thermal denaturation of DNA. I. Theoretical and practical considerations for the resolution of thermal subtransitions. Biopolymers. 1976;15(1):153-174. doi:10.1002/bip.1976.360150111

74. Ririe KM, Rasmussen RP and Wittwer C T. Product differentiation by analysis of DNA during melt curves during the polymetrase chain reaction. Anal Biochem. 1997;245(2):154-160.

75. Wittwer CT. High-resolution DNA melting analysis: Advancements and limitations. Hum Mutat. 2009;30(6):857-9. doi: 10.1002/humu.20951

76. Ahmed FE, Hussein LA, Gouda MM, Vos PW and Ahmed NC. Melt Curve Analysis in Interpretation of Nutrigenomics' MicroRNA Expression Data. Trends Res. 2018;1 (1):1-10.

77. Ahmed FE, Gouda MM, Hussein LA, Ahmed NC, Vos PW, Mohammad M. Role of Melt Curve Analysis in Interpretation of Nutrigenomics' MicroRNA Expression Data. Cancer Genom. Proteom. 2017;14(6):469-481.

78. Freier SM, Kierzek R, Jaeger JA, Sugimoto N M H Caruthers, T Neilson, D H Turner, et al. Improved free-energy parameters for prefictions

of RNA duplex stability. Proc Natl Acad Sci USA. 1986;83(24):9373-9377.

79. Lay MJ and Wittwer CT. Real-time fluorescence genotyping of factor V Leiden during rapid-cycle PCR. Clin Chem. 1997;43(12):2262-2267.

80. Wienken C J, Baaske P, Duhr S, Braun D. Thermophoretic melting curve quantify the conformation and stability of RNA and DNA. Nucleic Acids Res. 2011;39(8):e52. doi: 10.1093/nar/gkr035

81. Dwight Z, Palais R and Wittwer CT. µMELT:prediction of high-resolution melting curves and dynamic melting profiles of PCR products in a rich web application. Bioinformatics. 2011;27(7):1019-1020. doi: 10.1093/bioinformatics/btr065

82. Gudnason H, Dufva M, Bang D D, Wolff A. Comparison of multiple DNA dyes for real-time PCR:effects of dye concentration and sequence composition on DNA amplification and melting temperature. Nucleic Acids Res. 2007;35(19):e127. doi: 10.1093/nar/gkm671

83. Oscorbin IP, Belousova EA, Zakabunin AI, Boyarskikh UA and Filipenko M L. Comparison of fluorescent intercalating dyes for quantitative loop-mediated isothermal amplification. BioTechniques. 2016;61(1):20-25. doi: 10.2144/000114432

84. Horakova H, Polakovicova I, Shaik GM, Eitler J, Bugajev V, Draberova L, Draber P, et al. 1,2-propanediol-trehalose mixture as a potent quantitative real-time PCR enhancer. MBC Biotechnol. 2011;11:41

85. Venables WN and Ripley BD (Eds.). Modern and Applied Stastistics. Fourth Edition. Springer, New York. 2002.

86. R Core Team R. A language and environment for statistical computing. R Foundation for Statistical Computing, Vienna, Australia. 2015.

87. Ahmed F E. Statistical Analysis of microRNA as markers for Screening of Colon Cancer. Biostatistics and Biometrics J (BBOAJ). 2018;6(4):1-4. doi: 10.19080/BBOAJ.2018.06.555691

88. Benjamini Y and Yekutieli D. The controil of the false discovery rate in multiple testing under dependency. Annals Stat. 2001;29(4).1165-1188.

89. Huang da W, Sherman BT, Lempicki RA. Systematic and integrative analysis of large gene lists using DAVID bioinformatics resources. Nat Protocol. 2009;4(1).44-57. doi. 10.1038/nprot.2008.211

90. Thompson S M, Ufkin JA, Sathyanarayana M L, Liaw P, et al. Common features of micro RNA target prediction tools. Front Genet. 2014; 5:23. doi: 10.3389/fgene.2014.00023

91. Al-Soud WA and Radstrom P. Capacity of nine thermostable DNA polymerases to mediateDNA amplification in the presence of PCR-inhibiting samples. Appl Env Microbiol. 1998;64(10).3748-3753.

92. Wilson I G. Inhibition and facilitation of nucleic acid amplification. Appl Env Microbiol. 1997;63(10).3741-3751.

93. Montiero I, Bonnemaison D, Vekris A, Petry KG, J Bonnet, R Vidal, et al. Comples polysaccharides as PCR inhibitors in feces. Heliobacter pylori model. J Clin Microbiol. 1997;35(4):995–998.

94. Schrader C, Schielke A, Ellerbroek L and Johne R. PCR inhibitors- occurrence, properties and removal. J Appl Microbiol. 2012;113(5).1014-1026. dci. 10.1111/j.1365-2672.2012.05384.x.

95. Ahmed FE, Ahmed NC, Vos P, Bonnerup C, Atkins JN, Casey M. Diagnostic microRNA markers to screen for sporadic human colon cancer in blood. Cancer Genom Proteom. 2012;9(4).179-192.

96. Ahmed F E, Ahmed N C, Gouda M and Bonnerup C. MicroRNAs as Molecular Markers for Screening of Colon Cancer. Case Rep Surg Invasive Proced. 2017;1(2):14-17.

97. Ahmed F E, Ahmed N C, Gouda M and Vos P W. MiRNAs for the Diagnostic Screening of Early Stages of Colon Cancer in Stool or Blood. Surgical Case Reports and Reviewss. 2017;1(1):1-19. doi: 10.a5761/SCRR.1000103

98. Ahmed F E, Vos P W, Ijames S, Lysle DT, Flake G, Sinar DR, Naziri W, et al. Standardization for transceiptomic molecular markers to screen human colon cancer. Cancer Genom Proteom. 2007; 4(6): 419-431.

99. Davidson LA, Lupton JR, Miskovsky E, Fields AP. Quantification of human intestinal gene expression profiling using exfoliated colonocytes. a pilot study. Biomarkers. 2003;8(1).51-61.

100. Rådström P, Knutsson R, Wolffs P, Lovenklev M. Pre-PCR processing. strategies to generate PCR-compatible samples. Mol Biotechnol. 2004;26(2).133-146.

101. Scipioni A, Mauroy A, Ziant D, Saegerman C et al. A SYBR Green RT-PCR assay in single tube to detect human and bovine noroviruses and control for inhibition. Virology J. 2008;5.94. doi. 10.1186/1743-422X-5-94

102. Ahmed F E. MicroRNAs as Molecular Markers for Colon Cancer Diagnostic Screening in Stool & Blood. Int Med Rev . 2017;9:124. doi:10.3390/cancers9090124

103. Ahmed FE. Use of Chip-Based PCR for 3D Absolute Digital Quantification of microRNAs Molecules for The Non-Invasive Diagnostic Screening of Human Colon Cancer in Stool. *Integrative Care Sci Therapeutics.* 2018;5(2):1-1.

104. Ahmed, F E, Gouda M M, Ahmed N C. Chip-Based Digital PCR for Absolute Quantification of Colon Cancer miRNAs with 3D digital, chip-based PCR. Arch Oncol Cancer Ther. 2018;1(1):1-24.

21

Isolation and Characterization of Haloalkaliphilic Bacteria Isolated from the Rhizosphere of *Dichanthium annulatum*

Salma Mukhtar[1*], Kauser Abdulla Malik[1] and Samina Mehnaz[1]

[1]*Department of Biological Sciences, Forman Christian College (A Chartered University), Ferozepur Road, Lahore 54600, Pakistan*

*Corresponding author: Salma Mukhtar, Department of Biological Sciences, Forman Christian College (A Chartered University),Ferozepur Road, Lahore 54600, Pakistan, E-Mail: salmamukhtar85@gmail.com

Abstract

Diversity of haloalkaliphilic bacteria from the rhizosphere of halophytes is a crucial determinant of plant health and productivity. The main objective of this study is the identification and characterization of haloalkaliphilic bacteria from the rhizosphere, rhizoplane and root endosphere of *D. annulatum* collected from Khewra Salt Mine, Pakistan. A total of 41 bacterial strains were isolated and identified on the basis of morphological and biochemical characterization. Twenty two strains were selected for phylogenetic analysis based on 16S rRNA gene sequences. About 41% bacterial strains were identified as different species of *Bacillus*. *Exiguobacterium, Kocuria, Citricoccus* and *Staphylococcus* were dominant genera identified in this study. Most of the bacterial strains characterized in this study were alkaliphilic, moderately halophilic and mesophilic in nature. Mostly strains were considered as a good source of hydrolytic enzymes because of their ability to degrade proteins, carbohydrates and lipids. Results for screening of hydrolytic enzymes showed that more than 90% strains had ability to produce at least three enzymes screened in this study. These results showed that haloalkaliphilic bacterial diversity identified in this study had great biotechnological potential.

Keywords: Haloalkaliphilic bacteria; rhizosphere; 16S rRNA gene; *Dichanthium annulatum*; hydrolytic enzymes

Introduction

Hypersaline environments are widely distributed across the globe as salt mines, saline lakes, salt marshes and marine water [1, 2]. Halophytes such as *Atriplex, Salsola, Dichanthium,* kallar grass and para grass may contribute significantly to the developing world's supply of food, fiber, fuel and fodder. For areas where farm land has been salinized by poor irrigation practices or that overlie reservoirs of brackish water or for coastal desert regions, these plants could be successfully grown [3, 4].

The rhizosphere of halophytes harbors an impressive array of halophilic and alkaliphilic microorganisms. Poly extremophilic organisms have ability to tolerate two or more extreme conditions, such as haloalkaliphiles, halothermophiles and alkalithermophiles [5]. Haloalkaliphiles are organisms that require high salinity (3-30%) and an alkaline pH (pH 9-13) for their growth [6]. These organisms have been isolated and characterized from a number of environments such as saline-sodic lakes, acid mines, hypersaline saline soils, salt mines, marine environments and salt marshes [7, 8]. Haloalkaliphiles usually use small organic molecules (osmolytes, e.g., ectoine, betaine and proline) and intracellular enzymes (α-galactosidase) to maintain their osmotic balance and pH ranges near neutral to survive under extreme saline and basic environments [9, 10]

Haloalkaliphiles have a wide range of applications in biodefense, bioenzymes and biofuel production [11, 12]. These organisms provide a good source of novel alkaliphilic and halophilic enzymes such as proteases, gelatinase, amylases, lipases, cellulases and xylanases [13, 14]. Enzymes isolated and characterized from haloalkaliphiles have ability to function properly even at high pH and salinity. These enzymes can be used in industrial applications such as detergent industry, food stuffs, paper and pulp and pharmaceuticals industry [15, 16].

Though a number of studies have been reported on the isolation of haloalkaliphiles from different environments but this study is the first report on characterization of haloalkaliphilic bacteria from the rhizosphere of *Dichanthium annulatum* (halophyte) collected from Khewra Salt Mines, Pakistan. In the present study, haloalkaliphilic bacteria were isolated from rhizospheric soil and roots and identified on the basis of 16S rRNA gene sequence analysis. Selected haloalkaliphilic bacterial strains were further characterized for their biotechnological potential and ability to produce different industrially important enzymes (cellulases, proteases, amylases, xylanases and lipases).

Material and Methods

Sampling site

Khewra Salt Mine is the world second largest salt mine, It is located near Jhelum District, Punjab, Pakistan (32° 38′ North latitude, 73°10′ East longitude). It is classified as thalassic hypersaline environment because it is derived from evaporation of sea water [17]. It has Na+ and Cl- dominating ions and the pH is near neutral to slightly alkaline. Vegetation of this area is classified as sub-tropical dry evergreen forest. Plants like *Suaeda,*

Salsola, Atriplex, Dichanthium, Justica, Lantana, and *Chrysopogon* are dominant genera found here.

Sample collection

Rhizospheric soil samples were collected by gently removing the plants and obtaining the soil attached the roots. Soil and root samples were collected four sites from different sites of Khewra Salt Mine. At each site, soil samples of approximately 500 g were collected in black sterile polythene bags. These samples were stored at 4°C for further analysis.

Soil physicochemical parameters

Each soil sample (300 g) was thoroughly mixed and sieved through a pore size of 2 mm. Physical properties (pH, moisture content, salinity and temperature) of soil samples from different plants and non-rhizospheric soils samples were determined. Moisture (%); temperature and texture class were measured by Anderson method [18]; pH was measured by 1:2.5 (w/v) soil to water mixture and electrical conductivity (dS/m) was measured by 1:1 (w/v) soil to water mixture at 25°C [19]. Organic matter (Corg) was calculated by Walkley-Black method [20]. Cation exchange capacity (CEC) is capacity to retain and release cations (Ca^{2+}, Mg^{2+}, K^+ and Na^+) and sodium adsorption ratio (SAR) is the measure of the sodicity of soil which is calculated as the ratio of the sodium to the magnesium and calcium.

Isolation of haloalkaliphilic bacteria

Haloalkaliphilic Medium (HaP) (Tryptone 5 g/l, Yeast Extract 1 g/l, NaCl 30 g/l, 5 g/l KCl, 10 g/l $MgSO_4$, 2 g/l K_2HPO_4 and pH 9.2) was used for the isolation and purification of bacteria present in saline environments [21]. Rhizosphere was fractionated into rhizosphere fraction (RS), rhizoplane fraction (RP) and root endosphere or histoplane bacterial fraction (HP) according to the method described by Malik et al. [22]. RS fraction indicates the soil adhering with the roots; RP fraction is the root surface and HP is the interior of roots. In case of RS, the soil was mixed thoroughly, sieved and then one gram representative soil sample was taken. Bacterial fraction from RP was isolated by shifting one gram of washed root to a falcon tube containing 9 ml saline along with some pebbles and incubated in a shaker for 30 minutes. For the isolation of HP bacteria roots was sealed at both ends with wax after washing with water. Sealed roots were surface sterilized by using 10% bleach for 10 min. After sterilization waxed ends of roots were removed and roots were macerated by using FastPrep® instrument (MP Biomedicals). The soil from each non-rhizospheric soil samples and brine lake-bank soil samples was mixed thoroughly, sieved and then one gram representative soil sample was taken. Serial dilutions (10^{-1}-10^{-10}) were made for all samples [23]. Dilutions from 10-3 to 10-6 were inoculated on HaP plates for counting colony forming units (CFU) per gram of dry weight. Plates were incubated at 30°C until the appearance of bacterial colonies. Bacterial colonies were counted and number of bacteria per gram sample was calculated. The bacteria were purified by repeated sub-culturing of single colonies. Single colonies were selected, grown in HaP broth and stored in 33% glycerol at −80°C for subsequent characterization.

Morphological and biochemical characterization of haloalkaliphilic bacterial isolates

For morphological characterization, colony morphology (colour, shape, elevation, size and margin) and cell morphology (shape, size, motility and Gram-staining) were studied. Halophilic bacterial strains were biochemically characterized to detect different enzymes (β-galactosidase, arginine deaminase, lysine decarboxylase, tryptophan deaminase, gelatinase, catalase and oxidase) and carbon sources (glucose, sucrose, mannitol, maltose, arabinose, lactose and sorbitol) utilization by using QTS 24 strips (DESTO Laboratories, Karachi, Pakistan).

Molecular characterization of haloalkaliphilic bacterial isolates

Genomic DNA was isolated from different bacterial isolates by CTAB method [24]. PCR amplification of 16S rRNA were performed by using universal forward and reverse primers P15(5'-GAGAGTTTGATCCTGGTCAGAACGAAC-3'),P65 (5'CGTACGGCTACCTTGTTACGACTTCACC-3') for prokaryotes [25]. A PCR reaction of 25 μl was prepared by using Taq polymerase (5U) 0.5 μl, Taq buffer (10X) 1 μl, MgCl2 (25 mM) 1.5 μl, dNTPS (2.5 mM) 2 μl, 2 μl each of forward and reverse primer (10 pmol), 16 μl of dd.H_2O and 2 μl of template DNA. First denaturation step at 95°C for 5 min followed by 30 cycles of 94°C for 1 min, 54°C for 1 min and 72°C for 2 min and a final extension step was at 72°C for 10 min. PCR products were analyzed by using 1% agarose gel. PCR products were purified by using GeneJET PCR Purification Kit (K0702 - Thermo Fisher Scientific). Purified PCR products were sequenced by using forward and reverse primers (Eurofins, Germany).

Sequences of 16S rRNA gene were assembled and analyzed with the help of Chromus Lite 2.01 sequence analysis software (Technelysium Pty Ltd. Australia). The gene sequences were compared to those deposited in the GenBank nucleotide database using the NCBI BLAST program. Sequences were aligned using Clustal X 2.1 program and phylogenetic tree was constructed using neighbor-joining method. Bootstrap confidence analysis was performed on 1000 replicates to determine the reliability of the distance tree topologies obtained [26]. The evolutionary distances were computed using the Neighbor-joining method [27]. Phylogenetic analyses were conducted in MEGA7 [28]. There were a total of 1434 positions in the final dataset. Sequence of 16S rRNA gene from *Micrococcus luteus* was sued as outgroup. Bacterial strains identified in this study were submitted in GenBank under the accession numbers MH489029-MH489050.

Screening of haloalkaliphilic bacterial strains with respect to their salt, pH and temperature

Tolerance Ability

Bacterial isolates were grown in the presence of different salt concentrations (3-12% NaCl), pH ranges of 4-12 and temperature ranges of 4-42°C by using HaP broth medium. Isolates were cultured in 250 ml flasks at 30°C with continuous rotatory agitation at 150 rpm for 72 h (hours) [29]. During incubation, bacterial growth in terms of optical density (OD 600) was measured after different time intervals (3, 6, 12, 24, 48 and 72 h).

Enzyme assays for haloalkaliphilic bacterial strains

Cellulose and amylase activities were identified by using 2% iodine solution and spotting single colony of the bacterial strains on CMC (carboxymethyl cellulose 1%) and starch (1%) supplemented LB agar plates respectively [30]. Protease activity was tested on the medium described by Kumar et al. [31]. Test for gelatin hydrolysis was performed by using the method described by Pitt and Dey [32]. Lipase activity was tested by using HaP medium with 1% butyrin and Tween 80 hydrolysis assay as described by Sierra [33]. Xylanase activity was tested by using HaP medium supplemented with 1% xylan [34]. The clear zones around the bacterial colonies after 4-12 days of incubation at 30°C were considered as a positive result of protease, cellulose, xylan and lipase activities.

Results

Soil Physicochemical Analysis

Rhizospheric soil samples of four D. annulatum plants were analyzed and characterized on the basis of physicochemical properties such as soil pH, salinity, moisture, temperature, organic matter, NPK, CEC and SAR (Table 1). Soil pH ranged from 8.11 to 8.56 with the highest value in plant 3 and the lowest value in plant 1, electrical conductivity (EC1:1) ranged from 3.77 to 4.65 dS/m, values for soil moisture content ranged from 24.15 to 27.32%, temperature ranged from 29.23 to 32.52°C (Table 1). The value for total organic matter was maximum in soil sample 1 (35.77) and minimum in soil sample 4 (32.29). The amounts of available P, K, Ca and Mg were maximum in soil sample 1 as compared to other soil samples. CEC and SAR values were maximum (73.61 mg.dm-3 and 13.51) for soil sample 2 (Table 1).

Table 1: Physical and chemical properties of rhizospheric soil samples of D. annulatum

Parameters	D. annulatum 1	D. annulatum 2	D. annulatum 3	D. annulatum 4
pH	8.11[a]	8.29[ab]	8.56[b]	8.35[ab]
$EC_{1:1}$ (dS/m)	4.14[ab]	3.77[a]	4.19[ab]	4.65[b]
Moisture (%)	25.83[ab]	24.15[a]	27.32[b]	25.52[ab]
Temperature (°C)	29.23[a]	32.52[b]	31.01[ab]	30.82[ab]
Texture class	Silty loam	Silty loam	Silty loam	Silty loam
OM (g.Kg^{-1})	35.77[b]	33.15[a]	34.55[ab]	32.59[a]
P (mg.kg^{-1})	3.99[ab]	3.26[a]	3.82[ab]	3.59[ab]
K (mg.kg^{-1})	0.76[a]	0.58[b]	0.65[b]	0.49[a]
Ca (mg.kg^{-1})	1.70[b]	1.67[b]	1.51[a]	1.48[a]
Mg (mg.kg^{-1})	1.28[b]	1.15[a]	1.26[b]	1.19[a]
NO^{-3} (mg.kg^{-1})	12.76[b]	13.12[b]	10.21[a]	10.87[a]
H+Al (mg.kg^{-1})	67.55[b]	59.32[a]	61.24[a]	65.87[b]
V (mg.kg^{-1})	4.13[b]	3.87[a]	4.18[b]	3.76[a]
CEC (mg.dm^{-3})	68.45[a]	73.61b	72.73[b]	67.78[a]
SAR	10.24[b]	13.51a	11.15[a]	12.42[b]

Note: EC (Electrical conductivity); OM (Organic matter); P (Phosphorous); K (Potassium); Ca (Calcium); Mg (Magnesium); NO^{-3} (Nitrate ion); H+Al (potential acidity); V (base saturation index); CEC (Cation exchange capacity) and SAR (Sodium adsorption ratio). Letters represent statistically significant values at 5% level.

Morphological and Biochemical Characterization of Haloalkaliphilic Bacterial Isolates

A total of 41 bacterial strains were isolated from the rhizosphere and roots of D. annulatum by using Hap medium with high salt concentration (3% NaCl) and pH (9.2). These isolates were identified on the basis of morphological and biochemical characterization. Out of 41, 40% bacterial isolates were identified as members of genus Bacillus, 16% isolates were related to Kocuria, 12% isolates were belonging to Exiguobacterium, 8% isolates were related to Citricoccus, 8% isolates were identified as Staphylococcus and 4% isolates were realted to Micrococcus (Fig. 1).

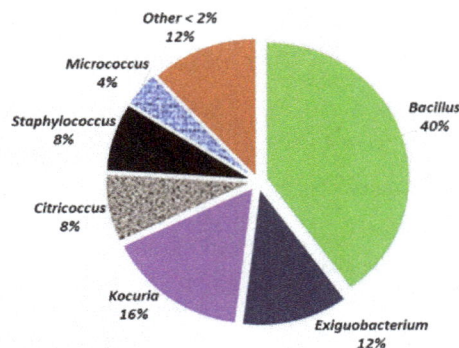

Figure 1: Relative abundance of haloalkaliphilic bacterial isolates from the rhizosphere, rhizoplane and root endosphere of *D. annulatum*

Phylogenetic Analysis of Haloalkaliphilic Bacterial Strains

On the basis of morphological and biochemical characterization, 22 bacterial isolates were selected for molecular characterization and phylogenetic analysis. Sequence analysis of 16S rRNA gene showed that 9 bacterial strains, PGRS2, PGRS7, PGRS9 and PGRS10 from the rhizosphere, PGRP3, PGRP6 and PGRP7 from the rhizoplane and PGHP2 and PGHP8 from the root endosphere of *D. annulatum* were identified as different species

of Bacillus (Table 2 and Fig. 2). Three bacterial strains (PGRS1, PGRS3 and PGHP1) had 99% similarity with *Exiguobacterium mexicanum*, 3 bacterial strains (PGRS5, PGRP4 and PGHP9) were related to *Kocuria* (*K. rosea and K. polaris*), 2 bacterial strains (PGRP2 and PGHP4) showed 99% similarity with *Citricoccus alkalitolerans* and one strain (PGHP5) had 99% similarity with Staphylococcus equorum. Bacterial strains related to *Oceanobacillus, Enterococcus, Virgibacillus* and *Micrococcus* were also identified in this study (Table 2 and Fig. 2).

Table 2: Identification of haloalkaliphilic bacterial isolates from the rhizosphere, rhizoplane and root endosphere of D. annulatum

Isolate code	Organism identified	Accession No.	Closest type strain in NCBI data base	Sequence length(bp)	Sequence similarity (%)
PGRS1	Exiguobacterium	MH489029	E. mexicanum DSM 6208 (JF505980)	1406	99
PGRS2	Bacillus	MH489030	B. pseudofirmus ATCC 700159 (NR_026137)	1425	100
PGRS3	Exiguobacterium	MH489031	E. mexicanum DSM 16483 (JF505982)	1365	99
PGRS5	Kocuria	MH489032	K. rosea ATCC 186 (KM114943)	1412	99
PGRS6	Oceanobacillus	MH489033	O. oncorhynchi DSM 16557 (KJ145755)	1305	100
PGRS7	Bacillus	MH489034	B. cohnii DSM 6307 (JF689927)	1465	100
PGRS9	Bacillus	MH489035	B. alcalophilus JCM 5262 (NR_036894)	1345	99
PGRS10	Bacillus	MH489036	B. polygoni NCIMB 14282 (NR_041571)	1513	99
PGRS11	Enterococcus	MH489037	E. durans ATCC 19432 (NR_036922)	1362	99
PGRS12	Virgibacillus	MH489038	V. halodenitrificans DSM 10037 (HG931337)	1473	100
PGRP2	Citricoccus	MH489039	C. alkalitolerans KCTC 19012 (KF322100)	1434	99

PGRP3	Bacillus	MH489040	B. alcalophilus ATCC 27647 (NR_036889)	1421	99
PGRP4	Kocuria	MH489041	K. polaris CIP 107764 (KF876845)	1464	99
PGRP6	Bacillus	MH489042	B. halodurans NRRL B-3881 (HQ446864)	1384	99
PGRP7	Bacillus	MH489043	B. alkalinitrilicus DSM 22532 (NR_044204)	1298	99
PGHP1	Exiguobacterium	MH489044	E. mexicanum DSM 16483 (JF505982)	1469	99
PGHP2	Bacillus	MH489045	B. clarkii DSM 8720 (KY849416)	1405	99
PGHP4	Citricoccus	MH489046	C. alkalitolerans DSM 15665 (KF322104)	1478	99
PGHP5	Staphylococcus	MH489047	S. equorum ATCC 43958 (AB975354)	1394	100
PGHP6	Micrococcus	MH489048	M. luteus CCM 169 (KJ843153)	1443	99
PGHP8	Bacillus	MH489049	B. pseudofirmus DSM 8715 (NR_026139)	1356	99
PGHP9	Kocuria	MH489050	K. rosea DSM 11630 (KF177263)	1429	99

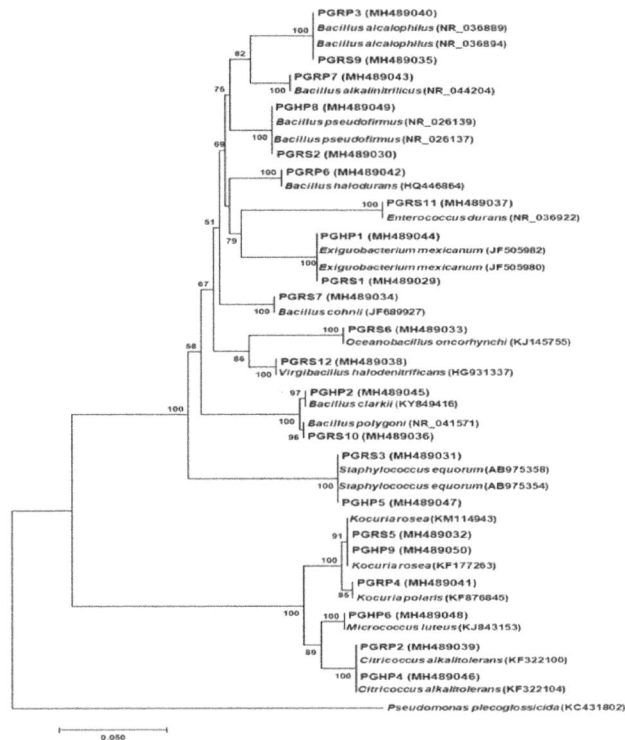

Figure 2: Phylogenetic tree based on 16S rRNA gene sequences of haloalkaliphilic bacterial strains from the rhizosphere, rhizosplane and root endosphere of *D. annulatum*. The percentage of replicate trees in which the associated taxa clustered together in the bootstrap test (1,000 replicates) is shown next to the branches.

Phenotypic characterization of haloalkaliphilic bacterial strains

All the strains had ability to grow at pH range from 8 to 12, but only few strains were able to survive at pH 4 and 6 (Fig. 3A). Mostly strains were able to grow at salt concentrations of 3-10% NaCl but only few strains (28%) especially members of *Bacillus* had ability to grow at 12% NaCl concentration (Fig. 3B). All the strains could grow well at temperature 28 and 37°C but only 38 % bacterial strains could tolerate at 4 and 62% strains were able to grow at 42°C (Fig. 3C).

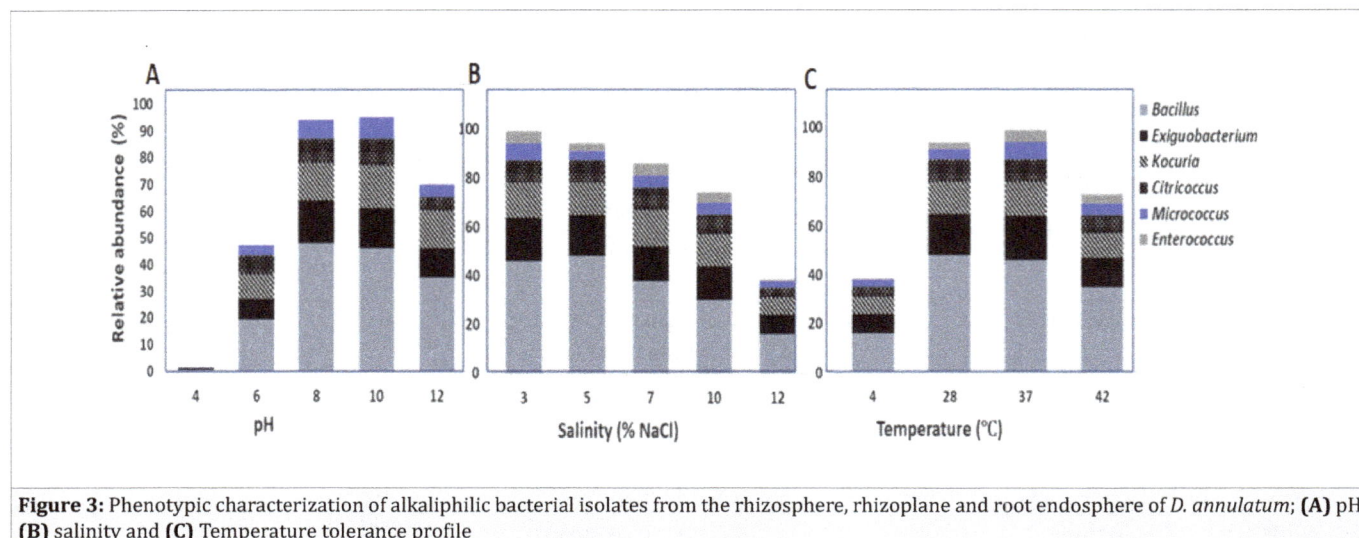

Figure 3: Phenotypic characterization of alkaliphilic bacterial isolates from the rhizosphere, rhizoplane and root endosphere of *D. annulatum*; **(A)** pH **(B)** salinity and **(C)** Temperature tolerance profile

Enzyme producing ability of haloalkaliphilic bacterial strains

Mostly haloalkaliphilic bacterial strains showed ability to degrade carbohydrates, lipids, proteins and gelatin at high salinity and pH (Table 3 and Fig. 4). Out of 22, sixteen bacterial strains showed positive results for protease activity, 20 strains had ability to degrade lipids, and 16 strains showed positive activity for amylase enzyme, 16 strains showed positive results for gelatinase, 14 strains were positive for cellulase activity and 14 strains showed positive results for xylanase activity (Table 3 and Fig. 4).

Table 3: Screening of hydrolytic enzymes produced by haloalkaliphilic bacterial strains from the   rhizosphere, rhizoplane and root endosphere of D. annulatum

Bacterial strains	Protease	Lipase	Amylase	Cellulase	Gelatinase	Xylanase
PGRS1	-	++	-	-	+	++
PGRS2	+++	-	++	+	++	-
PGRS3	++	+	-	-	+	++
PGRS5	-	+	+++	-	++	+
PGRS6	++	++	++	++	++	-
PGRS7	+++	+	-	++	-	++
PGRS9	-	++	++	+	+++	-
PGRS10	++	+	++	++	++	+
PGRS11	++	-	-	-	+++	-
PGRS12	+++	+	-	++	+	++
PGRP2	-	+	++	++	-	+
PGRP3	++	++	++	+	+++	+
PGRP4	+	+	+	-	+	-
PGRP6	++	++	++	++	-	+
PGRP7	+++	++	+	+	+++	-

PGHP1	+	+	-	-	+	++
PGHP2	++	++	+	++	++	++
PGHP4	-	+	++	+	-	-
PGHP5	+	+	+	+	-	+
PGHP6	++	++	++	-	-	+++
PGHP8	++	++	+	++	++	-
PGHP9	-	+	++	-	+	++

Note: -, no activity; +, low activity; ++, medium activity; +++, high activity

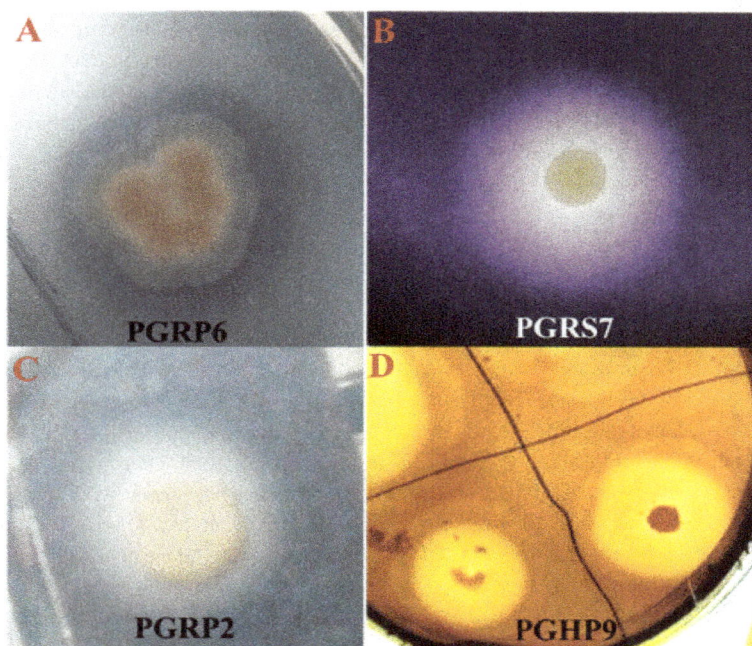

Figure 4: Enzyme assays for haloalkaliphilic bacterial strains from the rhizosphere, rhizoplane and root endosphere of D. annulatum using a drop spot technique; **(A)** Protease **(B)** Amylase **(C)** Lipase and **(D)** Cellulase

Discussion

High pH and salinity present a multifold challenge to all organisms in terms of ionic disequilibria and perturbed osmotic balance. Microorganisms that isolated and characterized from highly saline and saline-sodic soils have adapted special genetic and morphological modifications to survive under such extreme conditions [35, 36]. Here, we reported haloalkaliphilic bacterial diversity from the rhizosphere, rhizoplane and root endosphere of a halophyte (D. annulatum). The isolated bacterial strains were also screened for production of industrially important enzymes such as amylases, proteases, lipases, cellulases and gelatinase.

A total of 22 haloalkaliphilic bacterial strains have been identified from the rhizosphere and roots of D. annulatum. Phylogenetic analysis showed that these isolates were related to nine different bacterial genera Bacillus, Exiguobacterium, Kocuria, Citricoccus, Staphylococcus, Enterococcus, Oceanobacillus, Virgibacillus and Micrococcus (Table 2). Previous studies on the

isolation of haloalkaliphilic bacterial strains from Soda Lake Magadi (Kenya) showed that Bacillus, Exiguobacterium and Halomonas were the dominant genera identified from these environments [5, 37]. The abundance of Gram positive bacteria (Bacillus, Exiguobacterium, Kocuria, and Staphylococcus) is attributed to their cell wall and endospore formation in Bacillus enable them to survive in hypersaline and saline-sodic environments. Members of Actinobacteria Kocuria, Citricoccus and Micrococcus identified in this study have been previously reported from hypersaline soil of halophytes and Texcoco Lake [38, 39].

Bacterial isolates characterized in this study were alkaliphilic and moderately halophilic in nature. More than 87% strains were able to grow at pH more 10, salt concentrations 5-10% and temperature 28-42°C. Previous studies also reported that alkaliphiles, moderately halophiles and mesophiles are more abundant as compare to extremely halophilic and thermophilic bacteria in different soils [38, 40]. Halophilic strains from the groups, *Virgibacillus* and *Oceanobacillus* show optimum growth at salt concentration 5-10% NaCl and 28-37 °C [41].

Most of the bacterial strains showed ability to degrade different organic compounds such as carbohydrates, lipids, proteins and gelatin. More than 90% bacterial strains showed lipase activity, 73% bacterial strains showed proteolytic activity, 72% strains had ability to degrade carbohydrates, 73% strains showed positive results for gelatinase activity, 64% strains had ability to degrade cellulose and 63% strains showed xylanase activity (Table 3). Alkaliphilic, halophilic and mesophilic bacteria isolated from different saline and saline-sodic environments showed their ability to produce different industrially important enzymes such as amylases, proteases, lipases, gelatinase and xylanase [9, 11]. Enzymes produced by haloalkaliphilic bacteria have structural and catalytic properties to function properly even at high salinity, pH and temperature [42]. Lipase and protease producing alkaliphilic and halophilic bacteria have been previously isolated marine environment and food sources such as fish sauce [37, 43]. Halophilic bacterial strains related to *Bacillus and Oceanobacillus* are known to be a good source of different hydrolytic enzymes such as α-amylases, lipase, protease and xylanases [14, 42]. Members of Actinobacteria *Citricoccus, Kocuria* and *Micrococcus* have been well known for production of lipases, cellulases, amylases and gelatinase [39, 44]. Haloalkaliphilic cellulases and xylanases have been produced by different alkaliphilic and halophilic bacteria such as *Kocuria, Bacillus* and *Staphylococcus* [45]. Members of *Exiguobacterium* have been isolated from hypersaline tropical soils. These bacteria are able to grow at high pH and considered as good source of alkaliphilic enzymes such as proteases, lipases, cellulases and gelatinase [42, 46].

Conclusion

This study was the first report of its kind that deals with characterization of haloalkaliphilic bacteria from the rhizosphere and roots of *D. annulatum*. Twenty two haloalkaliphilic bacterial strains were identified on the basis of 16S rRNA gene analysis from the rhizosphere, rhizoplane and root endosphere. Nine strains showed more than 99% similarity with different species of Bacillus. Other dominant bacterial genera included *Kocuria, Exiguobacterium, Citricoccus, Oceanobacillus* and *Staphylococcus* was identified in this study. Most of the bacterial strains showed positive results for industrially important enzymes such as amylases, cellulases, proteases, lipases, gelatinase and xylanases. The ability of these bacterial strains to survive at high salinity, pH and temperature showed their potential biotechnological applications especially as a source of various enzymes.

Acknowledgement

We are highly thankful to **Higher Education Commission** [Project # HEC (FD/2012/1843)] for research grant.

References

1. Litchfield CD, Gillevet PM. Microbial diversity and complexity in hypersaline environments: a preliminary assessment. *J Ind Microbiol Biotechnol.* 2002;28(1):48-55.
2. Mukhtar S, Ishaq A, Hassan S, Mehnaz S, Mirza MS, Malik KA. Comparison of microbial communities associated with halophyte (*Salsola stocksii*) and non-halophyte (*Triticum aestivum*) using culture-independent approaches. *Pol J Microbiol.* 2017;66(3):375–386.
3. Ahmad R. Halophytes in Agriculture. DRIP Newsletter, Drainage and Reclamation Institute of Pakistan.1993;14(3).
4. Khan MA, Ansari R, Ali H, Gul B, Nielsen BL. *Panicum turgidum*, potentially sustainable cattle feed alternative to maize for saline areas. *Agri Ecosys Environ.* 2009;129(4):542-546.
5. Govender L, Naidoo L, Setati ME. Isolation of hydrolase producing bacteria from Sua pan solar salterns and the production of endo-1, 4-b-xylanase from a newly isolated haloalkaliphilic *Nesterenkonia* sp. *Afr J Biotechnol.* 2009; 8(20):5458–5466.
6. Hidri DE, Guesmi A, Najjari A, Hanen Cherif, Besma Ettoumi, Chadlia Hamdi, Abdellatif Boudabous, et al. Cultivation-dependant assessment, diversity, and ecology of haloalkaliphilic bacteria in Arid Saline Systems of Southern Tunisia. *Bio Med Research Inte.* 2013;1:1-16.
7. Takami H, Inoue A, Fuji F, Horikoshi K. Microbial flora in the deepest sea mud of the Mariana Trench. *FEMS Microbiol Lett.* 1997;152(2):279-285.
8. Horikoshi K. Alkalophiles: Some applications of their products for biotechnology. *Microbiol Mol Biol Rev.* 1999; 63(4): 735-750.
9. Jones BE, Grant WD. Microbial diversity and ecology of alkaline environments. Adaptation to Exotic Environments. Dordrecht: Kluwer Academic Publishers. 2000:177-190.
10. Bowers KJ, Mesbah NM, Wiegel J. Biodiversity of poly-extremophilic Bacteria: Does combining the extremes of high salt, alkaline pH and elevated temperature approach a physico-chemical boundary for life? *Saline Sys.* 2009;5:1-9. doi: 10.1186/1746-1448-5-9
11. Lundberg DS, Lebeis SL, Paredes SH, Yourstone S, Gehring J, Malfatti S, Tremblay J, Engelbrektson A, Kunin V. Defining the core *Arabidopsis thaliana* root microbiome. *Nature.* 2012;488:86-90.
12. Liu M, Cui Y, Chen Y, Lin X, Huang H, Bao S. Diversity of *Bacillus*-like bacterial community in the sediments of the Bamenwan mangrove wetland in Hainan, China. *Can J Microbiol.* 2017;63(3):238-245. doi: 10.1139/cjm-2016-0449
13. Taprig T, Akaracharanya A, Sitdhipol J, Visessanguan W, Tanasupawat S. Screening and characterization of protease-producing *Virgibacillus, Halobacillus* and *Oceanobacillus* strains from Thai fermented fish. *J appl pharm sci.* 2013; 3(2):25-30.
14. Horikoshi K. Enzymes isolated from alkaliphiles. Extremophiles Handbook. 2011:163-181.
15. Fujinami S, Fujisawa, M. Industrial application of alkaliphiles and their enzymes – past, present and future. *Environ Technol.* 2010;31(8-9):845-856. doi: 10.1080/09593331003762807
16. Krishna P, Arora A, Reddy MS. An Alkaliphilic and Xylanolytic strain of Actinomyctes. *World J Microbiol Biotechnol.* 2008;24(12):3079-3085.

17. Ahmad K, Hussain M, Ashraf M, Luqman M, Ashraf MY, Khan ZI. Indigenous vegetation of Soon valley at the risk of extinction. *Pak J Bot.* 2007;39(3):679-690.

18. Anderson JM, Ingram JS. Tropical Soil Biology and Fertility: A Handbook of Methods. 2nd ed. CAB International, Wallingford, UK. 1993:93-94.

19. Adviento-Borbe MA, Doran JW, Drijber RA, Dobermann, A. Soil electrical conductivity and water content affect nitrous oxide and carbon dioxide emissions in intensively managed soils. *J Environ Qual.* 2006;35(6):1999-2010.

20. Walkley A, Black IA. An examination of degtjareff method for determining soil organic matter and a proposed modification of the chromic acid titration method. Soil Sci. 1934;37(1):29-37.

21. Schneegurt MA. Media and conditions for the growth of halophilic and halotolerant bacteria and archaea. In: Vreeland RH (ed) Advances in understanding the biology of halophilic microorganisms. Springer, Dordrecht. 2012;35-58.

22. Malik KA, Bilal R, Mehnaz S, Rasool G, Mirza MS, Ali S. Association of nitrogen-fixing, plant growth promoting rhizobacteria (PGPR) with kallar grass and rice. Plant Soil. 1997;194(1-2):37-44.

23. Somasegaran P. Handbook for Rhizobia: Methods in Legume Rhizobium Technology. Springer-Verlag, cop. New York. 1994.

24. Winnepenninckx B, Backeljau T, de Wachter R. Extraction of high molecular weight DNA from molluscs. Trends Genet. 1993;9(12):407-412.

25. Tan ZY, Xu XD, Wan ET, Gao JL, Romer EM, Chen WX. Phylogenetic and genetic relationships of Mesorhizobium tianshanense and related Rhizobia. Int J Sys Bacteriol. 1997;47(3):874-879.

26. Varian H. Bootstrap tutorial. Math J. 2005;9:768-775.

27. Tamura K, Nei M, Kumar S. Prospects for inferring very large phylogenies by using the neighbor-joining method. Proc Natl Acad Sci USA. 2004;101(30):11030-11035.

28. Kumar S, Stecher G, Tamura K. MEGA7: Molecular Evolutionary Genetics Analysis version 7.0 for bigger datasets. Mol Biol Evol. 2016;33(7):1870-1874. doi: 10.1093/molbev/msw054

29. Bhadekar RK, Jadhav VV, Yadav A, Shouche YS. Isolation and cellular fatty acid composition of psychrotrophic Halomonas strains from Antarctic sea water. Eur Asia J BioSci. 2010;4:33-40.

30. Gupta P, Samant K, Sahu A. Isolation of cellulose-degrading bacteria and determination of their cellulolytic potential. Inter J Microbiol. 2012;20:28-35.

31. Kumar KV, Srivastava S, Singh N, Behl HM. Role of metal resistant plant growth promoting bacteria in ameliorating fly ash to the growth of Brassica juncea. J Haz Mat. 2009;170(1):51-57. doi: 10.1016/j.jhazmat.2009.04.132

32. Pitt TL, Dey D. A method for the detection of gelatinase production by bacteria. J Appl Microbiol. 1970;33(4):687-691.

33. Sierra G. A simple method for the detection of lipolytic activity of micro-organisms and some observations on the influence of the contact between cells and fatty acid substrates. A Van Leeuw J Microbiol. 1957;23(1):15-22.

34. Ghio S, DiLorenzo GS, Lia V, Talia, Cataldi A, Grasso D, Campos E. Isolation of Paenibacillus sp. and Variovorax sp. strains from decaying woods and characterization of their potential for cellulose deconstruction. Int J Biochem Mol Biol. 2012;3(4):352-364.

35. Sharma A, Singh P, Kumar S, Kashyap PL, Srivastava AK, Chakdar H, et al. Deciphering Diversity of Salt-Tolerant Bacilli from Saline Soils of Eastern Indo-gangetic Plains of India. Geomicrobiol J. 2015;32(2):170-180.

36. Mukhtar S, Mirza MS, Mehnaz S, Mirza BS, Malik KA. Diversity of Bacillus-like bacterial community in the rhizospheric and non-rhizospheric soil of halophytes (Salsola stocksii and Atriplex amnicola) and characterization of osmoregulatory genes in halophilic Bacilli. Can J Microbiol. 2018;64(8):567-579. DOI: 10.1139/cjm-2017-0544

37. Nyakeri EM, Mwirichia R, Boga H. Isolation and characterization of enzyme producing bacteria from Soda Lake Magadi, an extreme soda lake in Kenya. J Microbiol Exp. 2018;6(2):57-68.

38. Mukhtar S, Mirza MS, Awan HA, Maqbool A, Mehnaz S, Malik KA. Microbial diversity and metagenomic analysis of the rhizosphere of Para Grass (Urochloa mutica) growing under saline conditions. Pak J Bot. 2016;48(2):779-791.

39. Marisela YSP, Valenzuela-Encinas C, Dendooven L, Marsch R, Gortáres-Moroyoqui P, Estrada-Alvarado MI. Isolation and phylogenic identification of soil haloalkaliphilic strains in the former Texcoco Lake. Inte J Environ Health Res. 2014;24(1):82-90. doi: 10.1080/09603123.2013.800957

40. Irshad A, Ahmad I, Kim SB. Culturable diversity of halophilic bacteria in foreshore soils. Braz J Microbiol. 2014;45(2):563-571.

41. DasSarma S, DasSarma P. Halophiles and their enzymes: negativity put to good use. Curr Opin Microbiol. 2015;25C: 120-126. doi: 10.1016/j.mib.2015.05.009

42. Kumar S, Karan R, Kapoor S, Singh SP, Khare SK. Screening and isolation of halophilic bacteria producing industrially important enzymes. Braz J Microbiol. 2012;43(4):1595-1603. doi: 10.1590/S1517-838220120004000044

43. Phrommao E, Rodtong S, Yongsawatdigul J. Identification of novel halotolerant bacillopeptidase F-like proteinases from a moderately halophilic bacterium, Virgibacillus sp. SK37. J Applied Microbiol. 2011;110(1):191-201. doi: 10.1111/j.1365-2672.2010.04871.x

44. Mukhtar S, Zaheer A, Aiysha D, Malik KA, Mehnaz S. Actinomycetes: A source of industrially important enzymes. J Proteomics Bioinform. 2017;10:316-319.

45. De Lourdes MM, Pérez D, García MT, Mellado E. Halophilic bacteria as a source of novel hydrolytic enzymes. Life. 2013;3(1):38-51.

46. Anbu P, Annadurai G, Hur BK. Production of alkaline protease from a newly isolated Exiguobacterium profundum BK-P23 evaluated using the response surface methodology. Biologia. 2013;68(2):186-193.

22

The Possibility of Diabetic Wound Healing using Electro spun PLA Nano Fibers

Saeed Ahmadi Majd[1*], Mohammad Rabbani Khorasgani[1], Mahna Mapar[1], Mohammad Asadollahi[1], Nahid Zarini Mehr[1], Ardeshir Talebi[2], Hamed Karimi[3]

[1]University of Isfahan, Isfahan, Iran

[2]School of Medicine, Isfahan University of Medical Sciences, Isfahan, Iran

[3]Islamic Azad University, Science and Research Branch, Tehran, Iran

Corresponding author: *Saeed Ahmadi Majd, University of Isfahan, Isfahan, Iran; E-mail: saeedmajd68@gmail.com*

Abstract

Diabetes is a metabolic disorder demonstrated by hyperglycemia and is typically resulted from defects in insulin secretion and impaired pancreatic function. In general, wound infection is a common complication of diabetes and finding effective remedies (medications) and methods of diabetic wound healing has recently been of interests of researchers. In the present research study, PLA NANO fiber scaffolds are prepared by electro spinning in two different concentrations of 2% and 4%. Indeed, electron microscopic scanning of this NANO fiber represents the achievement of NANO fibers with average diameter of 190 nm ± 44/4 nm and 460 nm ± 75 nm that encompass a uniform morphology. Given the fact that the PLA NANO fiber scaffolding with concentration of 2% has less average diameter, smaller pores and higher biodegradability, hence, it has been selected to study wound healing. In this study, 18 male Wistar rats weighing 250-280g are divided into three groups. First, two groups of rats become diabetic with streptozotocin (55mg / kg i.p.) and another group is monitored as a control type (healthy rats). After observing the symptoms of diabetes and hyperglycemia (250-300 mg /dl in diabetic group), a wound is made on the back of neck of all four groups with diameter of 1 cm through surgical scissors. Then, wound is treated in the experimental group on a daily basis for 14 days by PLA NANO fiber scaffolding. Meanwhile, macroscopic and microscopic studies are adopted on 0, 4, 7, 14 and 7, 14 days respectively. The results revealed wound size reduction, epidermal gap decrement and reduction in wound treatment area in the treatment group compared to it in healthy groups. This, in turn, implies the positive impact of the NANO fibers in the process of wound healing.

Keywords: Diabetic; Wound healing; PLA; Nano fibers; Electro spinning

Abbreviations: PLA, poly lactic acid

Background

Generally, diabetic foot and diabetic foot ulcers are still considered as huge problems among diabetic patients. Diabetic foot problems are still largely unsolved despite many advances made in diagnosis and treatment of diabetes (1). Accordingly, wound healing depends on its blood supply as ulcers in diabetic patients are intensified due to vascular dysfunction and bacterial infection caused by diabetic ulcers that are not healed completely (2). When a wound is created, a series of complex biochemical functions are quickly involved to rebuild damaged tissue but this restoration cannot be fully fulfilled because of the scar tissue (3). Therefore, adopting wound coverages to protect the wound, extract excess fluid from the wound surrounding, sterilize external microorganisms, and improve the appearance of accelerating healing process is required. In order to meet such requirements, coatings on wound healing must create a physical barrier while they are permeable against the passage of moisture and oxygen.

A number of methods have been used to cover wounds each having special features. In fact, special features should be considered to realize a proper wound covering remedy such as biocompatibility, biodegradability, accelerating recovery and healing process, preventing of secondary infection associated to ulcers or trauma, and preventing of wound symptom generation (scar)(4). Traditional and even common methods have shortcomings because they often lack at least one of the above-mentioned features. The scaffold used wound covering should encircle biodegradability feature and degradation speed must be coordinated with generation speed of tissue. More importantly, these scaffolds should be highly porous to facilitate cell growth and penetration of nutrients (5). During the time period when the cell structure makes its own natural matrix, scaffold should provide a flawless structure and finally is released from new generated tissue as small molecules (6). Synthetic polymers used for acellular scaffold fabrication can be categorized as absorbable synthetic polymers, such as PCL, PLA, PEG, etc., and non absorbable synthetic polymers including polyurethane, nylon, polytetrafluoroethylene (PTFE), etc (7-9).

Poly-lactic acid is a biodegradable polymer with long substantial strings (10). Solid polymers are decomposed into available acids in the human body through delivering water to long polymer strands (11). The generated scaffolding in wound healing must be biodegradable, biocompatible and porous in addition to encompassing proper functioning feature. In this paper, we aimed to fabricate and study the characteristics mentioned in scaffolding PLA in order to apply it on skin wound healing in diabetic rats with streptozotocin (STZ) procedure.

Methods

Identification of Poly Lactic Acid

In general, gel permeation chromatography (GPC) is used to determine molecular weight and polymer dispersion index by the Central Laboratory of Isfahan University. The obtained results of GPC experiment revealed that the average molecular weight of poly lactic acid was about 300 KDa 300 with diffusion index of 1.575.

Structure of Polymer Scaffolds PLA using electro spinning method

After determining PLA molecular weight through Gel permeation chromatography, polymer solutions are prepared to make electro spun NANO fibers. In order to prepare polymer solutions, PLA is dissolved in a solvent and is located in an ultrasonic bath for one hour (Figure.1a). For this purpose, electro spinning device is used that includes a syringe pump to inject the polymer solution and a source of high voltage between collector surface and feeder pump (12). Initially, PLA polymer is dissolved within solvents tri flora ethanol (TFE). Then, electro spinning solution is deployed on 5 ml syringe including a needle with internal diameter of (0.337 ± 0.019). Moreover, clamp connected to power supply is joined to needle heading while rotating cylinder is connected to ground. In addition, solution injection rate is considered as 0.3-0.4 ml/h and operating distance is set as 9cm. Especially, NANO fibers formed on aluminum plate are collected which is located at a distance of 9 cm from the tip of the needle. Moreover, Electro spinning is executed in a lab environment with a temperature of 25 °C for 6 hours (Figure.1b).

NANO fiber scaffold morphology scanning within electron microscopy

For providing electron microscopy, web NANO fiber sample with size of (5 x 10) is cut and deployed on the disc. The samples are then coated with gold material for 5 minutes and after that, NANO fibers morphology is examined by electron microscopy under 20 kv voltage (Figure.1c).

Determination of NANO fibers porosity

Indirect calculation is adopted to determine porosity by applying polymer and NANO fiber scaffolds density. Furthermore, total porosity as percentage is obtained from porosity equation as follows.

Porosity percentage equation

$$\text{Porosity \%} = \left\{1 - \frac{d_s}{d_p}\right\} \times 100$$

$$= \left[1 - \left\{\frac{m_s / \upsilon_s}{d_p}\right\}\right] \times 100$$

$$= \left[1 - \left\{\frac{m_s / \pi d^2 t}{d_p}\right\}\right] \times 100$$

m_s = scaffolds mass)g(' v_s = scaffolds volume)cm³(' d=scaffolds diameter (cm)

d_s =scaffolds density) g/cm³) 'dp=polymer density)g/cm³)

NANO fibers diameter and pore size measurements by Image J software

In this section, images obtained from electron microscopy image scanning are used to image analysis and measure the diameter of the NANO fibers and pores size. This measurement is derived via Image J software (Java Version). There is a scale bar within each image that must be calibrated through the Set scale option in initialization phase of image analysis. Moreover, NANO fibers diameter and pore size are analyzed manually and randomly. Before starting the process of image analysis, contrast and threshold of each image are optimized to identify top layer of fiber scaffolds. This is adopted to ensure that only top layer scaffold porosity will be measured. Information obtained from the measurement is then extracted and transferred to Excel for data analysis. Entirely, porous scaffolds are evaluated through this program and compared with theoretical data.

Biodegradability test

The concept of NANO fiber scaffold degradation refers to loss of scaffolds mass during chemical and biological reactions that occurs in the environment. In most cases, this deterioration is appeared due to hydrolysis of the main chain. For analysis of generated

NANO fiber scaffold degradation, samples are cut out in forms of circular pieces with a diameter of 1.88 cm and also weighed separately. Initially, NANO fiber scaffolds are rinsed with 70% ethanol and each side of scaffolds are sterilized for 20 minutes under the UV rays. Then third samples of each scaffold are maintained in cell cultivation container for 6 weeks within a solution of PBS 10 m-Molar with PH=7. Meanwhile, scaffolds are immersed in a solution of PBS. Moreover, cell cultivation containers are kept in a shaking incubator at 37 °C temperature. Regularly, PBS solution is replaced with fresh solution every 2-3 days and subsequently 3 samples are removed from each scaffold after a week and then washed 3 times with deionized water and dried in a vacuum oven for 24 hours at ambient temperature weighing phase is executed.

Animal model

In this study, 18 outdoor male wistar rats weighing 280-250g are purchased from medical science of Isfahan University and are kept in biology laboratory Group of Isfahan University under 12 hours illumination and 12 hours dark condition in temperature of 2 ± 22 degree in order to adapt them with new environment. These animals are then randomly assigned to experimental and controlled and negative control groups.

Induction of diabetes in animal models

In this section, streptozotocin (STZ) medicine manufactured by sigma firm is applied for induction of diabetes with dose of 55 mg/kg as injection to intraperitoneal. However, animal's blood glucose is measured three days after injection and animals with blood glucose greater than 250 mg/dl are considered as diabetic rats.

Scarring in animal models

In order to make scar, mice were first anesthetized by 50 mg/kg of ketamine and 10 mg/kg of Xylazin. Moreover, their hairs on the back of the neck and between their shoulders are carefully shaved and wounds are created with dimensions of 1*1 cm² (Figure.1d).

Macroscopic studies

According to this study, wounds are evaluated in terms of macroscopic model to determine the area of ulcer and wound healing trend for days of 0, 4, 7 and 14. It should however be noted that 7 and 14 days reflect significant importance in terms of wound healing process. Therefore, wound area in these days is determined with imaging and images analysis using the image processing toolbox of Photoshop software Cs6.

Microscopic studies

Specific samples of wound were released from each group on days of 7 and 14 for histological studies. Meanwhile, sample tissue to be stabilized is placed in container including 10% formalin and is transferred to pathology laboratory to prepare a tissue section slides. After complete stabilization of samples, tissue sections with a thickness of 5 microns are prepared and utilized Hematoxylin and eosin coloration with conventional methods. Finally, slide colored by a microscope (4Nikon, Japan) is investigated with zoom magnification of 10 and 40 (Figure.1e).

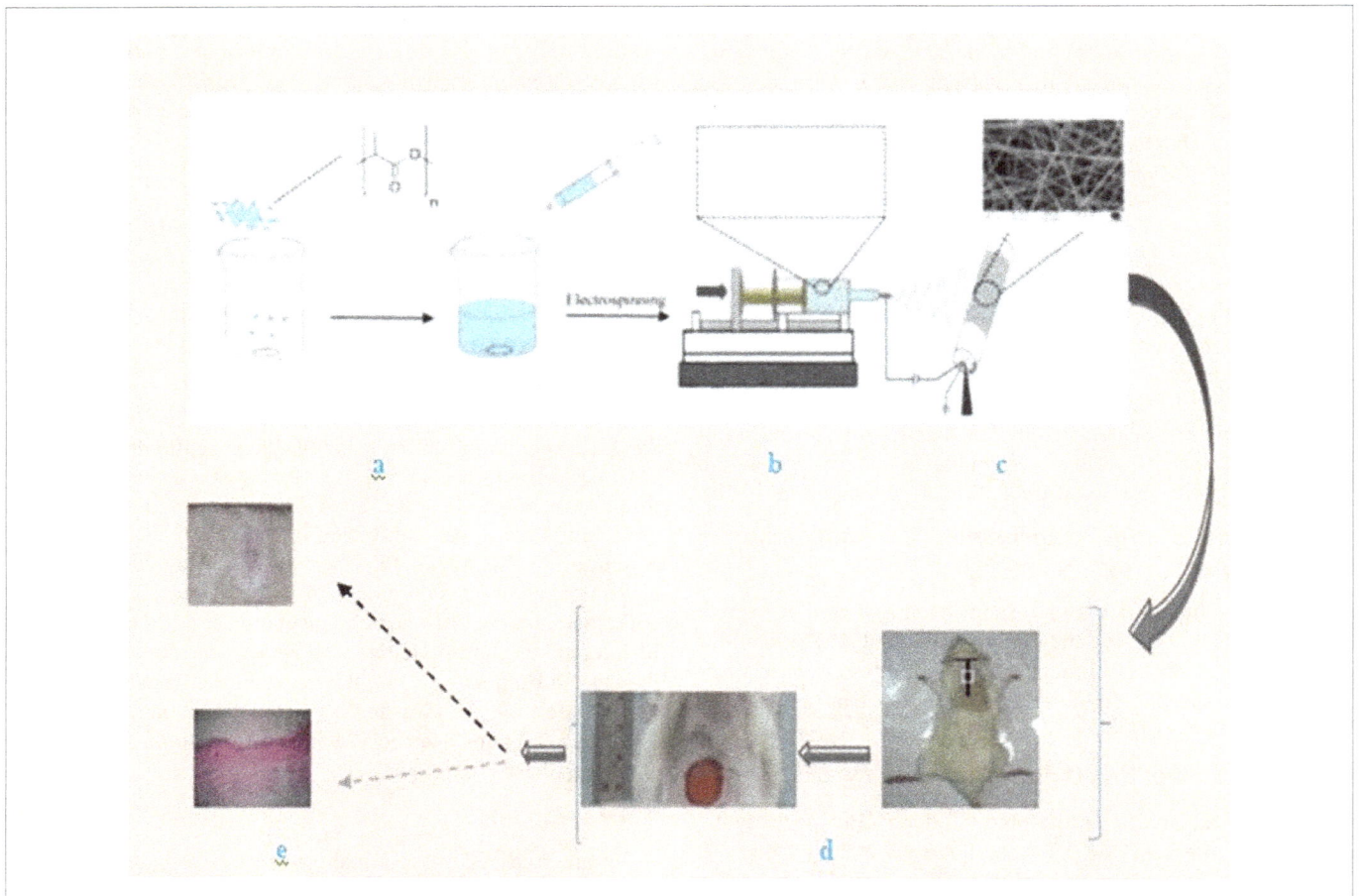

Figure 1: Schematic depicting, **a:** PLA solution, **b:** Electro spinning setup and phenomenon of electro spinning, **c:** electro spun fibers of PLA deposited on aluminum foil, **d:** Ulceration, **e:** Histological evaluation

Biopsy of normal and diabetic rats pancreas tissue

For the study and comparison of pancreas Langerhans islet beta cells in streptozotocin-induced diabetic rats, and normal rats, pancreatic biopsy of normal and diabetic rats were done and tissue samples were fixed in 10% formalin, then were stained by Hematoxylin & Eosin and finally were evaluated by Leitz microscope with 400 times enlargement.

Results

Electron microscopy

Electron microscope and AFM images prepared from PLA scaffold surface is shown in (Figure.2). In this study, PLA polymer with concentrations of 2% is used fiber diameter at a concentration of 2% is determined 190 nm ± 44/4.

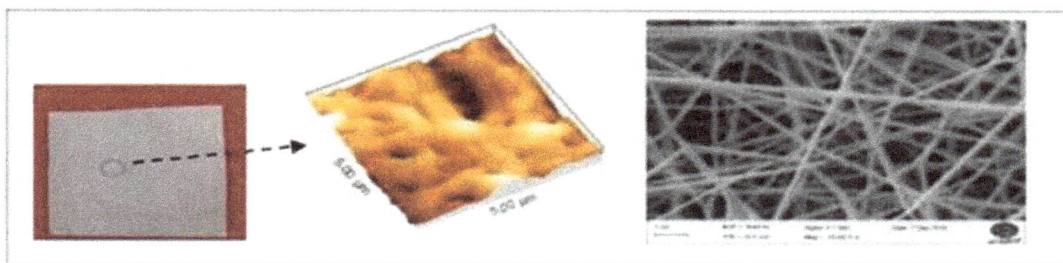

Figure 2: SEM image of poly lactic acid (PLA) Nano fibers prepared by electro spinning from with poly lactic acid concentrations of 2%

Porosity and pore size

Porosity and pore size is reported in the table 1. The theoretical porosity is reported based on average weight of diameter and thickness of the spun fibers and apparent porosity is obtained based on SEM images analysis.

Table 1: Porosity and pore size distribution of electro spun scaffolds

	Apparent porosity%	Theoretical Porosity %	The pore size
Scaffolds of NANO fibers2%	89	99/17±0/15	10 μm±4 μm

Biodegradability

In this study, degradation speed is only dependent to morphology structure (size and diameter of the NANO fibers) due to utilization of PLA polymer. The PLA scaffold degradation speed with 2% concentrations is given in figures; as it is evident from figures, PLA scaffold degradation speed with 2% concentrations is higher which is induced as a result of lower diameter of NANO fibers produced through electro spinning process (Figure.3).

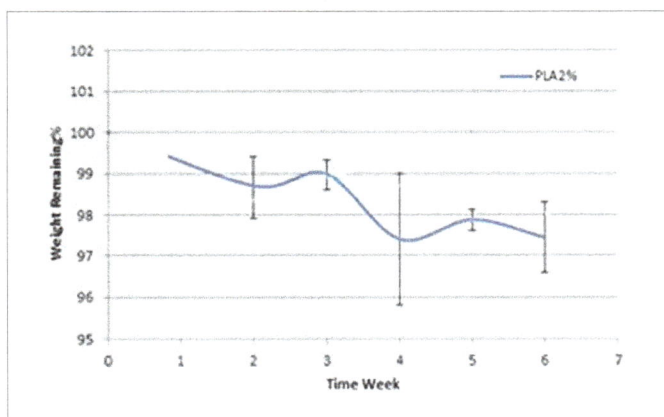

Figure 3: Weight remaining of PLA scaffolds after in vitro degradation for 6 weeks

Induction of diabetes by streptozotocin

Streptozotocin at dose 60 mg/kg, resulted in high mortality in test group (11death per 25 rats), Therefore the dose: 55 mg/kg was replaced it and diabetes developed within 3 days. The mortality rate significantly reduced (3death per 25 rats). Pancreatic biopsy of normal and diabetic rats confirmed that the islet and cells were destroyed due to the effect of Streptozotocin

in diabetic rats. The comparison of these pictures shows that the tissue of pancreatic Langerhans and the beta cells of diabetic rats have been degenerated irreversibly (figure.4). The results of measuring of rats body weights indicated that average of body weight in diabetic rats reveals loss of weight and thinness in diabetic adult rats (figure.5). The blood glucose measurement before and after diabetes revealed that: the levels of glucose in healthy adult rats was 101 ± 5 mg/dl, But in diabetic rats was measured as 374 ± 10 mg/dl (figure.6).

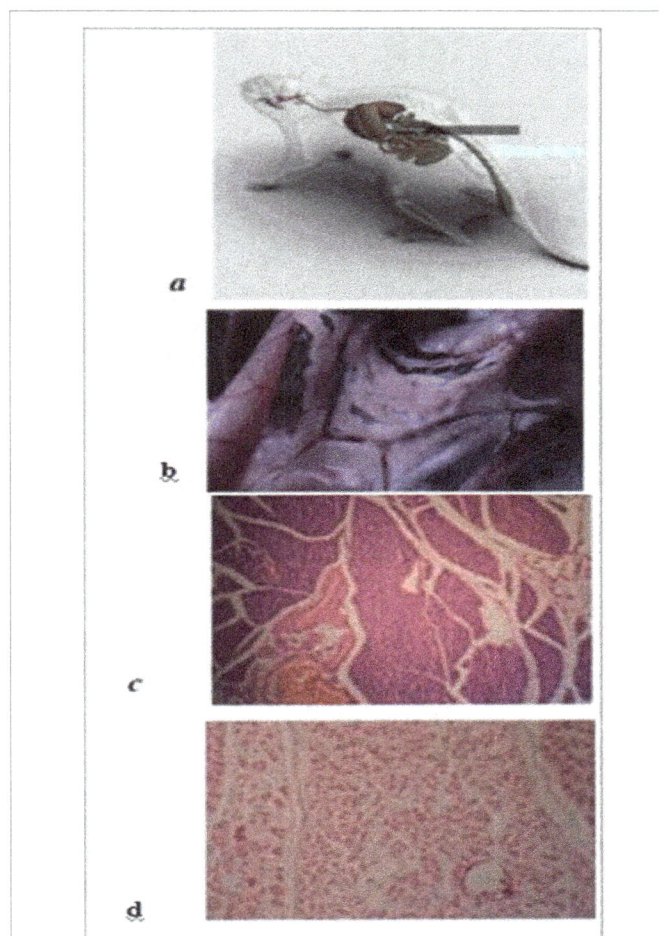

Figure 4: a: Anatomic relationships of the pancreas with surrounding organs and structures **b:** An isolated rat pancreas **c:** Pancreatic biopsy of normal rats **d:** Pancreatic biopsy of diabetic rats that confirms the necrosis of islets and cells due to the effect of streptozotocin

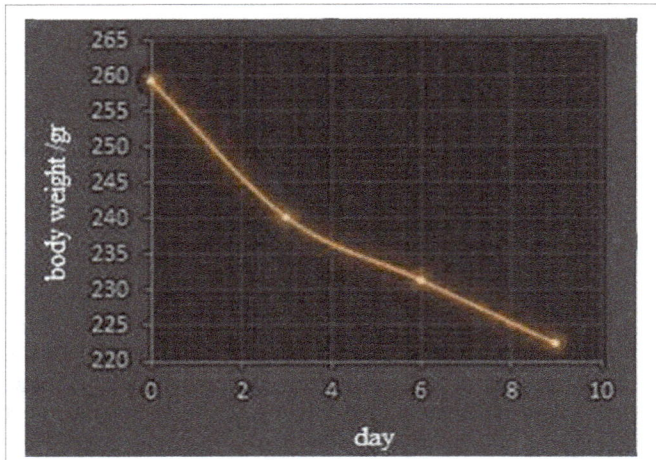

Figure 5: Shows continuous changes in average of body weight in diabetic rats in days 0, 3, 6 and 9

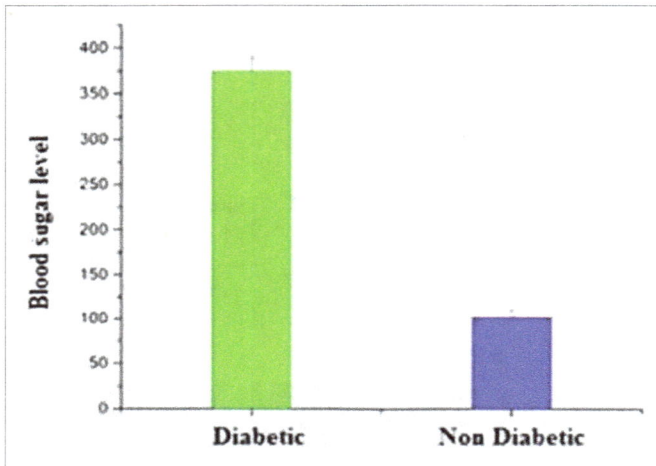

Figure 6: Shows the changes of average level of glucose in serum of diabetic and Non diabetic rats

Macroscopic assessment of wound

The average area of wound in diabetic group on days 7 and 14 is 93 ± 4.25 and 27.045 mm² respectively. Moreover, average area of wound in treated group diabetic group with PLA NANO fibers on days 7 and 14 is 64 ± 2.37 and 27.045 mm² respectively. According to comparison of average area of ulcers treated with diabetes control it can be concluded that there are significant differences between them *(P < 0.05)* (figure.7-8).

Microscopic evaluation of wounds healing

In the experiment group (PLA Nano fiber), histological findings indicate the reduction in the wound size and epidermis gap and dermis lesion surface length in the 7th and 14th days compared to the diabetic and healthy control groups (figure.7). This issue indicates its positive effect on the process of wound healing (figures.9-11).

The microscopic pilot group in the seventh day

There were no immediately apparent changes in the panniculus. On the seventh day, the wounds were open and featured with a pack of infiltrated inflammatory cells. The keratinocyte layers at the wound edges got thicker and the panniculus was not observed in the wound. In study of tissue sections marked with H & E gap (the epidermal gap is not restored and the epithelial lining is developing and the epidermal gap length and dermal area in the wound is less than the control group. It has been seen that in granulation tissue, the cell congestion is more and the blood vessels are more repeated than those in the control group (figures.9-11).

The microscopic pilot group in the fourteenth day

Re-epithelialization was perfect in most of the wounds on the fourteenth day when high cellularity was still obvious and certain new collagens could be seen in the wounds. By day 14, the wound areas became shorter and dense granulation tissue was present. By day 32, cellularity significantly decreased and scar tissue was seen at the wound site. In the investigation of tissue, sections stained with H & E gap (the gap) epidermal layer has been restored and the remaining scar was not seen on the wound. The epidermal gap length in the wound has been reached 0 and epidermal area in the wound area is less than that in the control group (figures.9-11).

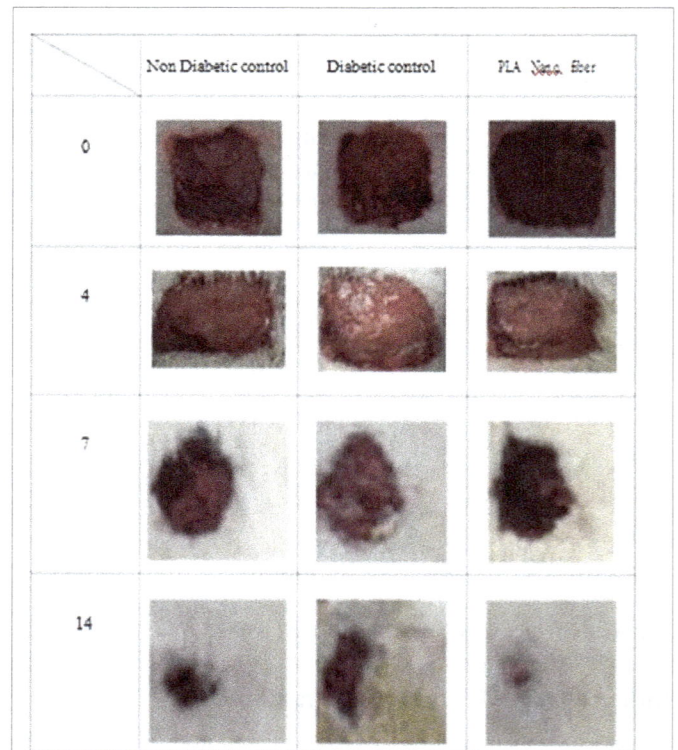

Figure 7: SPhotographs of macroscopic appearances of wound excised from untreated diabetic and non-diabetic rats (control groups) and experimentally diabetic rats treated with PLA Nano Fibers in days 7 and 14

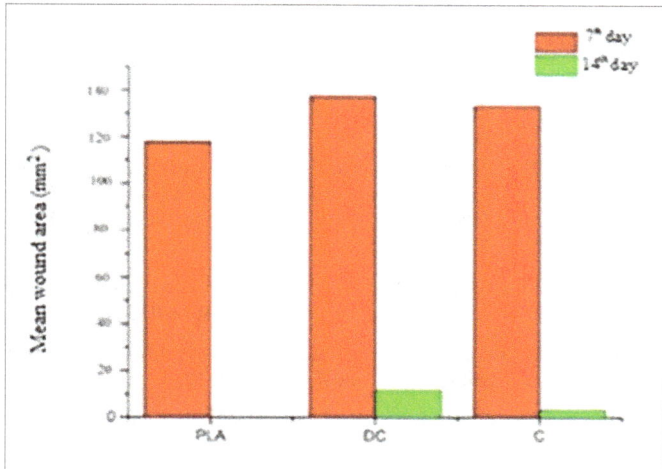

Figure 8: Wound areas in different groups on days 7 and 14C. Non diabetic control: DC, diabetic control: PLA, Poly Latctic and Nano fiber wound dressings

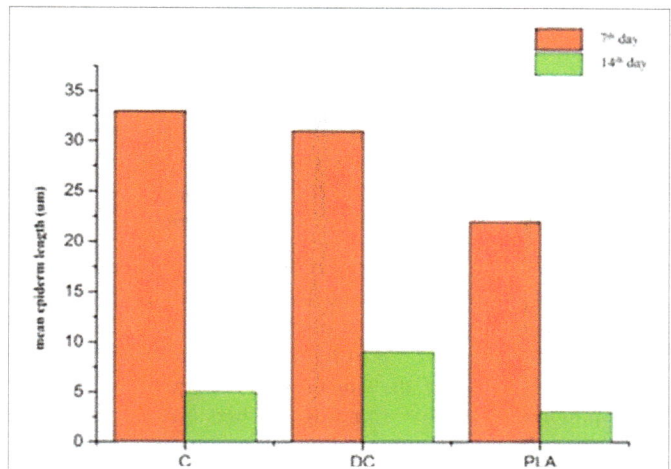

Figure 10: Length of epidermis gap in the study groups at day 7 and day 14

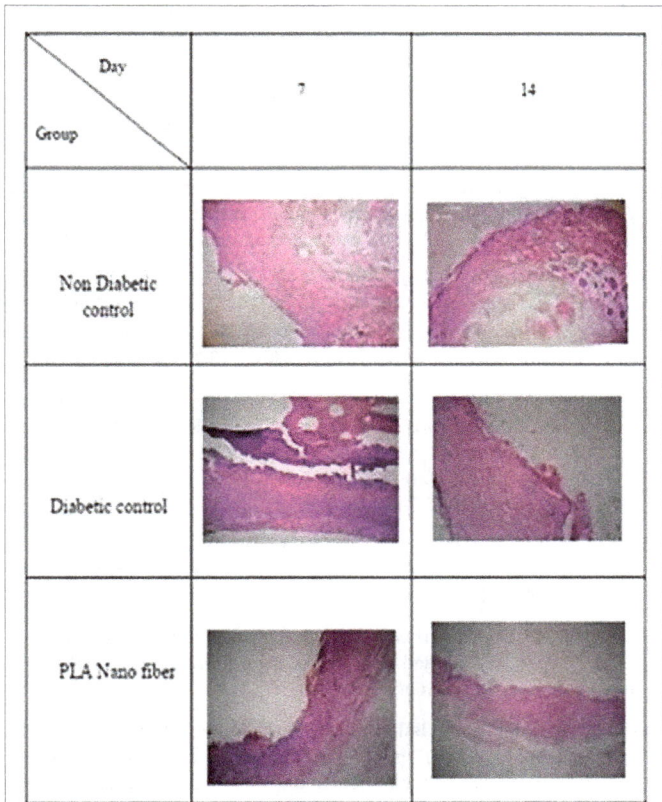

Figure 9: Histological study of wound healing in groups Non diabetic control, Diabetic control and PLA Nano fiber

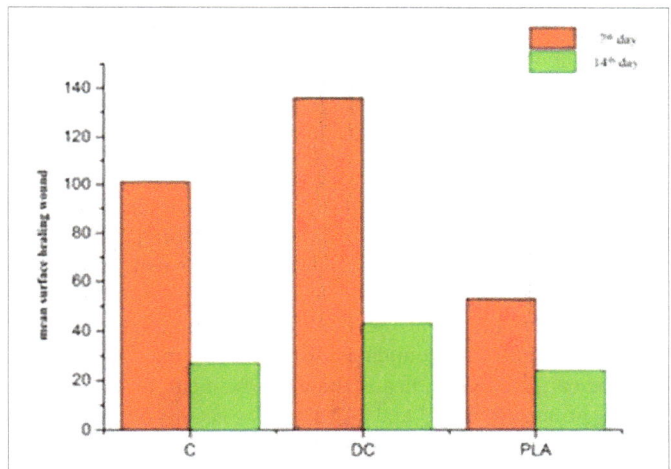

Figure 11: Healing Wound surface of dermis in the study groups at day 7 and 14

Discussion

Significant technological and scientific advances have been made in the field of electros pinning for the repair and regeneration of tissues including skin. Success in skin tissue engineering is mainly based on regulating cell behavior and tissue progress through the development of a scaffold similar to the natural extracellular matrix that can support cell culture. As a natural extracellular matrix provides an ideal environment chemically, electrically and topographically for cell adhesion and proliferation, tissue engineering also needs a biocompatible scaffold, immunological neutral, driving, biodegradable to support the growth and repair of skin lesions. In general, a suitable covering for the wound has features as follows: preventing the infiltration of external infectious agents (some are very small), failure to prevent the entry of oxygen and water, biocompatible, biodegradable, accelerating wound healing, preventing bleeding and failure to create ulcer syndrome. Our studies showed that PLA Nano-fibrous scaffold at a concentration of 2%, due to the small pore size, high biodegradability and wound regenerative properties has a proper feature to be used as a wound dressing.

Today, by imitating the in vivo conditions as much as possible, researchers are trying to improve the efficacy in vitro conditions. Tissue engineering is of great importance to achieve this goal by providing a variety of scaffolds in cell culture environment that can be an imitation of the body's extracellular space. Recent developments in electro spinning method have lead to the production of a continuous solid fiber with diameters in the range of several nanometers by controlling the surface inter-molecular structure. According to the survey, it is found that using electro spinning process is the best method of producing Nano fibers from polymeric materials where the production of Nano fibers is continuously done. Thus, in this study, we used the electro spinning method to prepare PLA Nano fiber scaffold.

Nguyen, et al suggested that Cur-loaded Nano fibers with appropriate Cur concentration are nontoxic and have potential as component of wound-healing patches (13).

Treatment with PLA Glass dressings significantly reduced the wound area, compared with the PLA- or commercial dressings-treated wounds.

The result of the present study revealed that the PLA Nano fibers might be properly used as a wound dressing. The wound healing was accelerated and the Nano fibers sheet was tightly adhered to the wound. In the experiment group PLA Nano fiber, some fragments of the PLA Nano fiber, equal as 1.5*1.5 cm^2 were put on the wounds as dressing. The macroscopic and histological findings indicate the improvement of the wound healing process in the 7th and 14th days. The results obtained from the present study indicate a significant reduction in the wound size and improvement of wound healing process of the wounds in artificially diabetic rats after application of PLA Nano fiber scaffold. Therefore, the application of the mentioned compound to help diabetic wound healing can be investigated in further studies particularly in clinical trials. In general, two main reasons might be mentioned for wound healing with Nano fiber PLA: the three-dimensional, network and porous structure of the Nano fiber scaffold which causes the pus and blood moisture absorption and high passage of oxygen over wounds and facilitating wound healing. 2. The chemical structure of Nano fiber PLA and its favorable biological properties such as its biocompatibility and antibacterial nature cause the assimilation of the biochemical environment of the natural tissue and absorption of fibroblasts

as well as facilitating wound healing. The instances of wounds are healed after 10 days (14).

References

1.Shahbazian H, Yazdanpanah L, Latifi SM. Risk assessment of patients with diabetes for foot ulcers according to risk classification consensus of International Working Group on Diabetic Foot (IWGDF). Pak J Med Sci. 2013;29(3):730-734.

2. O'Brien I, Corrall R. Epidemiology of diabetes and its complications. The New England journal of medicine. 1988;318(24):1619-20.

3. Romo T, Al Moutran H, Pearson J, Yalamanchili H, Pafford W, Zoumalan R. Skin wound healing. Medscape Reference. 2012.

4 .Gist S, Tio-Matos I, Falzgraf S, Cameron S, Beebe M. Wound care in the geriatric client. Clin Interv Aging. 2009;4(1):269-287.

5. Ma PX, Langer R. Fabrication of biodegradable polymer foams for cell transplantation and tissue engineering. Tissue engineering methods and protocols. 1999:47-56. doi: 10.1385/0-89603-516-6:47.

6. Shin M ,Ishii O, Sueda T, Vacanti J. Contractile cardiac grafts using a novel nanofibrous mesh. Biomaterials. 2004;25(17):3717-23. doi: 10.1016/j.biomaterials.2003.10.055

7. Vats A, Tolley N, Polak J, Gough J. Scaffolds and biomaterials for tissue engineering: a review of clinical applications. Clinical Otolaryngology. 2003;28(3):165-172.

8. Zhong S, Zhang Y, Lim C. Tissue scaffolds for skin wound healing and dermal reconstruction. Wiley Interdisciplinary Reviews: Nanomedicine and Nanobiotechnology. 2010;2(5):510-525.

9. Almany L, Seliktar D. Biosynthetic hydrogel scaffolds made from fibrinogen and polyethylene glycol for 3D cell cultures. Biomaterials. 2005;26(15):2467-2477. doi: 10.1016/j.biomaterials.2004.06.047

10. Lu X, Wei X, Huang J, Yang L, Zhang G, He G, et al. Supertoughened poly (lactic acid)/polyurethane blend material by in situ reactive interfacial compatibilization via dynamic vulcanization. Industrial & Engineering Chemistry Research. 2014;53(44):17386-17393.

11.You Y, Lee SW, Youk JH, Min B-M, Lee SJ, Park WH. In vitro degradation behaviour of non-porous ultra-fine poly (glycolic acid)/poly (L-lactic acid) fibres and porous ultra-fine poly (glycolic acid) fibres. Polymer Degradation and Stability. 2005;90(3):441-8. doi: 10.1016/j.polymdegradstab.2005.04.015

12.Jang J-H, Castano O, Kim H-W. Electrospun materials as potential platforms for bone tissue engineering. Advanced drug delivery reviews. 2009;61(12):1065-1083. doi: 10.1016/j.addr.2009.07.008

13. Nguyen TTT, Ghosh C, H wang S-G, Dai Tran L, Park JS. Characteristics of acid) curcumin-loaded poly (lactic nanofibers for wound healing. Journal of materials science. 2013;48(20):7125-7133

14.Gholipour-Kanani A, Bahrami SH, Samadikuchaksaraei A. Novel blend scaffolds from poly (caprolactone) chitosan-poly (vinyl alcohol): Physical, morphological and biological studies. Journal of Biomaterials and Tissue Engineering. 2014;4(3):245-52. doi: 10.1166/jbt.2014.1163

23

Further Developments on the (EG) Exponential-MIR Class of Distributions

Clement Boateng Ampadu*

31 Carrolton Road, Boston, MA 02132-6303, USA

Corresponding author: Clement Boateng Ampadu, 31 Carrolton Road, Boston, MA 02132-6303, USA, E-mail id: DrAmpadu@hotmail.com

Abstract

The Modified Inverse Rayleigh (MIR) distribution appeared in [Khan, M. S. (2014).Modified inverse Rayleigh distribution. International Journal of Computer Applications, 87(13):28–33] who got some theoretical properties of this distribution, and in[Nasiru, S., Mwita, P. N. and Ngesa, O. (2017). Exponentiated Generalized Exponential Dagum Distribution. Journal of King Saud University-Science, In Press] they introduced the (EG) Exponential-X class of distributions and obtained some theoretical properties with application. By assuming the random variable X follows the MIR distribution, some theoretical properties with application of the (EG) Exponential-MIR

Class of distributions appeared in [Nasiru, S., Mwita, P. N. and Ngesa, O. (2018). Discussion on Generalized Modified Inverse Rayleigh Distribution. Applied Mathematics and Information Sciences, 12(1):113-124]. In the present paper we propose some extensions of the (EG) Exponential-MIR class of distributions. The (EG) Exponential-MIR class of distributions is part of Chapter 5 [Nasiru, S. (2018). A New Generalization of Transformed-Transformer Family of Distributions. Doctor of Philosophy thesis in Mathematics (Statistics Option). Pan African University, Institute for Basic Sciences, Technology and Innovation, Kenya], where the naming convention "NEGMIR" is used

Keywords: *T-X (W)* family of distributions; Exponentiated Generalized distributions; Modified Inverse Rayleigh distribution; biological data; health data

Introduction

T-X (W) Family of Distributions

This family of distributions is a generalization of the beta-generated family of distribu-tions first proposed by Eugene et.al [Eugene, N, Lee, C, Famoye, F: The beta-normal dis-tribution and its applications. Communications in Statistics-Theory and Methods 31(4), 497–512 (2002)]. In particular, let r(t) be the PDF of the random variable $T \in [a, b]$, $-\infty \le a < b \le \infty$, and let $W(F(x))$ be a monotonic and absolutely continuous function of the *CDF F (x)* of any random variable X. The CDF of a new family of distributions defined by Alzaatreh et.al [Alzaatreh, A, Lee, C, Famoye, F: A new method for generating families of continuous distributions. Metron 71(1), 63–79 (2013b)] is given by

$$G(x) = \int_a^{W(F(x))} r(t)dt = R\{W(F(x))\}$$

Where R ($\bullet$) is the CDF of the random variable T and $a \ge 0$

Remark 1.1

The PDF of the T-X (W) family of distributions is obtained by differentiating the CDF above.

Remark 1.2

When we set $W(F(x)) := -\ln(1 - F(x))$ then we use the term "T-X Family of Distributions" to describe all sub-classes of the T-X (W) family of distributions induced by the weight function $W(x) = -\ln(1 - x)x$. A description of different weight functions that are appropriate given the support of the random variable T is discussed in [Alzaatreh, A, Lee, C, Famoye, F: A new method for generating families of continuous distributions. Metron 71(1), 63–79 (2013b)] A plethora of results studying properties and application of the T- X(W) family of distributions have appeared in the literature, and the research papers, assuming open access, can be easily obtained on the web via common search engines, like Google, etc.

The Exponentiated Generalized (EG) T-X family of distributions

This class of distributions appeared in [Suleman Nasiru, Peter N. Mwita and Oscar Ngesa, Exponentiated Generalized Transformed-Transformer Family of Distributions, Journal of Statistical and Econometric Methods, vol.6, no.4, 2017, 1-17] In particular the CDF Admits the following integral representation

$$G(x) = \int_0^{-\log[1-(1-\bar{F}(x)^d)^c]} r(t)dt$$

Where $c, d > 0$ and $\bar{F}(x) = 1 - F(x)$ and $F(x)$ is the CDF of a base distribution.

Remark 1.3

Note that if we set $L(x) := (1 - \bar{F}(x)^d)^c$ where c, d > 0 and $\bar{F}(x) = 1 - F(x)$, and F(x) is the CDF of a base distribution, then L(x) gives the CDF of the exponentiated generalized class of distributions [G.M. Cordeiro, E.M.M. Ortega and C.C.D. da Cunha, The exponentiated generalized class of distributions, Journal of Data Science,11(1), (2013),1-27]

The (EG) Exponential-MIR family of distributions

Here we assume the random variable X follows the MIR distribution with CDF

$$F_{\alpha,\theta(x)} = e^{-(\frac{\alpha}{x} + \frac{\theta}{x^2})}$$

And the random variable T follows the exponential distribution with PDF

$$r_\lambda(t) = \lambda e^{-\lambda t}$$

and CDF $\quad R_\lambda(t) = 1 - e^{-\lambda t}$

Now put $\quad L_{(\alpha,\theta,d,c)}(x) := (1 - \overline{F}_{(\alpha,\theta)}(x)^d)^c$

And observe the CDF of the (EG) Exponential-MIR family of distributions as proposed in [Nasiru, S, Mwita, P N and Ngesa, O.(2018) Discussion on Generalized Modified Inverse Rayleigh Distribution. Applied Mathematics and Information Sciences, 12(1):113-124; Nasiru, S. (2018). A New Generalization of Transformed-Transformer Family of Distributions. Doctor of Philosophy thesis in Mathematics (Statistics Option) Pan African University, Institute for Basic Sciences, Technology and Innovation, Kenya] is given by

$$G_{(\alpha,\theta,d,c,\lambda)}(x) = 1 - (1 - L_{(\alpha,\theta,d,c)}(x))^\lambda$$

Further Developments

In this section we present some new generalizations of the (EG) Exponential-MIR family of distributions which are induced by the other weight functions introduced in [Alzaatreh, A, Lee, C, Famoye, F: A new method for generating families of continuous distributions. Metron 71(1), 63–79 (2013b)], when the random variable T in the T-X(W) class of distributions has support $[a, \infty)$, where $a \geq 0$

Definition 2.1

A random variable Y1 will be called T − X distributed of type I if the weight function is given by

$$W(x) = -\log(1 - x)$$

Definition 2.2

A random variable Y2 will be called T − X distributed of type II if the weight function is given by

$$W(x) = \frac{x}{1 - x}$$

Definition 2.3

A random variable Y3 will be called T − X distributed of type III if the weight function is given by

$$W(x) = -\log(1 - x^\alpha)$$

For some $\alpha > 0$

Definition 2.4

A random variable Y4 will be called T − X distributed of type IV if the weight function is given by

$$W(x) = \frac{x^\alpha}{1 - x^\alpha}$$

For some $\alpha > 0$

Type II

Definition 2.5

The CDF of the (EG) Exponential-MIR class of distributions of type II has the following integral representation for $a \geq 0$

$$K_{(\alpha,\theta,d,c,\lambda)}(x) = \int_{a}^{\frac{L_{(\alpha,\theta,d,c)}(x)}{1 - L_{(\alpha,\theta,d,c)}(x)}} r_\lambda(t)dt$$

Note that $L_{(\alpha,,\theta,1,c)}(x)$ implies the following from Definition 2.5

Corollary 2.6

The CDF of the exponentiated Exponential-MIR class of distributions of type II is given by

$$K^*_{(\alpha,\theta,c,\lambda)}(x) = R_\lambda(\frac{F^c_{(\alpha,\theta)}(x)}{1 - F^c_{(\alpha,\theta)}(x)})$$

Where

$$R_\lambda(.) = 1 - e^{-\lambda(.)}$$

and

$$F_{(\alpha,\theta)}(x) = e^{-(\frac{\alpha}{x} + \frac{\theta}{x^2})}$$

Note that $L_{(\alpha,,\theta,1,1)}(x)$ implies the following from Definition 2.5

Corollary 2.7

The CDF of the Exponential-MIR class of distributions of type II

$$K^{**}_{(\alpha,\theta,\lambda)}(x) = R_\lambda(\frac{F_{(\alpha,\theta)}(x)}{1 - F_{(\alpha,\theta)}(x)})$$

Where $R_\lambda(.) = 1 - e^{-\lambda(.)}$

and $\quad F_{(\alpha,\theta)}(x) = e^{-(\frac{\alpha}{x} + \frac{\theta}{x^2})}$

Type III

Definition 2.8

The CDF of the (EG) Exponential-MIR class of distributions of type III has the following integral representation for $a \geq 0$ and $\xi > 0$

$$Z_{(\alpha,\theta,d,c,\lambda,\xi)}(x) = \int_{a}^{-\log(1 - L^\xi_{(\alpha,\theta,d,c)}(x))} r_\lambda(t)dt$$

Note that $L_{(\alpha,\theta,1,c)}(x)$ implies the following from Definition 2.8

Corollary 2.9

The CDF of the exponentiated Exponential-MIR class of distributions of type III is given by

$$Z^*_{(\alpha,\theta,c,\lambda,\xi)}(x) = R_{\lambda}(-\log(1 - F_{(\alpha,\theta)}(x)^{c\xi}(x)))$$

Where $R_{\lambda}(.) = 1 - e^{-\lambda(.)}$

and $F_{(\alpha,\theta)}(x) = e^{-(\frac{\alpha}{x}+\frac{\theta}{x^2})}$

Note that $L_{(\alpha,\theta,1,1)}(x)$ implies the following from Definition 2.8

Corollary 2.10

The CDF of the Exponential-MIR class of distributions of type III is given by

$$Z^{**}_{(\alpha,\theta,\lambda,\xi)}(x) = R_{\lambda}(-\log(1 - F_{(\alpha,\theta)}(x)^{\xi}(x)))$$

Where $R_{\lambda}(.) = 1 - e^{-\lambda(.)}$

and $F_{(\alpha,\theta)}(x) = e^{-(\frac{\alpha}{x}+\frac{\theta}{x^2})}$

Type IV

Definition 2.11

The CDF of the (EG) Exponential-MIR class of distributions of type IV has the following integral representation for $\xi > 0$ and $a \geq 0$

$$Q_{(\alpha,\theta,d,c,\lambda,\xi)}(x) = \int_{a}^{\frac{L^{\xi}_{(\alpha,\theta,d,c)}(x)}{1-L^{\xi}_{(\alpha,\theta,d,c)}(x)}} r_{\lambda}(t)dt$$

Note that $L_{(\alpha,\theta,1,c)}(x)$ implies the following from Definition 2.11

Corollary 2.12

The CDF of the exponentiated Exponential-MIR class of distributions of type IV is given by

$$Q^*_{(\alpha,\theta,c,\lambda,\xi)}(x) = R_{\lambda}(\frac{F_{(\alpha,\theta)}(x)^{\xi}(x)}{1 - F_{(\alpha,\theta)}(x)^{\xi}(x)})$$

Where $R_{\lambda}(.) = 1 - e^{-\lambda(.)}$

and $F_{(\alpha,\theta)}(x) = e^{-(\frac{\alpha}{x}+\frac{\theta}{x^2})}$

Note that $L_{(\alpha,\beta,\theta,1,1)}(x)$ implies the following from Definition 2.11

Corollary 2.13

The CDF of the Exponential-MIR class of distributions of type IV is given by

$$Q^{**}_{(\alpha,\theta,\lambda,\xi)}(x) = R_{\lambda}(\frac{F_{(\alpha,\theta)}(x)^{\xi}(x)}{1 - F_{(\alpha,\theta)}(x)^{\xi}(x)})$$

Where $R_{\lambda}(.) = 1 - e^{-\lambda(.)}$

and $F_{(\alpha,\theta)}(x) = e^{-(\frac{\alpha}{x}+\frac{\theta}{x^2})}$

Application

In this section we compare the Exponential-Weibull class of distributions of type II and the Exponential-Weibull class of distributions of type IV in modeling the aircraft data, Table 1 [Suleman Nasiru, Peter N. Mwita and Oscar Ngesa, Discussion on Generalized Modified Inverse Rayleigh, Appl. Math. Inf. Sci. 12, No. 1, 113-124 (2018)]

Remark 3.1

When a random variable X follows the Exponential-Weibull class of distributions of type II we write $X \sim EWII(a,b,c)$

Remark 3.2

When a random variable X follows the Exponential-Weibull class of distributions of type IV we write $X \sim EWIV(a,b,c,d)$ In this section we assume the CDF of the Weibull distribution is given by

$$F_{(b,c)}(x) = 1 - e^{-(\frac{x}{c})^b}$$

and the CDF of the Exponential distribution is given by

$$R_{(a)}(x) = 1 - e^{-ax}$$

Theorem 3.3

The CDF of the Exponential-Weibull class of distributions of type II is given by

$$K^{**}_{(a,b,c)}(x) = 1 - e^{-ae(\frac{x}{c})^b(1-e^{-(\frac{x}{c})^b})}$$

Proof in Corollary 2.7, let $\lambda := a$ and $F_{(\alpha,\theta)}(x) := F_{(b,c)}(x)$

Theorem 3.4

The CDF of the Exponential-Weibull class of distributions of type IV is given by

$$Q^{**}_{(a,b,c,d)}(x) = 1 - e^{-\frac{a(1-e^{-(\frac{x}{c})^b})^d}{1-(1-e^{-(\frac{x}{c})^b})^d}}}$$

Proof In Corollary 2.13, let $\xi := d$, $\lambda := a$ and $F_{(\alpha,\theta)}(x) := F_{(b,c)}(x)$ In order to compare the two distribution models, we used the following criteria: -2(Log likelihood) and AIC (Akaike information criterion) , AICC (corrected Akaike information criterion), and BIC (Bayesian information criterion) for the data set. The better distribution corresponds to the smaller -2(Log-likelihood) AIC, AICC, and BIC values:

$$AIC = 2k - 2l$$

$$AICC = AIC + \frac{2k(k+1)}{n-k-1}$$

$$BIC = k \log(n) - 2l$$

Where k is the number of parameters in the statistical model, n is the sample size, and l is the maximized value of the log-likelihood function under the considered model. From Table 1 above, it is clear the *EWII* (3975.82, 1.0863, 121474) distribution has the smallest AICC and BIC values, whilst the *EWIV* (216660, 1.47149, 1.21237_107, 0.681738) distribution has the smallest -2(Log-likelihood) and AIC values. When we compared the CDF's of the two distributions we obtained the following

On the other hand when we compared the PDF's of the two distributions we obtained the Following

The results from Table 1 and the Figures above, suggest the *EWIV* (216660, 1.47149, 1.21237_107, 0.681738) distribution is slightly better than the EWII (3975.82, 1.0863, 121474) distribution in modeling the aircraft data, Table 3 [Suleman Nasiru, Peter *N. Mwita* and *Oscar Ngesa*, Discussion on Generalized Modified Inverse Rayleigh, Appl. Math. Inf. Sci. 12, No. 1, 113-124 (2018)]

Concluding Remarks

Our hope is that the researchers will further develop the properties and applications of the new class of distributions presented in this paper. Finally we hope the new developments have practical significance in modeling biological data, health data, etc.

Note that $L_{(\alpha,\theta,1,c)}(x)$ implies the following from Section 1.3

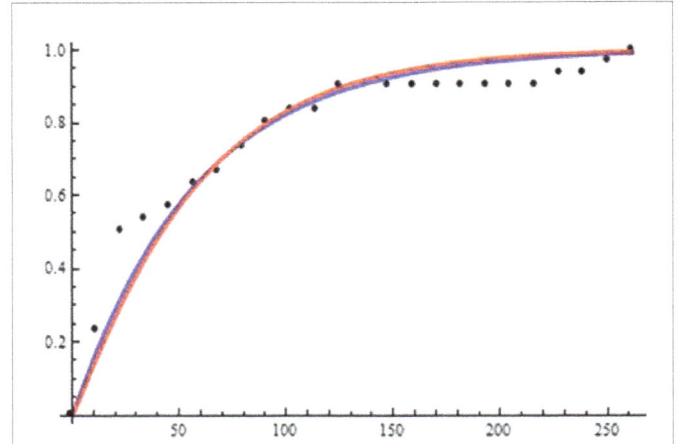

Figure 1: The CDF of EWII(3975.82, 1.0863, 121474) (red) and EWIV (216660, 1.47149, 1.21237 _ 107, 0.681738) (blue) fitted to the empirical distribution of the aircraft data, Table 3 [Suleman Nasiru, Peter N. Mwita and Oscar Ngesa, Discussion on Generalized Modified Inverse Rayleigh, Appl.Math. Inf. Sci. 12, No. 1, 113-124 (2018)]

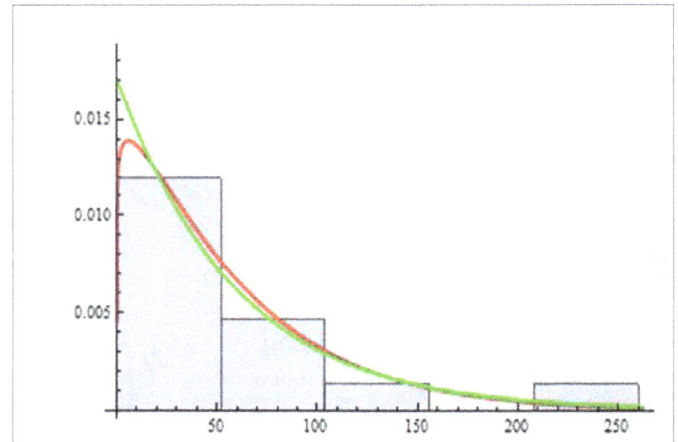

Figure 2: The PDF of *EWII* (3975.82, 1.0863, 121474) (red) and *EWIV* (216660, 1.47149, 1.21237 _ 107, 0.681738) (green) fitted to the empirical distribution of the aircraft data, Table 3 [Suleman Nasiru, Peter *N. Mwita* and *Oscar Ngesa*, Discussion on Generalized Modified Inverse Rayleigh, Appl. Math. Inf. Sci. 12, No. 1, 113-124 (2018)]

Table 1: Criteria for Comparison				
Model	**-2(Log-likelihood)**	**AIC**	**AICC**	**BIC**
EWII(3975.82, 1.0863, 121474)	307.351	313.351	314.274	317.554
EWIV (216660, 1.47149, 1.21237*10⁷, 0.681738)	305.338	313.338	314.938	318.943

Theorem 4.1

The CDF of the exponentiated Exponential-MIR class of distributions of type I is given by

$$G^{*}_{(\alpha,\theta,c,\lambda)}(x) = 1 - \left(1 - F_{(\alpha,\theta)}(x)^{c}(x)\right)^{\lambda}$$

Where $\quad F_{(\alpha,\theta)}(x) = e^{-(\frac{\alpha}{x} + \frac{\theta}{x^2})}$

Similarly, $L_{(\alpha,\theta,1,1)}(x)$ implies the Exponential-MIR class of distributions of type I from Section 1.3. Consequently several Corollaries can be deduced from Chapter 5[4], where they obtained several statistical/mathematical properties with application. For example, we have the following from Section 5.2 of Chapter 5[4]

Corollary 4.2

The survival function of the exponentiated Exponential-MIR class of distributions of type I is given by

$$S^{*}(x) = \{1 - F^{c}_{(\alpha,\theta)}(x)\}^{\lambda}$$

Proof Let $d = 1$ in eqn (5.8) contained in Section 5.2 of Chapter 5[4]

References

1. Khan,M S. Modified inverse Rayleigh distribution. International Journal of Computer Applications. 2014;87(13):28–33.

2. Nasiru S, Mwita P N, Ngesa O. Exponentiated Generalized Exponential Dagum Distribution. Journal of King Saud University-Science, In Press. 2017.

3. Nasiru S, Mwita P N, Ngesa O. Discussion on Generalized Modified Inverse Rayleigh Distribution. Applied Mathematics and Information Sciences. 2018;12(1):113-124.

4. Nasiru, S. A New Generalization of Transformed-Transformer Family of Distributions. Doctor of Philosophy thesis in Mathematics (Statistics Option). Pan African University, Institute for Basic Sciences, Technology and Innovation, Kenya. 2018.

5. Eugene, N, Lee, C, Famoye, F. The beta-normal distribution and its applications. Communications in Statistics-Theory and Methods. 2002;31(4):497–512.

6. Alzaatreh A, Lee C, Famoye F. A new method for generating families of continuous distributions. Metron. 2013;71(1): 63–79.

7. Suleman Nasiru, Peter N. Mwita and Oscar Ngesa. Exponentiated Generalized Transformed-Transformer Family of Distributions. Journal of Statistical and Econometric Methods. 2017;6(4):1-17.

8. G M Cordeiro, E M M Ortega and C C D da Cunha. The exponentiated generalized class of distributions. Journal of Data Science. 2013;11(1):1-27.

24

Biochemical and molecular characterization of cellulase producing bacterial isolates from cattle dung samples

Preeti Vyas and Ashwani Kumar*

Metagenomics and Secretomics Research Laboratory, Department of Botany, Dr. Harisingh Gour University (A Central University), Sagar-470003, (M.P.), India

***Corresponding author:** Ashwani Kumar, Assistant Professor, Metagenomics and Secretomics Research Laboratory, Department of Botany, Dr. Harisingh Gour University (A Central University), Sagar-470003, (M.P.), India, E-mail: ashwaniiitd@hotmail.com*

Abstract

Recently production of bio fuel and other platform chemicals from lignocellulosic biomass is gaining constant attention to meet the demand of energy all over the Globe. Utilizing agricultural residues for bio ethanol generation using novel bacterial isolates will provide a solution for energy problem. This study has been conducted with the following objectives; i) to search the novel bacterial isolates that can efficiently degrade the cellulose and ii) their biochemical and molecular characterization by using 16S rRNA sequencing. Almost, 15 bacteria's have been isolated from cattle dung out of which 10 isolates showed cellulase activity on CMC (carboxymethyl cellulose) screening media. These bacterial isolates with high cellulase activity were further characterized on the basis of gram staining, morphology and also by various biochemical utilization tests (Citrate, lysine, ornithine, urease, phenylalanine, H_2S production, nitrate reduction, glucose, lactose, adonitol, sorbitol, arabinose, and 35 different carbohydrate sources). Results showed that out of three isolates (AKPCD108, AKPCD109 and AKPBD1), the isolate number AKPBD1 showed 58.33 % utilization of biochemical's which is followed by AKPCD108 and AKPCD109, respectively. Among all carbohydrates utilized, AKPBD1 showed 68.57% utilization and other isolates (AKPCD108 & AKPCD109) were at par. These three isolates were molecularly identified using 16S rDNA sequencing. The obtained sequences were blast and phylogenetic analysis of their sequences showed that AKPCD108 has 99% similarity with *Pseudomonas otitidis* strain MCC10330, AKPCD109 has 99% similarity with *Stenotrophomonas koreensis* strain TR6-01 (*Flavobacterium lutescens*) and AKPBD1 has 99% similar with *Serratia marcescens* subsp. sakuensis strain KRED. In this study, we conclude that, these isolates are highly promising and have the ability to produce cellulase enzyme and can be used for efficient biological pretreatment of lignocellulosic biomass following production of biofuels and bioproducts.

Keywords: Cellulolytic bacteria; Lignocelluloses; 16S rDNA; Biomass; Biofuel

Introduction

Depletion of fossil fuels and worldwide increases of energy demand and environmental concern has lead in search for an alternative renewable energy resource like lignocellulosic biomass [1,2]. Lignocelluloses are renewable, cost efficient, eco-friendly potential feedstock for the production of bio fuels. However, the bioconversion of this lignocellulosic biomass into sugars is not easy due to the cellulose crystallinity, degree of polymerization, presence of lignin which poses hurdle for enzymatic hydrolysis. Thus, utilizing different pre-treatment methods for bioconversion process increases the overall digestibility of cellulose and hemicelluloses, and therefore removal of lignin from lignocellulosic biomass. However, among the pre-treatment, use of chemical pre-treatment is hazardous for the environment. Therefore, the biological pre-treatment method on the other hand is environmental friendly and gaining popularity now a day because this method is less toxic, requires very low energy input, and much higher sugar yield. The biomass of the plant is mainly composed of cellulose, hemicelluloses and lignin, which constitute an important material to be converted into various type of liquid fuels [3–5].

Present trends to produce cellulosic ethanol has been focused mainly on sorghum, wheat straw, sunflower, maize, sugarcane, and non-edible oil seed plants; however, use of agricultural residues that are not exploited or which could be used more efficiently for ethanol production. The waste material generated from various agricultural fields like stalks, straw, husk, stover, and stems, etc. are highly rich in cellulosic content that had generated great interests to use them as an energy source [6–9]. Among all the component of biomass, cellulose is the principal constituent of the cell wall and mainly composed of units of D-glucose that are linked together to form a linear chain by ß-1,4-glycosidic linkages. Cellulose is generally found as micro fibrils that are "2–20nm" in diameter and "100–40,000 nm" long in the biomass. The enzyme named cellulase has great potential to degrade cellulose and also convert cellulose into simple sugars. Cellulase mainly composed of three major components namely, cellobiohydrolase, endoglucanase or carboxymethylcellulase (CMCase), and beta-glucosidases [10]. Microorganism particularly bacteria and fungi produced this enzyme in nature and exploring different natural resources to isolate bacteria is imperative for higher production of cellulolytic enzymes [11,12].

Although, it is validated fact that proper degradation of lignocellulosic substrates requires a complex set of enzymes such as laccases, exoglucanases, endoglucanases, peroxidases, fucosidases, xylanases and β-glucosidases [13]. Variety of microbes that are present in nature like many bacterial and fungal species secretes different types of lignocellulolytic enzymes which help in degradation of lignocellulosic biomass into simple sugars. These sugars can be further converted into bio fuels and other platform chemicals. The present work was conducted to isolate cellulolytic bacteria from cattle dung and to identify them biochemically and molecularly by using 16S rRNA gene.

Materials and Methods

Isolation and selection of bacterial isolates for cellulase activity

Cattle dung samples were collected in sterile plastic bottles from Sagar, Madhya Pradesh. One gram of dung sample was used to grow bacteria in Nutrient Agar media (Hi Media, India) and incubated at 28°C for 48hrs in incubator and maintained at nutrient broth for further use. Pure culture of bacterial isolates was maintained in separate plate before testing them on selective media. Then 0.1 ml of the bacterial suspension was adjusted to 0.05 cell density by sterile 0.85% NaCl solution. Then these bacterial inoculums was transferred to Carboxymethyl Cellulose (CMC) plates for screening CMC activity (Shankar et al., 2011) and incubated at 37°C for 2 days. These plates were then flooded with 0.1% Congo red/Gram's Iodine solution for about 20 min and then washed by using 1 M NaCl for 15 min. Formation of the clear halos zones by these bacterial isolates was indicated by their cellulose degrading activity or cellulase activity [14]

Biochemical characterization of cellulase producing bacterial isolates

A combination of 12 biochemical tests (HiAssortedTM KB002, HIMEDIA, INDIA) and 35 tests for utilization of carbohydrates (HiCarboTM, HIMEDIA, INDIA) were used for biochemical characterization of cellulase producing bacterial isolates by following the instruction provided by the manufacturer. Tests used in this kit are presented in table 1 and table 2. These tests are based on the principle of color change, change in pH, and utilization of substrate by bacterial isolates. The bacterial isolates used for these test were isolated and purified. Only pure cultures were used and kit was opened aseptically. 50μl of the prepared inoculums was added to each well by using surface inoculation method. The kit were also be inoculated by stabbing each individual well with a loopful of inoculums and incubation at 35°C temp for 18-24 hours and changes in color were observed and recorded.

DNA Extraction and molecular identification

Total genomic DNA of bacterial isolates was extracted by using the Insta Gene TM Matrix Genomic DNA isolation kit. DNA concentration was determined by nano drop. The isolated bacterial DNA was further amplified by using universal forward primer 27F with primer sequence (AGAGTTTGATCMTGGCTCAG) and reverse primer 1492R with primer sequence (TACGGTACCTTGTTACGACTT). The PCR mixture contained 1 μL of DNA template, and 20 μL of PCR reaction mixture. The PCR amplification were performed by using automated thermal cycler (PTC-200, Biorad, Hercules, CA, USA) with the following PCR conditions: an initial denaturation at 94°C for about 5 min followed by 30 cycles of amplification at 94°C for 30 s, annealing at 52°C for about a min and extension at 72°C for 2 min, final elongation at 72°C for 10 min. The PCR products were purified using QIAquick PCR Purification Kit (Qiagen, Hilden, Germany) according to the manufacturer's instructions. Selected PCR products were then sequenced using an ABI Big Dye Terminator v3.1 cycle sequencing kit (Applied Biosystems, Grand Island, NY, USA) and these sequences were read on an Applied Biosystems 3130 genetic analyzer (Applied Biosystems). Sequence analysis was performed using Sequence Scanner v1.0 software. Amplified PCR products of 16S ribosomal gene were confirmed on 1% agarose gel, purified using QIAquick PCR purification kit (QIAGEN)

Bioinformatics protocol

These 16S rDNA sequences were then compared with other sequences available in Gene Bank databases using the NCBI BLAST at http://www.ncbi.n1m.nih.gov/blast/Blast.cgi.Sequences were submitted to NCBI GenBank data base for getting accession numbers. The phylogenetic analysis of sequences with the closely related sequence of NCBI blast results was performed by using the automated NCBI pipeline. A phylogenetic tree for these bacterial sequences were constructed by using iTOL (Interactive tree of life) after establishing relationship among the similar sequences analysis generated from Mega 5.05 software [15,16].

Results and Discussion

The large number of microorganisms from the diverse environment permits screening for more efficient lignocellulolytic enzymes to help overcome current challenges in bio fuel production [4,17–20]. Previous studies reported that cow/ buffalo dung is rich in cellulase producing bacterial isolates [21]. Therefore, in this study we have tried to isolate good cellulase producing bacteria from the cow dung collected from the Dr. Harisingh Gour University Campus, Sagar, Madhya Pradesh. Almost, 15 bacteria's have been isolated from cattle dung out of which 10 isolates showed cellulase activity on CMC (Carboxymethyl Cellulose) screening media (Figure 1). Cellulase enzyme activities were seen by the appearance of zones around the bacterial colonies on the screening media. Out of ten isolates we selected only three for further characterization; these were named as AKPCD108, AKPCD109 and AKPBD1. These isolates were found to be gram negative after gram staining test. These bacterial isolates with high cellulose activity were tested by gram staining, morphological ground and also by various biochemical utilization tests (Citrate, ornithine, lysine, phenylalanine, urease, H_2S production, nitrate reduction, glucose, lactose, adonitol, arabinose, sorbitol and 35 different carbon sources) (Table 1 & 2).

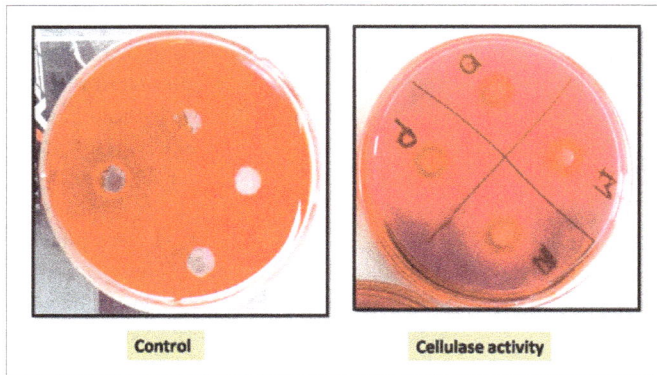

Figure 1: Bacterial isolates from buffalo dung showing cellulase activity

Biochemical utilization test	Strain AKPCD108	Strain AKPCD109	Strain AKPBD1
Citrate utilization			
Lysine utilization			
Ornithine utilization			
Urease			
Phenylalanine deamination			
Nitrate reduction			
H₂S production			
Glucose			
Adonitol			
Lactose			
Arabinose			
Sorbitol			
Control			
Utilization (%)	41.66	33.33	58.33

Table 1: Heat map of Biochemical Utilization by selected bacterial isolates from cattle dung
Note: (Dark blue colors indicates positive test for the respective source and grey indicates negative test result.)

Carbohydrates Utilization Test (35 Carbon Source)	Strain AKPCD108	Strain AKPCD109	Strain AKPBD1
Lactose			
Xylose			
Maltose			
Fructose			
Dextrose			
Galactose			
Raffinose			
Trehalose			
Melibiose			
Sucrose			
L-Arabinose			
Mannose			
Inulin			
Sodium gluconate			
Glycerol			
Salicin			
Glucosamine			
Dulcitol			
Inositol			
Sorbitol			
Mannnitol			
Adonitol			
α-Methyl-D-Gluconate			
Ribose			
Rhamnose			
Cellobiose			
Melezitose			
α-Methyl- D-mannoside			
Xylitol			
ONPG			
Esculin			
D-Arabinose			
Citrate			
Malonate			
Sorbose			
Control			
Utilization (%)	40.00	42.86	68.57

Table 2: Heat map of Carbohydrate utilization tests of selected bacterial isolates from cattle dung
Note: (Black colors indicates positive test for the respective source and grey indicates negative test result)

Results showed that among these three isolates (AKPCD108, AKPCD109 and AKPBD1), the isolate number AKPBD1 showed 58.33 % utilization of biochemical's (12 different biochemical's and one control) which is followed by AKPCD108 and AKPCD109. Among all carbohydrates (35 different carbon source and one control) utilized, AKPBD1 showed 68.57% utilization and other isolates (AKPCD108 & AKPCD109) were at par for the utilization of these carbohydrates (Table 1 and Table 2).

Among the biochemical (ornithine, nitrate and glucose) were utilized by all isolates and phenylalanine and arabinose have not been used by any isolates. On the other hand dextrose, sodium gluconate and inositol were used by all the isolates and lactose, melezitose, xylitol have not been used by any isolate. Our results are supported by Khianngam et al. (2014), in which they have screening and identified cellulase producing bacteria from oil palm meal. They obtained 8 isolates belongs to Bacillus, and one to *Paenibacillus* and *Lysinibacillus* by using 16S rDNA sequencing. The obtained sequences were blast and phylogenetic analysis of their sequences showed that AKPCD108 has 99% similarity with *Pseudomonas otitidis* strain MCC10330, AKPCD109 has 99% similarity with *Stenotrophomonas koreensis* strain TR6-01 (*Flavobacterium lutescens*) and AKPBD1 has 99% similar with *Serratia marcescens* subsp. sakuensis strain KRED (Table 3). The phylogenetic relationship of these identified cellulase producer are shown in Figures 2a-c. Among the bacterial isolates most common cellulase producer are belong to the following genera; *Cellulomonas, Cellvibrio, Clostridium, Paenibacillus, Pseudomonas, Ruminococcus, B. subtilis, B. cereus, B. altitudinis, Paenibacillus, Acetivibrio, Bacillus, Sporocytophaga, Bacteroides, Micrococcus* etc [14, 22]. But the isolates screened in this study for cellulase production were not reported by other workers.

Table 3: Different bacterial isolates from cattle dung identified using 16SrRNA sequencing

S.No.	Isolate number	Bacterial strain	Gram staining reaction	NCBI GENE BANK NO	NCBI database match	Percentage of identity /Accession
1.	AKPCD108	*Pseudomonas otitidis*	Gram negative	KX698105	*Pseudomonas otitidis* strain MCC10330	99%/NR_043289.1
2.	AKPCD109	*Flavobacterium lutescens*	Gram negative rod	KX698103	*Stenotrophomonas koreensis* strain TR6-01	99%/NR_041019.1
3.	AKPBD1	*Serratia marcescens*	Gram negative rod	KX698107	*Serratia marcescens* subsp. sakuensis strain KRED	99%/NR_036886.1

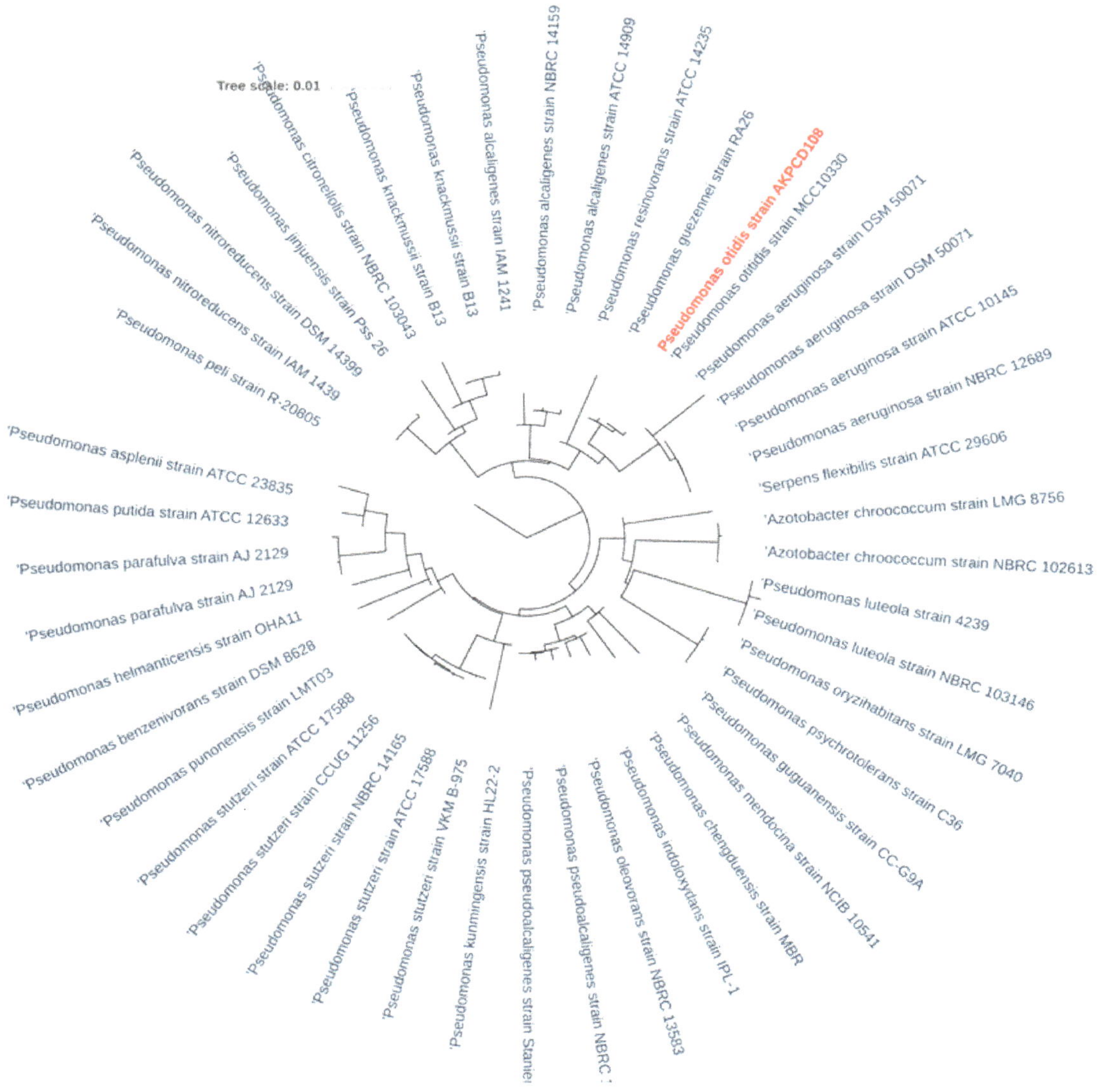

Figure2 (a): AKPCD108

Figure2 (b): AKPCD 109

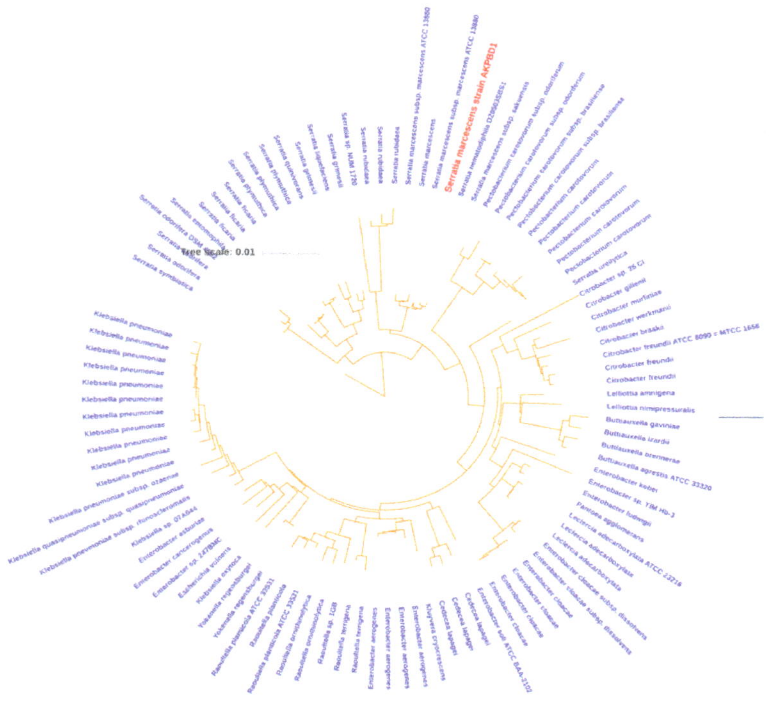

Figure2 (c): AKPBD1

Figure 2: Phylogenetic tree created by iTOL software to present relationship between closely related species

Conclusion

In conclusion, 10 cellulase producing bacteria were isolated from cattle dung samples from University Campus, Sagar Madhya Pradesh, India. They were screened for their cellulase activity by Congo red test and they showed hydrolysis capacity on plates. On the basis of their phenotypic characteristics and phylogenetic analyses, three isolates AKPCD108, AKPCD109 and AKPBD1 were closely related to *Pseudomonas, Stenotrophomonas* and *Serratia marcescens*, respectively. However, the further research on these isolates would be required utilize their full potential or to maximise the production of cellulase enzymes and degradation of agriculture residues. Finally based on the present research it can be concluded that cattle dung can be a very good source for the isolation of cellulase producing bacteria and their advance biotechnological application, mainly for biomass hydrolysis.

Contributory Statements

PV has conducted the experiment. AK proposed the experiments, prepared and edited the manuscript.

Acknowledgement

PV and AK would like to acknowledge the UGC Start-up grant (Awarded to AK) for the financial support. Authors would like to acknowledge Ms. Anamika Dubey (DST Inspire) for valuable suggestions.

References

1. Sharma HK, Xu C, Qin W. Biological Pretreatment of Lignocellulosic Biomass for Biofuels and Bioproducts:An Overview. Waste and Biomass Valorization. 2017;1–17.

2. Ahmad S, Pathak V V, Kothari R, Kumar A, Babu S, Krishna N. Optimization of nutrient stress using C pyrenoidosa for lipid and biodiesel production in integration with remediation in dairy industry wastewater using response surface methodology.3 Biotech. 2018.

3. Vyas P, Kumar A, Singh S. Biomass breakdown : A review on pretreatment, instrumentations and methods. Front Biosci. 2018;10:155–174.

4. Raghunandan K, Mchunu S, Kumar A, Kumar KS, Govender A, Permaul K, Singh S, et al. Biodegradation of glycerol using bacterial isolates from soil under aerobic conditions. J Environ Sci Heal Part A Toxic/Hazardous Subst Environ Eng. 2014;49(1):85–92. doi: 10.1080/10934529.2013.824733

5. Tiwari G, Sharma A, Dalela M, Gupta R, Sharma S, Kuhad RC. Microwave assisted alkali pretreatment of fruit peel wastes for enzymatic hydrolysis. Indian J Agric Sci. 2017;87(4):496–499.

6. Tiwari G, Sharma A, Kumar A, Sharma S. Assessment of microwave-assisted alkali pretreatment for the production of sugars from banana fruit peel waste. Biofuels. 2018

7. Gupta A, Kumar A, Sharma S, Vijay VK. Comparative evaluation of raw and detoxified mahua seed cake for biogas production. Applied Energy. 2013;102:1514–1521.

8. Kumar A, Sharma S. Potential non-edible oil resources as biodiesel feedstock: An Indian perspective. Renew Sustain Energy Rev. 2011;15(4):1791-1800.

9. Kothari R, Pandey A, Ahmad S, Kumar A, Pathak V V, Tyagi V V. Microalgal cultivation for value-added products: a critical enviro-economical assessment. 3 Biotech. 2017;7:243.

10. Bhat MK. Cellulases and related enzymes in biotechnology. Biotechnol Adv. 2000;18(5):355-383.

11. Raghunandan K, Kumar A, Kumar S, Permaul K, Singh S. Production of gellan gum, an exopolysaccharide, from biodiesel-derived waste glycerol by Sphingomonas spp. 3 Biotech. 2018;8(1):71. doi: 10.1007/s13205-018-1096-3

12. Singh JK, Vyas P, Dubey A, Upadhyaya CP, Kothari R, Tyagi VV, Kumar A, et al. Assessment of different pretreatment technologies for efficient bioconversion of lignocellulose to ethanol. FrontBiosci-Sch. 2018;10:350-371.

13. de Lima Brossi MJ, Jiménez DJ, Cortes-Tolalpa L, van Elsas JD. Soil-Derived Microbial Consortia Enriched with Different Plant Biomass Reveal Distinct Players Acting in Lignocellulose Degradation. Microb Ecol. 2016;71(3):616–627. doi: 10.1007/s00248-015-0683-7

14. Khianngam S, Pootaeng-on Y, Techakriengkrai T, Tanasupawat S. Screening and identification of cellulase producing bacteria isolated from oil palm meal. J. Appl. Pharm. Sci. 2014;4(4):90–96.

15. Kimura M. A simple method for estimating evolutionary rates of base substitutions through comparative studies of nucleotide sequences. J Mol Evol. 1980;16(2):111–120.

16. Letunic I, Bork P. Interactive tree of life (iTOL) v3 : an online tool for the display and annotation of phylogenetic and other trees. Nucleic Acid Res. 2016;44(1):242–245. doi: 10.1093/nar/gkw290

17. Lokhande, S and Pethe AS. Isolation and screening of cellulolytic bacteria from soil and optimization of cellulase production and activity. Int J Life Sci. 2017;5(2):277–282.

18. Gothwal R, Gupta A, Kumar A, Sharma S, Alappat BJ. Feasibility of dairy waste water (DWW) and distillery spent wash (DSW) effluents in increasing the yield potential of Pleurotus flabellatus (PF 1832) and Pleurotus sajor-caju (PS 1610) on bagasse. 3 Biotech. 2012;2(3):249–257.

19. Singh NB, Kumar A, Rai S. Potential production of bioenergy from biomass in an Indian perspective. Renew Sustain Energy Rev. 2014;39:65–78.

20. Kumar A, Kumar K, Kaushik N, Sharma S, Mishra S. Renewable energy in India: Current status and future potentials. Renew Sustain Energy Rev. 2010;14(8):2434–2442.

21. Maki ML, Broere M, Leung KT, Qin W. Characterization of some efficient cellulase producing bacteria isolated from paper mill sludges and organic fertilizers. Int J Biochem Mol Biol. 2011;2(2):146–154.

22. Sethi S, Datta A, Gupta BL, Gupta S, Sethi S, Datta A, et al. Optimization of Cellulase Production from Bacteria Isolated from Soil. Int Sch Res Not. 2013:e985685.

25

Bactericidal potential of Agrochemicals against bacterial leaf spot pathogen *Xanthomonas campestris* pv. *Vesicatoria* of tomato prevalent in Nashik region, Maharashtra and ability of bacteria to form pesticide resistant mutant

Ajayasree T S*, S G Borkar And B G Barhate

Department of Plant Pathology and Agricultural Microbiology, Mahatma Phule Agriculture University, Rahuri, 413 722, Dist- Ahmednagar, Maharashtra state

Corresponding author: *AJAYASREE T S, Department of Plant Pathology and Agricultural Microbiology, Mahatma Phule Agriculture University, Rahuri, 413 722, Dist- Ahmednagar, Maharashtra state; E-mail id: ajayasree128@gmail.com*

Abstract

Selection of suitable agrochemicals/pesticides plays an important role in the management of bacterial plant pathogen prevalent in a given area, may be due to the presence of pesticide resistance available in the pathogen. In this scenario, the assessment of bactericidal potential of agrochemical to be used in the management of bacterial plant pathogen is very necessary. To manage the bacterial leaf spot and blight pathogen *Xanthomonas campestris* pv. Vesicatoria on tomato prevalent in Nashik region, the bactericidal potential of various agrochemicals were assayed.

The bacterial potential of antibiotic Streptomycin sulphate, and Streptocycline; the bactericide Bactericin-100, and the Bordeaux mixture; the copper fungicide Copper-ox chloride and the di-thiocarbamate group of fungicide Mancozeb; the combination of Copper ox chloride + Streptocycline were against the tomato leaf spot and blight pathogen *Xanthomonas campestris* pv. *Vesicatoria* under *in vitro* condition. The minimum inhibitory concentration of these agrochemicals against *Xanthomonas campestris* pv. *Vesicatoria* varies indicating the variable bactericidal potential of these agrochemicals against the said bacterium prevalent in Nashik areas.

The bacterial population of *Xanthomonas campestris* pv. *Vesicatoria* was resistant to the antibiotic Terramycin, Streptomycin sulphate and to the fungicide Copper ox chloride, Mancozeb and a combination of Copper ox chloride + Streptomycin sulphate and induced the pesticide resistant mutant in the bacterium with various mutation frequencies. The mutation rate for Terramycin and Streptomycin sulphate was 3.5 x 10^{-4} and 2 x 10^{-4} respectively, whereas, for Dithane Z-78, Mancozeb and a combination of Copper ox chloride+ Streptomycin sulphate was 2.6 x 10^{-4}, 8 x 10^{-5} and 5 x 10^{-5}.respectively. The agrochemicals which were completely inhibitory to the bacterium and did not allow the formation of antibiotic or antibiotic + fungicide resistance mutant were Streptocycline, Streptocycline+ Copper ox chloride and Streptocycline + Bordeaux mixture.

Keywords: Antibiotics; Bactericides; Fungicides; Mutant

Introduction

Bacterial plant pathogens cause considerable losses due to their infection and spread under favorable climatic conditions (Borkar and Yumlembam 2016). The success of management of bacterial plant pathogens depends on the selection of suitable and effective agrochemicals/pesticides to control them. Most of the times, the bacterial plant pathogen develops the pesticide resistance to certain pesticides under field conditions. The development of this pesticide resistance can be a location specified in the crop region/areas depending upon the bacterial strains of the pathogen. Therefore, it is very necessary to know the efficacy of agrochemicals/pesticides against the bacterial pathogen in the given location.

Bacterial leaf spot and blight is a serious disease of tomato crop in tomato growing areas of Nashik district in Maharashtra state, India. The bacterial pathogen *Xanthomonas campestris* pv. *Vesicatoria* is prevalent in this region for over two decades (Borkar and Yumlembam 2016) in spite of much application of the pesticides on the crop to manage this bacterial disease. This may be due to the development of pesticide resistance in the bacterium, thereby making the pesticide/agrochemicals ineffective in its management. The rate of formation of pesticide-resistant mutant against the pesticide also varies with the pesticide and the strain of the pathogen (Anderson 2006; Araujo et al. 2012). Therefore, in the present investigation, the bactericidal potential of agrochemicals used by the farmers against the bacterial leaf spot pathogen *Xanthomonas campestris* pv. *vesicatoria* of tomato prevalent in Nashik region, Maharashtra and the ability of the bacterial population to form pesticide resistant mutant against these agrochemicals was studied.

Material and Methods

Isolation of bacterial pathogen from infected tomato leaves

The isolation of bacteria responsible for leaf spot and blight in tomato leaves, collected from the infected tomato crop in Nashik district of Maharashtra, was done on nutrient agar media by routine bacteriological laboratory techniques of isolation of plant pathogenic bacteria (Borkar 2018). The bacterium was identified as *Xanthomonas campestris* PV. *Vesicatoria* based on the identification test (Borkar 2018).

The purified bacterial culture was assayed under *in vitro* condition for its sensitivity to different agrochemicals used by the farmers in the tomato field. The routine poison food technique was used to determine the bactericidal potential of these agrochemicals against the bacterium *Xanthomonas campestris* PV. *Vesicatoria*

The bactericidal potential of different agrochemicals against tomato bacterium *Xanthomonas campestris* PV. *Vesicatoria*

The bactericidal potential of antibiotics

The antibiotics viz. Streptomycin sulphate, Streptocycline, and Terramycin were evaluated for its efficacy against the tomato bacterium at a concentration of 25, 50, 75, 100, 250, 500 and 1000 ppm.

The bactericidal potential of bactericides

Two bactericides viz. Bactericin-100 and Bordeaux mixture was evaluated for its efficacy against the bacterium. Bactericin-100 was evaluated at a concentration of 25, 50, 75, 100, 250 and 500 ppm; whereas the Bordeaux mixture was evaluated at 0.01, 0.025, 0.05, 0.1 and 0.2% concentration.

The bactericidal potential of fungicides

Three fungicides viz. Copper ox chloride, Mancozeb, and Zineb were evaluated for its efficacy against the bacterium as these fungicides are commonly used in the tomato field by the farmer for control of different fungal diseases including bacterial leaf spot.

The fungicides Copper ox chloride, Mancozeb, and Zineb were evaluated at 0.01, 0.025, 0.05, 0.1 and 0.2 per cent concentration.

Bactericidal potential of combination of fungicides + antibiotics

Three combinations particularly copper ox chloride + Streptocycline, Bordeaux mixture + Streptocycline and Copper ox chloride + Streptomycin sulphate were evaluated for their efficacy against the bacterium by poison food technique. A combination of Copper ox chloride + Streptocycline was evaluated at 0.025 % + 50 ppm, 0.05 % + 50 ppm and 0.2 % + 50 ppm. A combination of Bordeaux mixture + Streptocycline was evaluated at 0.025 % + 50 ppm and 0.05 % + 50 ppm and a combination of Copper ox chloride + Streptomycin sulphate was evaluated at 0.025 % + 50 ppm, 0.05 % + 75 ppm and 0.2 % + 500 ppm respectively.

Studies on pesticide resistant mutant inducing ability in the bacterium

A bacterial suspension of Xanthomonas campestris PV. Vesicatoria was prepared from the 24 hrs old bacterial growth in sterile distilled water and the optical density of bacterial suspension was adjusted to 0.1 which content 107cfu/ml (Borkar 2018). 1 ml of this suspension was mixed in the 25 ml poison nutrient medium of above agrochemicals of minimum inhibitory concentration for the bacteria and poured in the sterile Petri plate. The medium in the plates was solidified in the slanting position and the plates were incubated at 280C in BOD incubator. The formation of pesticide-resistant mutant colonies was observed after a week in the plates and the mutant population was calculated to determine the rate of mutation.

Result and Discussion

The bactericidal potential of different agrochemicals against tomato bacterium *Xanthomonas campestris* pv. *Vesicatoria*

Table 1: The bactericidal potential of antibiotics on tomato leaf spot pathogen *Xanthomonas campestris* pv. *Vesicatoria*

Sr.No	Antibiotics	Concentration (ppm)	Bactericidal potential	MIC
1.	Streptomycin sulphate	1000	–	1000 ppm
		500	+	
		250	+	
		100	+	
		75	+	
		50	+	
		25	+	
2.	Streptocycline	500	–	25 ppm
		250	–	
		100	–	
		75	–	
		50	–	
		25	–	
3.	Terramycin	500	–	75 ppm
		250	–	
		100	–	
		75	–	
		50	+	
		25	+	

+ = Bacterial growth;   – = inhibition of bacterial growth

The bactericidal potential of antibiotics

The antibiotic Streptocycline was effective at a very low concentration of 25 ppm whereas the antibiotic Terramycin was effective at a concentration of 75 ppm and Streptomycin sulphate at a concentration of 1000 ppm to inhibit the bacterial growth of *Xanthomonas campestris* pv. *Vesicatoria*

The minimum inhibitory concentration (MIC) for Streptomycin sulphate was 1000 ppm, for Streptocycline 25 ppm and for Terramycin 75 ppm. Below the MIC, there was a bacterial growth in the poison food plates (Table 1).

The bactericidal potential of bactericides

The bactericide Bactericin-100 was effective at a concentration of 50ppm and above while it was not effective at 25 ppm. The Bordeaux mixture was effective at a concentration of 0.1 percent but it was not effective at 0.05 % to inhibit the growth of the bacterium. The MIC for Bactericin-100 was 50 ppm whereas for Bordeaux mixture it was 0.1 percent (Table 2).

Table 2: The bactericidal potential of bactericides on tomato leaf spot pathogen *Xanthomonas campestris* pv. *Vesicatoria*

Sr.No	Bactericides	Concentration	Bactericidal potential	MIC
1.	Bacteriocin -100	500	–	50 ppm
		250	–	
		100	–	
		75	–	
		50	–	
		25	+	
2.	Bordeaux mixture	0.2	–	0.1 %
		0.1	–	
		0.05	+	
		0.025	+	
		0.01	+	

+ = Bacterial growth; – = inhibition of bacterial growth

The bactericidal potential of fungicides

The fungicide Copper ox chloride was effective at 0.1 percent and above concentration. The fungicide Mancozeb was effective at 0.05 percent and Zineb was effective at 0.025 percent concentration to inhibit the bacterial growth (Table 3).

The bactericidal potential of a combination of fungicides + antibiotics

A combination of Copper ox chloride + Streptocycline was effective at 0.025 % + 50 ppm indicating that the concentration of Copper ox chloride can be lowered down when used with Streptocycline. A combination of Bordeaux mixture + Streptocycline was effective at 0.025 % + 50 ppm indicating that the concentration of the Bordeaux mixture can also be lowered

Table 3: The bactericidal potential of copper and di-thiocarbamate group fungicides on tomato leaf spot pathogen *Xanthomonas campestris* pv. *Vesicatoria*

Sr.No	Fungicide	Concentration (%)	Bactericidal potential	MIC
1.	Copper ox chloride	0.2	–	0.1%
		0.1	–	
		0.05	+	
		0.025	+	
		0.01	+	
2.	Mancozeb	0.2	–	0.05%
		0.1	–	
		0.05	–	
		0.025	+	
		0.01	+	
3.	Zineb	0.2	–	0.025%
		0.1	–	
		0.05	–	
		0.025	–	
		0.01	+	

+ = Bacterial growth; – = inhibition of bacterial growth

Table 4: The bactericidal potential of fungicide + antibiotics on tomato leaf spot pathogen *Xanthomonas campestris* pv. *Vesicatoria*

Sr.No	Combination	Concentration	Bactericidal potential	MIC
1.	Copper ox chloride + Streptocycline	0.2 % + 50 ppm	–	0.025 % + 50 ppm
		0.05 % + 50 ppm	–	
		0.025 % + 50 ppm	–	
2.	Bordeaux mixture + Streptocycline	0.05 % + 50 ppm	–	0.025 % + 50 ppm
		0.025 % + 50 ppm	–	
3.	Copper ox chloride + Streptomycin sulphate	0.2 % + 500 ppm	–	0.2 % + 500 ppm
		0.05 % + 75 ppm	+	
		0.25% + 50 ppm	+	

+ = Bacterial growth; – = inhibition of bacterial growth

down when mixed with Streptocycline. The combination of Copper ox chloride + Streptomycin sulphate was effective at

0.2 % + 500 ppm and not below this concentration. The MIC for Copper ox chloride with Streptocycline combination was 0.025 % + 50 ppm, for Bordeaux mixture with Streptocycline combination it was 0.025 % + 50 ppm and for Copper ox chloride, with Streptomycin sulphate combination it was 0.2 % + 500 ppm (Table 4).

Pesticide resistant mutant inducing ability in the population of *Xanthomonas campestris* PV. *Vesicatoria*

The bacterial population of *Xanthomonas campestris* PV. *Vesicatoria* was resistant to the antibiotic Terramycin, streptomycin sulphate and to the fungicide Copper ox chloride, Mancozeb and a combination of Copper ox chloride + Streptomycin sulphate and induced the pesticide resistant mutant in the bacterium. The mutation rate for Terramycin and Streptomycin sulphate was 3.5×10^{-4} and 2×10^{-4} respectively, whereas, for Dithane Z-78, Mancozeb and a combination of Copper ox chloride+ Streptomycin sulphate was $2.6 \times 10^{-4,}$ 8×10^{-5} and 5×10^{-5}.respectively. The agrochemicals which were completely inhibitory to the bacterium and did not allow the formation of antibiotic or antibiotic + fungicide resistance mutant were Streptocycline, Streptocycline+ Copper ox chloride and Streptocycline + Bordeaux mixture.

Several workers had studied the potential of agrochemicals as bactericide (Patyka et al. 2012; Hulloli et al. 1998; Verma et al. 1992; Nafde and Verma 1984; Verma and Singh 1976) and formation of pesticide resistance in bacterial plant pathogens (Marques et al. 2009; Cazrola et al. 2002; Ritchie and Dittapongpitch 1991; Adaskaveg and Hine 1985).

Jones et al. (1991) monitored the populations of copper-resistant (Cur) strains of *Xanthomonas campestris* PV. *Vesicatoria* in the field on non-symptomatic tomato leaflets treated with copper or with a copper and Mancozeb mixture over three and four seasons, respectively. In a greenhouse study, where a Cur strain of *Xanthomonas campestris* PV. *Vesicatoria* was applied to tomato foliage; bacterial populations were significantly less on plants treated with copper or with a copper and Mancozeb mixture than on untreated plants. However, leaflets treated with the copper and Mancozeb combination had significantly lower Cu^r populations than leaflets treated with copper alone. Bouzar et al. (1999) found *Xanthomonas* isolates resistant to both Streptomycin and copper in the Caribbean and Central America. Buonaurio et al. (1994) reported few strains of race1, race 2 and race 3 of *Xanthomonas campestris* PV. *Vesicatoria* tolerant to Copper sulphate (200 µg/ml), while susceptible to Streptomycin sulphate (10 µg/ml)

Chand et al. (1994) reported widespread resistance in isolates of Xanthomonas campestris PV. Vesicatoria to copper and zinc collected in India during 1991-92. Stall and Thayer (1962) stated that resistance to streptomycin in Xanthomonas was found a century ago in Florida (USA).

Schroth et al. (1979) reported the increased concentrations of Streptomycin in media up to 1,000µg/ml increased the generation times of Streptomycin resistant strains but did not prevent growth. Virulence among Streptomycin resistant and Streptomycin susceptible strains varied but there was no consistent difference between the two groups. The Streptomycin resistant strains appeared to be relatively stable and were detected in orchards 6 years after termination of Streptomycin application.

Marco and Stall (1983) also reported the copper resistance in bacterial leaf spot diseases in Florida. Adaskaveg and Hine (1985) isolated the copper-sensitive strains of *Xanthomonas campestris* PV. *Vesicatoria* from infected pepper plants from two locations in Arizona where there was limited use of copper bactericides. Three copper tolerant strains of the bacterium were also isolated from diseased plants from the West Coast and Central Mexico, where copper bactericides have been used for more than 30 years. The Arizona strains were sensitive to various copper formulations (Copper hydroxide, Copper sulphate, Copper ammonium carbonate, and Basic copper sulphate) with and without the addition of Mancozeb as determined by the presence of inhibition zone in disk assays.

Bender et al. (1990) evaluated the efficacy of copper bactericides for control of *Xanthomonas campestris* PV. *Vesicatoria* in eastern Oklahoma tomato fields. Copper bactericides did not provide adequate control, and copper resistant strains of the pathogen were isolated. Ritchie (2000) reported that chemical control of *Xanthomonas vesicatoria* is limited to copper or copper combined with Maneb sprays that provide only marginal success due to the formation of resistant mutant thus making the disease very difficult to control once the epidemic is underway.

References

1. Adaskaveg JE and Hine RB. Copper tolerance and zinc sensitivity of Mexican strains of Xanthomonas campestris PV. vesicatoria, the causal agent of bacterial spot of pepper. Plant Dis. 1985;69(11):993-996.

2. Anderson DI. The biological cost of mutational antibiotic resistance: any practical conclusion. Curr Opin Microbiol. 2006;9(5):461-465.

3. Araújo ER, Costa JR, Ferreira MA and Quezado-Duval AM. Simultaneous detection and identification of the Xanthomonas species complex associated with tomato bacterial spot using species-specific primers and multiplex PCR. J Appl Microbiol. 2012;113(6):1479-1490. doi: 10.1111/j.1365-2672.2012.05431.x

4. Bender L, Malvick K, Conway E, George S and Pratt P. Characterization of pXV10A, a copper resistance plasmid in Xanthomonas campestris pv. Vesicatoria. Appl Environ Microbiol. 1990;56(1):170-175.

5. Borkar SG. Laboratory technique in plant bacteriology. CRC Press, U S A. 2018;320.

6. Borkar SG and Yumlembam RA. Bacterial Diseases of Crop Plants. CRC Press, U S A. 2016;594.

7. Bouzar H, Jones JB, Stall RE, Louws FJ, Schneider M, Rademaker JL, Bruijn FJ, et al. Multiphasic analysis of Xanthomonads causing bacterial spot disease on tomato and pepper in the Caribbean and Central America: evidence for common lineages within and between countries. Phytopathology. 1999;89(4):328-335. doi: 10.1094/PHYTO.1999.89.4.328

8. Buonarurio R, Stravato VM and Scortichni M. Characterization of Xanthomonas campestris pv. Vesicatoria from Capsicum annum L. in Italy. Plant Dis. 1994;78(3):296-299.

9. Cazrola FM, Arrebola E, Sesma A, Perez-Garcia A, Codima JC, Murillo J and de Vicente A, et al. Copper resistant in Pseudomonas syringe strain isolated from mango is encoded mainly in plasmids. Phytopathology. 2002;92(8):909-916. doi: 10.1094/PHYTO.2002.92.8.909

10. Chand R, Singh R and Singh PK. Distribution of pathogenic groups and races in Xanthomonas campestris pv. vesicatoria in peninsular India. Indian Phytopath.1994;47(3):251-252.

11. Hulloli SS, Singh RP and Verma JP. Management of bacterial blight of cotton with the use of neem based formulations. Indian Phytopath. 1998;51(1):21-25.

12. Jones JB, Woltz SS, Jones JP and Portier KL. Population dynamics of Xanthomonas campestris pv. vesicatoria on tomato leaflets treated with copper bactericides. Phytopathology. 1991;81(7):714-719.

13. Marco GM and Stall RE. Control of bacterial spot of pepper initiated by strains of Xanthomonas campestris pv. Vesicatoria that differ in sensitivity to copper. Plant Dis. 1983;67(7):779-781.

14. Marques E, Uesugi CH, and Ferreira MASV. Sensitivity to copper in Xanthomonas campestris pv. vesicatoria. Trop Plant Pathol. 2009;34(6):406-411.

15. Nafde SD and Verma JP. Effect of chemicals on translocation of streptomycin in the cotton seedling. Indian Phytopath. 1984;37:524-528.

16. Patyka V, Buletsa N, Pasichnyk L, Zhitkevish N, Kalinichenko A, Gnatiuk T and Butsenko L, et al. Specifics of pesticide effects on the phytopathogenic bacteria. Ecol Chem Eng S. 2012;23(2):311-331.

17. Ritchie DF and Dittapongpitch V. Copper and streptomycin resistant strains and host differentiated races of Xanthomonas campestris pv. vesicatoria. Plant Dis.1991;75:733-736.

18. Ritchie DF. Bacterial spot of pepper and tomato. The Plant Health Instructor.2000; doi: 10.1094/PHI-I-2000-1027-01

19. Schroth MN, Thomson SV and Moller WJ. Streptomycin resistance in Erwinia amylovora. Phytopathology.1979;69(1):565-568.

20. Stall RE and Thayer PL. Streptomycin resistance of the bacterial spot pathogen and control with streptomycin. Pl Dis Rep. 1962;46(1):389-392.

21. Verma J P and Singh RP. Chemical control of bacterial diseases of plants. Chemical Concept.1976;4:31-50.

22. Verma JP, Singh RP, Jindal JK and Trivedi BM. Bacterial plant pathogens and their management. Fusion Asia.1992;8:29-36.

26

Chitinase and Glucanase Activities of Antagonistic Streptomyces Spp Isolated from Fired Plots under Shifting Cultivation in Northeast India

Mukesh K Malviya[1], Pankaj Trivedi[2] and Anita Pandey[3*]

[1]*Crop Genetics Improvement and Biotechnology Lab, Guangxi Academy of Agricultural Sciences, Nanning, 530007, China*

[2]*Department of Bioagricultural Sciences and Pest Management, Colarado State University, C034 Plant Sciences, Fort Collins - CO - 80523-1177*

[*3]*Biotechnological Applications, G.B. Pant National Institute of Himalayan Environment and Sustainable Development, Kosi-Katarmal, Almora, 263 643 Uttarakhand, India*

***Corresponding author:** Anita Pandey, Biotechnological Applications, G. B. Pant National Institute of Himalayan Environment and Sustainable Development, Kosi-Katarmal, Almora, 263 643 Uttarakhand, India; E-mail: anita@gbpihed.nic.in*

Abstract

Antagonistic *Streptomyces* spp (*Streptomyces* sp NEA55 and *Streptomyces cavourensis* NEA5), isolated from fired plots under shifting cultivation in northeast India, are studied for their chitinase and glucanase activities. The species showed strong antagonism against test fungi (*Rhizoctonia solani* and *Cladosporium* sp.) in plate assays. Maximum % inhibition was observed due to the effect of diffusible compounds produced by these species. *Streptomyces* sp. NEA55 showed 54.83 % inhibition against *R. solani* while *S. cavourensis* NEA5 showed up to 50.00 % inhibition against *Cladosporium* sp. The inhibitory effect of volatile compounds by *Streptomyces* sp NEA55 was recorded up to 50.7 % against *R. solani* and 37.50 % against *Cladosporium* sp. While *S. cavourensis* NEA5 showed 49.23 % inhibition against R. solani and 34.37 % inhibition against *Cladosporium* sp. *S. cavourensis* NEA5 and *Streptomyces* sp. NEA55 produced 0.138±0.006 µg/ml and 0.15±0.004 µg/ml chitinase, 0.22±0.001 µg/ml and 0.25±0.002 µg/ml β 1,3 glucanase, respectively. Both the species showed maximum chitinase activity at pH 6 and temperature 50 ºC, while minimum enzyme activity was observed at pH 10 and temperature 20 ºC. Both the species showed glucanase activity maximum at pH 7 and temperature 40 ºC and minimum activity at pH 10 and temperature 20 ºC. Both the species hydrolyzed glycol–chitin as a substrate in denaturing conditions showing variable amount of different isoforms. This study demonstrates that the antagonistic species of *Streptomyces* survive the fire operations under shifting cultivation.

Keywords: *Streptomyces*; Antagonism; Chitinase; Glucanase; Shifting cultivation

Introduction

Actinobacteria are most widely distributed and distinct group of microorganisms in nature. *Actinobacteria*, *Streptomyces* species in particular, has been a broadly exploited group of microorganisms for the production of important secondary metabolites and enzymes in the field of medicine and agriculture

[1, 2]. *Streptomyces* are well known as antifungal biocontrol agents that inhibit several plant pathogenic fungi [3]. The antagonistic activity of *Streptomyces* to fungal pathogens is usually related to the production of antifungal compounds and extracellular hydrolytic enzymes [4, 5]. Many species of *Streptomyces* are well known as antifungal biocontrol agents that inhibit several plant pathogenic fungi e.g., *Phytophthora capsici Sclerotinia rolfsii, Fusarium sporotrichiodes, Rhizoctonia solani* and *Sclerotium rolfsii, Alternaria alternata* and *Phomopsis archeri* [6-9]. Furthermore, *Streptomyces* produce bioactive compounds such as antimicrobial, antiparasitic and immune-suppressing compounds via secondary metabolism. *Streptomycetes* have been found in beneficial associations with plants where they improve plant growth and protect against pests; this has attracted the attention of researchers worldwide [10].

Chitinase and β-1,3-glucanase are considered to be important hydrolytic enzymes in the lysis of fungal cell walls [11]. Chitinolytic enzymes have been identified in several *Streptomyces* spp. including *Streptomyces* sp. M-20, *S. venezuelae* P10, and *S. anulatus* CS242 [12-14]. Glucanase has been known to be produced by several microorganisms and playing important role in biocontrol [15]. Several *Streptomyces* have been studied for antifungal properties along with the production of glucanase, some of the examples are *Streptomyces* sp. S27 and *Streptomyces* sp. Mo [16, 17]. Shifting cultivation, refers to 'slash and burn', is a predominant form of agricultural practice in hills of northeast India. The microbiological aspects, basically survival of bacterial, fungal and actinobacterial communities after fire events have been studied in recent times [18-20]. The focus of present study is on the antagonistic potential of two *Streptomyces* species that were isolated after the fire events. These species have been studied with respect to production of diffusible and volatile compounds against test fungi along with the chitinase and β-1, 3-glucanase activities. In addition, both the species are also studied with hydrolyzed glycol–chitin as a substrate in denaturing conditions.

Material and Methods

Study site and isolation of actinobacteria

Actinobacteria were isolated from the soil samples that were collected after the completion of fire events in Papumpare District, Itanagar, Arunachal Pradesh under shifting cultivation. The details of the study sites have been described in Pandey et al. [18]. Among the two actinobacterial isolates used in the present study, isolate no. NEA5 showed maximum similarity with *Streptomyces cavourensis* NR_043851 and NEA55 with *Streptomyces* sp. YIM8 AF389344 [19]. The test fungi (*Rhizoctonia solani* and *Cladosporium* sp.) were also originally isolated from shifting cultivation site [20].

Scanning electron microscopy

In order to see the deep morphological pattern (substrate and aerial mycelia) of *Streptomyces* spp. scanning electron microscopy was performed. Glutaraldehyde (2.5 %) was added to the culture and then centrifuged. After centrifugation, washing was given with phosphate buffer saline (PBS) twice and centrifuged. The sample was dehydrated with CPD (Critical dry point), and then the samples were coated with gold by auto fine coater (JFC-1600). After coating, the sample was viewed under scanning electron microscope (JSM-6610LV).

Dual culture technique for determination of the production of diffusible antifungal compounds

To test the ability of the *Streptomyces* spp. to inhibit the phytopathogens, the test fungal culture and *Streptomyces* sp. was spot inoculated off-center on each potato carrot agar (PCA) plate. After 7 days of incubation at 28 ºC, the zone of inhibition was measured. Per cent growth inhibition was calculated using the following formula: (R1-R2)/R1 × 100 (where R1 represents the radius of the test fungus in the direction with no bacterial colony and R2 is the radius of the fungal colony in the direction of the bacterial colony) as described in Trivedi et al. [21].

Sealed double Petri plate technique for determination of the production of volatile antifungal compounds

Each *Streptomyces* sp. was streaked on Petri plate containing PCA and a 6.0 mm disc of 4 days old culture of the test fungus was placed in the middle of another PCA plate. The lid of both the plates were replaced with the plates of same diameter, placed face to face, and sealed by parafilm, preventing any physical contact between the pathogen and the *Streptomyces* sp. Petri plates were incubated at 28 ºC for 7 days and growth of the pathogen was measured and compared to control developed in the absence of the antagonist (mocked inoculation with a 6.0 mm disc of PCA). Results were expressed as means of per cent inhibition of the growth of each fungus in the presence or absence of the antagonistic *Streptomyces* sp. Per cent growth inhibition was

calculated by the following formula: (R1-R2)/R1 × 100, where R1 represents the diameter of the test fungus on control plate and R2 is the diameter of the growth on inoculated plate.

Quantitative estimation of production of chitinase and glucanase on different temperature and pH

Quantification of chitinase and glucanase has been done following prescribed procedures [22]. The *Streptomyces* species were cultured at 28 ºC for 5 days on a rotary shaker in 250.0 ml of chitin peptone medium for chitinase production and peptone medium containing laminarin (0.2 %) (from Laminaria digitata, Sigma) for glucanase production [23]. The cultures were centrifuged at 12,000 g for 20 min at 4 ºC and the supernatants were used as enzyme source. The reaction mixture contained 0.25 ml of enzyme solution, 0.3 ml of pH buffer (1.0 M citrate buffer pH 5, phosphate buffer pH 6 to 8, and glycine buffer pH 9). The reaction mixtures were incubated at different temperatures (20, 30, 40, 50 and 60 ºC) for 4 h in a water bath. One unit of chitinase was determined as 1 µmol of N acetyl glucosamine (GlcNAc) released min-1 and β-1, 3-glucanase activity was determined as 1 µmol of glucose released min-1. Protein content was determined as described by Lowry method [24].

SDS-PAGE analysis for chitinase

Streptomyces spp. was grown as shake culture in chitin-peptone medium at 28 ºC for 7 days. The supernatant was centrifuged at 12000 g for 20 min at 4 ºC and filtered through 0.22 µl sterile filter (Millipore) and collected in conical flask. Protein content was analyzed by SDS-PAGE [25]. Polyacrylamide slab gel consisted of 4 % stacking gel and 10 % separating gel containing 0.01 % glycol–chitin. Electrophoresis was carried out at a constant voltage of 65 V; gel was stained with coomassie brilliant blue R-250 and analyzed under gel documentation system (Alpha Imager 2200). Glycol chitin was prepared as described by Trudel and Asselin [26].

Results and Discussion

The scanning electron microscopy of both the species is presented in Figure: 1A and Figure: 1B. Scanning electron microscopy revealed definite structures of mycelium and spores of both the *Streptomyces* spp. Two of the isolates that were used for detailed studies based on their antagonistic properties were subjected to scanning electron microscopy. The scanning electron microscopy has been referred to provide perfect characterization of *streptomycetes* [27]. The effect of diffusible compounds produced by the *Streptomyces* spp., evaluated in terms of reduction in radial growth of two test fungi viz. *Rhizoctonia solani* and *Cladosporium* sp., following 7 days of incubation at 28 ºC, are presented in Figure:2 & Table 1. The results showed that volatile compounds produced by *Streptomyces* spp. were inhibitory to the growth of the test fungi, viz. *R. solani* and *Cladosporium* sp. Inhibition (%) in fungal growth by Streptomyces species is presented in Table 1. Microscopic observations revealed that the diffusible as well as volatile compounds produced by

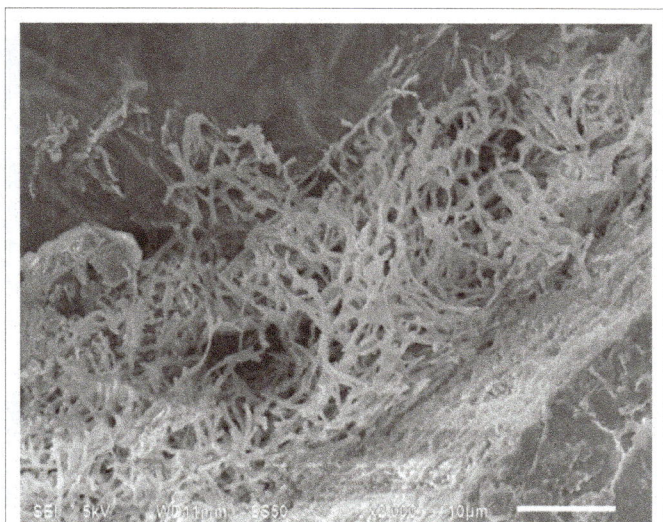

Figure 1A: Scanning electron micrograph of *Streptomyces cavourensis* NEA5

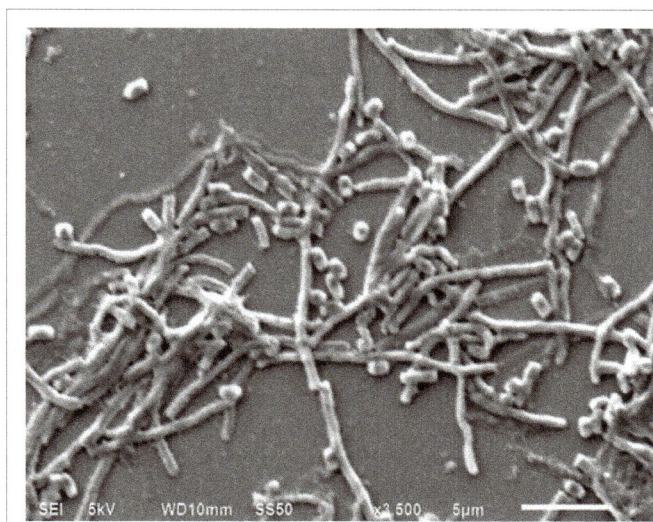

Figure 1B: Scanning electron micrograph of *Streptomyces* sp. NEA55

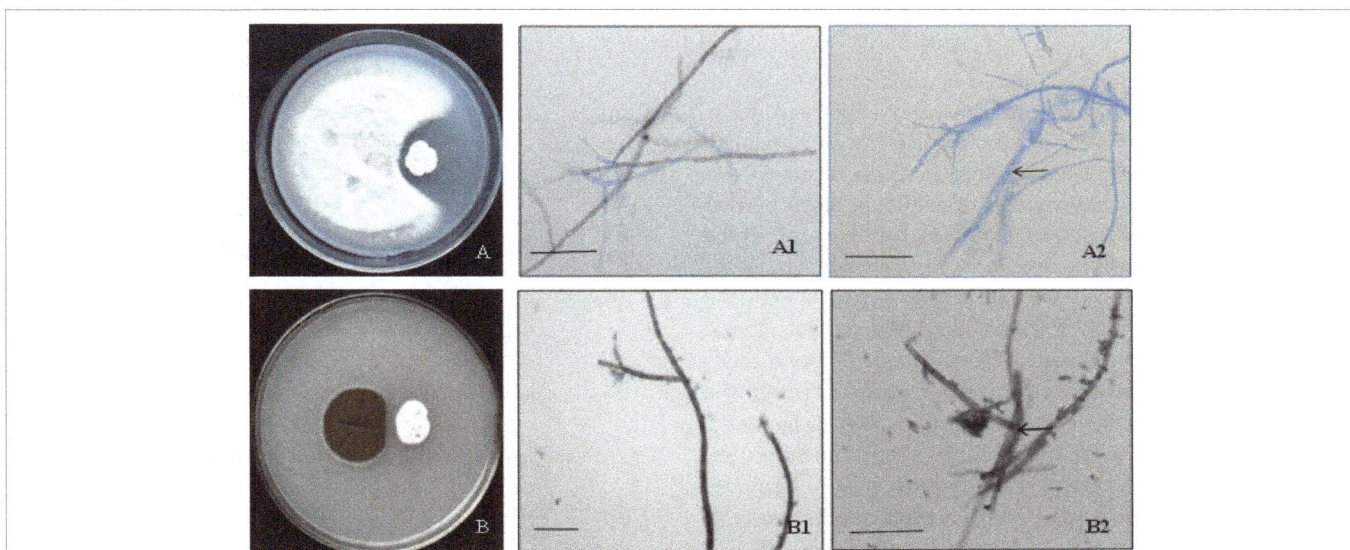

Figure 2: Production of antifungal compounds by *Streptomyces cavourensis* NEA5 & *Streptomyces* sp. NEA55 (A&B) respectively: Inhibition of *R. solani* and *Cladosporium* sp. due to diffusible compounds produced. Morphological deformities: (A1&B1) Normal structures of *R. soloni* and *Cladosporium* sp. (A2&B2) deformed structures of respective fungus. (Bar = 5 μm)

Table 1: In vitro percent *inhibitory* effect of diffusible and volatile metabolites of *Streptomyces* species on growth of test fungi

Pathogenic test fungi	% Inhibition in fungal growth by *Streptomyces* spp. after 7 days of incubation			
	Streptomyces cavourensis NEA5		*Streptomyces* sp NEA55	
Rhizoctonia solani	51.61	49.23	54.83	50.07
Cladosporium sp.	50	34.37	47.7	37.5

Table 2: Comparative account of structures of normal and antagonized pathogen

Test fungi	Structure	Normal	Antagonized
Rhizoctonia	Mycelium	Somatic and fertile hypha well distinguished	Irregular
Solani	Somatic hypha	Septate, dia. 2.5- 3.5 μm	Irregular
			Septate, dia 3.0-6.0
	Fertile hypha	Septate and dark in colour, dia. 3.0-4.5 μm	Irregular septate, dia 3.0-7.5 μm
	Hyphal wall	Present	Lysed
Cladosporium sp.	Mycelium	Somatic hypha septate, dia. 2-3.5 μm	Somatic hypha septate, dia. 2-5 μm
	Conidia	Curved or straight width 2.0-2.5 μm	Irregular width 2-4 μm

Streptomyces species induced morphological abnormalities in the fungal structures (Table 2). Deformation was observed in *mycelial, hyphal* or *conidial* structures. The longitudinal septae completely disappeared and the conidia became thick walled and spherical or irregular in shape. Lysis of fungal hyhae and vacuolization as well as granulation in mycelium was observed in *R. solani*. Size of the somatic and fertile hypha of *R. solani* increased due to the antagonistic effect of actinobacteria. Due to the antifungal activity of *Streptomyces* sp., hyphal wall of *R. solani* was completely lysed. In case of *Cladosporium* sp., size of mycelium and conidia increased with irregular shape of conidia. The inhibition of growth of the test pathogenic fungi continued to increase with increasing incubation time. A clear inhibitory area and dense sporulation ring was observed near the growth of the isolate which antagonized respective test fungi. No physical contact was observed between the *Streptomyces* species and the test fungi that were observed to be antagonized. Moreover, the formation of the inhibitory halo suggested the presence of fungicidal metabolites secreted by *Streptomyces*. Similar observations have been reported by Aghigni et al. [28] from a number of actinobacterial isolates from Iranian soil. These isolates formed inhibition zones in dual culture based assays inhibiting the growth of *Alternaria solani, A. alternata, Fusarium solani, Phytophthora megaperma, Verticillium dahlia* and *Saccharomyces cerevisiae*. The antifungal potential of actinobacteria against *Colletotrichum gloeosporioides* and *Sclerotium rolfsii* was assessed by dual culture technique [10]. Al-Askar et al. [29] reported that *Streptomyces spororaveus* RDS28 produces antifungal compounds against some phytopathogenic fungi, viz., *Rhizoctonia solani, Fusarium solani, F. verticillioides, Alternaria alternata* and *Botrytis cinerea*. There was a change in colour of the test fungi in inoculated plates, indicative of the inhibitory effect of volatile compounds. The test fungi, inhibited by the *Streptomyces*, under present study exhibited morphological abnormalities due to the production of diffusible and volatile antifungal compounds. Abnormal hyphal swelling, degradation and lysis of mycelia were observed by Joo [6] when *Phtophthora capsici* was grown with the high or low molecular fraction of *Streptomyces halstedii* AJ-7.

Quantitative estimation of production of chitinase and β 1, 3 glucanase activities is presented in Figure 3 (A&B) showing maximum activity of chitinase and glucanase. Effect of different

pH and temperature on chitinase activity of both the species is presented in Figure 4 (A&B). Both the species showed maximum activity of chitinase at pH 6 and temperature 50 °C, while minimum enzyme activity was observed at pH 10 and temperature 20 °C. In case of glucanase activity both the species showed maximum enzyme activity at pH 7 and temperature 40 °C and minimum activity at pH 10 and temperature 20 °C (Figure 5A&B). Kim et al. [12] reported maximum activity of chitinase

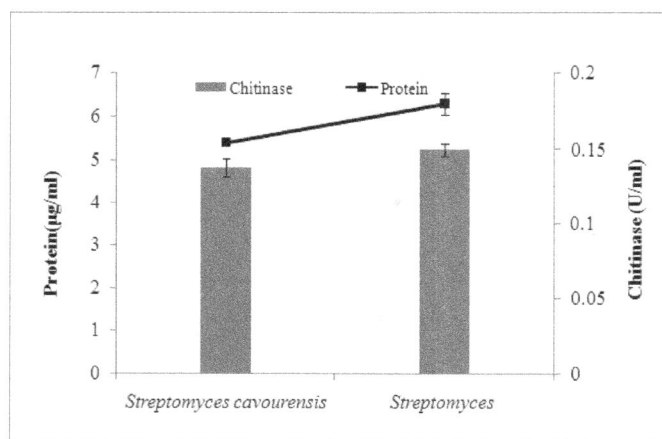

Figure 3A: Quantitative estimation of chitinase activity

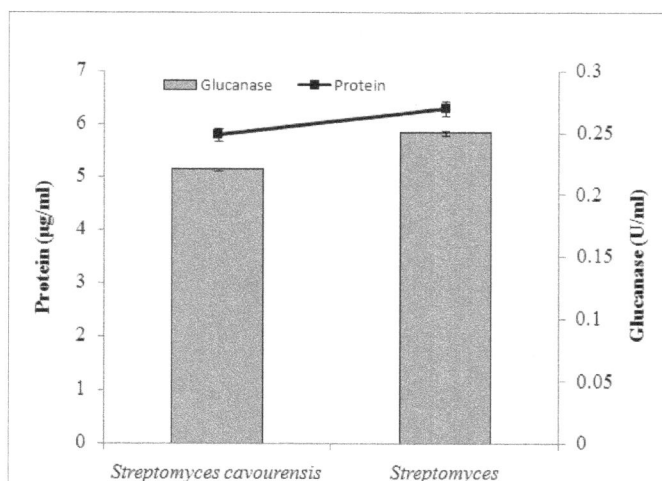

Figure 3B: Quantitative estimation of glucanase activity

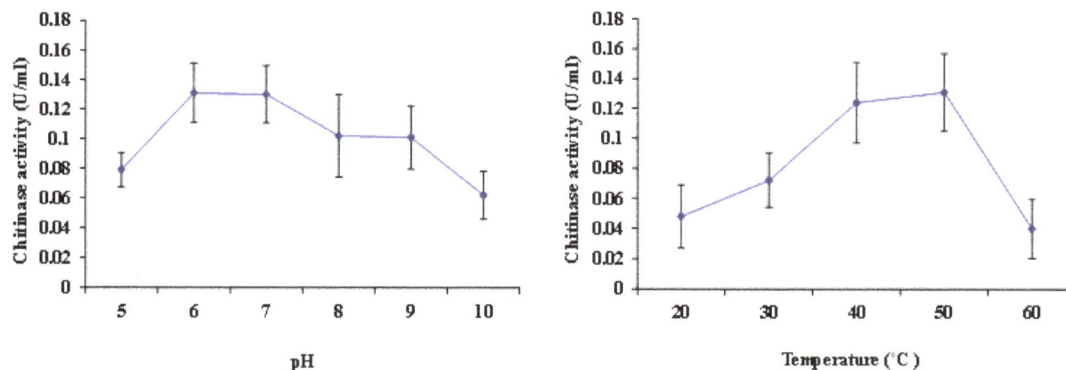

Figure 4A: Chitinase activity of *Streptomyces cavourensis* NEA5 at different pH & temperature

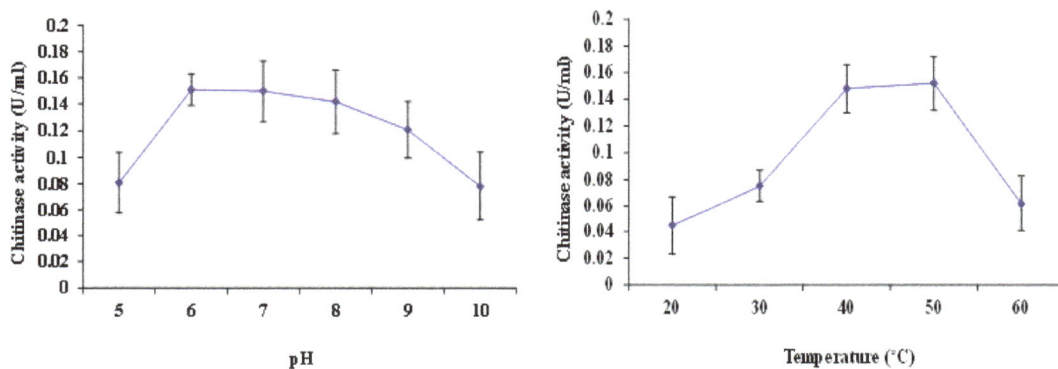

Figure 4B: Chitinase activity of *Streptomyces* sp NEA55 at different pH & temperature

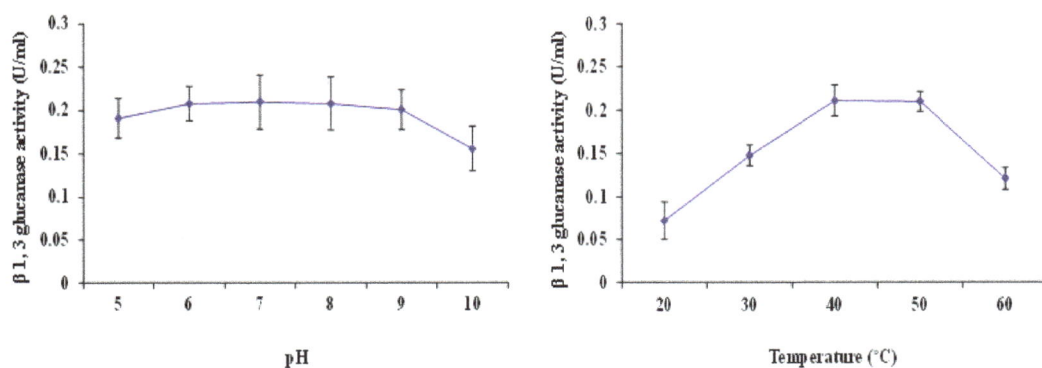

Figure 5A: Glucanase activity of *Streptomyces cavourensis* NEA5 at different pH & temperature

between pH 5.0 to 6.0 and it was relatively stable at pH 4.0 to 8.0 when kept at 4 ºC. Beyond this pH range there was a rapid loss in the activity. Glucanase activity of the two isolates was found maximum at pH 7 and temperature 40 ºC. Highest activity was found between pH 5 to 8 and temperature 30 to 50 ºC, in previous studies. The optimum activity for short term incubation is often seen at temperature in the range of 30 to 50 ºC, while many fungal glucanases appear stable between 50 to 60 ºC [30].

SDS-PAGE exhibited the chitinase activity of the two *Streptomyces* species (NEA5 & NEA55). *Streptomyces* species hydrolyzed glycol-chitin as a substrate in denaturing conditions. Species showing variable amount of different isoforms with molecular weights between 31 to 40 kDa are presented in Figure 6. Majority of bacterial chitinases has been reported to be in the range of 20 to 60 kDa. Chitinases from various *Streptomyces* were found to possess molecular weights as 20 kDa from *Streptomyces*

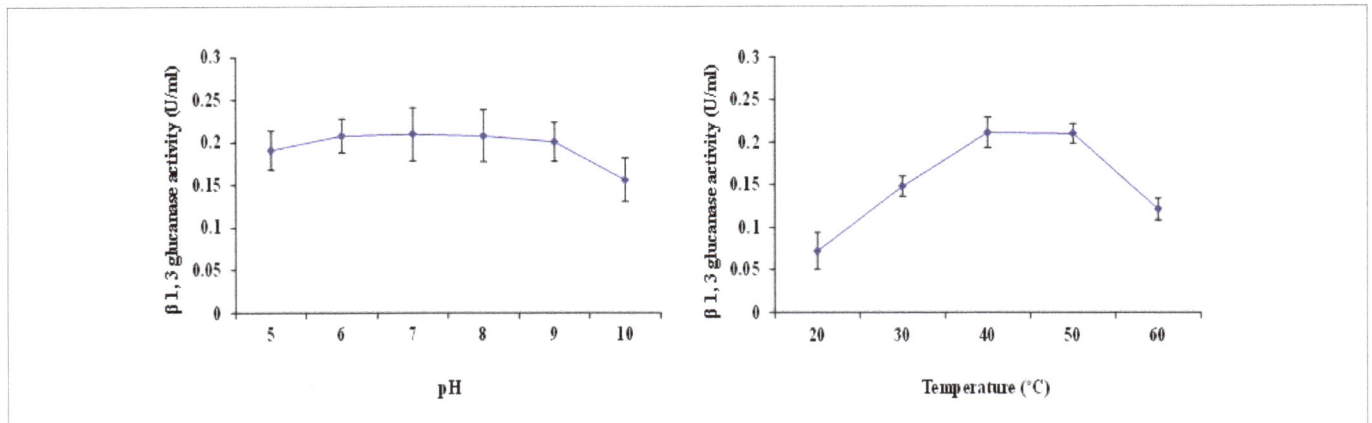

Figure 5B: Glucanase activity of *Streptomyces* sp NEA55 at different pH & temperature

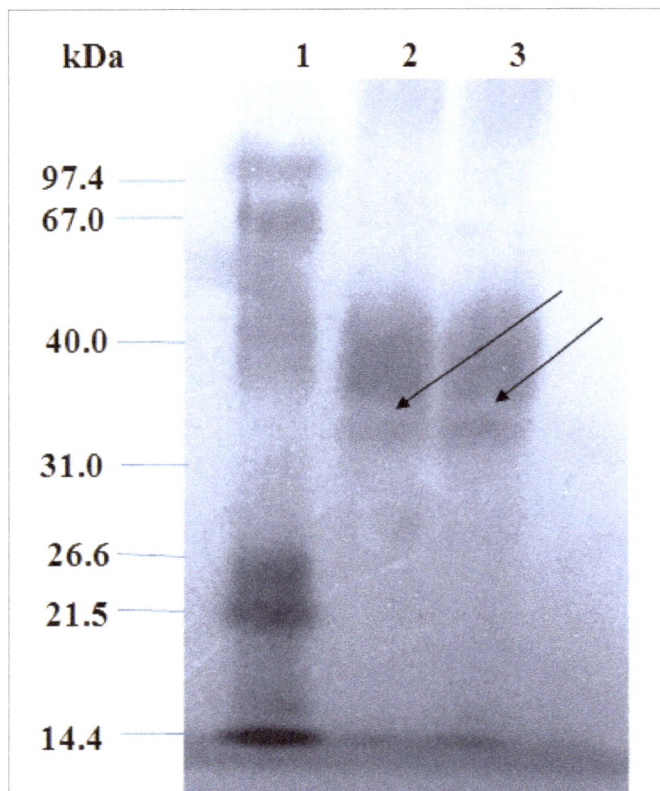

Figure 6: Detection of chitinase activity after SDS-PAGE using glycol chitin as substrate, lane 1 (Molecular weight markers), lane 2 and 3 sample *Streptomyces cavourensis* NEA5 and *Streptomyces* sp NEA55, respectively

sp. M-20 [13] and some strains of *Streptomyces* varying from approximately 25 kDa to 200 kDa [31].

The aim of this work is to prove the potential of *Streptomyces* species to survive the fire operations retaining their potential to reduce or eliminate the plant pathogenic fungi. Such robust isolates may prove to be the potential biocontrol agents for field applications. Further studies are needed in order to determinate the nature of *Streptomyces* species metabolites and their mechanism of action.

Acknowledgment

Director (GBPNIHESD) for extending the facilities and Ministry of Environment, Forest and Climate Change, Govt. of India, New Delhi for financial support.

References

1. Kumar D, Gupta RK. Biocontrol of wood rotting fungi. Ind J Biotechnol. 2006; 5:20-25.
2. Narayana KJP, Vijayalakshmi M. Chitinase production by *streptomyces* sp. anu 6277. Braz J Microbiol. 2009; 40(4): 725-733.
3. El-Tarabily KA, Soliman MH, Nassar AH, Al-Hassani HA, Sivasithamparam K. Biocontrol of sclerotinia minor using a chitinolytic bacterium and actinomycetes. Plant Pathol. 2000; 49: 573-583.
4. Fourati-Ben Fguira L, Fotso S, Ben Ameur-Mehdi R, Mellouli L, Laatsch H. Purification and structure elucidation of antifungal and antibacterial activities of newly isolated *Streptomyces* sp. strain US80. Res Microbiol. 2005; 156(3): 341-347.
5. Trejo-Estrada SR, Sepulveda IR, Crawford DL. In vitro and in vivo antagonism of *Streptomyces violaceusniger* YCED9 against fungal pathogens of turfgrass. World J Microbiol Biotechnol. 1998;14(6):865-872.
6. Joo GJ. Production of an anti-fungal substance for biological control of *Phytopthora capsici causing* blight in red-peppers by *Streptomyces halstedii*. Biotechnol Lett. 2005;27(3):201-205.
7. Errakhi R, Bouteau F, Lebrihi A, Barakate M. Evidences of biological control capacities of *Streptomyces* spp. against *Sclerotium rolfsii* responsible for damping-off disease in sugar beet (*Beta vulgaris L.*). World J Microbiol Biotechnol. 2007;23(11):1503-1509.
8. Rugthaworn P, Uraiwan Dilokkunanant, Sangchote S, Piadang N, Kitpreechavanich V. Search and improvement of actinomycete strains for biological control of plant pathogens. Kasetsart J (Nat Sci). 2007;41:248-254.
9. Malviya MK, Pandey A, Trivedi P, Gupta G, Kumar B. Chitinolytic activity of cold tolerant antagonistic species of *streptomyces* isolated from glacial sites of Indian Himalaya. Curr Microbiol. 2009;59(5):502-508. doi: 10.1007/s00284-009-9466-z
10. Sousa JAJ, Olivares FL. Plant growth promotion by streptomycetes: ecophysiology, mechanisms and applications. Chem Biol Technol Agric. 2016; 3:24.

11. Prapagdee B, Kuekulvong C, Mongkolsuk S. Antifungal potential of extracellular metabolites produced by *Streptomyces hygroscopicus* against phytopathogenic fungi. Int J Biol Sci. 2008;4(5):330-337.

12. Kim KJ, Yang YJ, Kim JG. Purification and characterization of chitinase from *Streptomyces* sp. M-20. J Biochem Mole Biol. 2003;36(2):185-189.

13. Mukherjee G, Sen SK. Purification, characterization, and antifungal activity of chitinase from *Streptomyces venezuelae* P10. Curr Microbiol. 2006;53(4):265-269.

14. Mander P, Cho SS, Choi YH, Panthi S, Choi YS, Kim HM, Yoo JC et al. Purification and characterization of chitinase showing antifungal and biodegradation properties obtained from *Streptomyces anulatus* CS242. Arch Pharm Res. 2016;39(7):878-886. doi: 10.1007/s12272-016-0747-3

15. El-Katatny MH, Gudelj M, Robra KH, Elnaghy MA, Gubitz GM. Characterization of a chitinase and an endo- β-1,3-glucanase from *Trichoderma harzianum* Rifai T24 involved in control of the phytopathogen *Sclerotium rolfsii*. Appl Microbiol Biotechnol. 2001; 56(1-2):137-143.

16. Shi P, Yao G, Li N, Luo H, Bai Y, Wang Y, Yao Bin et al. Cloning, characterization, and antifungal activity of an endo-1,3-β-D:-glucanase from *Streptomyces sp.* S27. Appl Microbiol Biotechnol. 2010;85(5):1483-1490. doi: 10.1007/s00253-009-2187-1

17. Kurakake M, Yamanouchi Y, Kinohara K, Moriyama S. Enzymatic properties of β-1,3-glucanase from *Streptomyces* sp Mo. J Food Sci. 2013;78(4):502-506. doi: 10.1111/1750-3841.12076

18. Pandey A, Chaudhry S, Sharma A, Choudhary VS, Malviya MK, Chamoli S, Rinu K, Trivedi P, Palni LM et al. Recovery of *Bacillus* and *Pseudomonas* spp from the 'Fired Plots' under Shifting Cultivation in Northeast India. Curr Microbiol. 2011; 62(1): 273-280. doi: 10.1007/s00284-010-9702-6

19. Malviya MK, Pandey A, Sharma A, Tiwari SC. Characterization and identification of actinomycetes isolated from 'fired plots' under shifting cultivation in northeast Himalaya, India. Ann Microbiol. 2013;63(2):561-569.

20. Jain R, Chaudhary D, Dhakar K, Pandey A. A consortium of fungal survivors after fire operations under shifting cultivation in Northeast Himalaya, India. Nat Acad Sci Lett. 2016;39(5):343-346.

21. Trivedi P, Pandey A, Palni LM. In vitro evaluation of antagonistic properties of *Pseudomonas corrugata*. Microbiol Res. 2008;163(3):329-336.

22. Nagarajkumar M, Bhaskaran R, Velazhahan R. Involvement of secondary metabolites and extracellular lytic enzymes produced by Pseudomonas fluorescens in inhibition of *Rhizoctonia solani*, the rice sheath blight pathogen. Microbiol Res. 2006;159(1):73-81.

23. Lim H, Kim Y, Kim S. *Pseudomonas stutzeri* YLP-1 genetic transformation and antifungal mechanism against *Fusarium solani*, an agent of plant root rot. Appl Environ Microbiol. 1991;57(2):510-516.

24. Bollag DM, Rozycki MD, Edelstein SJ. Protein methods, 2nd Ed. Wiley New York. 1996:432.

25. Laemmli UK. Cleavage of structural proteins during the assembly of head of bacteriophage T4. Nature (London). 1970;277:680-685.

26. Trudel J, Asselin A. Detection of chitinase activity after polyacrylamide gel electrophoresis. Ann Biochem. 1989; 178(2):362-366.

27. Dietz A, Mathews J. Scanning Electron Microscopy of Selected Members of the Streptomyces hygroscopicus Group. Appl Microbiol. 1969;18(4):694-696.

28. Aghigni S, Boniar GHS, Rowashdeh R, Batayneh S, Saadoun L. First report on antifungal spectra of activity of Iranian actinomycetes strains against *Alternaria solani, Alternaria alternata, Fusarium solani, Phytophthora megasperma, Verticillium dahlia and Saccharomyces cerevisiae*. Asian J Plant Sci. 2004;3(4):463-471. doi: 10.3923/ajps.2004.463.471

29. Al-Askar AA, Abdul Khair WM, Rashad YM. In vitro antifungal activity of *Streptomyces spororaveus* RDS28 against some phytopathogenic fungi. Afr J Agric Res. 2011;6(12):2835-2842.

30. Muskhazli M, Ramli SA, Ithni A, Tohfah NF. Purification and characterisation of β-l, 3-giucanase from *Trichodenna harzianum* BIO 10671. Pertanika J Trop Agric Sci. 2005; 28(1): 23-31.

31. Gomes RC, Semeˆdo LTAS, Soares RMA, Alviano CS, Linhares LF, Coelho RR et al. Chitinolytic activity of actinomycetes from a cerrado soil and their potential in biocontrol. Lett Appl Microbiol. 2000; 30(2):146-150.

27

Comparison of the anti inflammatory capacities of erythropoietin and U-74389G

C Tsompos[1*], C Panoulis[2], K Toutouzas[3], A Triantafyllou[4], G C Zografos[5] and A Papalois[6]

[*1]Constantinos Tsompos: Consultant A, Department of Gynecology, General Hospital of Thessaloniki "St. Dimitrios" Thessaloniki, Greece

[2]Constantinos Panoulis: Assistant Professor, Department of Obstetrics & Gynecology, Aretaieion Hospital, Athens University, Athens, Attiki, Greece

[3]Konstantinos Toutouzas: Assistant Professor, Department of Surgery, Ippokrateion General Hospital, Athens University, Athens, Attiki, Greece

[4]Aggeliki Triantafyllou: Associate Professor, Department of Biologic Chemistry, Athens University, Athens, Attiki, Greece

[5]George C. Zografos: Professor, Department of Surgery, Ippokrateion General Hospital, Athens University, Athens, Attiki, Greece

[6]Apostolos Papalois: Director, Experimental Research Centre ELPEN Pharmaceuticals, S.A. Inc., Co., Pikermi, Attiki, Greece

*Corresponding author: Tsompos Constantinos, Department of Gynecology, General Hospital of Thessaloniki "St. Dimitrios", 2 Elenis Zografou street, Thessaloniki 54634, Greece, E-mail: Tsomposconstantinos@gmail.com

Abstract

Aim: This study compared the anti inflammatory effects of erythropoietin (Epo) and antioxidant drug U-74389G based on 2 preliminary studies. The provided results at white blood cells count (wbc) count restoration, were co-evaluated in a hypoxia reoxygenation protocol of an animal model.

Materials and methods: Wbc count were evaluated at the 60th reoxygenation min (for groups A, C and E) and at the 120th reoxygenation min (for groups B, D and F) in 60 rats. Groups A and B received no drugs, rats from groups C and D were administered with Epo; whereas rats from groups E and F were administered with U-74389G.

Results: The first preliminary study of Epo kept significantly increased the wbc count by 14.64%±5.40% (p-value=0.0080). The second preliminary study of U-74389G also kept significantly increased the wbc count by 23.64%±6.32% (p-value=0.0004). These 2 studies were co-evaluated since they came from the same experimental setting. The outcome of the co-evaluation was that U-74389G is at least 1.6-fold less anti inflammatory than Epo (p-value=0.0000).

Conclusions: Epo is at least 1.6-fold more anti inflammatory than the antioxidant drug U-74389G (p-value=0.0000).

Key words: hypoxia; erythropoietin; U-74389G; white blood cells count; reoxygenation

Introduction

The short-term anti inflammatory 1action of U-74389G is not satisfactory (p-value=0.0004). U-74389G is a novel antioxidant factor. It implicates just only 255 known biomedical studies at present. 4.31% of these studies concern tissue hypoxia and reoxygenation (HR) experiments. The promising effect of U-74389G in tissue protection has been noted in these HR studies. U-74389G or also known as 21-[4-(2,6-di-1-pyrrolidinyl-4-pyrimidinyl)-1-piperazinyl]-pregna-1,4,9(11)-triene-3,20-dione maleate salt is an antioxidant which prevents both arachidonic acid-induced and iron-dependent lipid peroxidation. It protects against HR injury in animal heart, liver and kidney models. These membrane-associating antioxidants are particularly effective in preventing permeability changes in brain microvascular endothelial cells monolayers. Some biochemical capacities of U-74389G are summarized as activating attenuation of leukocytes; proinflammatory gene down-regulation; endotoxin shock treatment; cytokine production; mononuclear immunoenhancement; antishock and endothelial protection.

However, the anti inflammatory capacity of U-74389G gets more comprehensible whether is compared with the same capacity of a standard known drug. Such one of the more well studied drug; also without satisfactory anti inflammatory action (p-value=0.0080) is erythropoietin (Epo). Actually, Epo implicates over 29,735 known biomedical studies at present. 10.47% at least of these studies concern tissue hypoxia and reoxygenation (HR) experiments. Certainly, the concept has been moved away from the original action of Epo as a glycoprotein cytokine secreted by the kidney in response to cellular hypoxia; which stimulates red blood cell production (erythropoiesis) in the bone marrow. However, just few related reports were found, not covering completely the specific matter with white blood cells count (wbc).

The special aim of this experimental work was to compare the anti inflammatory effects of U-74389G and Epo on a rat model and mainly in an HR protocol. Their effects were tested by measuring the serum wbc counts.

Materials and methods

Animal preparation

The Vet licenses of the research were provided under 3693/12-11- 2010 & 14/10-1-2012 decisions. The granting company and the place of experiment are mentioned in related references1,2. Accepted standards of human animal care were adopted for Albino female Wistar rats. 7 days pre-experimental normal housing included ad libitum diet in laboratory. Continuous intra-experimental anesthesiologic techniques, oxygen supply, electrocardiogram and acidometry were provided. Post-experimental euthanasia excluded awakening and preservation of animals. Rats 16 – 18 weeks old were randomly delivered to four (6) groups (n=10), using the following protocols of HR: Hypoxia for 45 min followed by reoxygenation for 60 min (group A); hypoxia for 45 min followed by reoxygenation for 120 min (group B); hypoxia for 45 min followed by immediate Epo intravenous (IV) administration and reoxygenation for 60 min (group C); hypoxia for 45 min followed by immediate Epo IV administration and reoxygenation for 120 min (group D); hypoxia for 45 min followed by immediate U-74389G intravenous (IV) administration and reoxygenation for 60 min (group E); hypoxia for 45 min followed by immediate U-74389G IV administration and reoxygenation for 120 min (group F). The dose height selection criteria of Epo and U-74389G were assessed at preliminary studies as 10 mg/Kg body mass of animals for both drugs.

Hypoxia was caused by laparotomic clamping inferior aorta over renal arteries with forceps for 45 min. Reoxygenation was induced by removing the clamp and restoration the inferior aorta patency. After exclusion of the blood flow, the protocol of HR was applied, as described above for each experimental group. The drugs were administered at the time of reperfusion; through catheterized inferior vena cava. The wbc count were determined at 60th min of reoxygenation (for A, C and E groups) and at 120th min of reoxygenation (for B, D and F groups).

Statistical analysis

Table 1 presents the (%) restoration influence of Epo regarding reoxygenation time. Also, Table 2 presents the (%) restoration influence of U-74389G regarding reoxygenation time. Chi-square tests was applied using the ratios which produced the (%) results per endpoint. The outcomes of chi-square tests are depicted at Table 3. The statistical analysis was performed by Stata 6.0 software [Stata 6.0, StataCorp LP, Texas, USA].

Table 1: The (%) influence of erythropoietin on wbc count restoration in connection with reoxygenation time

Restoration	+SD	Reoxygenation time	p-values
24.01%	+13.38%	1h	0.1012
22.09%	+9.11%	1.5h	0.0163
20.17%	+12.94%	2h	0.0902
14.55%	+9.53%	reoxygenation time	0.0883
14.64%	+5.40%	interaction	0.008

Table 2: The (%) influence of U-74389G on wbc count restoration in connection with reoxygenation time.

Restoration	+SD	Reoxygenation time	p-values
22.99%	+12.45	1h	0.0914
30.85%	+11.14	1.5h	0.0045
38.70%	+17.39	2h	0.0185
24.97%	+11.55	reoxygenation time	0.0272
23.45%	+6.28	interaction	0.0004

Table 3: The U-74389G / erythropoietin efficacies ratios on wbc counts restoration after chi-square tests application

Odds ratio	[95% Conf. Interval]		p-values	Endpoint
0.957451	0.869207	1.054654	0.3782	1h
1.396122	1.394892	1.397353	0.0000	1.5 h
1.918237	1.763902	2.086076	0.0000	2h
1.71622	1.714481	1.717962	0.0000	Reperfusion time
1.601887	1.60025	1.603525	0.0000	interaction

Results

The successive application of chi-square tests revealed that the restoring capacity of U-74389G was superior than that of erythropoietin by 0.9574511-fold [0.8692073 - 1.054654, p-value=0. 3782] at 1h, by 1.396122-fold [1.394892 - 1.397353] at 1.5h, by 1.918237-fold [1.763902 - 2.086076] at 2h, by 1.71622-fold [1.714481 - 1.717962] without drugs and by 1.601887-fold [1.60025 - 1.603525] whether all variables have been considered (p-value=0.0000).

Discussion

The same authors summarized1 6 IR studies for the effect of U-74389G leading to consistent results in humans or animals. They recorded lower wbc count in 2 studies, general anti inflammatory properties in 3 studies and reducing leukopenia in 1 study. Even during reperfusion phase, a reperfusion syndrome occurs which seamlessly carries on the vicious cycle of leukocytosis. mRNAs expression of inflammatory (TNFa) and anti inflammatory (IL-10) cytokines were up-regulated still 1 hour after ischemia removal. The macromolecular permeability and adhesiveness of capillaries for wbc are due to oxygen free radicals. Reperfusion with consecutive re-entry of molecular oxygen into microvasculature, provokes the formation of oxygen-radicals and accumulation of leukocytes adhering to endothelium of post-capillary venules. Targeted release of reactivate oxygen metabolites, hydrolytic enzymes, additional oxygen-radicals and aggressive mediators delivery by activated neutrophils, such as proteases, cytokines and eicosanoids, which have chemotactic influence on wbc, result in a vicious cycle during reperfusion phase of tissue injury. Leukocyte-generated oxygen free radical are implicated as mediators of reperfusion-associated cellular membrane injury in IR tissues. Whole body systemic extension becomes through these activated inflammatory cells and possibly, results in secondary detectable tissue damage in endothelial cells of the systemic circulation inducing prolonged DNA damage even in early reperfusion period. A vicious cycle of wbc trapping, activation and tissue damage is engaged. The assumption is whether U-74389G administration which has oxygen free radical scavenging properties is a promising new anti inflammatory drug for the treatment of IR injury.

The same authors summarized2 24 IR studies for the effect of Epo leading to inconsistent results in humans or rats. They recorded no change of wbc count in 13 trials; significant decrease of wbc count in 5 trials and significant increase of wbc count in 6 trials. Stevenson JL et al determined3 no significant changes in wbc count but increase of Epo levels for echinacea-based dietary supplement treatment doses groups in endurance-trained men. Ren Y et al characterized4 abnormally increased mean wbc count and higher Epo level mainly in wild-type JAK2 V617F group (P<0.05) at diagnosis of polycythemia vera. Shen W et al found5 that Epo stimulated the production and recruitment of wbc count and CD34(+) cells along with effective mobilization of CD34(+)/VEGF-R2(+) cells into the retina in Royal College of Surgeons rats. Thiel A et al reported6 that piperine significantly decreased the wbc count adding mechanistic endpoints including Epo level in mice. Benders MJ et al found7 no adverse effects on wbc count after rhEpo total 3000 IU/kg administration in neonates with perinatal arterial ischemic stroke. Ofori-Acquah SF et al suggested8 that SDF-1α produced by ischemic tissues mobilizes significantly at least twice higher circulating progenitor cells; total wbc count; many mononuclear cell colonies and plasma Epo concentrations in hemoglobin SS subjects 5-18 years old compared with control subjects. Yan D et al noticed9 an increase of Epo levels and wbc count in Jak2V617F mice expressing all features of human polycythemia vera. Tentori F et al associated10 lower serum wbc counts with longer hemodialysis (HD) treatment time for the same Epo dose. Powers A et al found11 less Epo use therapy and potential complications of neutropenia; pneumonia diagnoses and decreased wbc count in younger myelodysplastic

syndrome patients than in older ones (p ≤0.034). Sugiura Y et al reported12 a secondary polycythemia due to normal wbc count and non increased Epo level in a 67-year-old patient with smokers' polycythemia and lung adenocarcinoma. Li Q et al found13 that adenovirus-mediated human hepatocyte growth factor (HGF) gene transfer could increase significantly the wbc count, the Epo levels enhancing immune function in irradiated C57BL/6 mice. Alsaran K et al assessed14 the mean wbc count dropped by 26.54% (p < 0.001), but the Epo dose increased by 28.27% (p = 0.776) after 48 weeks of peginterferon α-2b (12 kDa) plus ribavirin treatment in HD of chronic HCV patients. Rumi E et al measured15 higher wbc count and lower mutant allele burden and serum Epo levels in essential thrombocythemia JAK2 (V617F) patients than those with CALR mutation. Szygula Z et al found16 higher number of wbc count and Epo concentration after 10 and 20 whole-body cryostimulation treatments (-130°C, treatment duration: 3 minutes) in 45 men than baseline and control group. Zhang H et al accelerated17 the recovery of wbc count and the Epo secretion stimulation after the rhizome of Panax japonicus administration in blood deficiency model mice. Chiu YH et al showed18 a statistically significant rise of blood Epo values and wbc count in the immediate post-race values but a rapid drop in values at 24 hours post-race for Epo values compared with pre-race values in recruited runners. Fauchère JC et al found19 significantly higher reticulocyte and wbc counts at day 7-10 in the rhEpo group after high dose rhEpo administration shortly after birth and subsequently over the first 2 days for neuroprotection in very preterm infants. Jeong G et al demonstrated 20 leukocytosis and decreased Epo in a 61-year-old female with a history of transient ischemic attack and follicular lymphoma.

According to above, table 3 shows that U-74389G has at least 1.6-fold less anti inflammatory capacity than Epo (p-value=0.0000). Perhaps, a longer study time or a higher U-74389G dose may reveal more effective anti inflammatory property. A meta-analysis of these ratios from the same experiment, for 6 other hematologic variables, provides comparable results (table 4).

Table 4: A U-74389G / erythropoietin efficacies ratios meta-analysis on 6 hematologic variableç (4 variables with balancing efficacies and 2 variables with opposite efficacies)

Endpoint Variable	1h	p-value	1.5h	p-value	2h	p-value	Reperfusion time	p-value	interaction	p-value
Hematocrit	38.424	0.0000	9.076658	0.0000	6.222898	0.0000	1.001356	0.2184	12.66419	0.0000
Hemoglobin	1.268689	0.0000	1.839035	0.0000	13.1658	0.0000	1.252422	0.0000	1.94889	0.0000
Platelet DW	0.694023	0.0000	0.0000	0.0000	2.206972	0.0000	2.248401	0.0000	2.458888	0.0000
Creatinine	168.9034	0.0000	4.872332	0.0000	3.039572	0.0000	1.026202	0.0000	5.005523	0.0000
Mean	8.694459		3.2183563		4.8418607		1.3042516		4.1748246	

Endpoint Variable	1h	p-value	1.5h	p-value	2h	p-value	Reperfusion time	p-value	interaction	p-value
Mean corpuscular hemoglobin concentrations	-0.27742	0	-0.55047	0	-0.85224	0	+3.044774	0	-0.77932	0
Platelet crit	-0.2312	0	-0.67194	0	-1.33076	0.0886	5.620077	0	-0.97715	0
Mean	-0,2532076		-0,6081795		-1,0649544		4,1366488		-0,8726499	

Conclusion

The anti-oxidant capacities of U-74389G cannot provide satisfactory short-term anti inflammatory properties; whereas the cytocine capacities of Epo are proved more anti inflammatory at that certain setting. Otherwise, U-74389G was found 62.42% [62.40% - 62.44%] less anti inflammatory than epo (p-value=0.0000).

Acknowledgements

Ackowledged in preliminary studies

References

1. Tsompos C, Panoulis C, Toutouzas K, Triantafyllou A, Zografos G, Papalois A. The effect of the antioxidant drug "U-74389G" on white blood cells levels during hypoxia reoxygenation injury in rats. Asian Journal of Pharmacology and Toxicology. 2016;43(2):22-32. doi:10.1515/amb-2016-0012

2. Tsompos C, Panoulis C, Toutouzas K, Triantafyllou A, Zografos G, Papalois A. The Effect of Erythropoietin on White Blood Cell Count during Ischemia Reperfusion. Sci Chronicles 2013;18(2): 92 - 103.

3. Stevenson JL, Krishnan S, Inigo MM, Stamatikos AD, Gonzales JU, Cooper JA. Echinacea-Based Dietary Supplement Does Not Increase Maximal Aerobic Capacity in Endurance-Trained Men and Women. J Diet Suppl. 2016;13(3):324-338. doi: 10.3109/19390211.2015

4. Ren Y, Fu R, Qu W, Ruan E, Wang X, Wang G, et al. Clinical analysis of 70 cases of polycythemia vera. Zhonghua Yi Xue Za Zhi. 2015;95(18):1378-1381.

5. Shen W, Chung SH, Irhimeh MR, Li S, Lee SR, Gillies MC. Systemic administration of erythropoietin inhibits retinopathy in RCS rats. PLoS One. 2014;9(8):e104759. doi.org/10.1371/journal.pone.0104759

6. Thiel A, Buskens C, Woehrle T, Etheve S, Schoenmakers A, Fehr M, et al. Black pepper constituent piperine: genotoxity studies in vitro and in vivo. Food Chem Toxicol. 2014 ;66:350-357. doi:10.1016/j.fct.2014.01.056

7. Benders MJ, van der Aa NE, Roks M, van Straaten HL, Isgum I, Viergever MA, et al. Feasibility and safety of erythropoietin for neuroprotection after perinatal arterial ischemic stroke. J Pediatr. 2014;164(3):481-6.e1-2. doi: 10.1016/j.jpeds.2013.10.084

8. Ofori-Acquah SF, Buchanan ID, Osunkwo I, Manlove-Simmons J, Lawal F, Quarshie A, et al. Elevated circulating angiogenic progenitors and white blood cells are associated with hypoxia-inducible angiogenic growth factors in children with sickle cell disease. Anemia. 2012;2012:156598. doi: 10.1155/2012/156598

9. Yan D, Hutchison RE, Mohi G. Critical requirement for Stat5 in a mouse model of polycythemia vera. Blood. 2012 ;119(15):3539-3549. doi:10.1182/blood-2011-03-345215

10. Tentori F, Zhang J, Li Y, Karaboyas A, Kerr P, Saran R, et al. Longer dialysis session length is associated with better intermediate outcomes and survival among patients on in-center three times per week hemodialysis: results from the Dialysis Outcomes and Practice Patterns Study (DOPPS). Nephrol Dial Transplant. 2012 ;27(11):4180-4188. doi: 10.1093/ndt/gfs021

11. Powers A, Faria C, Broder MS, Chang E, Cherepanov D. Hematologic complications, healthcare utilization, and costs in commercially insured patients with myelodysplastic syndrome receiving supportive care. Am Health Drug Benefits. 2012 ;5(7):455-465.

12. Sugiura Y, Nemoto E, Shinoda H, Nakamura N, Kaseda S. Surgery for lung adenocarcinoma with smokers' polycythemia: a case report. BMC Res Notes. 2013;6:38. doi:10.1186/1756-0500-6-38

13. Li Q, Sun H, Xiao F, Wang X, Yang Y, Liu Y, et al. Protection against radiation-induced hematopoietic damage in bone marrow by hepatocyte growth factor gene transfer. Int J Radiat Biol. 2014;90(1):36-44. doi: 10.3109/09553002.2014.847294

14. Alsaran K, Sabry A, Molhem A. Treatment of chronic hepatitis C with peginterferon alfa-2b, plus ribavirin in end stage renal disease patients treated by hemodialysis: single Saudi center experience. Ren Fail. 2013;35(10):1305-9. doi: 10.3109/0886022X.2013.826136

15. Rumi E, Pietra D, Ferretti V, Klampfl T, Harutyunyan AS, Milosevic JD, et al. JAK2 or CALR mutation status defines subtypes of essential thrombocythemia with substantially different clinical course and outcomes. Blood. 2014 ;123(10):1544-51. doi: 10.1182/blood-2013-11-539098

16. Szygula Z, Lubkowska A, Giemza C, Skrzek A, Bryczkowska I, Dołęgowska B. Hematological parameters, and hematopoietic growth factors: EPO and IL-3 in response to whole-body cryostimulation (WBC) in military academy students. PLoS One. 2014;9(4):e93096. doi: 10.1371/

journal.pone.0093096

17. Zhang H, Wang HF, Liu Y, Huang LJ, Wang ZF, Li Y. The haematopoietic effect of Panax japonicus on blood deficiency model mice. J Ethnopharmacol. 2014;154(3):818-824. doi:10.1016/j.jep.2014.05.008

18. Chiu YH, Lai JI, Wang SH, How CK, Li LH, Kao WF, et al. Early changes of the anemia phenomenon in male 100-km ultramarathoners. J Chin Med Assoc. 2015;78(2):108-113. doi:10.1016/j.jcma.2014.09.004

19. Fauchère JC, Koller BM, Tschopp A, Dame C, Ruegger C, Bucher HU, et al. Safety of Early High-Dose Recombinant Erythropoietin for Neuroprotection in Very Preterm Infants. J Pediatr. 2015;167(1):52-7.e1-3. doi: 10.1016/j.jpeds.2015.02.052

20. Jeong G, Kim J, Han S, Lee J, Park K, Pak C, et al. Coexistence of follicular lymphoma and an unclassifiable myeloproliferative neoplasm in a treatment-naïve patient: A case report. Oncol Lett. 2016;11(2):1469-1473. doi:10.3892/ol.2015.4040

21. C. Tsompos, C. Panoulis, K Toutouzas, A. Triantafyllou, G. Zografos, A. Papalois. Comparison of the Widening Capacities of Erythropoietin and U-74389g Concerning Platelet Distribution Width Levels. Journal of Biotechnology and Bioengineering. 1(1) 2017:1-4.

PERMISSIONS

The contributors of this book come from diverse backgrounds, making this book a truly international effort. This book will bring forth new frontiers with its revolutionizing research information and detailed analysis of the nascent developments around the world.

We would like to thank all the contributing authors for lending their expertise to make the book truly unique. They have played a crucial role in the development of this book. Without their invaluable contributions this book wouldn't have been possible. They have made vital efforts to compile up to date information on the varied aspects of this subject to make this book a valuable addition to the collection of many professionals and students.

This book was conceptualized with the vision of imparting up-to-date information and advanced data in this field. To ensure the same, a matchless editorial board was set up. Every individual on the board went through rigorous rounds of assessment to prove their worth. After which they invested a large part of their time researching and compiling the most relevant data for our readers.

The editorial board has been involved in producing this book since its inception. They have spent rigorous hours researching and exploring the diverse topics which have resulted in the successful publishing of this book. They have passed on their knowledge of decades through this book. To expedite this challenging task, the publisher supported the team at every step. A small team of assistant editors was also appointed to further simplify the editing procedure and attain best results for the readers.

Apart from the editorial board, the designing team has also invested a significant amount of their time in understanding the subject and creating the most relevant covers. They scrutinized every image to scout for the most suitable representation of the subject and create an appropriate cover for the book.

The publishing team has been an ardent support to the editorial, designing and production team. Their endless efforts to recruit the best for this project, has resulted in the accomplishment of this book. They are a veteran in the field of academics and their pool of knowledge is as vast as their experience in printing. Their expertise and guidance has proved useful at every step. Their uncompromising quality standards have made this book an exceptional effort. Their encouragement from time to time has been an inspiration for everyone.

The publisher and the editorial board hope that this book will prove to be a valuable piece of knowledge for researchers, students, practitioners and scholars across the globe.

LIST OF CONTRIBUTORS

Si Young Cho, Su Hwan Kim, Sunmi Kim, Chan-Woong Park, Hyun Woo Park, Dae Bang Seo and Song Seok Shin
R&D Unit, Amore Pacific Corporation, Yongin-si, Gyeonggi-do 446-729, Republic of Korea, Japan

Juewon Kim
R&D Unit, Amore Pacific Corporation, Yongin-si, Gyeonggi-do 446-729, Republic of Korea, Japan
Department of Integrated Biosciences, University of Tokyo, Chiba 277-8562, Japan

Sarika Pawar, Bela Dhamangaonkar and Abhishek Cukkemane
Bijasu Agri Research Laboratory LLP, Sr. No. 37, Kondhwa Industrial Estate, Khadi Machine Chowk, Kondhwa, Pune-411048, India

Vidya Kalyankar and Shobha Waghmode
Department of Chemistry, M. E. S Abasaheb Garware College, Pune - 411 007, Maharashtra, India

Sharada Dagade
Department of Chemistry, Y. M. College, Bharathi Vidyapeeth, Pune-411038, Maharashtra, India

Mandalaywala HP and Ratna Trivedi
Department of Environmental Sciences, Shree Ramkrishna Institute of Computer Education & Applied Sciences

Shah MP
Division of Applied & Environmental Microbiology, Enviro Technology Limited, Industrial Waste Water Research Laboratory, India

Thuy Trang Nguyen
Department of Pharmacy, Ho Chi Minh City University of Technology (HUTECH), Ho Chi Minh City, Vietnam
Department of Bionano Technology, Gachon Medical Research Institute,Gachon University, Seongnam, South Korea

Vo Van Giau
Deparment of Faculty of Food Technology, Ho Chi Minh City University of Food Industry (HUFI), 140 Le Trong Tan, Tan Phu district, Ho Chi Minh City, Vietnam
Department of Bionano Technology, Gachon Medical Research Institute,Gachon University, Seongnam, South Korea

Tuong Kha Vo
Vietnam Sports Hospital, Ministry of Culture, Sports and Tourism, Do Xuan Hop Road, My Dinh I Ward, Nam Tu Liem District, Hanoi City, Vietnam

Shailesh R. Dave, Monal B. Shah and Devayani R. Tipre
Department of Microbiology and Biotechnology, University School of Sciences, Gujarat University, Ahmedabad 380 009, Gujarat, India

Cristiano José de Andrade and Lidiane Maria de Andrade
Chemical Engineering Department of Polytechnic School of the University of São Paulo

Abhishek Cukkemane
Bijasu Agri Research Laboratory LLP, Kondhwa, Pune-411048, Maharashtra, India

Fei Gui, Xuan Jing, Fu Sheng Huang, Hong Yan Chai, Chunzi Liang and Jian Cheng Tu
Department of Clinical Laboratory & Center for Gene Diagnosis, Zhongnan Hospital of Wuhan University, Wuhan, china

Mei Jun Wang
Department of Clinical Laboratory & Center for Gene Diagnosis, Zhongnan Hospital of Wuhan University, Wuhan, china
Department of Clinical laboratory, Shiyan People's Hospital, Affiliated to Hubei Medical College, Shiyan China

Weipeng Wang and Haiqing Xu
Women and Children's hospital of Hubei province, Wuhan China

Sajad Majeed Zargar
Centre for Plant Biotechnology, Division of Biotechnology, Sher-e-Kashmir University of Agricultural Sciences & Technology of Kashmir, Shalimar, Srinagar, J&K, India

Nancy Gupta
School Biotechnology, Sher-e-Kashmir University of Agricultural Sciences & Technology of Jammu, Chatha, Jammu, J&K, India

Rakeeb A Mir
School of Bio resources & Biotechnology, BGSB University, Rajouri, J&K, India

Vandna Rai
NRCPB, New Delhi, India

Poornananda M. Naik and Jameel M. Al-Khayri
Department of Agricultural Biotechnology, College of Agriculture and Food Sciences, King Faisal University, P.O. Box 420, Al-Hassa 31982, Saudi Arabia

Dr. Sci. Tsuneo. ISHIDA
Life and Environment Science Research

Ganesh Chandra Jagetia
Department of Zoology, Mizoram University, Aizawl-796 004, India

Prakash Chandra Shetty
Department of Anatomy, Melaka Manipal Medical College, Manipal-576 004, India

Gorla V. Reddy
K Scientific Solutions Private Limited, Gachibowli, Hyderabad 500 032, India

Maulin P Shah
Industrial Waste Water Research Lab, Division of Applied & Environmental Microbiology, Environ Technology Limited, India

Shabnam S Shaikh and Meenu S Saraf
Department of Microbiology and Biotechnology, School of sciences, Gujarat University, Ahmedabad

Spandan Chaudhary
Department of Medical Genetics, Xcelris Labs Limited, Old Premchandnagar Road, Opp. Satyagrah Chhavani, Bodakdev, Ahmedabad-380015, Gujarat, India

Pooja S. Chaudhary
NGS department, Xcelris Labs Limited, Old Premchandnagar Road, Opp. Satyagrah Chhavani, Bodakdev, Ahmedabad-380015, Gujarat, India

Toral A. Vaishnani
Bioinformatics department, Xcelris Labs Limited, Old Premchandnagar Road, Opp. Satyagrah Chhavani, Bodakdev, Ahmedabad-380015, Gujarat, India

D. O. Akeredolu
Department of Microbiology, University of Benin, Benin-City, Edo State
Ambrose Alli University, Ekpoma, Edo State

AO. Ekundayo
Ambrose Alli University, Ekpoma, Edo State

Preeti Vyas, Anamika Dubey and Ashwani Kumar
Metagenomics & Secretomics Research laboratory, Department of Botany, Dr.Harisingh Gour University (A Central University), Sagar (M.P.), India

Damendra Kumar
Department of Biotechnology, Dr. Harisingh Gour University (A Central University), Sagar (M.P.), India

Shiwangi Morya and Gauri Aeron
Department of Biotechnology, SD College of Engineering &Technology, Muzaffarnagar-25100, India

Farid E Ahmed and Nancy C Ahmed
GEM Tox Labs, Institute for Research in Biotechnology, 2905 South Memorial Drive, Greenville, NC 27834, USA

Laila Hussein
Department of Nutrition & Food Science, National Research Center, El-Bohooth Street, Dokki, Cairo, Egypt

Mostafa M Gouda
Department of Nutrition & Food Science, National Research Center, El-Bohooth Street, Dokki, Cairo, Egypt
National Research and Development Center for Egg Processing, College of Food Science and Technology, Huazhong Agricultural University, Wuhan, Hubei, PR, China

Paul W Vos
Department of Biostatistics, College of Allied Health Sciences, East Carolina University, 600 Moye Boulevard, Greenville, NC 27858, USA

Mohamed Mahmoud
USDA/ARS Children's Nutrition Research, 1100 Bates Street, Houston, TX 77030, USA

Salma Mukhtar, Kauser Abdulla Malik and Samina Mehnaz
Department of Biological Sciences, Forman Christian College (A Chartered University), Ferozepur Road, Lahore 54600, Pakistan

Saeed Ahmadi Majd, Mohammad Rabbani Khorasgani, Mahna Mapar, Mohammad Asadollahi and Nahid Zarini Mehr
University of Isfahan, Isfahan, Iran

Ardeshir Talebi
School of Medicine, Isfahan University of Medical Sciences, Isfahan, Iran

Hamed Karimi
Islamic Azad University, Science and Research Branch,
Tehran, Iran

Clement Boateng Ampadu
31 Carrolton Road, Boston, MA 02132-6303, USA

Preeti Vyas and Ashwani Kumar
Metagenomics and Secretomics Research Laboratory,
Department of Botany, Dr. Harisingh
Gour University (A Central University), Sagar-470003,
(M.P.), India

Ajayasree T S, S G Borkar and B G Barhate
Department of Plant Pathology and Agricultural
Microbiology, Mahatma Phule Agriculture University,
Rahuri, 413 722, Dist- Ahmednagar, Maharashtra state

Mukesh K Malviya
Crop Genetics Improvement and Biotechnology Lab,
Guangxi Academy of Agricultural Sciences, Nanning,
530007, China

Pankaj Trivedi
Department of Bioagricultural Sciences and Pest
Management, Colarado State University, C034 Plant
Sciences, Fort Collins - CO - 80523-1177

Anita Pandey
Biotechnological Applications, G.B. Pant National
Institute of Himalayan Environment and Sustainable
Development, Kosi-Katarmal, Almora, 263 643
Uttarakhand, India

C Tsompos
Department of Gynecology, General Hospital of
Thessaloniki "St. Dimitrios" Thessaloniki, Greece

C Panoulis
Department of Obstetrics & Gynecology,
Aretaieion Hospital, Athens University, Athens,
Attiki, Greece

K Toutouzas
Department of Surgery, Ippokrateion General
Hospital, Athens University, Athens, Attiki, Greece

A Triantafyllou
Department of Biologic Chemistry, Athens
University, Athens, Attiki, Greece

G C Zografos
Department of Surgery, Ippokrateion General
Hospital, Athens University, Athens, Attiki, Greece

A Papalois
Experimental Research Centre ELPEN
Pharmaceuticals, S.A. Inc., Co., Pikermi, Attiki, Greece

Index